Foundations of Linear and Generalized Linear Models

Foundations of Linear and Generalized Linear Models

ALAN AGRESTI

Distinguished Professor Emeritus
University of Florida
Gainesville, FL

Visiting Professor
Harvard University
Cambridge, MA

WILEY

Published by John Wiley & Sons, Inc., Hoboken, New Jersey.
Published simultaneously in Canada.

Library of Congress Cataloging-in-Publication Data

Agresti, Alan, author.
Foundations of linear and generalized linear models / Alan Agresti.
pages cm. – (Wiley series in probability and statistics)
Includes bibliographical references and index.
ISBN 978-1-118-73003-4 (hardback)
1. Mathematical analysis–Foundations. 2. Linear models (Statistics) I. Title.
QA299.8.A37 2015
003′.74–dc23

2014036543

10 9 8 7 6 5 4 3 2 1

To my statistician friends in Europe

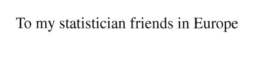

Contents

Website

Data sets for the book arc at www.stat.ufl.edu/~aa/glm/data

Preface

PURPOSE OF THIS BOOK

Why yet another book on linear models? Over the years, a multitude of books have already been written about this well-traveled topic, many of which provide more comprehensive presentations of linear modeling than this one attempts. *My book is intended to present an overview of the key ideas and foundational results of linear and generalized linear models.* I believe this overview approach will be useful for students who lack the time in their program for a more detailed study of the topic. This situation is increasingly common in Statistics and Biostatistics departments. As courses are added on recent influential developments (such as "big data," statistical learning, Monte Carlo methods, and application areas such as genetics and finance), programs struggle to keep room in their curriculum for courses that have traditionally been at the core of the field. Many departments no longer devote an entire year or more to courses about linear modeling.

Books such as those by Dobson and Barnett (2008), Fox (2008), and Madsen and Thyregod (2011) present fine overviews of both linear and generalized linear models. By contrast, my book has more emphasis on the theoretical foundations—showing how linear model fitting projects the data onto a model vector subspace and how orthogonal decompositions of the data yield information about effects, deriving likelihood equations and likelihood-based inference, and providing extensive references for historical developments and new methodology. In doing so, my book has less emphasis than some other books on practical issues of data analysis, such as model selection and checking. However, each chapter contains at least one section that applies the models presented in that chapter to a dataset, using R software. The book is not intended to be a primer on R software or on the myriad details relevant to statistical practice, however, so these examples are relatively simple ones that merely convey the basic concepts and spirit of model building.

The presentation of linear models for continuous responses in Chapters 1–3 has a geometrical rather than an algebraic emphasis. More comprehensive books on linear models that use a geometrical approach are the ones by Christensen (2011) and by

Seber and Lee (2003). The presentation of generalized linear models in Chapters 4–9 includes several sections that focus on discrete data. Some of this significantly abbreviates material from my book, *Categorical Data Analysis* (3rd ed., John Wiley & Sons , 2013). Broader overviews of generalized linear modeling include the classic book by McCullagh and Nelder (1989) and the more recent book by Aitkin et al. (2009). An excellent book on statistical modeling in an even more general sense is by Davison (2003).

USE AS A TEXTBOOK

This book can serve as a textbook for a one-semester or two-quarter course on linear and generalized linear models. It is intended for graduate students in the first or second year of Statistics and Biostatistics programs. It also can serve programs with a heavy focus on statistical modeling, such as econometrics and operations research. The book also should be useful to students in the social, biological, and environmental sciences who choose Statistics as their minor area of concentration.

As a prerequisite, the reader should be familiar with basic theory of statistics, such as presented by Casella and Berger (2001). Although not mandatory, it will be helpful if readers have at least some background in applied statistical modeling, including linear regression and ANOVA. I also assume some linear algebra background. In this book, I recall and briefly review fundamental statistical theory and matrix algebra results where they are used. This contrasts with the approach in many books on linear models of having several chapters on matrix algebra and distribution theory before presenting the main results on linear models. Readers wanting to improve their knowledge of matrix algebra can find on the Web (e.g., with a Google search of "review of matrix algebra") overviews that provide more than enough background for reading this book. Also helpful as background for Chapters 1–3 on linear models are online lectures, such as the MIT linear algebra lectures by G. Strang at `http://ocw.mit.edu/courses/mathematics` on topics such as vector spaces, column space and null space, independence and a basis, inverses, orthogonality, projections and least squares, eigenvalues and eigenvectors, and symmetric and idempotent matrices. By not including separate chapters on matrix algebra and distribution theory, I hope instructors will be able to cover most of the book in a single semester or in a pair of quarters.

Each chapter contains exercises for students to practice and extend the theory and methods and also to help assimilate the material by analyzing data. Complete data files for the text examples and exercises are available at the text website, `http://www.stat.ufl.edu/~aa/glm/data/`. Appendix A contains supplementary data analysis exercises that are not tied to any particular chapter. Appendix B contains solution outlines and hints for some of the exercises.

I emphasize that this book is not intended to be a complete overview of linear and generalized linear modeling. Some important classes of models are beyond its scope; examples are transition (e.g., Markov) models and survival (time-to-event) models. I intend merely for the book to be an overview of the *foundations* of this subject—that is, core material that should be part of the background of any statistical scientist. I invite readers to use it as a stepping stone to reading more specialized books that focus on recent advances and extensions of the models presented here.

ACKNOWLEDGMENTS

This book evolved from a one-semester course that I was invited to develop and teach as a visiting professor for the Statistics Department at Harvard University in the fall terms of 2011–2014. That course covers most of the material in Chapters 1–9. My grateful thanks to Xiao-Li Meng (then chair of the department) for inviting me to teach this course, and likewise thanks to Dave Harrington for extending this invitation through 2014. (The book's front cover, showing the Zakim bridge in Boston, reflects the Boston-area origins of this book.) Special thanks to Dave Hoaglin, who besides being a noted statistician and highly published book author, has wonderful editing skills. Dave gave me detailed and helpful comments and suggestions for my working versions of all the chapters, both for the statistical issues and the expository presentation. He also found many errors that otherwise would have found their way into print!

Thanks also to David Hitchcock, who kindly read the entire manuscript and made numerous helpful suggestions, as did Maria Kateri and Thomas Kneib for a few chapters. Hani Doss kindly shared his fine course notes on linear models (Doss 2010) when I was organizing my own thoughts about how to present the foundations of linear models in only two chapters. Thanks to Regina Dittrich for checking the R code and pointing out errors. I owe thanks also to several friends and colleagues who provided comments or datasets or other help, including Pat Altham, Alessandra Brazzale, Jane Brockmann, Phil Brown, Brian Caffo, Leena Choi, Guido Consonni, Brent Coull, Anthony Davison, Kimberly Dibble, Anna Gottard, Ralitza Gueorguieva, Alessandra Guglielmi, Jarrod Hadfield, Rebecca Hale, Don Hedeker, Georg Heinze, Jon Hennessy, Harry Khamis, Eunhee Kim, Joseph Lang, Ramon Littell, I-Ming Liu, Brian Marx, Clint Moore, Bhramar Mukherjee, Dan Nettleton, Keramat Nourijelyani, Donald Pierce, Penelope Pooler, Euijung Ryu, Michael Schemper, Cristiano Varin, Larry Winner, and Lo-Hua Yuan. James Booth, Gianfranco Lovison, and Brett Presnell have generously shared materials over the years dealing with generalized linear models. Alex Blocker, Jon Bischof, Jon Hennessy, and Guillaume Basse were outstanding and very helpful teaching assistants for my Harvard Statistics 244 course, and Jon Hennessy contributed solutions to many exercises from which I extracted material at the end of this book. Thanks to students in that course for their comments about the manuscript. Finally, thanks to my wife Jacki Levine for encouraging me to spend the terms visiting Harvard and for support of all kinds, including helpful advice in the early planning stages of this book.

ALAN AGRESTI

Brookline, Massachusetts, and Gainesville, Florida

June 2014

CHAPTER 1

Introduction to Linear and Generalized Linear Models

This is a book about *linear models* and *generalized linear models*. As the names suggest, the linear model is a special case of the generalized linear model. In this first chapter, we define generalized linear models, and in doing so we also introduce the linear model.

Chapters 2 and 3 focus on the linear model. Chapter 2 introduces the *least squares* method for fitting the model, and Chapter 3 presents statistical inference under the assumption of a *normal* distribution for the response variable. Chapter 4 presents analogous model-fitting and inferential results for the generalized linear model. This generalization enables us to model non-normal responses, such as categorical data and count data.

The remainder of the book presents the most important generalized linear models. Chapter 5 focuses on models that assume a *binomial* distribution for the response variable. These apply to binary data, such as "success" and "failure" for possible outcomes in a medical trial or "favor" and "oppose" for possible responses in a sample survey. Chapter 6 extends the models to multicategory responses, assuming a *multinomial* distribution. Chapter 7 introduces models that assume a *Poisson* or *negative binomial* distribution for the response variable. These apply to count data, such as observations in a health survey on the number of respondent visits in the past year to a doctor. Chapter 8 presents ways of weakening distributional assumptions in generalized linear models, introducing *quasi-likelihood* methods that merely focus on the mean and variance of the response distribution. Chapters 1–8 assume *independent* observations. Chapter 9 generalizes the models further to permit *correlated* observations, such as in handling *multivariate* responses. Chapters 1–9 use the traditional *frequentist* approach to statistical inference, assuming probability distributions for the response variables but treating model parameters as fixed, unknown values. Chapter 10 presents the *Bayesian* approach for linear models and generalized linear models, which treats the model parameters as random variables having their

Foundations of Linear and Generalized Linear Models, First Edition. Alan Agresti.

own distributions. The final chapter introduces extensions of the models that handle more complex situations, such as *high-dimensional* settings in which models have enormous numbers of parameters.

1.1 COMPONENTS OF A GENERALIZED LINEAR MODEL

The ordinary linear regression model uses linearity to describe the relationship between the mean of the response variable and a set of explanatory variables, with inference assuming that the response distribution is normal. *Generalized linear models* (GLMs) extend standard linear regression models to encompass non-normal response distributions and possibly nonlinear functions of the mean. They have three components.

- *Random component*: This specifies the response variable y and its probability distribution. The observations[1] $\boldsymbol{y} = (y_1, \dots, y_n)^{\text{T}}$ on that distribution are treated as independent.
- *Linear predictor*: For a *parameter vector* $\boldsymbol{\beta} = (\beta_1, \beta_2, \dots, \beta_p)^{\text{T}}$ and a $n \times p$ *model matrix* $\boldsymbol{X}$ that contains values of p explanatory variables for the n observations, the linear predictor is $\boldsymbol{X\beta}$.
- *Link function*: This is a function g applied to each component of $E(\boldsymbol{y})$ that relates it to the linear predictor,

$$g[E(\boldsymbol{y})] = \boldsymbol{X\beta}.$$

Next we present more detail about each component of a GLM.

1.1.1 Random Component of a GLM

The *random component* of a GLM consists of a response variable y with independent observations $(y_1, \dots, y_n)$ having probability density or mass function for a distribution in the *exponential family*. In Chapter 4 we review this family of distributions, which has several appealing properties. For example, $\sum_i y_i$ is a sufficient statistic for its parameter, and regularity conditions (such as differentiation passing under an integral sign) are satisfied for derivations of properties such as optimal large-sample performance of maximum likelihood (ML) estimators.

By restricting GLMs to exponential family distributions, we obtain general expressions for the model likelihood equations, the asymptotic distributions of estimators for model parameters, and an algorithm for fitting the models. For now, it suffices to say that the distributions most commonly used in Statistics, such as the normal, binomial, and Poisson, are exponential family distributions.

[1]The superscript T on a vector or matrix denotes the transpose; for example, here $\boldsymbol{y}$ is a column vector. Our notation makes no distinction between random variables and their observed values; this is generally clear from the context.

1.1.2 Linear Predictor of a GLM

For observation i, $i = 1, \dots, n$, let x_{ij} denote the value of explanatory variable x_j, $j = 1, \dots, p$. Let $\boldsymbol{x}_i = (x_{i1}, \dots, x_{ip})$. Usually, we set $x_{i1} = 1$ or let the first variable have index 0 with $x_{i0} = 1$, so it serves as the coefficient of an intercept term in the model. The *linear predictor* of a GLM relates parameters $\{\eta_i\}$ pertaining to $\{E(y_i)\}$ to the explanatory variables $x_1, \dots, x_p$ using a linear combination of them,

$$\eta_i = \sum_{j=1}^{p} \beta_j x_{ij}, \quad i = 1, \dots, n.$$

The labeling of $\sum_{j=1}^{p} \beta_j x_{ij}$ as a *linear* predictor reflects that this expression is *linear in the parameters*. The explanatory variables themselves can be nonlinear functions of underlying variables, such as an interaction term (e.g., $x_{i3} = x_{i1}x_{i2}$) or a quadratic term (e.g., $x_{i2} = x_{i1}^2$).

In matrix form, we express the linear predictor as

$$\boldsymbol{\eta} = \boldsymbol{X}\boldsymbol{\beta},$$

where $\boldsymbol{\eta} = (\eta_1, \dots, \eta_n)^{\mathrm{T}}$, $\boldsymbol{\beta}$ is the $p \times 1$ column vector of model parameters, and $\boldsymbol{X}$ is the $n \times p$ matrix of explanatory variable values $\{x_{ij}\}$. The matrix $\boldsymbol{X}$ is called the *model matrix*. In experimental studies, it is also often called the *design matrix*. It has n rows, one for each observation, and p columns, one for each parameter in $\boldsymbol{\beta}$. In practice, usually $p \leq n$, the goal of *model parsimony* being to summarize the data using a considerably smaller number of parameters.

GLMs treat y_i as random and $\boldsymbol{x}_i$ as fixed. Because of this, the linear predictor is sometimes called the *systematic component*. In practice $\boldsymbol{x}_i$ is itself often random, such as in sample surveys and other observational studies. In this book, we condition on its observed values in conducting statistical inference about effects of the explanatory variables.

1.1.3 Link Function of a GLM

The third component of a GLM, the *link function*, connects the random component with the linear predictor. Let $\mu_i = E(y_i)$, $i = 1, \dots, n$. The GLM links η_i to μ_i by $\eta_i = g(\mu_i)$, where the link function $g(\cdot)$ is a monotonic, differentiable function. Thus, g links μ_i to explanatory variables through the formula:

$$g(\mu_i) = \sum_{j=1}^{p} \beta_j x_{ij}, \quad i = 1, \dots, n. \tag{1.1}$$

In the exponential family representation of a distribution, a certain parameter serves as its *natural parameter*. This parameter is the mean for a normal distribution, the log of the odds for a binomial distribution, and the log of the mean for a Poisson distribution. The link function g that transforms μ_i to the natural parameter is called the *canonical link*. This link function, which equates the natural parameter with the linear predictor, generates the most commonly used GLMs. Certain simplifications result when the GLM uses the canonical link function. For example, the model has a concave log-likelihood function and simple sufficient statistics and likelihood equations.

1.1.4 A GLM with Identity Link Function is a "Linear Model"

The link function $g(\mu_i) = \mu_i$ is called the *identity link function*. It has $\eta_i = \mu_i$. A GLM that uses the identity link function is called a *linear model*. It equates the linear predictor to the mean itself. This GLM has

$$\mu_i = \sum_{j=1}^{p} \beta_j x_{ij}, \quad i = 1, \dots, n.$$

The standard version of this, which we refer to as the *ordinary linear model*, assumes that the observations have constant variance, called *homoscedasticity*. An alternative way to express the ordinary linear model is

$$y_i = \sum_{j=1}^{p} \beta_j x_{ij} + \epsilon_i,$$

where the "error term" ϵ_i has $E(\epsilon_i) = 0$ and $\text{var}(\epsilon_i) = \sigma^2$, $i = 1, \dots, n$. This is natural for the identity link and normal responses but not for most GLMs.

In summary, ordinary linear models equate the linear predictor directly to the mean of a response variable y and assume constant variance for that response. The normal linear model also assumes normality. By contrast, a GLM is an extension that equates the linear predictor to a link-function-transformed mean of y, and assumes a distribution for y that need not be normal but is in the exponential family. We next illustrate the three components of a GLM by introducing three of the most important GLMs.

1.1.5 GLMs for Normal, Binomial, and Poisson Responses

The class of GLMs includes models for continuous response variables. Most important are ordinary normal linear models. Such models assume a normal distribution for the random component, $y_i \sim N(\mu_i, \sigma^2)$ for $i = 1, ..., n$. The natural parameter for a normal distribution is the mean. So, the canonical link function for a normal GLM is the identity link, and the GLM is then merely a linear model. In particular, standard regression and analysis of variance (ANOVA) models are GLMs assuming a normal random component and using the identity link function. Chapter 3 develops statistical inference for such normal linear models. Chapter 2 presents model fitting for linear models and shows this does not require the normality assumption.

Many response variables are binary. We represent the "success" and "failure" outcomes, such as "favor" and "oppose" responses to a survey question about legalizing same-sex marriage, by 1 and 0. A *Bernoulli trial* for observation i has probabilities $P(y_i = 1) = \pi_i$ and $P(y_i = 0) = 1 - \pi_i$, for which $\mu_i = \pi_i$. This is the special case of the binomial distribution with the number of trials $n_i = 1$. The natural parameter for the binomial distribution is $\log[\mu_i/(1 - \mu_i)]$. This is the log odds of response outcome 1, the so-called *logit* of μ_i. The logit is the canonical link function for binary random components. GLMs using the logit link have the form:

$$\log\left(\frac{\mu_i}{1-\mu_i}\right) = \sum_{j=1}^{p} \beta_j x_{ij}, \quad i = 1, \dots, n.$$

They are called *logistic regression models*, or sometimes simply *logit models*. Chapter 5 presents such models. Chapter 6 introduces generalized logit models for multinomial random components, for handling categorical response variables that have more than two outcome categories.

Some response variables have counts as their possible outcomes. In a criminal justice study, for instance, each observation might be the number of times a person has been arrested. Counts also occur as entries in contingency tables. The simplest probability distribution for count data is the Poisson. It has natural parameter $\log \mu_i$, so the canonical link function is the log link, $\eta_i = \log \mu_i$. The model using this link function is

$$\log \mu_i = \sum_{j=1}^{p} \beta_j x_{ij}, \quad i = 1, \dots, n.$$

Presented in Chapter 7, it is called a *Poisson loglinear model*. We will see there that a more flexible model for count data assumes a negative binomial distribution for y_i.

Table 1.1 lists some GLMs presented in Chapters 2–7. Chapter 4 presents basic results for GLMs, such as likelihood equations, ways of finding the ML estimates, and large-sample distributions for the ML estimators.

1.1.6 Advantages of GLMs versus Transforming the Data

A traditional way to model data, introduced long before GLMs, transforms y so that it has approximately a normal conditional distribution with constant variance. Then, the least squares fitting method and subsequent inference for ordinary normal linear

Table 1.1 Important Generalized Linear Models for Statistical Analysis

Random Component	Link Function	Model	Chapters
Normal	Identity	Regression	2 and 3
		Analysis of variance	2 and 3
Exponential family	Any	Generalized linear model	4
Binomial	Logit	Logistic regression	5
Multinomial	Generalized logits	Multinomial response	6
Poisson	Log	Loglinear	7

Chapter 4 presents an overview of GLMs, and the other chapters present special cases.

models presented in the next two chapters are applicable on the transformed scale. For example, with count data that have a Poisson distribution, the distribution is skewed to the right with variance equal to the mean, but $\sqrt{y}$ has a more nearly normal distribution with variance approximately equal to 1/4. For most data, however, it is challenging to find a transformation that provides both approximate normality and constant variance. The best transformation to achieve normality typically differs from the best transformation to achieve constant variance.

With GLMs, by contrast, the choice of link function is separate from the choice of random component. If a link function is useful in the sense that a linear model with the explanatory variables is plausible for that link, it is not necessary that it also stabilizes variance or produces normality. This is because the fitting process maximizes the likelihood for the choice of probability distribution for y, and that choice is not restricted to normality.

Let g denote a function, such as the log function, that is a link function in the GLM approach or a transformation function in the transformed-data approach. An advantage of the GLM formulation is that the model parameters describe $g[E(y_i)]$, rather than $E[g(y_i)]$ as in the transformed-data approach. With the GLM approach, those parameters also describe effects of explanatory variables on $E(y_i)$, after applying the inverse function for g. Such effects are usually more relevant than effects of explanatory variables on $E[g(y_i)]$. For example, with g as the log function, a GLM with $\log[E(y_i)] = \beta_0 + \beta_1 x_{i1}$ translates to an exponential model for the mean, $E(y_i) = \exp(\beta_0 + \beta_1 x_{i1})$, but the transformed-data model[2] $E[\log(y_i)] = \beta_0 + \beta_1 x_{i1}$ does not translate to exact information about $E(y_i)$ or the effect of x_{i1} on $E(y_i)$. Also, the preferred transform is often not defined on the boundary of the sample space, such as the log transform with a count or a proportion of zero.

GLMs provide a unified theory of modeling that encompasses the most important models for continuous and discrete response variables. Models studied in this text are GLMs with normal, binomial, or Poisson random component, or with extended versions of these distributions such as the multinomial and negative binomial, or multivariate extensions of GLMs. The ML parameter estimates are computed with an algorithm that iteratively uses a weighted version of least squares. The same algorithm applies to the entire exponential family of response distributions, for any choice of link function.

1.2 QUANTITATIVE/QUALITATIVE EXPLANATORY VARIABLES AND INTERPRETING EFFECTS

So far we have learned that a GLM consists of a random component that identifies the response variable and its distribution, a linear predictor that specifies the explanatory variables, and a link function that connects them. We now take a closer look at the form of the linear predictor.

[2]We are not stating that a model for log-transformed data is never relevant; modeling the mean on the original scale may be misleading when the response distribution is very highly skewed and has many outliers.

1.2.1 Quantitative and Qualitative Variables in Linear Predictors

Explanatory variables in a GLM can be

- quantitative, such as in simple linear regression models.
- qualitative factors, such as in analysis of variance (ANOVA) models.
- mixed, such as an interaction term that is the product of a quantitative explanatory variable and a qualitative factor.

For example, suppose observation i measures an individual's annual income y_i, number of years of job experience x_{i1}, and gender x_{i2} ($1 =$ female, $0 =$ male). The linear model with linear predictor

$$\mu_i = \beta_0 + \beta_1 x_{i1} + \beta_2 x_{i2} + \beta_3 x_{i1} x_{i2}$$

has quantitative x_{i1}, qualitative x_{i2}, and mixed $x_{i3} = x_{i1}x_{i2}$ for an interaction term. As Figure 1.1 illustrates, this model corresponds to straight lines $\mu_i = \beta_0 + \beta_1 x_{i1}$ for males and $\mu_i = (\beta_0 + \beta_2) + (\beta_1 + \beta_3)x_{i1}$ for females. With an interaction term relating two variables, the effect of one variable changes according to the level of the other. For example, with this model, the effect of job experience on mean annual income has slope β_1 for males and $\beta_1 + \beta_3$ for females. The special case, $\beta_3 = 0$, of a lack of interaction corresponds to parallel lines relating mean income to job experience for females and males. The further special case also having $\beta_2 = 0$ corresponds to identical lines for females and males. When we use the model to compare mean incomes for females and males while accounting for the number of years of job experience as a covariate, it is called an *analysis of covariance* model.

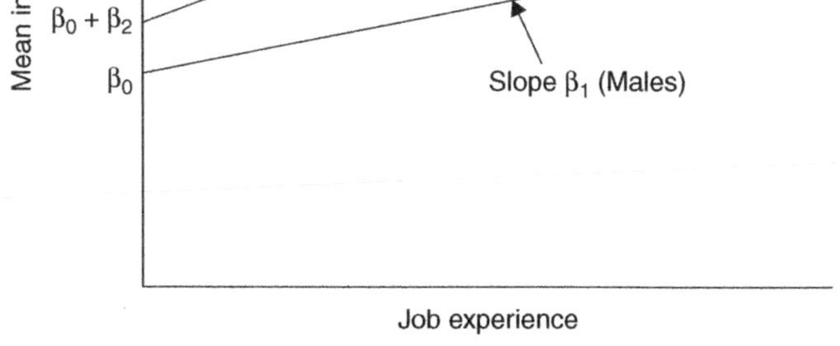

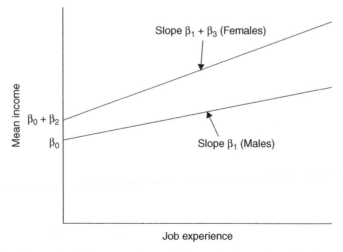

Figure 1.1 Portrayal of linear predictor with quantitative and qualitative explanatory variables.

A quantitative explanatory variable x is represented by a single βx term in the linear predictor and a single column in the model matrix $\boldsymbol{X}$. A qualitative explanatory variable having c categories can be represented by $c - 1$ indicator variables and terms in the linear predictor and $c - 1$ columns in the model matrix $\boldsymbol{X}$. The R software uses as default the "first-category-baseline" parameterization, which constructs indicators for categories $2, \ldots, c$. Their parameter coefficients provide contrasts with category 1. For example, suppose racial–ethnic status is an explanatory variable with $c = 3$ categories, (black, Hispanic, white). A model relating mean income to racial–ethnic status could use

$$\mu_i = \beta_0 + \beta_1 x_{i1} + \beta_2 x_{i2}$$

with $x_{i1} = 1$ for Hispanics and 0 otherwise, $x_{i2} = 1$ for whites and 0 otherwise, and $x_{i1} = x_{i2} = 0$ for blacks. Then β_1 is the difference between the mean income for Hispanics and the mean income for blacks, β_2 is the difference between the mean income for whites and the mean income for blacks, and $\beta_1 - \beta_2$ is the difference between the mean income for Hispanics and the mean income for whites. Some other software, such as SAS, uses an alternative "last-category-baseline" default parameterization, which constructs indicators for categories $1, \ldots, c - 1$. Its parameters then provide contrasts with category c. All such possible choices are equivalent, in terms of having the same model fit.

Shorthand notation can represent terms (variables and their coefficients) in symbols used for linear predictors. A quantitative effect βx is denoted by X, and a qualitative effect is denoted by a letter near the beginning of the alphabet, such as A or B. An interaction is represented[3] by a product of such terms, such as $A.B$ or $A.X$. The period represents forming component-wise product vectors of constituent columns from the model matrix. The crossing operator $A{*}B$ denotes $A + B + A.B$. *Nesting* of categories of B within categories of A (e.g., factor A is states, and factor B is counties within those states) is represented by $A/B = A + A.B$, or sometimes by $A + B(A)$. An intercept term is represented by 1, but this is usually assumed to be in the model unless specified otherwise. Table 1.2 illustrates some simple types of linear predictors and lists the names of normal linear models that equate the mean of the response distribution to that linear predictor.

Table 1.2 Types of Linear Predictors for Normal Linear Models

Linear Predictor	Name of Model
$X_1 + X_2 + X_3 + \cdots$	Multiple regression
A	One-way ANOVA
$A + B$	Two-way ANOVA, no interaction
$A + B + A.B$	Two-way ANOVA, interaction
$A + X$ or $A + X + A.X$	Analysis of covariance

[3]In R, a colon is used, such as $A{:}B$.

1.2.2 Interval, Nominal, and Ordinal Variables

Quantitative variables are said to be measured on an *interval scale*, because numerical intervals separate levels on the scale. They are sometimes called *interval variables.* A qualitative variable, as represented in a model by a set of indicator variables, has categories that are treated as unordered. Such a categorical variable is called a *nominal variable.*

By contrast, a categorical variable whose categories have a natural ordering is referred to as *ordinal*. For example, attained education might be measured with the categories (<high school, high school graduate, college graduate, postgraduate degree). Ordinal explanatory variables can be treated as qualitative by ignoring the ordering and using a set of indicator variables. Alternatively, they can be treated as quantitative by assigning monotone scores to the categories and using a single βx term in the linear predictor. This is often done when we expect $E(y)$ to progressively increase, or progressively decrease, as we move in order across those ordered categories.

1.2.3 Interpreting Effects in Linear Models

How do we interpret the β coefficients in the linear predictors of GLMs? Suppose the response variable is a college student's math achievement test score y_i, and we fit the linear model having x_{i1} = the student's number of years of math education as an explanatory variable, $\mu_i = \beta_0 + \beta_1 x_{i1}$. Since β_1 is the slope of a straight line, we might say, "If the model holds, a one-year increase in math education corresponds to a change of β_1 in the expected math achievement test score." However, this may suggest the inappropriate causal conclusion that if a student attains another year of math education, her or his math achievement test score is expected to change by β_1. To validly make such a conclusion, we would need to conduct an experiment that adds a year of math education for each student and then observes the results. Otherwise, a higher mean test score at a higher math education level (if $\beta_1 > 0$) could at least partly reflect the correlation of several other variables with both test score and math education level, such as parents' attained educational levels, the student's IQ, GPA, number of years of science courses, etc. Here is a more appropriate interpretation: If the model holds, when we compare the subpopulation of students having a certain number of years of math education with the subpopulation having one fewer year of math education, the difference in the means of their math achievement test scores is β_1.

Now suppose the model adds x_{i2} = age of student and x_{i3} = mother's number of years of math education,

$$\mu_i = \beta_0 + \beta_1 x_{i1} + \beta_2 x_{i2} + \beta_3 x_{i3}.$$

Since $\beta_1 = \partial\mu_i/\partial x_{i1}$, we might say, "The difference between the mean math achievement test score of a subpopulation of students having a certain number of years of math education and a subpopulation having one fewer year of math education equals β_1, when we keep constant the student's age and the mother's math education." Controlling variables is possible in designed experiments. But it is unnatural and possibly inconsistent with the data for many observational studies to envision increasing one explanatory variable while keeping all the others fixed. For example, x_1 and x_2 are likely to be positively correlated, so increases in x_1 naturally tend to occur with increases in x_2. In some datasets, one might not even observe a 1-unit range in an explanatory variable when the other explanatory variables are all held constant. A better interpretation is this: "The difference between the mean math achievement test score of a subpopulation of students having a certain number of years of math education and a subpopulation having one fewer year equals β_1, when both subpopulations have the same value for $\beta_2 x_{i2} + \beta_3 x_{i3}$." More concisely we might say, "The effect of the number of years of math education on the mean math achievement test score equals β_1, *adjusting*[4] for student's age and mother's math education." When the model also has a qualitative factor, such as x_{i4} = gender (1 = female, 0 = male), then β_4 is the difference between the mean math achievement test scores for female and male students, adjusting for the other explanatory variables in the model. Analogous interpretations apply to GLMs for a link-transformed mean.

The effect β_1 in the equation with a sole explanatory variable is usually not the same as β_1 in the equation with multiple explanatory variables, because of factors such as confounding. The effect of x_1 on $E(y)$ will usually differ if we ignore other variables than if we adjust for them, especially in observational studies containing "lurking variables" that are associated both with y and with x_1. To highlight such a distinction, it is sometimes helpful to use different notation[5] for the model with multiple explanatory variables, such as

$$\mu_i = \beta_0 + \beta_{y1\cdot 23} x_{i1} + \beta_{y2\cdot 13} x_{i2} + \beta_{y3\cdot 12} x_{i3},$$

where $\beta_{yj\cdot k\ell}$ denotes the effect of x_j on y after adjusting for x_k and x_ℓ.

Some other caveats: In practice, such interpretations use an *estimated* linear predictor, so we replace "mean" by "estimated mean." Depending on the units of measurement, an effect may be more relevant when expressed with changes other than one unit. When an explanatory variable also occurs in an interaction, then its effect should be summarized separately at different levels of the interacting variable. Finally, for GLMs with nonidentity link function, interpretation is more difficult because β_j refers to the effect on $g(\mu_i)$ rather than μ_i. In later chapters we will present interpretations for various link functions.

1.3 MODEL MATRICES AND MODEL VECTOR SPACES

For the data vector $\boldsymbol{y}$ with $\boldsymbol{\mu} = E(\boldsymbol{y})$, consider the GLM $\boldsymbol{\eta} = \boldsymbol{X\beta}$ with link function g and transformed mean values $\boldsymbol{\eta} = g(\boldsymbol{\mu})$. For this GLM, $\boldsymbol{y}$, $\boldsymbol{\mu}$, and $\boldsymbol{\eta}$ are points in n-dimensional Euclidean space, denoted by $\mathbb{R}^n$.

[4]For linear models, Section 2.5.6 gives a technical definition of *adjusting*, based on removing effects of x_2 and x_3 by regressing both y and x_1 on them.

[5]Yule (1907) introduced such notation in a landmark article on regression modeling.

1.3.1 Model Matrices Induce Model Vector Spaces

Geometrically, model matrices of GLMs naturally induce *vector spaces* that determine the possible $\boldsymbol{\mu}$ for a model. Recall that a vector space S is such that if $\boldsymbol{u}$ and $\boldsymbol{v}$ are elements in S, then so are $\boldsymbol{u} + \boldsymbol{v}$ and $c\boldsymbol{u}$ for any constant c.

For a particular $n \times p$ model matrix $\boldsymbol{X}$, the values of $\boldsymbol{X\beta}$ for all possible vectors $\boldsymbol{\beta}$ of model parameters generate a vector space that is a linear subspace of $\mathbb{R}^n$. For all possible $\boldsymbol{\beta}$, $\boldsymbol{\eta} = \boldsymbol{X\beta}$ traces out the vector space spanned by the columns of $\boldsymbol{X}$, that is, the set of all possible linear combinations of the columns of $\boldsymbol{X}$. This is the *column space* of $\boldsymbol{X}$, which we denote by $C(\boldsymbol{X})$,

$$C(\boldsymbol{X}) = \{\boldsymbol{\eta} : \text{ there is a } \boldsymbol{\beta} \text{ such that } \boldsymbol{\eta} = \boldsymbol{X\beta}\}.$$

In the context of GLMs, we refer to the vector space $C(\boldsymbol{X})$ as the *model space*. The $\boldsymbol{\eta}$, and hence the $\boldsymbol{\mu}$, that are possible for a particular GLM are determined by the columns of $\boldsymbol{X}$.

Two models with model matrices $\boldsymbol{X}_a$ and $\boldsymbol{X}_b$ are equivalent if $C(\boldsymbol{X}_a) = C(\boldsymbol{X}_b)$. The matrices $\boldsymbol{X}_a$ and $\boldsymbol{X}_b$ could be different because of a change of units of an explanatory variable (e.g., pounds to kilograms), or a change in the way of specifying indicator variables for a qualitative predictor. On the other hand, if the model with model matrix $\boldsymbol{X}_a$ is a special case of the model with model matrix $\boldsymbol{X}_b$, for example, with $\boldsymbol{X}_a$ obtained by deleting one or more of the columns of $\boldsymbol{X}_b$, then the model space $C(\boldsymbol{X}_a)$ is a vector subspace of the model space $C(\boldsymbol{X}_b)$.

1.3.2 Dimension of Model Space Equals Rank of Model Matrix

Recall that the *rank* of a matrix $\boldsymbol{X}$ is the number of vectors in a *basis* for $C(\boldsymbol{X})$, which is a set of linearly independent vectors whose linear combinations generate $C(\boldsymbol{X})$. Equivalently, the rank is the number of linearly independent columns (or rows) of $\boldsymbol{X}$. The *dimension* of the model space $C(\boldsymbol{X})$ of $\boldsymbol{\eta}$ values, denoted by $\dim[C(\boldsymbol{X})]$, is defined to be the rank of $\boldsymbol{X}$. In all but the final chapter of this book, we assume $p \leq n$, so the model space has dimension no greater than p. We say that $\boldsymbol{X}$ has *full rank* when $\text{rank}(\boldsymbol{X}) = p$.

When $\boldsymbol{X}$ has less than full rank, the columns of $\boldsymbol{X}$ are linearly dependent, with any one column being a linear combination of the other columns. That is, there exist linear combinations of the columns that yield the $\boldsymbol{0}$ vector. There are then nonzero $p \times 1$ vectors $\boldsymbol{\zeta}$ such that $\boldsymbol{X\zeta} = \boldsymbol{0}$. Such vectors make up the *null space* of the model matrix,

$$N(\boldsymbol{X}) = \{\boldsymbol{\zeta} : \boldsymbol{X\zeta} = \boldsymbol{0}\}.$$

When $\boldsymbol{X}$ has full rank, then $\dim[N(\boldsymbol{X})] = 0$. Then, no nonzero combinations of the columns of $\boldsymbol{X}$ yield $\boldsymbol{0}$, and $N(\boldsymbol{X})$ consists solely of the $p \times 1$ zero vector, $\boldsymbol{0} = (0, 0, \ldots, 0)^{\mathrm{T}}$. Generally,

$$\dim[C(\boldsymbol{X})] + \dim[N(\boldsymbol{X})] = p.$$

When $\boldsymbol{X}$ has less than full rank, we will see that the model parameters $\boldsymbol{\beta}$ are not well defined. Then there is said to be *aliasing* of the parameters. In one way this can happen, called *extrinsic aliasing*, an anomaly of the data causes the linear dependence, such as when the values for one predictor are a linear combination of values for the other predictors (i.e., perfect *collinearity*). Another way, called *intrinsic aliasing*, arises when the linear predictor contains inherent redundancies, such as when (in addition to the usual intercept term) we use an indicator variable for each category of a qualitative predictor. The following example illustrates.

1.3.3 Example: The One-Way Layout

Many research studies have the central goal of comparing response distributions for different groups, such as comparing life-length distributions of lung cancer patients under two treatments, comparing mean crop yields for three fertilizers, or comparing mean incomes on the first job for graduating students with various majors. For c groups of independent observations, let y_{ij} denote response observation j in group i, for $i = 1, \dots, c$ and $j = 1, \dots, n_i$. This data structure is called the *one-way layout*.

We regard the groups as c categories of a qualitative factor. For $\mu_{ij} = E(y_{ij})$, the GLM has linear predictor,

$$g(\mu_{ij}) = \beta_0 + \beta_i.$$

Let μ_i denote the common value of $\{\mu_{ij}, j = 1, \dots, n_i\}$, for $i = 1, \dots, c$. For the identity link function and an assumption of normality for the random component, this model is the basis of the *one-way ANOVA* significance test of H_0: $\mu_1 = \cdots = \mu_c$, which we develop in Section 3.2. This hypothesis corresponds to the special case of the model in which $\beta_1 = \cdots = \beta_c$.

Let $\boldsymbol{y} = (y_{11}, \dots, y_{1n_1}, \dots, y_{c1}, \dots, y_{cn_c})^{\mathrm{T}}$ and $\boldsymbol{\beta} = (\beta_0, \beta_1, \dots, \beta_c)^{\mathrm{T}}$. Let $\mathbf{1}_{n_i}$ denote the $n_i \times 1$ column vector consisting of n_i entries of 1, and likewise for $\mathbf{0}_{n_i}$. For the one-way layout, the model matrix $\boldsymbol{X}$ for the linear predictor $\boldsymbol{X\beta}$ in the GLM expression $g(\boldsymbol{\mu}) = \boldsymbol{X\beta}$ that represents $g(\mu_{ij}) = \beta_0 + \beta_i$ is

$$\boldsymbol{X} = \begin{pmatrix} \mathbf{1}_{n_1} & \mathbf{1}_{n_1} & \mathbf{0}_{n_1} & \cdots & \mathbf{0}_{n_1} \\ \mathbf{1}_{n_2} & \mathbf{0}_{n_2} & \mathbf{1}_{n_2} & \cdots & \mathbf{0}_{n_2} \\ \vdots & \vdots & \vdots & \ddots & \vdots \\ \mathbf{1}_{n_c} & \mathbf{0}_{n_c} & \mathbf{0}_{n_c} & \cdots & \mathbf{1}_{n_c} \end{pmatrix}.$$

This matrix has dimension $n \times p$ with $n = n_1 + \cdots + n_c$ and $p = c + 1$.

Equivalently, this parameterization corresponds to indexing the observations as y_h for $h = 1, \dots, n$, defining indicator variables $x_{hi} = 1$ when observation h is in group i and $x_{hi} = 0$ otherwise, for $i = 1, \dots, c$, and expressing the linear predictor for the link function g applied to $E(y_h) = \mu_h$ as

$$g(\mu_h) = \beta_0 + \beta_1 x_{h1} + \cdots + \beta_c x_{hc}.$$

In either case, the indicator variables whose coefficients are $\{\beta_1, \dots, \beta_c\}$ add up to the vector $\mathbf{1}_n$. That vector, which is the first column of $\boldsymbol{X}$, has coefficient that is the intercept term β_0. The columns of $\boldsymbol{X}$ are linearly dependent, because columns 2 through $c+1$ add up to column 1. Here β_0 is intrinsically aliased with $\sum_{i=1}^{c} \beta_i$. The parameter β_0 is *marginal* to $\{\beta_1, \ldots, \beta_c\}$, in the sense that the column space for the coefficient of β_0 in the model lies wholly in the column space for the vector coefficients of $\{\beta_1, \ldots, \beta_c\}$. So, β_0 is redundant in any explanation of the structure of the linear predictor.

Because of the linear dependence of the columns of $\boldsymbol{X}$, this matrix does not have full rank. But we can achieve full rank merely by dropping one column of $\boldsymbol{X}$, because we need only $c-1$ indicators to represent a c-category explanatory variable. This model with one less parameter has the same column space for the reduced model matrix.

1.4 IDENTIFIABILITY AND ESTIMABILITY

In the one-way layout example, let d denote any constant. Suppose we transform the parameters $\boldsymbol{\beta}$ to a new set,

$$\boldsymbol{\beta}^* = (\beta_0^*, \beta_1^*, \ldots, \beta_c^*)^{\mathrm{T}} = (\beta_0 + d, \beta_1 - d, \ldots, \beta_c - d)^{\mathrm{T}}.$$

The linear predictor with this new set of parameters is

$$g(\mu_{ij}) = \beta_0^* + \beta_i^* = (\beta_0 + d) + (\beta_i - d) = \beta_0 + \beta_i.$$

That is, the linear predictor $\boldsymbol{X\beta}$ for $g(\boldsymbol{\mu})$ is exactly the same, for any value of d. So, for the model as specified with $c+1$ parameters, the parameter values are not unique.

1.4.1 Identifiability of GLM Model Parameters

For this model, because the value for $\boldsymbol{\beta}$ is not unique, we cannot estimate $\boldsymbol{\beta}$ uniquely even if we have an infinite amount of data. Whether we assume normality or some other distribution for $\boldsymbol{y}$, the likelihood equations have infinitely many solutions. When the model matrix is not of full rank, $\boldsymbol{\beta}$ is not *identifiable*.

Definition. For a GLM with linear predictor $\boldsymbol{X\beta}$, the parameter vector $\boldsymbol{\beta}$ is *identifiable* if whenever $\boldsymbol{\beta}^* \neq \boldsymbol{\beta}$, then $\boldsymbol{X\beta}^* \neq \boldsymbol{X\beta}$.

Equivalently, $\boldsymbol{\beta}$ is identifiable if $\boldsymbol{X\beta}^* = \boldsymbol{X\beta}$ implies that $\boldsymbol{\beta}^* = \boldsymbol{\beta}$, so this definition tells us that if we know $g(\boldsymbol{\mu}) = \boldsymbol{X\beta}$ (and hence if we know $\boldsymbol{\mu}$ satisfying the model), then we can also determine $\boldsymbol{\beta}$.

For the parameterization just given for the one-way layout, $\boldsymbol{\beta}$ is not identifiable, because $\boldsymbol{\beta} = (\beta_0, \beta_1, \ldots, \beta_c)^{\mathrm{T}}$ and $\boldsymbol{\beta}^* = (\beta_0 + d, \beta_1 - d, \ldots, \beta_c - d)^{\mathrm{T}}$ do not have different linear predictor values. In such cases, we can obtain identifiability and eliminate the intrinsic aliasing among the parameters by redefining the linear predictor with fewer parameters. Then, different $\boldsymbol{\beta}$ values have different linear predictor values $\boldsymbol{X\beta}$, and estimation of $\boldsymbol{\beta}$ is possible.

For the one-way layout, we can either drop a parameter or add a linear constraint. That is, in $g(\mu_{ij}) = \beta_0 + \beta_i$, we might set $\beta_1 = 0$ or $\beta_c = 0$ or $\sum_i \beta_i = 0$ or $\sum_i n_i\beta_i = 0$. With the first-category-baseline constraint $\beta_1 = 0$, we express the model as $g(\boldsymbol{\mu}) = \boldsymbol{X\beta}$ with

$$\boldsymbol{X\beta} = \begin{pmatrix} \mathbf{1}_{n_1} & \mathbf{0}_{n_1} & \mathbf{0}_{n_1} & \cdots & \mathbf{0}_{n_1} \\ \mathbf{1}_{n_2} & \mathbf{1}_{n_2} & \mathbf{0}_{n_2} & \cdots & \mathbf{0}_{n_2} \\ \mathbf{1}_{n_3} & \mathbf{0}_{n_3} & \mathbf{1}_{n_3} & \cdots & \mathbf{0}_{n_3} \\ \vdots & \vdots & \vdots & \ddots & \vdots \\ \mathbf{1}_{n_c} & \mathbf{0}_{n_c} & \mathbf{0}_{n_c} & \cdots & \mathbf{1}_{n_c} \end{pmatrix} \begin{pmatrix} \beta_0 \\ \beta_2 \\ \vdots \\ \beta_c \end{pmatrix}.$$

When used with the identity link function, this expression states that $\mu_1 = \beta_0$ (from the first n_1 rows of $\boldsymbol{X}$), and for $i > 1$, $\mu_i = \beta_0 + \beta_i$ (from the n_i rows of $\boldsymbol{X}$ in set i). Thus, the model parameters then represent $\beta_0 = \mu_1$ and $\{\beta_i = \mu_i - \mu_1\}$. Under the last-category-baseline constraint $\beta_c = 0$, the parameters are $\beta_0 = \mu_c$ and $\{\beta_i = \mu_i - \mu_c\}$. Under the constraint $\sum_i n_i\beta_i = 0$, the parameters are $\beta_0 = \bar{\mu}$ and $\{\beta_i = \mu_i - \bar{\mu}\}$, where $\bar{\mu} = (\sum_i n_i\mu_i)/n$.

A slightly more general definition of identifiability refers instead to linear combinations $\boldsymbol{\ell}^{\mathrm{T}}\boldsymbol{\beta}$ of parameters. It states that $\boldsymbol{\ell}^{\mathrm{T}}\boldsymbol{\beta}$ is identifiable if whenever $\boldsymbol{\ell}^{\mathrm{T}}\boldsymbol{\beta}^* \neq \boldsymbol{\ell}^{\mathrm{T}}\boldsymbol{\beta}$, then $\boldsymbol{X\beta}^* \neq \boldsymbol{X\beta}$. This definition permits a subset of the terms in $\boldsymbol{\beta}$ to be identifiable, rather than treating the entire $\boldsymbol{\beta}$ as identifiable or nonidentifiable. For example, suppose we extend the model for the one-way layout to include a quantitative explanatory variable taking value x_{ij} for observation j in group i, yielding the analysis of covariance model

$$g(\mu_{ij}) = \beta_0 + \beta_i + \gamma x_{ij}.$$

Then, without a constraint on $\{\beta_i\}$ or β_0, according to this definition $\{\beta_i\}$ and β_0 are not identifiable, but γ *is* identifiable. Here, taking $\boldsymbol{\ell}^{\mathrm{T}}\boldsymbol{\beta} = \gamma$, different values of $\boldsymbol{\ell}^{\mathrm{T}}\boldsymbol{\beta}$ yield different values of $\boldsymbol{X\beta}$.

1.4.2 Estimability in Linear Models

In a non-full-rank model specification, some quantities are unaffected by the parameter nonidentifiability and can be estimated. In a linear model, the adjective *estimable* refers to certain quantities that can be estimated in an unbiased manner.

Definition. In a linear model $E(\boldsymbol{y}) = \boldsymbol{X\beta}$, the quantity $\boldsymbol{\ell}^T\boldsymbol{\beta}$ is *estimable* if there exist coefficients $\boldsymbol{a}$ such that $E(\boldsymbol{a}^T\boldsymbol{y}) = \boldsymbol{\ell}^T\boldsymbol{\beta}$.

That is, some linear combination of the observations estimates $\boldsymbol{\ell}^{\mathrm{T}}\boldsymbol{\beta}$ unbiasedly.

We show now that if $\boldsymbol{\ell}^{\mathrm{T}}\boldsymbol{\beta}$ can be expressed as a linear combination of means, it is estimable. Recall that $\boldsymbol{x}_i$ denotes row i of the model matrix $\boldsymbol{X}$, corresponding to observation y_i, for which $E(y_i) = \boldsymbol{x}_i\boldsymbol{\beta}$. Letting $\boldsymbol{\ell}^{\mathrm{T}} = \boldsymbol{x}_i$ and taking $\boldsymbol{a}$ to be identically 0 except for a 1 in position i, we have $E(\boldsymbol{a}^{\mathrm{T}}\boldsymbol{y}) = E(y_i) = \boldsymbol{x}_i\boldsymbol{\beta} = \boldsymbol{\ell}^{\mathrm{T}}\boldsymbol{\beta}$ for all $\boldsymbol{\beta}$. So $E(y_i) =$

$\boldsymbol{x}_i\boldsymbol{\beta}$ is estimable. More generally, for any particular $\boldsymbol{a}$, since $E(\boldsymbol{a}^{\mathrm{T}}\boldsymbol{y}) = \boldsymbol{a}^{\mathrm{T}}E(\boldsymbol{y}) = \boldsymbol{a}^{\mathrm{T}}\boldsymbol{X}\boldsymbol{\beta}$, the quantity $\boldsymbol{\ell}^{\mathrm{T}}\boldsymbol{\beta}$ is estimable with $\boldsymbol{\ell}^{\mathrm{T}} = \boldsymbol{a}^{\mathrm{T}}\boldsymbol{X}$. That is, the estimable quantities are linear functions $\boldsymbol{a}^{\mathrm{T}}\boldsymbol{\mu}$ of $\boldsymbol{\mu} = \boldsymbol{X}\boldsymbol{\beta}$. This is not surprising, since $\boldsymbol{\beta}$ affects the response variable only through $\boldsymbol{\mu} = \boldsymbol{X}\boldsymbol{\beta}$.

To illustrate, for the one-way layout, consider the over-parameterization $\mu_{ij} = \beta_0 + \beta_i$. Then, $\beta_0 + \beta_i = \mu_i$ as well as contrasts such as $\beta_h - \beta_i = \mu_h - \mu_i$ are estimable. Any sole element in $\boldsymbol{\beta}$ is not estimable.

When $\boldsymbol{X}$ has full rank, $\boldsymbol{\beta}$ is identifiable, and then *all* linear combinations $\boldsymbol{\ell}^{\mathrm{T}}\boldsymbol{\beta}$ are estimable. (We will see how to form the appropriate $\boldsymbol{a}^{\mathrm{T}}\boldsymbol{y}$ for the unbiased estimator in Chapter 2 when we learn how to estimate $\boldsymbol{\beta}$.) The estimates do not depend on which constraints we employ, if necessary, to obtain identifiability. When $\boldsymbol{X}$ does not have full rank, $\boldsymbol{\beta}$ is not identifiable. Also in that case, for the more general definition of identifiability in terms of linear combinations $\boldsymbol{\ell}^{\mathrm{T}}\boldsymbol{\beta}$, at least one component of $\boldsymbol{\beta}$ is not identifiable. In fact, for that definition, $\boldsymbol{\ell}^{\mathrm{T}}\boldsymbol{\beta}$ is estimable if and only if it is identifiable. Then *the estimable quantities are merely the linear functions of* $\boldsymbol{\beta}$ *that are identifiable* (Christensen 2011, Section 2.1).

Nonidentifiability of $\boldsymbol{\beta}$ is irrelevant as long as we focus on $\boldsymbol{\mu} = \boldsymbol{X}\boldsymbol{\beta}$ and other estimable characteristics. In particular, when $\boldsymbol{\ell}^{\mathrm{T}}\boldsymbol{\beta}$ is estimable, the values of $\boldsymbol{\ell}^{\mathrm{T}}\hat{\boldsymbol{\beta}}$ are the same for every solution $\hat{\boldsymbol{\beta}}$ of the likelihood equations. So, just what is the set of linear combinations $\boldsymbol{\ell}^{\mathrm{T}}\boldsymbol{\beta}$ that are estimable? Since $E(\boldsymbol{a}^{\mathrm{T}}\boldsymbol{y}) = \boldsymbol{\ell}^{\mathrm{T}}\boldsymbol{\beta}$ with $\boldsymbol{\ell}^{\mathrm{T}} = \boldsymbol{a}^{\mathrm{T}}\boldsymbol{X}$, the linear space of such $p \times 1$ vectors $\boldsymbol{\ell}$ is precisely the set of linear combinations of rows of $\boldsymbol{X}$. That is, it is the row space of the model matrix $\boldsymbol{X}$, which is equivalently $C(\boldsymbol{X}^{\mathrm{T}})$. This is not surprising, since each mean is the inner product of a row of $\boldsymbol{X}$ with $\boldsymbol{\beta}$.

1.5 EXAMPLE: USING SOFTWARE TO FIT A GLM

General-purpose statistical software packages, such as R, SAS, Stata, and SPSS, can fit linear models and GLMs. In each chapter of this book, we introduce an example to illustrate the concepts of that chapter. We show R code and output, but the choice of software is less important than understanding how to interpret the output, which is similar with different packages.

In R, the `lm` function fits and performs inference for normal linear models, and the `glm` function does this for GLMs[6]. When the `glm` function assumes the normal distribution for y and uses the identity link function, it provides the same fit as the `lm` function.

1.5.1 Example: Male Satellites for Female Horseshoe Crabs

We use software to specify and fit linear models and GLMs with data from a study of female horseshoe crabs[7] on an island in the Gulf of Mexico. During spawning season,

[6] For "big data," the `biglm` package in R has functions that fit linear models and GLMs using an iterative algorithm that processes the data in chunks.

[7] See `http://en.wikipedia.org/wiki/Horseshoe_crab` and `horseshoecrab.org` for details about horseshoe crabs, including pictures of their mating.

Table 1.3 Number of Male Satellites (y) by Female Crab's Characteristics

y	C	S	W	Wt	y	C	S	W	Wt	y	C	S	W	Wt
8	2	3	28.3	3.05	0	3	3	22.5	1.55	9	1	1	26.0	2.30
4	3	3	26.0	2.60	0	2	3	23.8	2.10	0	3	2	24.7	1.90
0	3	3	25.6	2.15	0	3	3	24.3	2.15	0	2	3	25.8	2.65
0	4	2	21.0	1.85	14	2	1	26.0	2.30	8	1	1	27.1	2.95

Source: The data are courtesy of Jane Brockmann, University of Florida. The study is described in *Ethology* **102**: 1–21 (1996). Complete data ($n = 173$) are in file `Crabs.dat` at the text website, `www.stat.ufl.edu/~aa/glm/data`.

C, color (1, medium light; 2, medium; 3, medium dark; 4, dark); S, spine condition (1, both good; 2, one worn or broken; 3, both worn or broken); W, carapace width (cm); Wt, weight (kg).

a female migrates to the shore to breed. With a male attached to her posterior spine, she burrows into the sand and lays clusters of eggs. The eggs are fertilized externally, in the sand beneath the pair. During spawning, other male crabs may cluster around the pair and may also fertilize the eggs. These male crabs are called *satellites*.

The response outcome for each of the $n = 173$ female crabs is her y = number of satellites. Explanatory variables are the female crab's color, spine condition, weight, and carapace width.Table 1.3 shows a small portion of the data and the categories for color and spine condition. As you read through the discussion below, we suggest that you download the data from the text website and practice data analysis by replicating these analyses and conduct others that occur to you (including additional plots) using R or your preferred software.

We now fit some linear models and GLMs to these data. Since the data are counts, the Poisson might be the first distribution you would consider for modeling y.

```
-----------------------------------------------------------------------
> Crabs <- read.table("Crabs.dat", header=T)
> attach(Crabs)
> mean(y); var(y)
[1] 2.9191
[1] 9.9120
> hist(y) # Provides a histogram display
> table(y) # Shows frequency distribution for y values
 0  1 2  3  4  5  6 7 8 9 10 11 12 14 15
62 16 9 19 19 15 13 4 6 3  3  1  1  1  1
> fit.pois <- glm(y ~ 1, family = poisson(link=identity), data=Crabs)
> summary(fit.pois) # y ~ 1 puts only an intercept in model
Coefficients:
            Estimate  Std. Error  z value  Pr(>|z|)
(Intercept)   2.9191      0.1299    22.47    <2e-16
-----------------------------------------------------------------------
```

Fitting the Poisson distribution with a GLM containing only an intercept and using the identity link function gives an estimated Poisson mean that is the sample mean 2.92, for reasons we will see in Chapter 7 on models for count data. However, the Poisson mean equals its variance, and the mode is the integer part of the mean. The sample variance of 9.92 and the strong mode at 0 shown by the frequency distribution suggest that a Poisson assumption is inappropriate for the marginal distribution of y. We study more appropriate distributions for the counts in Chapter 7.

1.5.2 Linear Model Using Weight to Predict Satellite Counts

Of the explanatory variables, two are quantitative (width and weight) and two are ordinal categorical (color and spine condition). We begin by illustrating the use of a quantitative explanatory variable. Weight and width are very highly positively correlated, and for illustrative purposes we will use weight, in kilograms, as an explanatory variable. We first find some simple descriptive statistics:

```
----------------------------------------------------------------------
> mean(weight); sd(weight); quantile(weight, c(0, 0.25, 0.50, 0.75, 1))
[1] 2.4372
[1] 0.5770
  0%   25%   50%   75%  100% # minimum, quartiles, and maximum
1.20 2.00 2.35 2.85 5.20
> plot(weight, y) # Scatterplot of y and x = weight
----------------------------------------------------------------------
```

The quantiles reveal a relatively large maximum weight, which the scatterplot in Figure 1.2 of the number of satellites against weight also highlights. That plot shows there is not a clear trend in the relation.

We next fit the linear model having a straight-line relationship between $E(y)$ and $x =$ weight.

```
---------------------------------------------------------------------
> fit.weight <- lm(y ~ weight, data=Crabs)
> summary(fit.weight)
            Estimate  Std. Error   t value    Pr(>|t|)
(Intercept)  -1.9911      0.9710    -2.050      0.0418
weight        2.0147      0.3878     5.196    5.75e-07
---
> fit.weight2 <- glm(y ~  weight, family=gaussian(link=identity),
+ data=Crabs)
> summary(fit.weight2)
            Estimate  Std. Error   t value    Pr(>|t|)
(Intercept)  -1.9911      0.9710    -2.050      0.0418
weight        2.0147      0.3878     5.196    5.75e-07
> abline(lm(y ~ weight)) # puts fitted line on the scatterplot
---------------------------------------------------------------------
```

The fit of an ordinary linear model is the same as the fit of the GLM using normal (Gaussian family) random component with identity link function. The fit $\hat{\mu}_i = -1.991 + 2.015x_i$, with positive estimated slope, suggests that heavier female crabs tend to have more satellites. Figure 1.2 shows the fitted line superimposed on the scatterplot.

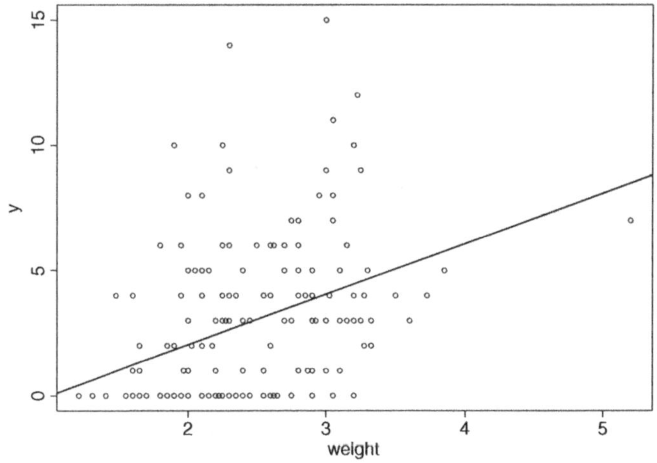

Figure 1.2 Scatterplot of y = number of crab satellites against x = crab weight.

For linear modeling, it is most common to assume a normal response distribution, with constant variance. This is not ideal for the horseshoe crab satellite counts, since they are discrete and since count data usually have variability that increases as the mean does. However, the normal assumption has the flexibility, compared with the Poisson, that the variance is not required to equal the mean. In any case, Chapter 2 shows that the linear model fit does not require an assumption of normality.

1.5.3 Comparing Mean Numbers of Satellites by Crab Color

To illustrate the use of a qualitative explanatory variable, we next compare the mean satellite counts for the categories of color. Color is a surrogate for the age of the crab, as older crabs tend to have a darker color. It has five categories, but no observations fell in the "light" color. Let us look at the category counts and the sample mean and variance of the number of satellites for each color category.

```
-----------------------------------------------------------------
> table(color)
color # 1 = medium light, 2 = medium, 3 = medium dark, 4 = dark
 1  2  3  4
12 95 44 22
> cbind(by(y,color,mean), by(y,color,var))
     [,1]      [,2]
1  4.0833   9.7197 # color 1 crabs have mean(y) = 4.08, var(y) = 9.72
2  3.2947  10.2739
3  2.2273   6.7378
4  2.0455  13.0931
-----------------------------------------------------------------
```

The majority of the crabs are of medium color, and the mean response decreases as the color gets darker. There is evidence of too much variability for a Poisson distribution to be realistic for y, conditional on color.

We next fit the linear model for a one-way layout with color as a qualitative explanatory factor. By default, without specification of a distribution and link function, the R `glm` function fits the normal linear model:

```
-------------------------------------------------------------------
> fit.color <- glm(y ~ factor(color)) # normal dist. is default
> summary(fit.color)
               Estimate   Std. Error   t value   Pr(>|t|)
(Intercept)      4.0833       0.8985     4.544   1.05e-05
factor(color)2  -0.7886       0.9536    -0.827     0.4094
factor(color)3  -1.8561       1.0137    -1.831     0.0689
factor(color)4  -2.0379       1.1170    -1.824     0.0699
-------------------------------------------------------------------
```

The output does not report a separate estimate for the first category of color, because that parameter is aliased with the other color parameters. To achieve identifiability, R specifies first-category-baseline indicator variables (i.e., for all but the first category). In fact, $\hat{\beta}_0 = \bar{y}_1$, $\hat{\beta}_2 = \bar{y}_2 - \bar{y}_1$, $\hat{\beta}_3 = \bar{y}_3 - \bar{y}_1$, and $\hat{\beta}_4 = \bar{y}_4 - \bar{y}_1$.

If we instead assume a Poisson distribution for the conditional distribution of the response variable, we find:

```
-------------------------------------------------------------------
> fit.color2 <- glm(y ~ factor(color), family=poisson(link=identity))
> summary(fit.color2)
                Estimate   Std. Error   z value   Pr(>|t|)
(Intercept)       4.0833       0.5833     7.000   2.56e-12
factor(color)2   -0.7886       0.6123    -1.288    0.19780
factor(color)3   -1.8561       0.6252    -2.969    0.00299
factor(color)4   -2.0379       0.6582    -3.096    0.00196
-------------------------------------------------------------------
```

The estimates are the same, because the Poisson distribution also has sample means as ML estimates of $\{\mu_i\}$ for a model with a single factor predictor. However, the standard error values are much smaller than under the normal assumption. Why do you think this is? Do you think they are trustworthy?

Finally, we illustrate the simultaneous use of quantitative and qualitative explanatory variables by including both weight and color in the normal model's linear predictor.

```
-------------------------------------------------------------------
> fit.weight.color <- glm(y ~ weight + factor(color))
> summary(fit.weight.color)
                Estimate   Std. Error   t value   Pr(>|t|)
(Intercept)      -0.8232       1.3549    -0.608      0.544
weight            1.8662       0.4018     4.645   6.84e-06
factor(color)2   -0.6181       0.9011    -0.686      0.494
factor(color)3   -1.2404       0.9662    -1.284      0.201
factor(color)4   -1.1882       1.0704    -1.110      0.269
-------------------------------------------------------------------
```

Let us consider the model for this analysis and its model matrix. For response y_i for female crab i, let x_{i1} denote weight, and let $x_{ij} = 1$ when the crab has color j and $x_{ij} = 0$ otherwise, for $j = 2, 3, 4$. Then, the model has linear predictor

$$\mu_i = \beta_0 + \beta_1 x_{i1} + \beta_2 x_{i2} + \beta_3 x_{i3} + \beta_4 x_{i4}.$$

The model has the form $\boldsymbol{\mu} = E(\boldsymbol{y}) = \boldsymbol{X}\boldsymbol{\beta}$ with, using some of the observations shown in Table 1.3,

$$\boldsymbol{y} = \begin{pmatrix} 8 \\ 0 \\ 9 \\ 4 \\ \vdots \end{pmatrix}, \quad \boldsymbol{X}\boldsymbol{\beta} = \begin{pmatrix} 1 & 3.05 & 1 & 0 & 0 \\ 1 & 1.55 & 0 & 1 & 0 \\ 1 & 2.30 & 0 & 0 & 0 \\ 1 & 2.60 & 0 & 1 & 0 \\ \vdots & \vdots & \vdots & \vdots & \vdots \end{pmatrix} \begin{pmatrix} \beta_0 \\ \beta_1 \\ \beta_2 \\ \beta_3 \\ \beta_4 \end{pmatrix}.$$

From $\hat{\beta}_1 = 1.866$, for crabs of a particular color that differ by a kilogram of weight, the estimated mean number of satellites is nearly 2 higher for the heavier crabs. As an exercise, construct a plot of the fit and interpret the color coefficients.

We could also introduce an interaction term, letting the effect of weight vary by color. However, even for the simple models fitted, we have ignored a notable outlier—the exceptionally heavy crab weighing 5.2 kg. As an exercise, you can redo the analyses without that observation to check whether results are much influenced by it. We'll develop better models for these data in Chapter 7.

CHAPTER NOTES

Section 1.1: Components of a Generalized Linear Model

1.1 **GLM**: Nelder and Wedderburn (1972) introduced the class of GLMs and the algorithm for fitting them, but many models in the class were in practice by then.

1.2 **Transform data**: For the transforming-data approach to attempting normality and variance stabilization of y for use with ordinary normal linear models, see Anscombe (1948), Bartlett (1937, 1947), Box and Cox (1964), and Cochran (1940).

1.3 **Random $\boldsymbol{x}$ and measurement error**: When $\boldsymbol{x}$ is random, rather than conditioning on $\boldsymbol{x}$, one can study how the bias in estimated effects depends on the relation between $\boldsymbol{x}$ and the unobserved variables that contribute to the error term. Much of the econometrics literature deals with this (e.g., Greene 2011). Random $\boldsymbol{x}$ is also relevant in the study of errors of measurement of explanatory variables (Buonaccorsi 2010). Such error results in attenuation, that is, biasing of the effect toward zero.

1.4 **Parsimony**: For a proof of the result that a parsimonious reduction of the data to fewer parameters results in improved estimation, see Altham (1984).

Section 1.2: Quantitative/Qualitative Explanatory Variables and Interpreting Effects

1.5 **GLM effect interpretation**: Hoaglin (2012, 2015) discussed appropriate and inappropriate interpretations of parameters in linear models. For studies that use a nonidentity link function g, $\partial\mu_i/\partial x_{ij}$ has value depending on g and μ_i as well as β_j. For sample data and a GLM fit, one way to summarize partial effect j, adjusting for the other explanatory variables, is by $\frac{1}{n}\sum_i(\partial\hat{\mu}_i/\partial x_{ij})$, averaging over the n sample settings. For example, for a Poisson loglinear model, $\frac{1}{n}\sum_i(\partial\hat{\mu}_i/\partial x_{ij}) = \hat{\beta}_j\bar{y}$ (Exercise 7.9).

1.6 **Average causal effect**: Denote two groups to be compared by $x_1 = 0$ and $x_1 = 1$. For GLMs, an alternative effect summary is the *average causal effect*,

$$\frac{1}{n}\sum_{i=1}^{n}\left[E(y_i|x_{i1} = 1, x_{i2}, \dots x_{ip}) - E(y_i|x_{i1} = 0, x_{i2}, \dots, x_{ip})\right].$$

This uses, for each observation i, the expected response for its values of $x_{i2}, \dots, x_{ip}$ if that observation were in group 1 and if that observation were in group 0. For a particular model fit, the sample version estimates the difference between the overall means if all subjects sampled were in group 1 and if all subjects sampled were in group 0. For observational data, this mimics a counterfactual measure to estimate if we could instead conduct an experiment and observe subjects under each treatment group, rather than have half the observations missing. See Gelman and Hill (2006, Chapters 9 and 10), Rubin (1974), and Rosenbaum and Rubin (1983).

EXERCISES

1.1 Suppose that y_i has a $N(\mu_i, \sigma^2)$ distribution, $i = 1, \dots, n$. Formulate the normal linear model as a special case of a GLM, specifying the random component, linear predictor, and link function.

1.2 Link function of a GLM:

a. Describe the purpose of the link function g.

b. The identity link is the standard one with normal responses but is not often used with binary or count responses. Why do you think this is?

1.3 What do you think are the advantages and disadvantages of treating an ordinal explanatory variable as (**a**) quantitative, (**b**) qualitative?

1.4 Extend the model in Section 1.2.1 relating income to racial–ethnic status to include education and interaction explanatory terms. Explain how to interpret parameters when software constructs the indicators using (**a**) first-category-baseline coding, (**b**) last-category-baseline coding.

1.5 Suppose you *standardize* the response and explanatory variables before fitting a linear model (i.e., subtract the means and divide by the standard deviations). Explain how to interpret the resulting *standardized regression coefficients*.

1.6 When $\boldsymbol{X}$ has full rank p, explain why the null space of $\boldsymbol{X}$ consists only of the $\boldsymbol{0}$ vector.

1.7 For any linear model $\boldsymbol{\mu} = \boldsymbol{X\beta}$, is the origin $\mathbf{0}$ in the model space $C(\boldsymbol{X})$? Why or why not?

1.8 A model M has model matrix $\boldsymbol{X}$. A simpler model M_0 results from removing the final term in M, and hence has model matrix $\boldsymbol{X}_0$ that deletes the final column from $\boldsymbol{X}$. From the definition of a column space, explain why $C(\boldsymbol{X}_0)$ is contained in $C(\boldsymbol{X})$.

1.9 For the normal linear model, explain why the expression $y_i = \sum_{j=1}^{p} \beta_j x_{ij} + \epsilon_i$ with $\epsilon_i \sim N(0, \sigma^2)$ is equivalent to $y_i \sim N(\sum_{j=1}^{p} \beta_j x_{ij}, \sigma^2)$.

1.10 GLMs normally use a hierarchical structure by which the presence of a higher-order term implies also including the lower-order terms. Explain why this is sensible, by showing that (**a**) a model that includes an x^2 explanatory variable but not x makes a strong assumption about where the maximum or minimum of $E(y)$ occurs, (**b**) a model that includes x_1x_2 but not x_1 makes a strong assumption about the effect of x_1 when $x_2 = 0$.

1.11 Show the form of $\boldsymbol{X\beta}$ for the linear model for the one-way layout, $E(y_{ij}) = \beta_0 + \beta_i$, using a full-rank model matrix $\boldsymbol{X}$ by employing the constraint $\sum_i \beta_i = 0$ to make parameters identifiable.

1.12 Consider the model for the *two-way layout* for qualitative factors A and B,

$$E(y_{ijk}) = \beta_0 + \beta_i + \gamma_j,$$

for $i = 1, \ldots, r, j = 1, \ldots, c$, and $k = 1, \ldots, n$. This model is *balanced*, having an equal sample size n in each of the rc cells, and assumes an absence of interaction between A and B in their effects on y.

a. For the model as stated, is the parameter vector identifiable? Why or why not?

b. Give an example of a quantity that is (i) not estimable, (ii) estimable. In each case, explain your reasoning.

1.13 Consider the model for the two-way layout shown in the previous exercise. Suppose $r = 2$, $c = 3$, and $n = 2$.

a. Show the form of a full-rank model matrix $\boldsymbol{X}$ and corresponding parameter vector $\boldsymbol{\beta}$ for the model, constraining $\beta_1 = \gamma_1 = 0$ to make $\boldsymbol{\beta}$ identifiable. Explain how to interpret the elements of $\boldsymbol{\beta}$.

b. Show the form of a full-rank model matrix and corresponding parameter vector $\boldsymbol{\beta}$ when you constrain $\sum_i \beta_i = 0$ and $\sum_j \gamma_j = 0$ to make $\boldsymbol{\beta}$ identifiable. Explain how to interpret the elements of $\boldsymbol{\beta}$.

c. In the full-rank case, what is the rank of $\boldsymbol{X}$?

1.14 For the model in the previous exercise with constraints $\beta_1 = \gamma_1 = 0$, generalize the model by adding an interaction term δ_{ij}.

a. Show the new full-rank model matrix. Specify the constraints that $\{\delta_{ij}\}$ satisfy. Indicate how many parameters the δ_{ij} term represents in $\boldsymbol{\beta}$.

b. Show how to write the linear predictor using indicator variables for the factor categories, with the model parameters as coefficients of those indicators and the interaction parameters as coefficients of products of indicators.

1.15 Refer to Exercise 1.12. Now suppose $r = 2$ and $c = 4$, but observations for the first two levels of B occur only at the first level of A, and observations for the last two levels of B occur only at the second level of A. In the corresponding model, $E(y_{ijk}) = \beta_0 + \beta_i + \gamma_{j(i)}$, B is said to be *nested* within A. Specify a full-rank model matrix $\boldsymbol{X}$, and indicate its rank.

1.16 Explain why the vector space of $p \times 1$ vectors $\boldsymbol{\ell}$ such that $\boldsymbol{\ell}^{\mathrm{T}}\boldsymbol{\beta}$ is estimable is $C(\boldsymbol{X}^{\mathrm{T}})$.

1.17 If $\boldsymbol{A}$ is a nonsingular matrix, show that $C(\boldsymbol{X}) = C(\boldsymbol{XA})$. (If two full-rank model matrices correspond to equivalent models, then one model matrix is the other multiplied by a nonsingular matrix.)

1.18 For the linear model for the one-way layout, Section 1.4.1 showed the model matrix that makes parameters identifiable by setting $\beta_1 = 0$. Call this model matrix $\boldsymbol{X}_1$.

a. Suppose we instead obtain identifiability by imposing the constraint $\beta_c = 0$. Show the model matrix, say $\boldsymbol{X}_c$.

b. Show how to obtain $\boldsymbol{X}_1$ as a linear transformation of $\boldsymbol{X}_c$.

1.19 Consider the analysis of covariance model without interaction, denoted by $1 + X + A$.

a. Write the formula for the model in such a way that the parameters are *not* identifiable. Show the corresponding model matrix.

b. For the model parameters in (**a**), give an example of a characteristic that is (i) estimable, (ii) not estimable.

c. Now express the model so that the parameters are identifiable. Explain how to interpret them. Show the model matrix when A has three groups, each containing two observations.

1.20 Show the first five rows of the model matrix for (**a**) the linear model for the horseshoe crabs in Section 1.5.2, (**b**) the model for a one-way layout in Section 1.5.3, (**c**) the model containing both weight and color predictors.

1.21 Littell et al. (2000) described a pharmaceutical clinical trial in which 24 patients were randomly assigned to each of three treatment groups (drug A, drug B, placebo) and compared on a measure of respiratory ability (FEV1 = forced expiratory volume in 1 second, in liters). The data file[8] `FEV.dat` at `www.stat.ufl.edu/~aa/glm/data` has the form shown in Table 1.4. Here, we let y be the response after 1 hour of treatment (variable *fev1* in the data file), x_1 = the baseline measurement prior to administering the drug (variable *base* in the data file), and x_2 = drug (qualitative with labels *a*, *b*, *p* in the data file). Download the data and fit the linear model for y with explanatory variables (**a**) x_1, (**b**) x_2, (**c**) both x_1 and x_2. Interpret model parameter estimates in each case.

Table 1.4 Part of FEV Clinical Trial Data File for Exercise 1.21

Patient	Base	fev1	fev2	fev3	fev4	fev5	fev6	fev7	fev8	Drug
01	2.46	2.68	2.76	2.50	2.30	2.14	2.40	2.33	2.20	a
02	3.50	3.95	3.65	2.93	2.53	3.04	3.37	3.14	2.62	a
03	1.96	2.28	2.34	2.29	2.43	2.06	2.18	2.28	2.29	a
...										
72	2.88	3.04	3.00	3.24	3.37	2.69	2.89	2.89	2.76	p

Complete data (file `FEV.dat`) are at the text website `www.stat.ufl.edu/~aa/glm/data`

1.22 Refer to the analyses in Section 1.5.3 for the horseshoe crab satellites.

a. With color alone as a predictor, why are standard errors much smaller for a Poisson model than for a normal model? Out of these two very imperfect models, which do you trust more for judging significance of the estimates of the color effects? Why?

b. Download the data (file `Crabs.dat`) from `www.stat.ufl.edu/~aa/glm/data`. When weight is also a predictor, identify an outlying observation. Refit the model with color and weight predictors without that observation. Compare results, to investigate the sensitivity of the results to this outlier.

1.23 Another horseshoe crab dataset[9] (`Crabs2.dat` at `www.stat.ufl.edu/~aa/glm/data`) comes from a study of factors that affect sperm traits of male crabs. A response variable, *SpermTotal*, is measured as the log of the total number of sperm in an ejaculate. It has mean 19.3 and standard deviation 2.0. Two explanatory variables are the crab's carapace width (in centimeters, with mean 18.6 and standard deviation 3.0) and color (1 = dark, 2 = medium,

[8]Thanks to Ramon Littell for making these data available.

[9]Thanks to Jane Brockmann and Dan Sasson for making these data available.

3 = light). Explain how to interpret the estimates in the following table. Is the model fitted equivalent to a GLM with the log link for the expected number of sperm? Why or why not?

```
-------------------------------------------------------------
> summary(lm(SpermTotal ~ CW + factor(Color))
Coefficients:
               Estimate  Std. Error  t value  Pr(>|t|)
(Intercept)      11.366       0.638   17.822   < 2e-16
CW                0.391       0.034   11.651   < 2e-16
factor(Color)2    0.809       0.246    3.292   0.00114
factor(Color)3    1.149       0.271    4.239  3.14e-05
-------------------------------------------------------------
```

1.24 For 72 young girls suffering from anorexia, the `Anorexia.dat` file at the text website shows their weights before and after an experimental period. Table 1.5 shows the format of the data. The girls were randomly assigned to receive one of three therapies during this period. A control group received the standard therapy, which was compared to family therapy and cognitive behavioral therapy. Download the data and fit a linear model relating the weight after the experimental period to the initial weight and the therapy. Interpret estimates.

Table 1.5 Weights of Anorexic Girls, in Pounds, Before and After Receiving One of Three Therapies

Cognitive Behavioral		Family Therapy		Control	
Weight Before	Weight After	Weight Before	Weight After	Weight Before	Weight After
80.5	82.2	83.8	95.2	80.7	80.2
84.9	85.6	83.3	94.3	89.4	80.1
81.5	81.4	86.0	91.5	91.8	86.4

Source: Thanks to Brian Everitt for these data. Complete data are at text website.

CHAPTER 2

Linear Models: Least Squares Theory

The next two chapters consider fitting and inference for the ordinary linear model. For n independent observations $\boldsymbol{y} = (y_1, \dots, y_n)^{\mathrm{T}}$ with $\mu_i = E(y_i)$ and $\boldsymbol{\mu} = (\mu_1, \dots, \mu_n)^{\mathrm{T}}$, denote the covariance matrix by

$$\boldsymbol{V} = \mathrm{var}(\boldsymbol{y}) = E[(\boldsymbol{y} - \boldsymbol{\mu})(\boldsymbol{y} - \boldsymbol{\mu})^{\mathrm{T}}].$$

Let $\boldsymbol{X} = \left(x_{ij}\right)$ denote the $n \times p$ model matrix, where x_{ij} is the value of explanatory variable j for observation i. In this chapter we will learn about model fitting when

$$\boldsymbol{\mu} = \boldsymbol{X}\boldsymbol{\beta} \quad \text{with} \quad \boldsymbol{V} = \sigma^2 \boldsymbol{I},$$

where $\boldsymbol{\beta}$ is a $p \times 1$ parameter vector with $p \leq n$ and $\boldsymbol{I}$ is the $n \times n$ identity matrix. The covariance matrix is a diagonal matrix with common value σ^2 for the variance. With the additional assumption of a normal random component, this is the *normal linear model*, which is a generalized linear model (GLM) with identity link function. We will add the normality assumption in the next chapter. Here, though, we will obtain many results about fitting linear models and comparing models that do not require distributional assumptions.

An alternative way to express the ordinary linear model is

$$\boldsymbol{y} = \boldsymbol{X}\boldsymbol{\beta} + \boldsymbol{\epsilon}$$

for an error term $\boldsymbol{\epsilon}$ having $E(\boldsymbol{\epsilon}) = \boldsymbol{0}$ and covariance matrix $\boldsymbol{V} = \mathrm{var}(\boldsymbol{\epsilon}) = \sigma^2 \boldsymbol{I}$. Such a simple additive structure for the error term is not natural for most GLMs, however, except for normal models and latent variable versions of some other models and their extensions with multiple error components. To be consistent with GLM formulas, we will usually express linear models in terms of $E(\boldsymbol{y})$.

Foundations of Linear and Generalized Linear Models, First Edition. Alan Agresti.

Section 2.1 introduces the *least squares* method for fitting linear models. Section 2.2 shows that the least squares model fit $\hat{\boldsymbol{\mu}}$ is a projection of the data $\boldsymbol{y}$ onto the model space $C(\boldsymbol{X})$ generated by the columns of the model matrix. Section 2.3 illustrates for a few simple linear models. Section 2.4 presents summaries of variability in a linear model. Section 2.5 shows how to use *residuals* to summarize how far $\boldsymbol{y}$ falls from $\hat{\boldsymbol{\mu}}$ and to estimate σ^2 and check the model. Following an example in Section 2.6, Section 2.7 proves the *Gauss–Markov theorem*, which specifies a type of optimality that least squares estimators satisfy. That section also generalizes least squares to handle observations that have nonconstant variance or are correlated.

2.1 LEAST SQUARES MODEL FITTING

Having formed a model matrix $\boldsymbol{X}$ and observed $\boldsymbol{y}$, how do we obtain parameter estimates $\hat{\boldsymbol{\beta}}$ and *fitted values* $\hat{\boldsymbol{\mu}} = \boldsymbol{X}\hat{\boldsymbol{\beta}}$ that best satisfy the linear model? The standard approach uses the *least squares* method. This determines the value of $\hat{\boldsymbol{\mu}}$ that minimizes

$$\|\boldsymbol{y} - \hat{\boldsymbol{\mu}}\|^2 = \sum_i (y_i - \hat{\mu}_i)^2 = \sum_{i=1}^{n} \left(y_i - \sum_{j=1}^{p} \hat{\beta}_j x_{ij} \right)^2 .$$

That is, the fitted values $\hat{\boldsymbol{\mu}}$ are such that

$$\|\boldsymbol{y} - \hat{\boldsymbol{\mu}}\| \leq \|\boldsymbol{y} - \boldsymbol{\mu}\| \quad \text{for all} \quad \boldsymbol{\mu} \in C(\boldsymbol{X}).$$

Using least squares corresponds to maximum likelihood when we add a normality assumption to the model. The logarithm[1] of the likelihood for independent observations $y_i \sim N(\mu_i, \sigma^2)$, $i = 1, \ldots, n$, is (in terms of $\{\mu_i\}$)

$$\log \left[\prod_{i=1}^{n} \left(\frac{1}{\sqrt{2\pi}\sigma} e^{-(y_i - \mu_i)^2 / 2\sigma^2} \right) \right] = \text{constant} - \left[\sum_{i=1}^{n} (y_i - \mu_i)^2 \right] / 2\sigma^2 .$$

To maximize the log-likelihood function, we must minimize $\sum_i (y_i - \mu_i)^2$.

2.1.1 The Normal Equations and Least Squares Solution

The expression $L(\boldsymbol{\beta}) = \sum_i (y_i - \mu_i)^2 = \sum_i (y_i - \sum_j \beta_j x_{ij})^2$ is quadratic in $\{\beta_j\}$, so we can minimize it by equating

$$\frac{\partial L}{\partial \beta_j} = 0, \quad j = 1, \ldots, p.$$

[1]In this book, we use the natural logarithm throughout.

These partial derivatives yield the equations:

$$\sum_i (y_i - \mu_i)x_{ij} = 0, \quad j = 1, \dots, p.$$

Thus, the least squares estimates satisfy

$$\sum_{i=1}^{n} y_i x_{ij} = \sum_{i=1}^{n} \hat{\mu}_i x_{ij}, \quad j = 1, \dots, p. \tag{2.1}$$

These are called[2] the *normal equations*. They occur naturally in more general settings than least squares. Chapter 4 shows that these are the likelihood equations for GLMs that use the canonical link function, such as the normal linear model, the binomial logistic regression model, and the Poisson loglinear model.

Using matrix algebra provides an economical expression for the solution of these equations in terms of the model parameter vector $\boldsymbol{\beta}$ for the linear model $\boldsymbol{\mu} = \boldsymbol{X\beta}$. In matrix form,

$$L(\boldsymbol{\beta}) = \|\boldsymbol{y} - \boldsymbol{X\beta}\|^2 = (\boldsymbol{y} - \boldsymbol{X\beta})^{\mathrm{T}}(\boldsymbol{y} - \boldsymbol{X\beta}) = \boldsymbol{y}^{\mathrm{T}}\boldsymbol{y} - 2\boldsymbol{y}^{\mathrm{T}}\boldsymbol{X\beta} + \boldsymbol{\beta}^{\mathrm{T}}\boldsymbol{X}^{\mathrm{T}}\boldsymbol{X\beta}.$$

We use the results for matrix derivatives that

$$\partial(\boldsymbol{a}^{\mathrm{T}}\boldsymbol{\beta})/\partial\boldsymbol{\beta} = \boldsymbol{a} \quad \text{and} \quad \partial(\boldsymbol{\beta}^{\mathrm{T}}\boldsymbol{A\beta})/\partial\boldsymbol{\beta} = (\boldsymbol{A} + \boldsymbol{A}^{\mathrm{T}})\boldsymbol{\beta},$$

which equals $2\boldsymbol{A\beta}$ for symmetric $\boldsymbol{A}$. So, $\partial L(\boldsymbol{\beta})/\partial\boldsymbol{\beta} = -2\boldsymbol{X}^{\mathrm{T}}(\boldsymbol{y} - \boldsymbol{X\beta})$. In terms of $\hat{\boldsymbol{\beta}}$, the normal equations (2.1) are

$$\boldsymbol{X}^{\mathrm{T}}\boldsymbol{y} = \boldsymbol{X}^{\mathrm{T}}\boldsymbol{X}\hat{\boldsymbol{\beta}}. \tag{2.2}$$

Suppose $\boldsymbol{X}$ has full rank p. Then, the $p \times p$ matrix $(\boldsymbol{X}^{\mathrm{T}}\boldsymbol{X})$ also has rank p and is nonsingular, its inverse exists, and the least squares estimator of $\boldsymbol{\beta}$ is

$$\hat{\boldsymbol{\beta}} = (\boldsymbol{X}^{\mathrm{T}}\boldsymbol{X})^{-1}\boldsymbol{X}^{\mathrm{T}}\boldsymbol{y}. \tag{2.3}$$

Since $\partial^2 L(\boldsymbol{\beta})/\partial\boldsymbol{\beta}^2 = 2\boldsymbol{X}^{\mathrm{T}}\boldsymbol{X}$ is positive definite, the minimum rather than maximum of $L(\boldsymbol{\beta})$ occurs at $\hat{\boldsymbol{\beta}}$.

2.1.2 Hat Matrix and Moments of Estimators

The fitted values $\hat{\boldsymbol{\mu}}$ are a linear transformation of $\boldsymbol{y}$,

$$\hat{\boldsymbol{\mu}} = \boldsymbol{X}\hat{\boldsymbol{\beta}} = \boldsymbol{X}(\boldsymbol{X}^{\mathrm{T}}\boldsymbol{X})^{-1}\boldsymbol{X}^{\mathrm{T}}\boldsymbol{y}.$$

[2]Here "normal" refers not to the normal distribution but to orthogonality of $(\boldsymbol{y} - \hat{\boldsymbol{\mu}})$ with each column of $\boldsymbol{X}$.

The $n \times n$ matrix $\boldsymbol{H} = \boldsymbol{X}(\boldsymbol{X}^{\mathrm{T}}\boldsymbol{X})^{-1}\boldsymbol{X}^{\mathrm{T}}$ is called[3] the *hat matrix* because it linearly transforms $\boldsymbol{y}$ to $\hat{\boldsymbol{\mu}} = \boldsymbol{H}\boldsymbol{y}$. The hat matrix $\boldsymbol{H}$ is a *projection matrix*, projecting $\boldsymbol{y}$ to $\hat{\boldsymbol{\mu}}$ in the model space $C(\boldsymbol{X})$. We define projection matrices and study their properties in Section 2.2.

Recall that for a matrix of constants $\boldsymbol{A}$, $E(\boldsymbol{A}\boldsymbol{y}) = \boldsymbol{A}E(\boldsymbol{y})$ and $\mathrm{var}(\boldsymbol{A}\boldsymbol{y}) = \boldsymbol{A}\mathrm{var}(\boldsymbol{y})\boldsymbol{A}^{\mathrm{T}}$. So, the mean and variance of the least squares estimator are

$$\begin{aligned} E(\hat{\boldsymbol{\beta}}) &= E[(\boldsymbol{X}^{\mathrm{T}}\boldsymbol{X})^{-1}\boldsymbol{X}^{\mathrm{T}}\boldsymbol{y}] = (\boldsymbol{X}^{\mathrm{T}}\boldsymbol{X})^{-1}\boldsymbol{X}^{\mathrm{T}}E(\boldsymbol{y}) = (\boldsymbol{X}^{\mathrm{T}}\boldsymbol{X})^{-1}\boldsymbol{X}^{\mathrm{T}}\boldsymbol{X}\boldsymbol{\beta} = \boldsymbol{\beta}, \\ \mathrm{var}(\hat{\boldsymbol{\beta}}) &= (\boldsymbol{X}^{\mathrm{T}}\boldsymbol{X})^{-1}\boldsymbol{X}^{\mathrm{T}}(\sigma^2\boldsymbol{I})\boldsymbol{X}(\boldsymbol{X}^{\mathrm{T}}\boldsymbol{X})^{-1} = \sigma^2(\boldsymbol{X}^{\mathrm{T}}\boldsymbol{X})^{-1}. \end{aligned} \tag{2.4}$$

For the ordinary linear model with normal random component, since $\hat{\boldsymbol{\beta}}$ is a linear function of $\boldsymbol{y}$, $\hat{\boldsymbol{\beta}}$ has a normal distribution with these two moments.

2.1.3 Bivariate Linear Model and Regression Toward the Mean

We illustrate least squares using the linear model with a single explanatory variable for a single response, that is, the "bivariate linear model"

$$E(y_i) = \beta_0 + \beta_1 x_i.$$

From (2.1) with $x_{i1} = 1$ and $x_{i2} = x_i$, the normal equations are

$$\sum_{i=1}^{n} y_i = n\beta_0 + \beta_1 \sum_{i=1}^{n} x_i, \qquad \sum_{i=1}^{n} x_i y_i = \beta_0\left(\sum_{i=1}^{n} x_i\right) + \beta_1 \sum_{i=1}^{n} x_i^2.$$

By straightforward solution of these two equations, you can verify that the least squares estimates are

$$\hat{\beta}_1 = \frac{\sum_{i=1}^{n}(x_i - \bar{x})(y_i - \bar{y})}{\sum_{i=1}^{n}(x_i - \bar{x})^2}, \qquad \hat{\beta}_0 = \bar{y} - \hat{\beta}_1\bar{x}. \tag{2.5}$$

From the solution for $\hat{\beta}_0$, the least squares fitted equation $\hat{\mu}_i = \hat{\beta}_0 + \hat{\beta}_1 x_i$ satisfies $\bar{y} = \hat{\beta}_0 + \hat{\beta}_1\bar{x}$. It passes through the center of gravity of the data, that is, the point $(\bar{x}, \bar{y})$. The analogous result holds for the linear model with multiple explanatory variables and the point $(\bar{x}_1, \dots, \bar{x}_p, \bar{y})$.

Denote the sample marginal standard deviations of x and y by s_x and s_y. From the Pearson product-moment formula, the sample *correlation*

$$r = \mathrm{corr}(x, y) = \frac{\sum_{i=1}^{n}(x_i - \bar{x})(y_i - \bar{y})}{\sqrt{[\sum_{i=1}^{n}(x_i - \bar{x})^2][\sum_{i=1}^{n}(y_i - \bar{y})^2]}} = \hat{\beta}_1\left(\frac{s_x}{s_y}\right).$$

[3]According to Hoaglin and Welsch (1978), John Tukey proposed the term "hat matrix."

One implication of this is that the correlation equals the slope when both variables are standardized to have $s_x = s_y = 1$. Another implication is that an increase of s_x in x corresponds to a change of $\hat{\beta}_1 s_x = r s_y$ in $\hat{\mu}$. This equation highlights the famous result of Francis Galton (1886) that there is *regression toward the mean*: When $|r| < 1$, a standard deviation change in x corresponds to a predicted change of less than a standard deviation in y.

In practice, explanatory variables are often *centered* before entering them in a model by taking $x_i^* = x_i - \bar{x}$. For the centered values, $\bar{x}^* = 0$, so

$$\hat{\beta}_0 = \bar{y}, \quad \hat{\beta}_1 = \left(\sum_{i=1}^{n} x_i^* y_i \right) \Big/ \sum_{i=1}^{n} (x_i^*)^2.$$

Under centering, $(X^T X)$ is a diagonal matrix with elements n and $\sum_i (x_i^*)^2$. Thus, the covariance matrix for $\hat{\beta}$ is then

$$\text{var}(\hat{\beta}) = \sigma^2 (X^T X)^{-1} = \sigma^2 \begin{pmatrix} 1/n & 0 \\ 0 & 1/[\sum_{i=1}^{n} (x_i - \bar{x})^2] \end{pmatrix}.$$

Centering the explanatory variable does not affect $\hat{\beta}_1$ and its variance but results in $\text{corr}(\hat{\beta}_0, \hat{\beta}_1) = 0$.

You can show directly from the expression for the model matrix X that the hat matrix for the bivariate linear model is

$$H = X(X^T X)^{-1} X^T = \begin{pmatrix} \frac{1}{n} + \frac{(x_1-\bar{x})^2}{\sum_i (x_i-\bar{x})^2} & \cdots & \frac{1}{n} + \frac{(x_1-\bar{x})(x_n-\bar{x})}{\sum_i (x_i-\bar{x})^2} \\ \vdots & \ddots & \vdots \\ \frac{1}{n} + \frac{(x_n-\bar{x})(x_1-\bar{x})}{\sum_i (x_i-\bar{x})^2} & \cdots & \frac{1}{n} + \frac{(x_n-\bar{x})^2}{\sum_i (x_i-\bar{x})^2} \end{pmatrix}.$$

In Section 2.5.4 we will see that each diagonal element of the hat matrix is a measure of the observation's potential influence on the model fit.

2.1.4 Least Squares Solutions When X Does Not Have Full Rank

When X does not have full rank, neither does $(X^T X)$ in the normal equations. A solution $\hat{\beta}$ of the normal equations then uses a *generalized inverse* of $(X^T X)$, denoted by $(X^T X)^-$. Recall that for a matrix A, G is a generalized inverse if and only if $AGA = A$. Generalized inverses always exist but may not be unique. The least squares estimate $\hat{\beta} = (X^T X)^- X^T y$ is not then unique, reflecting that β is not identifiable.

With rank$(X) < p$, the null space $N(X)$ has nonzero elements. For any solution $\hat{\beta}$ of the normal equations $X^T y = X^T X \hat{\beta}$ and any element $\gamma \in N(X)$, $\tilde{\beta} = \hat{\beta} + \gamma$ is also a solution. This follows because $X\gamma = 0$ and $X^T X(\hat{\beta} + \gamma) = X^T X \hat{\beta}$. Although there are multiple solutions $\tilde{\beta}$ for estimating β, $\hat{\mu} = X\tilde{\beta}$ is invariant to the solution (as are estimates of estimable quantities), because $X\tilde{\beta} = X(\hat{\beta} + \gamma)$ has the same fitted values as given by $\hat{\mu} = X\hat{\beta}$.

Likewise, if $\ell^T\beta$ is estimable, then $\ell^T\hat{\beta}$ is the same for all solutions to the normal equations. This follows because $\ell^T\hat{\beta}$ can be expressed as $a^TX\hat{\beta}$ for some a, and fitted values are identical for all $\hat{\beta}$.

2.1.5 Orthogonal Subspaces and Residuals

Section 1.3.1 introduced the model space $C(X)$ of $X\beta$ values for all the possible β values. This vector space is a linear subspace of n-dimensional Euclidean space, $\mathbb{R}^n$. Many results in this chapter relate to *orthogonality* for this representation, so let us recall a few basic results about orthogonality for vectors and for vector subspaces of $\mathbb{R}^n$:

- Two vectors u and v in $\mathbb{R}^n$ are *orthogonal* if $u^Tv = 0$. Geometrically, orthogonal vectors are perpendicular in $\mathbb{R}^n$.
- For a vector subspace W of $\mathbb{R}^n$, the subspace of vectors v such that for any $u \in W$, $u^Tv = 0$, is the *orthogonal complement* of W, denoted by $W^\perp$.
- Orthogonal complements W and $W^\perp$ in $\mathbb{R}^n$ satisfy $\dim(W) + \dim(W^\perp) = n$.
- For orthogonal complements W and $W^\perp$, any $y \in \mathbb{R}^n$ has a unique[4] *orthogonal decomposition* $y = y_1 + y_2$ with $y_1 \in W$ and $y_2 \in W^\perp$.

Figure 2.1 portrays the key result about orthogonal decompositions into components in orthogonal complement subspaces. In the decomposition $y = y_1 + y_2$, we will see in Section 2.2 that y_1 is the *orthogonal projection* of y onto W.

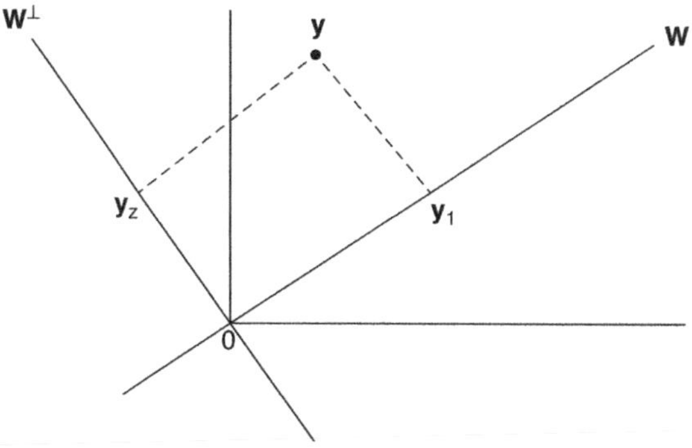

Figure 2.1 Orthogonal decomposition of y into components y_1 in subspace W plus y_2 in orthogonal complement subspace $W^\perp$.

[4] The proof uses the Gram–Schmidt process on a basis for $\mathbb{R}^n$ that extends one for W to construct an orthogonal basis of vectors in W and vectors in $W^\perp$; y_1 and y_2 are then linear combinations of these two sets of vectors. See Christensen (2011, pp. 414–416).

Now, suppose $\boldsymbol{W} = C(\boldsymbol{X})$, the model space spanned by the columns of a model matrix $\boldsymbol{X}$. Vectors in its orthogonal complement $C(\boldsymbol{X})^{\perp}$ in $\mathbb{R}^n$ are orthogonal with any vector in $C(\boldsymbol{X})$, and hence with each column of $\boldsymbol{X}$. So any vector $\boldsymbol{v}$ in $C(\boldsymbol{X})^{\perp}$ satisfies $\boldsymbol{X}^{\mathrm{T}}\boldsymbol{v} = \boldsymbol{0}$, and $C(\boldsymbol{X})^{\perp}$ is the null space of $\boldsymbol{X}^{\mathrm{T}}$, denoted $N(\boldsymbol{X}^{\mathrm{T}})$. We will observe next that $C(\boldsymbol{X})^{\perp}$ is an *error space* that contains differences between possible data vectors and model-fitted values for such data.

The normal equations $\boldsymbol{X}^{\mathrm{T}}\boldsymbol{y} = \boldsymbol{X}^{\mathrm{T}}\boldsymbol{X}\hat{\boldsymbol{\beta}}$ that the least squares estimates satisfy can be expressed as

$$\boldsymbol{X}^{\mathrm{T}}(\boldsymbol{y} - \boldsymbol{X}\hat{\boldsymbol{\beta}}) = \boldsymbol{X}^{\mathrm{T}}\boldsymbol{e} = \boldsymbol{0},$$

where $\boldsymbol{e} = (\boldsymbol{y} - \boldsymbol{X}\hat{\boldsymbol{\beta}})$. The elements of $\boldsymbol{e}$ are prediction errors when we use $\hat{\boldsymbol{\mu}} = \boldsymbol{X}\hat{\boldsymbol{\beta}}$ to predict $\boldsymbol{y}$ or $\boldsymbol{\mu}$. They are called *residuals*. The normal equations tell us that the residual vector $\boldsymbol{e}$ is orthogonal to each column of $\boldsymbol{X}$. So $\boldsymbol{e}$ is in the orthogonal complement to the model space $C(\boldsymbol{X})$, that is, $\boldsymbol{e}$ is in $C(\boldsymbol{X})^{\perp} = N(\boldsymbol{X}^{\mathrm{T}})$. Figure 2.2 portrays the orthogonality of $\boldsymbol{e}$ with $C(\boldsymbol{X})$.

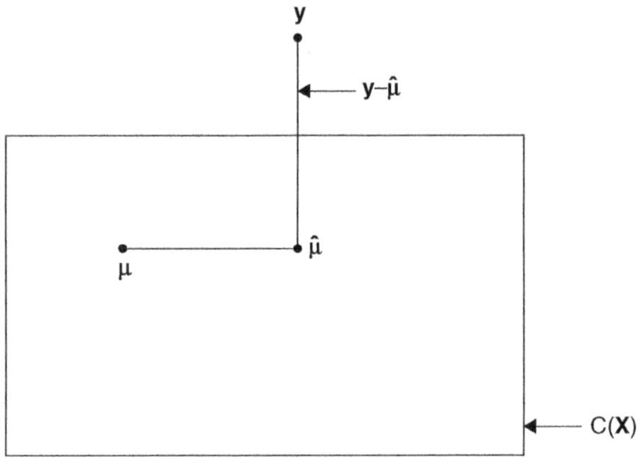

Figure 2.2 Orthogonality of residual vector $\boldsymbol{e} = (\boldsymbol{y} - \hat{\boldsymbol{\mu}})$ with vectors in the model space $C(\boldsymbol{X})$ for a linear model $\boldsymbol{\mu} = \boldsymbol{X}\boldsymbol{\beta}$.

Some linear model analyses decompose $\boldsymbol{y}$ into several orthogonal components. An orthogonal decomposition of $\mathbb{R}^n$ into k orthogonal subspaces $\{\boldsymbol{W}_i\}$ is one for which any $\boldsymbol{u} \in \boldsymbol{W}_i$ and any $\boldsymbol{v} \in \boldsymbol{W}_j$ have $\boldsymbol{u}^{\mathrm{T}}\boldsymbol{v} = 0$ for all $i \neq j$, and any $\boldsymbol{y} \in \mathbb{R}^n$ can be uniquely expressed as $\boldsymbol{y} = \boldsymbol{y}_1 + \cdots + \boldsymbol{y}_k$ with $\boldsymbol{y}_i \in \boldsymbol{W}_i$ for $i = 1, \ldots, k$.

2.1.6 Alternatives to Least Squares

In fitting a linear model, why minimize $\sum_i (y_i - \hat{\mu}_i)^2$ rather than some other metric, such as $\sum_i |y_i - \hat{\mu}_i|$? Minimizing a sum of squares is mathematically and computationally much simpler. For this reason, least squares has a long history, dating back to a published article by the French mathematician Adrien-Marie Legendre (1805), followed by German mathematician Carl Friedrich Gauss's claim in 1809 of priority[5] in having used it since 1795. Another motivation, seen at the beginning of this section, is that it corresponds to maximum likelihood when we add the normality assumption. Yet another motivation, presented in Section 2.7, shows that the least squares estimator is best in the class of estimators that are unbiased and linear in the data.

Recent research has developed alternatives to least squares that give sensible answers in situations that are unstable in some way. For example, instability may be caused by a severe outlier, because in minimizing a sum of squared deviations, a single observation can have substantial influence. Instability could also be caused by an explanatory variable being linearly determined (or nearly so) by the other explanatory variables, a condition called *collinearity* (Section 4.6.5). Finally, instability occurs in using least squares with datasets containing very large numbers of explanatory variables, sometimes even with $p > n$.

Regularization methods add an additional term to the function minimized, such as $\lambda \sum_j |\beta_j|$ or $\lambda \sum_j \beta_j^2$ for some constant λ. The solution then is a smoothing of the least squares estimates that shrinks them toward zero. This is highly effective when we have a large number of explanatory variables but expect few of them to have a substantively important effect. Unless n is extremely large, because of sampling variability the ordinary least squares estimates $\{\hat{\beta}_j\}$ then tend to be much larger in absolute value than the true values $\{\beta_j\}$. Shrinkage toward 0 causes a bias in the estimators but tends to reduce the variance substantially, resulting in their tending to be closer to $\{\beta_j\}$.

Regularization methods are increasingly important as more applications involve "big data." Chapter 11, which introduces extensions of the GLM, presents some regularization methods.

2.2 PROJECTIONS OF DATA ONTO MODEL SPACES

We have mentioned that the least squares fit $\hat{\boldsymbol{\mu}}$ is a projection of the data $\boldsymbol{y}$ onto the model space $C(\boldsymbol{X})$, and the hat matrix $\boldsymbol{H}$ that projects $\boldsymbol{y}$ to $\hat{\boldsymbol{\mu}}$ is a *projection matrix*. We now explain more precisely what is meant by projection of a vector $\boldsymbol{y} \in \mathbb{R}^n$ onto a vector subspace such as $C(\boldsymbol{X})$.

2.2.1 Projection Matrices

Definition. A square matrix $\boldsymbol{P}$ is a *projection matrix* onto a vector subspace $\boldsymbol{W}$ if

(1) for all $\boldsymbol{y} \in \boldsymbol{W}$, $\boldsymbol{P}\boldsymbol{y} = \boldsymbol{y}$.

(2) for all $\boldsymbol{y} \in \boldsymbol{W}^{\perp}$, $\boldsymbol{P}\boldsymbol{y} = \boldsymbol{0}$.

For a projection matrix $\boldsymbol{P}$, since $\boldsymbol{P}\boldsymbol{y}$ is a linear combination of the columns of $\boldsymbol{P}$, the vector subspace $\boldsymbol{W}$ onto which $\boldsymbol{P}$ projects is the column space $C(\boldsymbol{P})$. The projection

[5]See Stigler (1981, 1986, Chapters 1 and 4) for details.

matrix $\boldsymbol{P}$ onto $\boldsymbol{W}$ projects an arbitrary $\boldsymbol{y} \in \mathbb{R}^n$ to its component $\boldsymbol{y}_1 \in \boldsymbol{W}$ for the unique orthogonal decomposition of $\boldsymbol{y}$ into $\boldsymbol{y}_1 + \boldsymbol{y}_2$ using $\boldsymbol{W}$ and $\boldsymbol{W}^\perp$ (Recall Section 2.1.5). We next list this and some other properties of a projection matrix:

- If $\boldsymbol{y} = \boldsymbol{y}_1 + \boldsymbol{y}_2$ with $\boldsymbol{y}_1 \in \boldsymbol{W}$ and $\boldsymbol{y}_2 \in \boldsymbol{W}^\perp$, then $\boldsymbol{Py} = \boldsymbol{P}(\boldsymbol{y}_1 + \boldsymbol{y}_2) = \boldsymbol{Py}_1 + \boldsymbol{Py}_2 = \boldsymbol{y}_1 + \boldsymbol{0} = \boldsymbol{y}_1$. Since the orthogonal decomposition is unique, so too is the projection onto $\boldsymbol{W}$ unique[6].
- The projection matrix onto a subspace $\boldsymbol{W}$ is unique. To see why, suppose $\boldsymbol{P}^*$ is another one. Then for the orthogonal decomposition $\boldsymbol{y} = \boldsymbol{y}_1 + \boldsymbol{y}_2$ with $\boldsymbol{y}_1 \in \boldsymbol{W}$, $\boldsymbol{P}^*\boldsymbol{y} = \boldsymbol{y}_1 = \boldsymbol{Py}$ for all $\boldsymbol{y}$. Hence, $\boldsymbol{P} = \boldsymbol{P}^*$. (Recall that if $\boldsymbol{Ay} = \boldsymbol{By}$ for all $\boldsymbol{y}$, then $\boldsymbol{A} = \boldsymbol{B}$.)
- $\boldsymbol{I} - \boldsymbol{P}$ is the projection matrix onto $\boldsymbol{W}^\perp$. For an arbitrary $\boldsymbol{y} = \boldsymbol{y}_1 + \boldsymbol{y}_2$ with $\boldsymbol{y}_1 \in \boldsymbol{W}$ and $\boldsymbol{y}_2 \in \boldsymbol{W}^\perp$, we have $\boldsymbol{Py} = \boldsymbol{y}_1$ and $(\boldsymbol{I} - \boldsymbol{P})\boldsymbol{y} = \boldsymbol{y} - \boldsymbol{y}_1 = \boldsymbol{y}_2$. Thus,

$$\boldsymbol{y} = \boldsymbol{Py} + (\boldsymbol{I} - \boldsymbol{P})\boldsymbol{y}$$

  provides the orthogonal decomposition of $\boldsymbol{y}$. Also, $\boldsymbol{P}(\boldsymbol{I} - \boldsymbol{P})\boldsymbol{y} = \boldsymbol{0}$.
- $\boldsymbol{P}$ is a projection matrix if and only if it is symmetric and idempotent (i.e., $\boldsymbol{P}^2 = \boldsymbol{P}$).

We will use this last property often, so let us see why it is true. First, we suppose $\boldsymbol{P}$ is symmetric and idempotent and show that this implies $\boldsymbol{P}$ is a projection matrix. For any $\boldsymbol{v} \in C(\boldsymbol{P})$ (the subspace onto which $\boldsymbol{P}$ projects), $\boldsymbol{v} = \boldsymbol{Pb}$ for some $\boldsymbol{b}$. Then, $\boldsymbol{Pv} = \boldsymbol{P}(\boldsymbol{Pb}) = \boldsymbol{P}^2\boldsymbol{b} = \boldsymbol{Pb} = \boldsymbol{v}$. For any $\boldsymbol{v} \in C(\boldsymbol{P})^\perp$, we have $\boldsymbol{P}^{\mathrm{T}}\boldsymbol{v} = \boldsymbol{0}$, but this is also $\boldsymbol{Pv}$ by the symmetry of $\boldsymbol{P}$. So, we have shown that $\boldsymbol{P}$ is a projection matrix onto $C(\boldsymbol{P})$. Second, to prove the converse, we suppose $\boldsymbol{P}$ is the projection matrix onto $C(\boldsymbol{P})$ and show this implies $\boldsymbol{P}$ is symmetric and idempotent. For any $\boldsymbol{v} \in \mathbb{R}^n$, let $\boldsymbol{v} = \boldsymbol{v}_1 + \boldsymbol{v}_2$ with $\boldsymbol{v}_1 \in C(\boldsymbol{P})$ and $\boldsymbol{v}_2 \in C(\boldsymbol{P})^\perp$. Since

$$\boldsymbol{P}^2\boldsymbol{v} = \boldsymbol{P}(\boldsymbol{P}(\boldsymbol{v}_1 + \boldsymbol{v}_2)) = \boldsymbol{Pv}_1 = \boldsymbol{v}_1 = \boldsymbol{Pv},$$

we have $\boldsymbol{P}^2 = \boldsymbol{P}$. To show symmetry, let $\boldsymbol{w} = \boldsymbol{w}_1 + \boldsymbol{w}_2$ be any other vector in $\mathbb{R}^n$, with $\boldsymbol{w}_1 \in C(\boldsymbol{P})$ and $\boldsymbol{w}_2 \in C(\boldsymbol{P})^\perp$. Since $\boldsymbol{I} - \boldsymbol{P}$ is the projection matrix onto $C(\boldsymbol{P})^\perp$,

$$\boldsymbol{w}^{\mathrm{T}}\boldsymbol{P}^{\mathrm{T}}(\boldsymbol{I} - \boldsymbol{P})\boldsymbol{v} = \boldsymbol{w}_1^{\mathrm{T}}\boldsymbol{v}_2 = 0.$$

Since this is true for any $\boldsymbol{v}$ and $\boldsymbol{w}$, we have $\boldsymbol{P}^{\mathrm{T}}(\boldsymbol{I} - \boldsymbol{P}) = \boldsymbol{0}$, or $\boldsymbol{P}^{\mathrm{T}} = \boldsymbol{P}^{\mathrm{T}}\boldsymbol{P}$. Since $\boldsymbol{P}^{\mathrm{T}}\boldsymbol{P}$ is symmetric, so is $\boldsymbol{P}^{\mathrm{T}}$ and hence $\boldsymbol{P}$.

Next, here are two useful properties about the eigenvalues and the rank of a projection matrix.

- The eigenvalues of any projection matrix $\boldsymbol{P}$ are all 0 and 1.

[6]The projection defined here is sometimes called an *orthogonal projection*, because $\boldsymbol{Py}$ and $\boldsymbol{y} - \boldsymbol{Py}$ are orthogonal vectors. This text considers only orthogonal and not oblique projections, and we take "projection" to be synonymous with "orthogonal projection."

This follows from the definitions of a projection matrix and an eigenvalue. For an eigenvalue λ of $\boldsymbol{P}$ with eigenvector $\boldsymbol{v}$, $\boldsymbol{P}\boldsymbol{v} = \lambda\boldsymbol{v}$; but either $\boldsymbol{P}\boldsymbol{v} = \boldsymbol{v}$ (if $\boldsymbol{v} \in \boldsymbol{W}$) or $\boldsymbol{P}\boldsymbol{v} = \boldsymbol{0}$ (if $\boldsymbol{v} \in \boldsymbol{W}^{\perp}$), so $\lambda = 1$ or 0. In fact, this is a property of symmetric, idempotent matrices.

- For any projection matrix $\boldsymbol{P}$, rank($\boldsymbol{P}$) = trace($\boldsymbol{P}$), the sum of its main diagonal elements.

This follows because the trace of a square matrix is the sum of the eigenvalues, and for symmetric matrices the rank is the number of nonzero eigenvalues. Since the eigenvalues of $\boldsymbol{P}$ (which is symmetric) are all 0 and 1, the sum of its eigenvalues equals the number of nonzero eigenvalues.

Finally, we state a useful property[7] about decompositions of the identity matrix into a sum of projection matrices:

- Suppose $\{\boldsymbol{P}_i\}$ are symmetric $n \times n$ matrices such that $\sum_i \boldsymbol{P}_i = \boldsymbol{I}$. Then, the following three conditions are equivalent:
  1. $\boldsymbol{P}_i$ is idempotent for each i.
  2. $\boldsymbol{P}_i\boldsymbol{P}_j = \boldsymbol{0}$ for all $i \neq j$.
  3. $\sum_i \text{rank}(\boldsymbol{P}_i) = n$.

The aspect we will use is that symmetric idempotent matrices (thus, projection matrices) that satisfy $\sum_i \boldsymbol{P}_i = \boldsymbol{I}$ also satisfy $\boldsymbol{P}_i\boldsymbol{P}_j = \boldsymbol{0}$ for all $i \neq j$. The proof of this is a by-product of a key result of the next chapter (Cochran's theorem) about independent chi-squared quadratic forms.

2.2.2 Projection Matrices for Linear Model Spaces

Let $\boldsymbol{P}_X$ denote the projection matrix onto the model space $C(\boldsymbol{X})$ corresponding to a model matrix $\boldsymbol{X}$ for a linear model. We next present some properties for this particular case.

- If $\boldsymbol{X}$ has full rank, then $\boldsymbol{P}_X$ is the hat matrix, $\boldsymbol{H} = \boldsymbol{X}(\boldsymbol{X}^{\mathrm{T}}\boldsymbol{X})^{-1}\boldsymbol{X}^{\mathrm{T}}$.

This follows by noting that $\boldsymbol{H}$ satisfies the two parts of the definition of a projection matrix for $C(\boldsymbol{X})$:

- If $\boldsymbol{y} \in C(\boldsymbol{X})$, then $\boldsymbol{y} = \boldsymbol{X}\boldsymbol{b}$ for some $\boldsymbol{b}$. So

$$\boldsymbol{H}\boldsymbol{y} = \boldsymbol{X}(\boldsymbol{X}^{\mathrm{T}}\boldsymbol{X})^{-1}\boldsymbol{X}^{\mathrm{T}}\boldsymbol{y} = \boldsymbol{X}(\boldsymbol{X}^{\mathrm{T}}\boldsymbol{X})^{-1}\boldsymbol{X}^{\mathrm{T}}\boldsymbol{X}\boldsymbol{b} = \boldsymbol{X}\boldsymbol{b} = \boldsymbol{y}.$$

- Recall from Section 2.1.5 that $C(\boldsymbol{X})^{\perp} = N(\boldsymbol{X}^{\mathrm{T}})$. If $\boldsymbol{y} \in N(\boldsymbol{X}^{\mathrm{T}})$, then $\boldsymbol{X}^{\mathrm{T}}\boldsymbol{y} = \boldsymbol{0}$, and thus, $\boldsymbol{H}\boldsymbol{y} = \boldsymbol{X}(\boldsymbol{X}^{\mathrm{T}}\boldsymbol{X})^{-1}\boldsymbol{X}^{\mathrm{T}}\boldsymbol{y} = \boldsymbol{0}$.

[7] For a proof, see Bapat (2000, p. 60).

We have seen that $\boldsymbol{H}$ projects $\boldsymbol{y}$ to the least squares fit $\hat{\boldsymbol{\mu}} = \boldsymbol{X}\hat{\boldsymbol{\beta}}$.

- If $\boldsymbol{X}$ does not have full rank, then $\boldsymbol{P}_X = \boldsymbol{X}(\boldsymbol{X}^{\mathrm{T}}\boldsymbol{X})^{-}\boldsymbol{X}^{\mathrm{T}}$. Moreover, $\boldsymbol{P}_X$ is invariant to the choice for the generalized inverse $(\boldsymbol{X}^{\mathrm{T}}\boldsymbol{X})^{-}$.

The proof is outlined in Exercise 2.16. Thus, $\hat{\boldsymbol{\mu}}$ is invariant to the choice of the solution $\hat{\boldsymbol{\beta}}$ of the normal equations. In particular, if rank($\boldsymbol{X}$) $= r < p$ and if $\boldsymbol{X}_0$ is any matrix having r columns that form a basis for $C(\boldsymbol{X})$, then $\boldsymbol{P}_X = \boldsymbol{X}_0(\boldsymbol{X}_0^{\mathrm{T}}\boldsymbol{X}_0)^{-1}\boldsymbol{X}_0^{\mathrm{T}}$. This follows by the same proof just given for the full-rank case.

- If $\boldsymbol{X}$ and $\boldsymbol{W}$ are model matrices satisfying $C(\boldsymbol{X}) = C(\boldsymbol{W})$, then $\boldsymbol{P}_X = \boldsymbol{P}_W$.

To see why, for an arbitrary $\boldsymbol{y} \in \mathbb{R}^n$, we use the orthogonal decompositions $\boldsymbol{y} = \boldsymbol{P}_X\boldsymbol{y} + (\boldsymbol{I} - \boldsymbol{P}_X)\boldsymbol{y}$ and $\boldsymbol{y} = \boldsymbol{P}_W\boldsymbol{y} + (\boldsymbol{I} - \boldsymbol{P}_W)\boldsymbol{y}$. By the uniqueness of the decomposition, $\boldsymbol{P}_X\boldsymbol{y} = \boldsymbol{P}_W\boldsymbol{y}$. But $\boldsymbol{y}$ is arbitrary, so $\boldsymbol{P}_X = \boldsymbol{P}_W$. It follows that $\hat{\boldsymbol{\mu}}$ and $\boldsymbol{e} = \boldsymbol{y} - \boldsymbol{X}\hat{\boldsymbol{\beta}}$ are also the same for both models. For example, projection matrices and model fits are not affected by reparameterization, such as changing the indicator coding for a factor.

- **Nested model projections:** When model a is a special case of model b, with projection matrices $\boldsymbol{P}_a$ and $\boldsymbol{P}_b$ for model matrices $\boldsymbol{X}_a$ and $\boldsymbol{X}_b$, then $\boldsymbol{P}_a\boldsymbol{P}_b = \boldsymbol{P}_b\boldsymbol{P}_a = \boldsymbol{P}_a$ and $\boldsymbol{P}_b - \boldsymbol{P}_a$ is also a projection matrix.

When one model is a special case of another, we say that the models are *nested*. To show this result, for an arbitrary $\boldsymbol{y}$, we use the unique orthogonal decomposition $\boldsymbol{y} = \boldsymbol{y}_1 + \boldsymbol{y}_2$, with $\boldsymbol{y}_1 \in C(\boldsymbol{X}_a)$ and $\boldsymbol{y}_2 \in C(\boldsymbol{X}_a)^{\perp}$. Then, $\boldsymbol{P}_a\boldsymbol{y} = \boldsymbol{y}_1$, from which $\boldsymbol{P}_b(\boldsymbol{P}_a\boldsymbol{y}) = \boldsymbol{P}_b\boldsymbol{y}_1 = \boldsymbol{y}_1 = \boldsymbol{P}_a\boldsymbol{y}$, since the fitted value for the simpler model also satisfies the more complex model. So $\boldsymbol{P}_b\boldsymbol{P}_a = \boldsymbol{P}_a$. But $\boldsymbol{P}_b\boldsymbol{P}_a = \boldsymbol{P}_a\boldsymbol{P}_b$ because of their symmetry, so we have also that $\boldsymbol{P}_a\boldsymbol{P}_b = \boldsymbol{P}_a$. Since $\boldsymbol{P}_a\boldsymbol{y} = \boldsymbol{P}_a(\boldsymbol{P}_b\boldsymbol{y})$, we see that $\boldsymbol{P}_a\boldsymbol{y}$ is also the projection of $\boldsymbol{P}_b\boldsymbol{y}$ onto $C(\boldsymbol{X}_a)$. Since $\boldsymbol{P}_b\boldsymbol{P}_a = \boldsymbol{P}_a\boldsymbol{P}_b = \boldsymbol{P}_a$ and $\boldsymbol{P}_a$ and $\boldsymbol{P}_b$ are idempotent,

$$(\boldsymbol{P}_b - \boldsymbol{P}_a)(\boldsymbol{P}_b - \boldsymbol{P}_a) = \boldsymbol{P}_b - \boldsymbol{P}_a.$$

So $(\boldsymbol{P}_b - \boldsymbol{P}_a)$ is also a projection matrix. In fact, an extended orthogonal decomposition incorporates such difference projection matrices,

$$\boldsymbol{y} = \boldsymbol{I}\boldsymbol{y} = [\boldsymbol{P}_a + (\boldsymbol{P}_b - \boldsymbol{P}_a) + (\boldsymbol{I} - \boldsymbol{P}_b)]\boldsymbol{y} = \boldsymbol{y}_1 + \boldsymbol{y}_2 + \boldsymbol{y}_3.$$

Here $\boldsymbol{P}_a$ projects $\boldsymbol{y}$ onto $C(\boldsymbol{X}_a)$, $(\boldsymbol{P}_b - \boldsymbol{P}_a)$ projects $\boldsymbol{y}$ to its component in $C(\boldsymbol{X}_b)$ that is orthogonal with $C(\boldsymbol{X}_a)$, and $(\boldsymbol{I} - \boldsymbol{P}_b)$ projects $\boldsymbol{y}$ to its component in $C(\boldsymbol{X}_b)^{\perp}$.

2.2.3 Example: The Geometry of a Linear Model

We next illustrate the geometry that underlies the projections for linear models. We do this for two simple models for which we can easily portray projections graphically.

The first model has a single quantitative explanatory variable,

$$\mu_i = E(y_i) = \beta x_i, \quad i = 1, \dots, n,$$

but does not contain an intercept. Its model matrix $\boldsymbol{X}$ is the $n \times 1$ vector $(x_1, \dots, x_n)^{\mathrm{T}}$. Figure 2.3 portrays the model, the data, and the fit. The response values $\boldsymbol{y} = (y_1, \dots, y_n)^{\mathrm{T}}$ are a point in $\mathbb{R}^n$. The explanatory variable values $\boldsymbol{X}$ are another such point. The linear predictor values $\boldsymbol{X}\beta$ for all the possible real values for β trace out a line in $\mathbb{R}^n$ that passes through the origin. This is the model space $C(\boldsymbol{X})$. The model fit $\hat{\boldsymbol{\mu}} = \boldsymbol{P}_X\boldsymbol{y} = \hat{\beta}\boldsymbol{X}$ is the orthogonal projection of $\boldsymbol{y}$ onto the model space line.

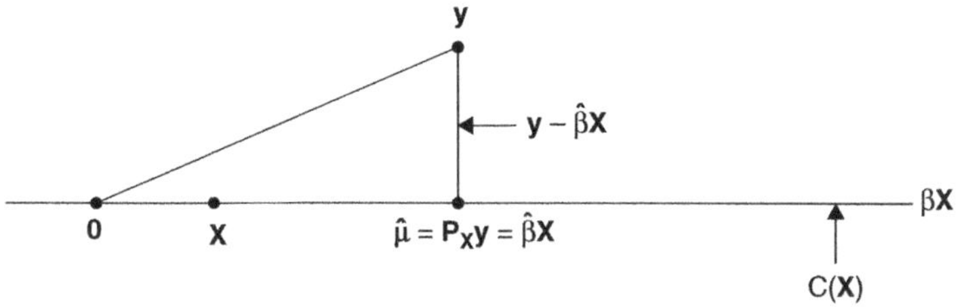

Figure 2.3 Portrayal of simple linear model with quantitative predictor x and no intercept, showing the observations $\boldsymbol{y}$, the model matrix $\boldsymbol{X}$ of predictor values, and the fit $\hat{\boldsymbol{\mu}} = \boldsymbol{P}_X\boldsymbol{y} = \hat{\beta}\boldsymbol{X}$.

Next, we extend the modeling to handle two quantitative explanatory variables. Consider the models

$$E(y_i) = \beta_{y1}x_{i1}, \quad E(y_i) = \beta_{y2}x_{i2}, \quad E(y_i) = \beta_{y1\cdot 2}x_{i1} + \beta_{y2\cdot 1}x_{i2}.$$

We use Yule's notation to reflect that $\beta_{y1\cdot 2}$ and $\beta_{y2\cdot 1}$ typically differ from β_{y1} and β_{y2}, as discussed in Section 1.2.3. Figure 2.4 portrays the data and the three model fits. When evaluated for all real $\beta_{y1\cdot 2}$ and $\beta_{y2\cdot 1}$, $\boldsymbol{\mu}$ traces out a plane in $\mathbb{R}^n$ that passes through the origin. The projection $\boldsymbol{P}_{12}\boldsymbol{y} = \hat{\beta}_{y1\cdot 2}\boldsymbol{X}_1 + \hat{\beta}_{y2\cdot 1}\boldsymbol{X}_2$ gives the least squares fit using both predictors together. The projection $\boldsymbol{P}_1\boldsymbol{y} = \hat{\beta}_{y1}\boldsymbol{X}_1$ onto the model space for $\boldsymbol{X}_1 = (x_{11}, \dots, x_{n1})^{\mathrm{T}}$ gives the least squares fit when x_1 is the sole predictor. The projection $\boldsymbol{P}_2\boldsymbol{y} = \hat{\beta}_{y2}\boldsymbol{X}_2$ onto the model space for $\boldsymbol{X}_2 = (x_{12}, \dots, x_{n2})^{\mathrm{T}}$ gives the least squares fit when x_2 is the sole predictor.

From the result in Section 2.2.2 that $\boldsymbol{P}_a\boldsymbol{P}_b = \boldsymbol{P}_a$ when model a is a special case of model b, $\boldsymbol{P}_1\boldsymbol{y}$ is also the projection of $\boldsymbol{P}_{12}\boldsymbol{y}$ onto the model space for $\boldsymbol{X}_1$, and $\boldsymbol{P}_2\boldsymbol{y}$ is also the projection of $\boldsymbol{P}_{12}\boldsymbol{y}$ onto the model space for $\boldsymbol{X}_2$. These ideas extend directly to models with several explanatory variables as well as an intercept term.

2.2.4 Orthogonal Columns and Parameter Orthogonality

Although $\hat{\beta}_{y1}$ in the reduced model $\boldsymbol{\mu} = \beta_{y1}\boldsymbol{X}_1$ is usually not the same as $\hat{\beta}_{y1\cdot 2}$ in the full model $\boldsymbol{\mu} = \beta_{y1\cdot 2}\boldsymbol{X}_1 + \beta_{y2\cdot 1}\boldsymbol{X}_2$, the effects are identical when $\boldsymbol{X}_1$ is orthogonal

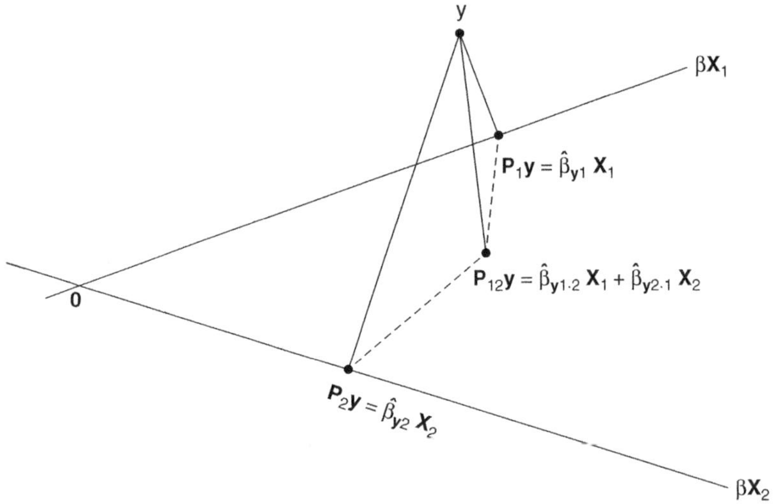

Figure 2.4 Portrayal of linear model with two quantitative explanatory variables, showing the observations $\boldsymbol{y}$ and the fits $\boldsymbol{P}_1\boldsymbol{y} = \hat{\beta}_{y1}\boldsymbol{X}_1$, $\boldsymbol{P}_2\boldsymbol{y} = \hat{\beta}_{y2}\boldsymbol{X}_2$, and $\boldsymbol{P}_{12}\boldsymbol{y} = \hat{\beta}_{y1\cdot 2}\boldsymbol{X}_1 + \hat{\beta}_{y2\cdot 1}\boldsymbol{X}_2$.

with $\boldsymbol{X}_2$. We show this for a more general context in which $\boldsymbol{X}_1$ and $\boldsymbol{X}_2$ may each refer to a *set* of explanatory variables.

We partition the model matrix and parameter vector for the full model into

$$\boldsymbol{X\beta} = (\boldsymbol{X}_1 : \boldsymbol{X}_2)\begin{pmatrix} \boldsymbol{\beta}_1 \\ \boldsymbol{\beta}_2 \end{pmatrix} = \boldsymbol{X}_1\boldsymbol{\beta}_1 + \boldsymbol{X}_2\boldsymbol{\beta}_2.$$

Then, $\boldsymbol{\beta}_1$ and $\boldsymbol{\beta}_2$ are said to be *orthogonal parameters* if each column from $\boldsymbol{X}_1$ is orthogonal with each column from $\boldsymbol{X}_2$, that is, $\boldsymbol{X}_1^{\mathrm{T}}\boldsymbol{X}_2 = \boldsymbol{0}$. In this case

$$\boldsymbol{X}^{\mathrm{T}}\boldsymbol{X} = \begin{pmatrix} \boldsymbol{X}_1^{\mathrm{T}}\boldsymbol{X}_1 & \boldsymbol{0} \\ \boldsymbol{0} & \boldsymbol{X}_2^{\mathrm{T}}\boldsymbol{X}_2 \end{pmatrix} \quad \text{and} \quad \boldsymbol{X}^{\mathrm{T}}\boldsymbol{y} = \begin{pmatrix} \boldsymbol{X}_1^{\mathrm{T}}\boldsymbol{y} \\ \boldsymbol{X}_2^{\mathrm{T}}\boldsymbol{y} \end{pmatrix}.$$

Because of this, $(\boldsymbol{X}^{\mathrm{T}}\boldsymbol{X})^{-1}$ also has block diagonal structure, and $\hat{\boldsymbol{\beta}}_1 = (\boldsymbol{X}_1^{\mathrm{T}}\boldsymbol{X}_1)^{-1}\boldsymbol{X}_1^{\mathrm{T}}\boldsymbol{y}$ from fitting the reduced model $\boldsymbol{\mu} = \boldsymbol{X}_1\boldsymbol{\beta}_1$ is identical to $\hat{\boldsymbol{\beta}}_1$ from fitting $\boldsymbol{\mu} = \boldsymbol{X}_1\boldsymbol{\beta}_1 + \boldsymbol{X}_2\boldsymbol{\beta}_2$. The same property holds if each column from $\boldsymbol{X}_1$ is orthogonal with each column from $\boldsymbol{X}_2$ after centering each column of $\boldsymbol{X}_1$ (i.e., from subtracting off the mean) or centering each column of $\boldsymbol{X}_2$. In that case, the correlation is zero for each such pair (Exercise 2.19), and the result is a consequence of a property to be presented in Section 2.5.6 showing that the same partial effects occur in regression modeling using two sets of residuals.

2.2.5 Pythagoras's Theorem Applications for Linear Models

The projection matrix plays a key role for linear models. The first important result is that the projection matrix projects the data vector $\boldsymbol{y}$ to the fitted value vector $\hat{\boldsymbol{\mu}}$ that is the *unique* point in the model space $C(\boldsymbol{X})$ that is closest to $\boldsymbol{y}$.

Data projection gives unique least squares fit: For each $\boldsymbol{y} \in \mathbb{R}^n$ and its projection $\boldsymbol{P_X y} = \hat{\boldsymbol{\mu}}$ onto the model space $C(\boldsymbol{X})$ for a linear model $\boldsymbol{\mu} = \boldsymbol{X\beta}$,

$$\|\boldsymbol{y} - \boldsymbol{P_X y}\| \leq \|\boldsymbol{y} - \boldsymbol{z}\| \quad \text{for all} \quad \boldsymbol{z} \in C(\boldsymbol{X}),$$

with equality if and only if $\boldsymbol{z} = \boldsymbol{P_X y}$.

To show why this is true, for an arbitrary $\boldsymbol{z} \in C(\boldsymbol{X})$ we express

$$\boldsymbol{y} - \boldsymbol{z} = (\boldsymbol{y} - \boldsymbol{P_X y}) + (\boldsymbol{P_X y} - \boldsymbol{z}).$$

Now $(\boldsymbol{y} - \boldsymbol{P_X y}) = (\boldsymbol{I} - \boldsymbol{P_X})\boldsymbol{y}$ is in $C(\boldsymbol{X})^{\perp} = N(\boldsymbol{X}^{\mathrm{T}})$, whereas $(\boldsymbol{P_X y} - \boldsymbol{z})$ is in $C(\boldsymbol{X})$ because each component is in $C(\boldsymbol{X})$. Since the subspaces $C(\boldsymbol{X})$ and $C(\boldsymbol{X})^{\perp}$ are orthogonal complements,

$$\|\boldsymbol{y} - \boldsymbol{z}\|^2 = \|\boldsymbol{y} - \boldsymbol{P_X y}\|^2 + \|\boldsymbol{P_X y} - \boldsymbol{z}\|^2,$$

because $\boldsymbol{u}^{\mathrm{T}}\boldsymbol{v} = 0$ for any $\boldsymbol{u} \in C(\boldsymbol{X})$ and $\boldsymbol{v} \in C(\boldsymbol{X})^{\perp}$. It follows from this application of Pythagoras's theorem that $\|\boldsymbol{y} - \boldsymbol{z}\|^2 \geq \|\boldsymbol{y} - \boldsymbol{P_X y}\|^2$, with equality if and only if $\boldsymbol{P_X y} = \boldsymbol{z}$.

The fact that the fitted values $\hat{\boldsymbol{\mu}} = \boldsymbol{P_X y}$ provide the unique least squares solution for $\boldsymbol{\mu}$ is no surprise, as Section 2.2.2 showed that the projection matrix for a linear model is the hat matrix, which projects the data to the least squares fit. Likewise, $(\boldsymbol{I} - \boldsymbol{P_X})$ is the projection onto $C(\boldsymbol{X})^{\perp}$, and the residual vector $\boldsymbol{e} = (\boldsymbol{I} - \boldsymbol{P_X})\boldsymbol{y} = \boldsymbol{y} - \hat{\boldsymbol{\mu}}$ falls in that error space.

Here is another application of Pythagoras's theorem for linear models.

True and sample residuals: For the fitted values $\hat{\boldsymbol{\mu}}$ of a linear model $\boldsymbol{\mu} = \boldsymbol{X\beta}$ obtained by least squares,

$$\|\boldsymbol{y} - \boldsymbol{\mu}\|^2 = \|\boldsymbol{y} - \hat{\boldsymbol{\mu}}\|^2 + \|\hat{\boldsymbol{\mu}} - \boldsymbol{\mu}\|^2.$$

This follows by decomposing $(\boldsymbol{y} - \boldsymbol{\mu}) = (\boldsymbol{y} - \hat{\boldsymbol{\mu}}) + (\hat{\boldsymbol{\mu}} - \boldsymbol{\mu})$ and using the fact that $(\hat{\boldsymbol{\mu}} - \boldsymbol{\mu})$, which is in $C(\boldsymbol{X})$, is orthogonal to $(\boldsymbol{y} - \hat{\boldsymbol{\mu}})$, which is in $C(\boldsymbol{X})^{\perp}$. In particular, the data tend to be closer to the model fit than to the true means, and the fitted values vary less than the data. From this result, a plot of $\|\boldsymbol{y} - \boldsymbol{\mu}\|^2$ against $\boldsymbol{\beta}$ shows a quadratic function that is minimized at $\hat{\boldsymbol{\beta}}$. Figure 2.5 portrays this for the case of a one-dimensional parameter β.

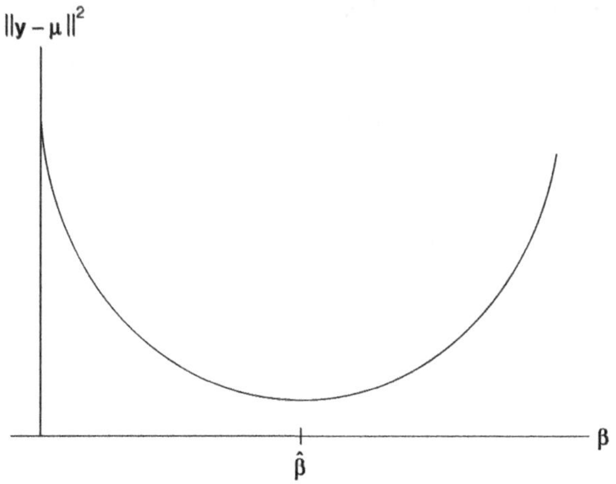

Figure 2.5 For a linear model $\boldsymbol{\mu} = \boldsymbol{X\beta}$, the sum of squares $\|\boldsymbol{y} - \boldsymbol{\mu}\|^2$ is minimized at the least squares estimate $\hat{\boldsymbol{\beta}}$.

Here is a third application of Pythagoras's theorem for linear models.

Data = fit + residuals: For the fitted values $\hat{\boldsymbol{\mu}}$ of a linear model $\boldsymbol{\mu} = \boldsymbol{X\beta}$ obtained by least squares,

$$\|\boldsymbol{y}\|^2 = \|\hat{\boldsymbol{\mu}}\|^2 + \|\boldsymbol{y} - \hat{\boldsymbol{\mu}}\|^2.$$

This uses the decomposition illustrated in Figure 2.6,

$$\boldsymbol{y} = \hat{\boldsymbol{\mu}} + (\boldsymbol{y} - \hat{\boldsymbol{\mu}}) = \boldsymbol{P_X y} + (\boldsymbol{I} - \boldsymbol{P_X})\boldsymbol{y}, \qquad \text{that is, } \textbf{data = fit + residuals},$$

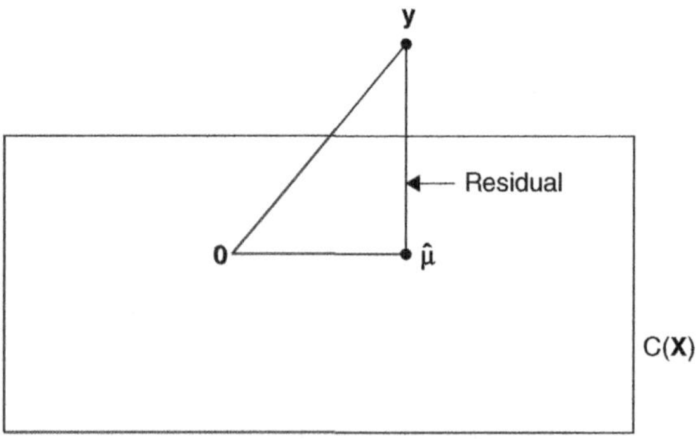

Figure 2.6 Pythagoras's theorem for a linear model applies to the data vector, the fitted values, and the residual vector; that is, *data* = *fit* + *residuals*.

with $\hat{\mu}$ in $C(X)$ orthogonal to $(y - \hat{\mu})$ in $C(X)^{\perp}$. It also follows using the symmetry and idempotence of projection matrices, from

$$\begin{aligned}\|y\|^2 &= y^{\mathrm{T}}y = y^{\mathrm{T}}[P_X + (I - P_X)]y = y^{\mathrm{T}}P_Xy + y^{\mathrm{T}}(I - P_X)y \\ &= y^{\mathrm{T}}P_X^{\mathrm{T}}P_Xy + y^{\mathrm{T}}(I - P_X)^{\mathrm{T}}(I - P_X)y = \hat{\mu}^{\mathrm{T}}\hat{\mu} + (y - \hat{\mu})^{\mathrm{T}}(y - \hat{\mu}).\end{aligned}$$

A consequence of $\hat{\mu}$ being the projection of y onto the model space is that the squared length of y equals the squared length of $\hat{\mu}$ plus the squared length of the residual vector. The orthogonality of the fitted values and the residuals is a key result that we will use often.

Linear model analyses that decompose y into several orthogonal components have a corresponding sum-of-squares decomposition. Let $P_1, P_2, \ldots, P_k$ be projection matrices satisfying an orthogonal decomposition:

$$I = P_1 + P_2 + \cdots + P_k.$$

That is, each projection matrix refers to a vector subspace in a decomposition of $\mathbb{R}^n$ using orthogonal subspaces. The unique decomposition of y into elements in those orthogonal subspaces is

$$y = Iy = P_1y + P_2y + \cdots + P_ky = y_1 + \cdots + y_k.$$

The corresponding sum-of-squares decomposition is

$$y^{\mathrm{T}}y = y^{\mathrm{T}}P_1y + \cdots + y^{\mathrm{T}}P_ky.$$

2.3 LINEAR MODEL EXAMPLES: PROJECTIONS AND SS DECOMPOSITIONS

We next use a few simple linear models to illustrate concepts introduced in this chapter. We construct projection matrices for the models and show corresponding sum-of-squares decompositions for the data.

2.3.1 Example: Null Model

The simplest model assumes independent observations with constant variance σ^2 but has only an intercept term,

$$E(y_i) = \beta, \quad i = 1, \ldots, n.$$

It is called the *null model*, because it has no explanatory variables. This is the relevant model for inference about the population marginal mean for a response variable. We also use it as a baseline for comparison with models containing explanatory variables to analyze whether those variables collectively have a significant effect.

With its model matrix $\boldsymbol{X}$, the null model is

$$E(\boldsymbol{y}) = \boldsymbol{X}\boldsymbol{\beta} \quad \text{for} \quad \boldsymbol{X} = \begin{pmatrix} 1 \\ 1 \\ \vdots \\ 1 \end{pmatrix} = \mathbf{1}_n.$$

The null model has projection matrix:

$$\boldsymbol{P}_X = \boldsymbol{X}(\boldsymbol{X}^{\mathrm{T}}\boldsymbol{X})^{-1}\boldsymbol{X}^{\mathrm{T}} = \boldsymbol{X}n^{-1}\boldsymbol{X}^{\mathrm{T}} = \frac{1}{n}\mathbf{1}_n\mathbf{1}_n^{\mathrm{T}},$$

which is a $n \times n$ matrix with $1/n$ in each entry. The fitted values for the model are therefore

$$\hat{\boldsymbol{\mu}} = \boldsymbol{P}_X\boldsymbol{y} = \frac{1}{n}\mathbf{1}_n\mathbf{1}_n^{\mathrm{T}}\boldsymbol{y} = \bar{y}\mathbf{1}_n,$$

the $n \times 1$ vector with the sample mean in each element. In the sum-of-squares decomposition $\boldsymbol{y}^{\mathrm{T}}\boldsymbol{y} = \boldsymbol{y}^{\mathrm{T}}\boldsymbol{P}_X\boldsymbol{y} + \boldsymbol{y}^{\mathrm{T}}(\boldsymbol{I} - \boldsymbol{P}_X)\boldsymbol{y}$,

$$\boldsymbol{y}^{\mathrm{T}}\boldsymbol{P}_X\boldsymbol{y} = \boldsymbol{y}^{\mathrm{T}}\bar{y}\mathbf{1}_n = \bar{y}\boldsymbol{y}^{\mathrm{T}}\mathbf{1}_n = \left(\sum_{i=1}^{n} y_i\right)^2 / n = n\bar{y}^2,$$

$$\boldsymbol{y}^{\mathrm{T}}(\boldsymbol{I} - \boldsymbol{P}_X)\boldsymbol{y} = \boldsymbol{y}^{\mathrm{T}}(\boldsymbol{I} - \boldsymbol{P}_X)^{\mathrm{T}}(\boldsymbol{I} - \boldsymbol{P}_X)\boldsymbol{y} = (\boldsymbol{y} - \bar{y}\mathbf{1}_n)^{\mathrm{T}}(\boldsymbol{y} - \bar{y}\mathbf{1}_n) = \sum_{i=1}^{n}(y_i - \bar{y})^2.$$

The sum-of-squares decomposition for "data = fit + residual" simplifies[8] to

$$\sum_{i=1}^{n} y_i^2 = n\bar{y}^2 + \sum_{i=1}^{n}(y_i - \bar{y})^2.$$

Let us visualize the model space, the projection matrix, and these sums of squares for a simple dataset—a sample of size $n = 2$ with $y_1 = 3$ and $y_2 = 4$. For it, $\hat{\beta} = \bar{y} = 3.5$. For two-dimensional Euclidean space, Figure 2.7 shows the model space for the null model. This is the straight line passing through the origin with slope 1, equating the two components. The figure shows the data point $\boldsymbol{y}$ having coordinates (3, 4) and its projection to the point $\hat{\boldsymbol{\mu}} = \boldsymbol{X}\hat{\boldsymbol{\beta}}$ in the model space having coordinates (3.5, 3.5) for

[8]Historical comment: Until the modern era of statistical software, introductory statistics textbooks suggested reducing complexity in by-hand computing of the numerator of the sample variance for an integer-valued response by instead computing $\sum_i y_i^2 - n\bar{y}^2$.

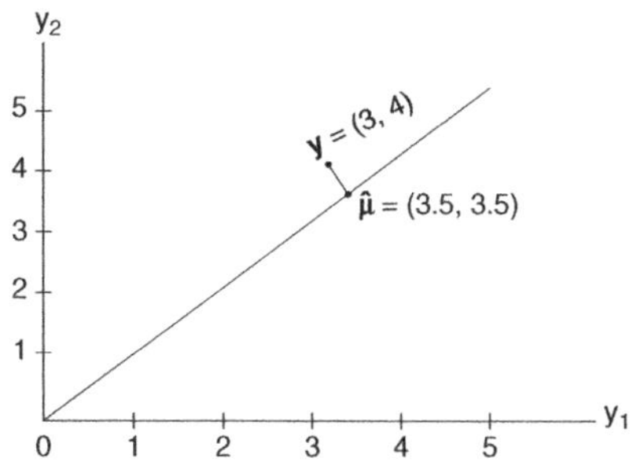

Figure 2.7 Data space and the projection of the data onto the null model space for $n = 2$ observations, $y_1 = 3$, and $y_2 = 4$.

which each component is the sample mean of the two observations. The projection matrix for the model space is

$$\boldsymbol{P_X} = \boldsymbol{X}(\boldsymbol{X}^{\mathrm{T}}\boldsymbol{X})^{-1}\boldsymbol{X}^{\mathrm{T}} = \frac{1}{n}\boldsymbol{1}_n\boldsymbol{1}_n^{\mathrm{T}} = \begin{pmatrix} 1/2 & 1/2 \\ 1/2 & 1/2 \end{pmatrix}.$$

The projection matrix for the error space is

$$\boldsymbol{I} - \boldsymbol{P_X} = \begin{pmatrix} 1/2 & -1/2 \\ -1/2 & 1/2 \end{pmatrix}.$$

This is the projection onto the orthogonal complement subspace spanned by the vector with coordinates $(-1, 1)$, that is, the line passing through the origin with slope -1. For these two observations, for example, $(\boldsymbol{I} - \boldsymbol{P_X})\boldsymbol{y} = (-1/2, 1/2)^{\mathrm{T}}$. The total variability is $\boldsymbol{y}^{\mathrm{T}}\boldsymbol{y} = 25$, which decomposes into $\boldsymbol{y}^{\mathrm{T}}\boldsymbol{P_X}\boldsymbol{y} = \hat{\boldsymbol{\mu}}^{\mathrm{T}}\hat{\boldsymbol{\mu}} = n\bar{y}^2 = 24.5$ plus $\boldsymbol{y}^{\mathrm{T}}(\boldsymbol{I} - \boldsymbol{P_X})\boldsymbol{y} = \sum_i (y_i - \bar{y})^2 = 0.5$.

2.3.2 Example: Model for the One-way Layout

We next extend the null model to the linear model for the *one-way layout*. This is the model for comparing means $\{\mu_i\}$ for c groups, first considered in a GLM context in Section 1.3.3. Let y_{ij} denote observation j in group i, for $j = 1, \dots, n_i$ and $i = 1, \dots, c$, with $n = \sum_i n_i$. With independent observations, an important case having this data format is the *completely randomized experimental design*: Experimental units are randomly assigned to c treatments, such as in a randomized clinical trial in which subjects with a particular malady are randomly assigned to receive drug A, drug B, or a placebo. The model for $\mu_i = E(y_{ij})$ has linear predictor

$$E(y_{ij}) = \beta_0 + \beta_i.$$

Identifiability requires a constraint, such as $\beta_1 = 0$. For now, we continue without a constraint. In Exercise 2.33 you can analyze how the following discussion simplifies by adding one.

As in Section 1.3.3, we express the linear predictor as $\boldsymbol{\mu} = \boldsymbol{X}\boldsymbol{\beta}$ with

$$\boldsymbol{X}\boldsymbol{\beta} = \begin{pmatrix} \mathbf{1}_{n_1} & \mathbf{1}_{n_1} & \mathbf{0}_{n_1} & \cdots & \mathbf{0}_{n_1} \\ \mathbf{1}_{n_2} & \mathbf{0}_{n_2} & \mathbf{1}_{n_2} & \cdots & \mathbf{0}_{n_2} \\ \vdots & \vdots & \vdots & \ddots & \vdots \\ \mathbf{1}_{n_c} & \mathbf{0}_{n_c} & \mathbf{0}_{n_c} & \cdots & \mathbf{1}_{n_c} \end{pmatrix} \begin{pmatrix} \beta_0 \\ \beta_1 \\ \vdots \\ \beta_c \end{pmatrix}.$$

Here, $\boldsymbol{\beta}$ has $p = c + 1$ elements, but $\boldsymbol{X}$ has rank c. For this model matrix,

$$\boldsymbol{X}^{\mathrm{T}}\boldsymbol{X} = \begin{pmatrix} n & n_1 & n_2 & \cdots & n_c \\ n_1 & n_1 & 0 & \cdots & 0 \\ n_2 & 0 & n_2 & \cdots & 0 \\ \vdots & \vdots & \vdots & \ddots & \vdots \\ n_c & 0 & 0 & \cdots & n_c \end{pmatrix}.$$

You can verify that a generalized inverse for this matrix is

$$(\boldsymbol{X}^{\mathrm{T}}\boldsymbol{X})^{-} = \begin{pmatrix} 0 & 0 & 0 & \cdots & 0 \\ 0 & 1/n_1 & 0 & \cdots & 0 \\ 0 & 0 & 1/n_2 & \cdots & 0 \\ \vdots & \vdots & \vdots & \ddots & \vdots \\ 0 & 0 & 0 & \cdots & 1/n_c \end{pmatrix},$$

by checking that $(\boldsymbol{X}^{\mathrm{T}}\boldsymbol{X})(\boldsymbol{X}^{\mathrm{T}}\boldsymbol{X})^{-}(\boldsymbol{X}^{\mathrm{T}}\boldsymbol{X}) = \boldsymbol{X}^{\mathrm{T}}\boldsymbol{X}$. Since

$$\boldsymbol{X}^{\mathrm{T}}\boldsymbol{y} = (n\bar{y}, n_1\bar{y}_1, \ldots, n_c\bar{y}_c)^{\mathrm{T}},$$

this generalized inverse yields the model parameter estimate, $\hat{\boldsymbol{\beta}} = (\hat{\beta}_0, \hat{\beta}_1, \ldots, \hat{\beta}_c)^{\mathrm{T}} = (0, \bar{y}_1, \ldots, \bar{y}_c)^{\mathrm{T}}$.

For this model matrix and generalized inverse, the corresponding projection matrix is

$$P_X = X(X^TX)^-X^T = \begin{pmatrix} \frac{1}{n_1}\mathbf{1}_{n_1}\mathbf{1}_{n_1}^T & 0 & \cdots & 0 \\ 0 & \frac{1}{n_2}\mathbf{1}_{n_2}\mathbf{1}_{n_2}^T & \cdots & 0 \\ \vdots & \vdots & \ddots & \vdots \\ 0 & 0 & \cdots & \frac{1}{n_c}\mathbf{1}_{n_c}\mathbf{1}_{n_c}^T \end{pmatrix}. \quad (2.6)$$

This yields the fitted values

$$\hat{\mu} = P_X y = (\bar{y}_1, \dots, \bar{y}_1, \bar{y}_2, \dots, \bar{y}_2, \dots, \bar{y}_c, \dots, \bar{y}_c)^T.$$

The same fitted values are generated by all generalized inverses and their corresponding $\hat{\beta}$, since they all have the same projection matrix. That is, the estimable quantities $\{\mu_i = \beta_0 + \beta_i\}$ from the linear predictor have $\{\bar{y}_i\}$ as their least squares estimates.

With this parameterization, any individual β_i in the linear predictor is not estimable. So which linear combinations $\sum_i a_i\beta_i$ of $\{\beta_i\}$ are estimable? In Section 1.4.2 we noted that $\ell^T\beta$ is estimable when $\ell \in C(X^T)$, that is, in the row space of X. Now, the null space $N(X)$ is the orthogonal complement for $C(X^T)$. Since $\dim[C(X)] + \dim[N(X)] = p$, with here $p = c + 1$, and since X has rank c, $N(X)$ has dimension 1. Now $(1, -1, \dots, -1)^T$ is in $N(X)$, since from the above expression for X, $X(1, -1, \dots, -1)^T = \mathbf{0}$ (i.e, the first column is the sum of the other c columns). Thus, this vector serves as a basis for $N(X)$ and is orthogonal to $C(X^T)$. So $\sum_i a_i\beta_i$ is estimable when $(0, a_1, \dots, a_c)$ is orthogonal to $(1, -1, \dots, -1)$. It follows that $\sum_i a_i\beta_i$ is estimable if and only if $\sum_i a_i = 0$, that is, when $\sum_i a_i\beta_i$ is a *contrast*. Then, since $\beta_i = \mu_i - \beta_0$, a contrast has form $\sum_i a_i\beta_i = \sum_i a_i\mu_i$, and its estimate is $\sum_i a_i\bar{y}_i$. An example of a contrast is $\beta_1 - \beta_2 = \mu_1 - \mu_2$, with estimate $\bar{y}_1 - \bar{y}_2$.

The normal equations (2.2), namely $X^Ty = X^TX\hat{\beta}$, are also satisfied by

$$\hat{\beta} = \begin{pmatrix} \hat{\beta}_0 \\ \hat{\beta}_1 \\ \vdots \\ \hat{\beta}_c \end{pmatrix} = \begin{pmatrix} 0 \\ \bar{y}_1 \\ \vdots \\ \bar{y}_c \end{pmatrix} - \lambda \begin{pmatrix} -1 \\ 1 \\ \vdots \\ 1 \end{pmatrix}$$

for an arbitrary real value of λ. In fact, these are the non-full-rank solutions corresponding to the various generalized inverses. For example, with $\lambda = \bar{y}_1$, we have $\hat{\beta} = (\bar{y}_1, 0, \bar{y}_2 - \bar{y}_1, \dots, \bar{y}_c - \bar{y}_1)^T$. This corresponds to the full-rank solution obtained by imposing the constraint $\beta_1 = 0$ to make the parameters identifiable.

2.3.3 Sums of Squares and ANOVA Table for One-Way Layout

For the linear model for the one-way layout, we next use an orthogonal decomposition of the data to induce a corresponding sum-of-squares decomposition and a table for displaying data analyses. We use

$$y_{ij} = \bar{y} + (\bar{y}_i - \bar{y}) + (y_{ij} - \bar{y}_i),$$

for $j = 1, \ldots, n_i$ and $i = 1, \ldots, c$. Let $\boldsymbol{P}_0 = \frac{1}{n}\boldsymbol{1}_n\boldsymbol{1}_n^{\mathrm{T}}$ denote the projection matrix for the null model. That model has $\boldsymbol{X} = \boldsymbol{1}_n$, which is the first column of $\boldsymbol{X}$ as given above for the one-way layout, and $\boldsymbol{P}_0$ projects $\boldsymbol{y}$ to the overall mean vector $\bar{y}\boldsymbol{1}_n$. When we view the data decomposition for the entire $n \times 1$ vector $\boldsymbol{y}$ expressed as $\boldsymbol{Iy}$ for the $n \times n$ identity matrix $\boldsymbol{I}$, it uses a decomposition of $\boldsymbol{I}$ into three projection matrices using $\boldsymbol{P}_0$ and $\boldsymbol{P}_X$ from (2.6),

$$\boldsymbol{y} = \boldsymbol{Iy} = [\boldsymbol{P}_0 + (\boldsymbol{P}_X - \boldsymbol{P}_0) + (\boldsymbol{I} - \boldsymbol{P}_X)]\boldsymbol{y}.$$

The corresponding sum-of-squares decomposition is

$$\boldsymbol{y}^{\mathrm{T}}\boldsymbol{y} = \boldsymbol{y}^{\mathrm{T}}\boldsymbol{Iy} = \boldsymbol{y}^{\mathrm{T}}[\boldsymbol{P}_0 + (\boldsymbol{P}_X - \boldsymbol{P}_0) + (\boldsymbol{I} - \boldsymbol{P}_X)]\boldsymbol{y}.$$

For the null model, we have already found that $\boldsymbol{y}^{\mathrm{T}}\boldsymbol{P}_0\boldsymbol{y} = n\bar{y}^2$. Using the block diagonal structure for the projection matrix $\boldsymbol{P}_X$ found above for the one-way layout, you can verify that $\boldsymbol{y}^{\mathrm{T}}\boldsymbol{P}_X\boldsymbol{y} = \sum_{i=1}^{c} n_i\bar{y}_i^2$. Therefore,

$$\boldsymbol{y}^{\mathrm{T}}(\boldsymbol{P}_X - \boldsymbol{P}_0)\boldsymbol{y} = \sum_{i=1}^{c} n_i\bar{y}_i^2 - n\bar{y}^2 = \sum_{i=1}^{c} n_i(\bar{y}_i - \bar{y})^2,$$

called the *between-groups sum of squares*. It represents variability explained by adding the group factor to the model as an explanatory variable. The final term in the sum-of-squares decomposition is

$$\boldsymbol{y}^{\mathrm{T}}(\boldsymbol{I} - \boldsymbol{P}_X)\boldsymbol{y} = \sum_{i=1}^{c}\sum_{j=1}^{n_i} y_{ij}^2 - \sum_{i=1}^{c} n_i\bar{y}_i^2 = \sum_{i=1}^{c}\sum_{j=1}^{n_i}(y_{ij} - \bar{y}_i)^2,$$

called the *within-groups sum of squares*. Since the fitted value corresponding to observation y_{ij} is $\hat{\mu}_{ij} = \bar{y}_i$, this is a sum of squared residuals for the model for the one-way layout.

An *analysis of variance* (ANOVA) *table* displays the components in the sum-of-squares decomposition. Table 2.1 shows the form of this table for a one-way layout. We have included in this table the corresponding projection matrices. The degrees of freedom (*df*) values listed determine specific sampling distributions for the normal linear model in the next chapter. Each *df* value equals the rank of the projection matrix and the dimension of the corresponding vector subspace to which that matrix

Table 2.1 ANOVA Table for Linear Model for One-Way Layout

Source	Projection Matrix	*df* (rank)	Sum of Squares	Name for Sum of Squares
Mean	$\boldsymbol{P}_0$	1	$n\bar{y}^2$	
Groups	$\boldsymbol{P}_X - \boldsymbol{P}_0$	$c - 1$	$\sum_{i=1}^{c} n_i(\bar{y}_i - \bar{y})^2$	Between-groups
Error	$\boldsymbol{I} - \boldsymbol{P}_X$	$n - c$	$\sum_{i=1}^{c} \sum_{j=1}^{n_i} (y_{ij} - \bar{y}_i)^2$	Within-groups
Total	$\boldsymbol{I}$	n	$\sum_{i=1}^{c} \sum_{j=1}^{n_i} y_{ij}^2$	

projects. Sometimes the first source (mean) is not shown, and the total row is replaced by a *corrected total sum of squares*:

$$\sum_{i=1}^{c} \sum_{j=1}^{n_i} y_{ij}^2 - n\bar{y}^2 = \sum_{i=1}^{c} \sum_{j=1}^{n_i} (y_{ij} - \bar{y})^2.$$

This sum of squares has $df = n - 1$.

2.3.4 Example: Model for Two-Way Layout with Randomized Block Design

The model for the one-way layout generalizes to models with two or more explanatory factors. We outline some basic ideas for two factors, with one observation at each combination of their categories. An important case having this data format is the *randomized block design*: The rows represent treatments whose means we would like to compare. The columns represent blocks such that experimental units are more homogeneous within blocks than between blocks. A classic example is comparing mean yields of some type of crop for r fertilizer treatments, using c fields as blocks. Within each field, the treatments are randomly assigned to r plots within that field. The experiment yields $n = rc$ observations.

Let y_{ij} be the observation for treatment i in block j. Consider the linear model with linear predictor

$$E(y_{ij}) = \beta_0 + \beta_i + \gamma_j, \quad i = 1, \dots, r, \quad j = 1, \dots, c,$$

with constraints such as $\beta_1 = \gamma_1 = 0$ for identifiability. Let $\bar{y}_{i.}$ be the sample mean observation for treatment i, $\bar{y}_{.j}$ the sample mean observation in block j, and $\bar{y}$ the overall sample mean. The orthogonal decomposition of the data for all i and j as

$$y_{ij} = \bar{y} + (\bar{y}_{i.} - \bar{y}) + (\bar{y}_{.j} - \bar{y}) + (y_{ij} - \bar{y}_{i.} - \bar{y}_{.j} + \bar{y})$$

corresponds to applying to $\boldsymbol{y}$ a decomposition of the $n \times n$ identity matrix into four projection matrices,

$$\boldsymbol{I} = \boldsymbol{P}_0 + (\boldsymbol{P}_r - \boldsymbol{P}_0) + (\boldsymbol{P}_c - \boldsymbol{P}_0) + (\boldsymbol{I} - \boldsymbol{P}_r - \boldsymbol{P}_c + \boldsymbol{P}_0).$$

Here, the null model projection matrix $\boldsymbol{P}_0 = \frac{1}{n}\mathbf{1}_n\mathbf{1}_n^{\mathrm{T}}$, when applied to $\boldsymbol{y} = (y_{11}, \dots, y_{1c}, \dots, y_{r1}, \dots, y_{rc})^{\mathrm{T}}$, generates $\bar{y}\mathbf{1}_n$. The matrix

$$\boldsymbol{P}_r = \begin{pmatrix} 1/c & \cdots & 1/c & \cdots & 0 & \cdots & 0 \\ \vdots & \ddots & \vdots & \cdots & \vdots & \ddots & \vdots \\ 1/c & \cdots & 1/c & \cdots & 0 & \cdots & 0 \\ \vdots & \vdots & \vdots & \ddots & \vdots & \vdots & \vdots \\ 0 & \cdots & 0 & \cdots & 1/c & \cdots & 1/c \\ \vdots & \ddots & \vdots & \cdots & \vdots & \ddots & \vdots \\ 0 & \cdots & 0 & \cdots & 1/c & \cdots & 1/c \end{pmatrix}$$

is a block diagonal matrix with r blocks of size $c \times c$, in which each element equals $1/c$. The projection $\boldsymbol{P}_r\boldsymbol{y}$ generates the vector $(\bar{y}_{1.}, \dots, \bar{y}_{1.}, \dots, \bar{y}_{r.}, \dots, \bar{y}_{r.})^{\mathrm{T}}$, where $\bar{y}_{i.}$ occurs in the locations for the c observations for treatment i. A corresponding projection matrix $\boldsymbol{P}_c$ has elements $1/r$ in appropriate locations to generate the block means $(\bar{y}_{.1}, \dots, \bar{y}_{.c}, \dots, \bar{y}_{.1}, \dots, \bar{y}_{.c})$, where $\bar{y}_{.j}$ occurs in the locations for the r observations for block j. You can check that $\boldsymbol{P}_r\boldsymbol{P}_c = \boldsymbol{P}_0$ and that each of the four matrices in the decomposition for $\boldsymbol{I}$ is symmetric and idempotent and hence a projection matrix.

Recall that the rank of a projection matrix equals its trace. For this decomposition of projection matrices, the trace of $(\boldsymbol{P}_r - \boldsymbol{P}_0)$ equals the difference of traces, $r - 1$, corresponding to the $(r - 1)$ nonredundant $\{\beta_i\}$. Applied to $\boldsymbol{y}$, it generates the sample treatment effects $(\bar{y}_{1.} - \bar{y}, \dots, \bar{y}_{1.} - \bar{y}, \dots, \bar{y}_{r.} - \bar{y}, \dots, \bar{y}_{r.} - \bar{y})^{\mathrm{T}}$. Likewise, $(\boldsymbol{P}_c - \boldsymbol{P}_0)$ has rank $c - 1$ and generates sample block effects. The projection matrix $(\boldsymbol{I} - \boldsymbol{P}_r - \boldsymbol{P}_c + \boldsymbol{P}_0)$ generates the term representing the residual error. Its rank is $rc - r - c + 1 = (r - 1)(c - 1)$.

The decomposition of observations and of projection matrices corresponds to the sum-of-squares decomposition shown in the ANOVA table (Table 2.2). The corrected total sum of squares is $\sum_i \sum_j y_{ij}^2 - rc\bar{y}^2$. The larger the between-treatments sum of squares relative to the error sum of squares, the stronger the evidence of a treatment

Table 2.2 ANOVA Table for Linear Model for Two-Way $r \times c$ Layout with One Observation Per Cell (as in Randomized Block Design)

Source	Projection Matrix	df (rank)	Sum of Squares
Mean	$\boldsymbol{P}_0$	1	$rc\bar{y}^2$
Treatments	$\boldsymbol{P}_r - \boldsymbol{P}_0$	$r - 1$	$c\sum_{i=1}^{r}(\bar{y}_{i.} - \bar{y})^2$
Blocks	$\boldsymbol{P}_c - \boldsymbol{P}_0$	$c - 1$	$r\sum_{j=1}^{c}(\bar{y}_{.j} - \bar{y})^2$
Error	$\boldsymbol{I} - \boldsymbol{P}_r - \boldsymbol{P}_c + \boldsymbol{P}_0$	$(r - 1)(c - 1)$	$\sum_{i=1}^{r}\sum_{j=1}^{c}(y_{ij} - \bar{y}_{i.} - \bar{y}_{.j} + \bar{y})^2$
Total	$\boldsymbol{I}$	$n = rc$	$\sum_{i=1}^{r}\sum_{j=1}^{c} y_{ij}^2$

effect. Inferential details for the normal linear model follow from results in the next chapter.

2.4 SUMMARIZING VARIABILITY IN A LINEAR MODEL

For a linear model $E(y) = X\beta$ with model matrix X and covariance matrix $V = \sigma^2 I$, in Section 2.2.5 we introduced the "data = fit + residuals" orthogonal decomposition using the projection matrix $P_X = X(X^T X)^{-1} X^T$ (i.e., the hat matrix H),

$$y = \hat{\mu} + (y - \hat{\mu}) = P_X y + (I - P_X) y.$$

This represents the orthogonality of the fitted values $\hat{\mu}$ and the raw residuals $e = (y - \hat{\mu})$. We have used $P_X y = \hat{\mu}$ to estimate μ. The other part of this decomposition, $(I - P_X)y = (y - \hat{\mu})$, falls in the error space $C(X)^{\perp}$ orthogonal to the model space $C(X)$. We next use it to estimate the variance σ^2 of the conditional distribution of each y_i, given its explanatory variable values. This variance is sometimes called the *error variance*, from the representation of the model as $y = X\beta + \epsilon$ with $\text{var}(\epsilon) = \sigma^2 I$.

2.4.1 Estimating the Error Variance for a Linear Model

To obtain an unbiased estimator of σ^2, we apply a result about $E(y^T A y)$, for a $n \times n$ matrix A. Since $E(y - \mu) = \mathbf{0}$,

$$E[(y - \mu)^T A (y - \mu)] = E(y^T A y) - \mu^T A \mu.$$

Using the commutative property of the trace of a matrix,

$$\begin{aligned} E[(y - \mu)^T A (y - \mu)] &= E\{\text{trace}[(y - \mu)^T A (y - \mu)]\} = E\{\text{trace}[A(y - \mu)(y - \mu)^T]\} \\ &= \text{trace}\{AE[(y - \mu)(y - \mu)^T]\} = \text{trace}(AV). \end{aligned}$$

It follows that

$$E(y^T A y) = \text{trace}(AV) + \mu^T A \mu. \tag{2.7}$$

For a linear model with full-rank model matrix X and projection matrix P_X, we now apply this result with $A = (I - P_X)$ and $V = \sigma^2 I$ for the $n \times n$ identity matrix I. The rank of X, which also is the rank of P_X, is the number of model parameters p. We have

$$\begin{aligned} E[y^T (I - P_X) y] &= \text{trace}[(I - P_X)\sigma^2 I] + \mu^T (I - P_X)\mu \\ &= \sigma^2 \text{trace}(I - P_X), \end{aligned}$$

because $(\boldsymbol{I} - \boldsymbol{P}_X)\boldsymbol{\mu} = \boldsymbol{\mu} - \boldsymbol{\mu} = \boldsymbol{0}$. Then, since

$$\text{trace}(\boldsymbol{P}_X) = \text{trace}[\boldsymbol{X}(\boldsymbol{X}\boldsymbol{X}^{\text{T}})^{-1}\boldsymbol{X}^{\text{T}}] = \text{trace}[\boldsymbol{X}^{\text{T}}\boldsymbol{X}(\boldsymbol{X}\boldsymbol{X}^{\text{T}})^{-1}] = \text{trace}(\boldsymbol{I}_p),$$

where $\boldsymbol{I}_p$ is the $p \times p$ identity matrix, we have $\text{trace}(\boldsymbol{I} - \boldsymbol{P}_X) = n - p$, and

$$E\left[\frac{\boldsymbol{y}^{\text{T}}(\boldsymbol{I} - \boldsymbol{P}_X)\boldsymbol{y}}{n-p}\right] = \sigma^2.$$

So $s^2 = [\boldsymbol{y}^{\text{T}}(\boldsymbol{I} - \boldsymbol{P}_X)\boldsymbol{y}]/(n-p)$ is an unbiased estimator of σ^2. Since $\boldsymbol{P}_X$ and $(\boldsymbol{I} - \boldsymbol{P}_X)$ are symmetric and idempotent, the numerator of s^2 is

$$\boldsymbol{y}^{\text{T}}(\boldsymbol{I} - \boldsymbol{P}_X)\boldsymbol{y} = \boldsymbol{y}^{\text{T}}(\boldsymbol{I} - \boldsymbol{P}_X)^{\text{T}}(\boldsymbol{I} - \boldsymbol{P}_X)\boldsymbol{y} = (\boldsymbol{y} - \hat{\boldsymbol{\mu}})^{\text{T}}(\boldsymbol{y} - \hat{\boldsymbol{\mu}}) = \sum_{i=1}^{n}(y_i - \hat{\mu}_i)^2.$$

In summary, an unbiased estimator of the error variance σ^2 in a linear model with full-rank model matrix is

$$s^2 = \frac{\sum_{i=1}^{n}(y_i - \hat{\mu}_i)^2}{n-p},$$

an average of the squared residuals. Here, the average is taken with respect to the dimension of the error space in which these residual components reside. When $\boldsymbol{X}$ has less than full rank $r < p$, the same argument holds with the $\text{trace}(\boldsymbol{P}_X) = r$. Then, s^2 has denominator $n - r$. The estimate s^2 is called[9] the *error mean square*, where error = residual, or *residual mean square*.

For example, for the null model (Section 2.3.1), the numerator of s^2 is $\sum_{i=1}^{n}(y_i - \bar{y})^2$ and the rank of $\boldsymbol{X} = \boldsymbol{1}_n$ is 1. An unbiased estimator of σ^2 is

$$s^2 = \frac{\sum_{i=1}^{n}(y_i - \bar{y})^2}{n-1}.$$

This is the sample variance and the usual estimator of the marginal variance of y.

There is nothing special about using an unbiased estimator. In fact s, which is on a more helpful scale for interpreting variability, is biased. However, s^2 occurs naturally in distribution theory for the ordinary linear model, as we will see in the next chapter. The denominator $(n-p)$ of the estimator occurs as a degrees of freedom measure in sampling distributions of relevant statistics.

[9]Not to be confused with the "mean squared error," which is $E(\hat{\theta} - \theta)^2$ for an estimator $\hat{\theta}$ of a parameter θ.

2.4.2 Sums of Squares: Error (SSE) and Regression (SSR)

The sum of squares $\sum_i (y_i - \hat{\mu}_i)^2$ in the numerator of s^2 is abbreviated by SSE, for "sum of squared errors," It is also referred to as the *residual sum of squares.*

The orthogonal decomposition of the data, $\boldsymbol{y} = \boldsymbol{P_X y} + (\boldsymbol{I} - \boldsymbol{P_X})\boldsymbol{y}$, expresses observation i as $y_i = \hat{\mu}_i + (y_i - \hat{\mu}_i)$. Correcting for the sample mean,

$$(y_i - \bar{y}) = (\hat{\mu}_i - \bar{y}) + (y_i - \hat{\mu}_i).$$

Using $(y_i - \bar{y})$ as the observation corresponds to adjusting y_i by including an intercept term before investigating effects of the explanatory variables. (For the null model $E(y_i) = \beta$, Section 2.3.1 showed that $\hat{\mu}_i = \bar{y}$.) This orthogonal decomposition into the component in the model space and the component in the error space yields the sum-of-squares decomposition:

$$\sum_i (y_i - \bar{y})^2 = \sum_i (\hat{\mu}_i - \bar{y})^2 + \sum_i (y_i - \hat{\mu}_i)^2.$$

We abbreviate this decomposition as

$$\text{TSS} = \text{SSR} + \text{SSE},$$

for the (corrected) *total sum of squares* TSS, the *sum of squares due to the regression model* SSR, and the *sum of squared errors* SSE. Here, TSS summarizes the total variation in the data after fitting the model containing only an intercept. The SSE component represents the variation in $\boldsymbol{y}$ "unexplained" by the full model, that is, a summary of prediction error remaining after fitting that model. The SSR component represents the variation in $\boldsymbol{y}$ "explained" by the full model, that is, the reduction in variation from TSS to SSE resulting from adding explanatory variables to a model that contains only an intercept term. For short, we will refer to SSR as the *regression sum of squares*. It is also called the *model sum of squares.*

We illustrate with the model for the one-way layout. From Section 2.3.3, TSS partitions into a between-groups SS = $\sum_{i=1}^{c} n_i(\bar{y}_i - \bar{y})^2$ and a within-groups SS = $\sum_{i=1}^{c} \sum_{j=1}^{n_i} (y_{ij} - \bar{y}_i)^2$. The between-groups SS is the SSR for the model, representing variability explained by adding the indicator predictors to the model. Since the fitted value corresponding to observation y_{ij} is $\hat{\mu}_{ij} = \bar{y}_i$, the within-groups SS is SSE for the model. For the model for the two-way layout in Section 2.3.4, SSR is the sum of the SS for the treatment effects and the SS for the block effects.

2.4.3 Effect on SSR and SSE of Adding Explanatory Variables

The least squares fit minimizes SSE. When we add an explanatory variable to a model, SSE cannot increase, because we could (at worst) obtain the same SSE value by setting $\hat{\beta}_j = 0$ for the new variable. So, SSE is monotone decreasing as the set of explanatory variables grows. Since TSS depends only on $\{y_i\}$ and is identical for every model fitted to a particular dataset, SSR = TSS − SSE is monotone increasing as variables are added.

Let $\text{SSR}(x_1, x_2)$ denote the regression sum of squares for a model with two explanatory variables and let $\text{SSR}(x_1)$ and $\text{SSR}(x_2)$ denote it for the two models having only one of those explanatory variables (plus, in each case, the intercept). We can partition $\text{SSR}(x_1, x_2)$ into $\text{SSR}(x_1)$ and the additional variability explained by adding x_2 to the model. Denote that additional variability explained by x_2, adjusting for x_1, by $\text{SSR}(x_2 \mid x_1)$. That is,

$$\text{SSR}(x_1, x_2) = \text{SSR}(x_1) + \text{SSR}(x_2 \mid x_1).$$

Equivalently, $\text{SSR}(x_2 \mid x_1)$ is the decrease in SSE from adding x_2 to the model.

Let $\{\hat{\mu}_{i1}\}$ denote the fitted values when x_1 is the sole explanatory variable, and let $\{\hat{\mu}_{i12}\}$ denote the fitted values when both x_1 and x_2 are explanatory variables. Then, from the orthogonal decomposition $(\hat{\mu}_{i12} - \bar{y}) = (\hat{\mu}_{i1} - \bar{y}) + (\hat{\mu}_{i12} - \hat{\mu}_{i1})$,

$$\text{SSR}(x_2 \mid x_1) = \sum_{i=1}^{n} (\hat{\mu}_{i12} - \hat{\mu}_{i1})^2.$$

To show that this application of Pythagoras's theorem holds, we need to show that $\sum_i (\hat{\mu}_{i1} - \bar{y})(\hat{\mu}_{i12} - \hat{\mu}_{i1}) = 0$. But denoting the projection matrices by $\boldsymbol{P}_0$ for the model containing only an intercept, $\boldsymbol{P}_1$ for the model that also has x_1 as an explanatory variable, and $\boldsymbol{P}_{12}$ for the model that has x_1 and x_2 as explanatory variables, this sum is

$$(\boldsymbol{P}_1\boldsymbol{y} - \boldsymbol{P}_0\boldsymbol{y})^{\text{T}}(\boldsymbol{P}_{12}\boldsymbol{y} - \boldsymbol{P}_1\boldsymbol{y}) = \boldsymbol{y}^{\text{T}}(\boldsymbol{P}_1 - \boldsymbol{P}_0)(\boldsymbol{P}_{12} - \boldsymbol{P}_1)\boldsymbol{y}.$$

Since $\boldsymbol{P}_a\boldsymbol{P}_b = \boldsymbol{P}_a$ when model a is a special case of model b, $(\boldsymbol{P}_1 - \boldsymbol{P}_0)(\boldsymbol{P}_{12} - \boldsymbol{P}_1) = \boldsymbol{0}$, so $\boldsymbol{y}^{\text{T}}(\boldsymbol{P}_1 - \boldsymbol{P}_0)(\boldsymbol{P}_{12} - \boldsymbol{P}_1)\boldsymbol{y} = 0$. This also follows from the result about decompositions of $\boldsymbol{I}$ into sums of projection matrices stated at the end of Section 2.2.1, whereby projection matrices that sum to $\boldsymbol{I}$ have pairwise products of $\boldsymbol{0}$. Here, $\boldsymbol{I} = \boldsymbol{P}_0 + (\boldsymbol{P}_1 - \boldsymbol{P}_0) + (\boldsymbol{P}_{12} - \boldsymbol{P}_1) + (\boldsymbol{I} - \boldsymbol{P}_{12})$.

2.4.4 Sequential and Partial Sums of Squares

Next we consider the general case with p explanatory variables, $x_1, x_2, \ldots, x_p$, and an intercept or centered value of y. From entering these variables in sequence into the model, we obtain the regression sum of squares and successive increments to it,

$$\text{SSR}(x_1), \text{SSR}(x_2 \mid x_1), \text{SSR}(x_3 \mid x_1, x_2), \ldots, \text{SSR}(x_p \mid x_1, x_2, \ldots, x_{p-1}).$$

These components are referred to as *sequential sums of squares*. They sum to the regression sum of squares for the full model, denoted by $\text{SSR}(x_1, \ldots, x_p)$. The sequential sum of squares corresponding to adding a term to the model can depend strongly on which other variables are already in the model, because of correlations among the predictors. For example, $\text{SSR}(x_p)$ often tends to be much larger than $\text{SSR}(x_p \mid x_1, \dots, x_{p-1})$ when x_p is highly correlated with the other predictors, as happens in many observational studies. We discuss this further in Section 4.6.5.

An alternative set[10] of increments to regression sums of squares, called *partial sums of squares*, uses the same set of p explanatory variables for each:

$$\text{SSR}(x_1 \mid x_2, \dots, x_p), \text{SSR}(x_2 \mid x_1, \dots, x_p), \dots, \text{SSR}(x_p \mid x_1, \dots, x_{p-1}).$$

Each of these represents the additional variability explained by adding a particular explanatory variable to the model, when *all* the other explanatory variables are already in the model. Equivalently, it is the drop in SSE when that explanatory variable is added, after all the others. These *partial* SS values may differ from all the corresponding *sequential* SS values $\text{SSR}(x_1), \text{SSR}(x_2 \mid x_1), \dots, \text{SSR}(x_p \mid x_1, x_2, \dots, x_{p-1})$, except for the final one.

2.4.5 Uncorrelated Predictors: Sequential SS = Partial SS = SSR Component

We have seen that the "data = fit + residuals" orthogonal decomposition $\boldsymbol{y} = \boldsymbol{P}_X\boldsymbol{y} + (\boldsymbol{I} - \boldsymbol{P}_X)\boldsymbol{y}$ implies the corresponding SS decomposition, $\boldsymbol{y}^{\text{T}}\boldsymbol{y} = \boldsymbol{y}^{\text{T}}\boldsymbol{P}_X\boldsymbol{y} + \boldsymbol{y}^{\text{T}}(\boldsymbol{I} - \boldsymbol{P}_X)\boldsymbol{y}$. When the values of y are centered, this is TSS = SSR + SSE. Now, suppose the p parameters are orthogonal (Section 2.2.4). Then, $\boldsymbol{X}^{\text{T}}\boldsymbol{X}$ and its inverse are diagonal. With the model matrix partitioned into $\boldsymbol{X} = (\boldsymbol{X}_1 : \boldsymbol{X}_2 : \cdots : \boldsymbol{X}_p)$,

$$\boldsymbol{P}_X = \boldsymbol{X}(\boldsymbol{X}^{\text{T}}\boldsymbol{X})^{-1}\boldsymbol{X}^{\text{T}} = \boldsymbol{X}_1(\boldsymbol{X}_1^{\text{T}}\boldsymbol{X}_1)^{-1}\boldsymbol{X}_1^{\text{T}} + \cdots + \boldsymbol{X}_p(\boldsymbol{X}_p^{\text{T}}\boldsymbol{X}_p)^{-1}\boldsymbol{X}_p^{\text{T}}.$$

In terms of the projection matrices for separate models, each with only a single explanatory variable, this is $\boldsymbol{P}_{X_1} + \cdots + \boldsymbol{P}_{X_p}$. Therefore,

$$\boldsymbol{y}^{\text{T}}\boldsymbol{y} = \boldsymbol{y}^{\text{T}}\boldsymbol{P}_{X_1}\boldsymbol{y} + \cdots + \boldsymbol{y}^{\text{T}}\boldsymbol{P}_{X_p}\boldsymbol{y} + \boldsymbol{y}^{\text{T}}(\boldsymbol{I} - \boldsymbol{P}_X)\boldsymbol{y}.$$

Each component of SSR equals the SSR for the model with that sole explanatory variable, so that

$$\text{SSR}(x_1, \dots, x_p) = \text{SSR}(x_1) + \text{SSR}(x_2) + \cdots + \text{SSR}(x_p). \tag{2.8}$$

When $\boldsymbol{X}_1 = \boldsymbol{1}_n$ is the coefficient of an intercept term, $\text{SSR}(x_1) = n\bar{y}^2$ and TSS = $\boldsymbol{y}^{\text{T}}\boldsymbol{y} - \text{SSR}(x_1)$ for the uncentered $\boldsymbol{y}$. The sum of squares that software reports as

[10]Alternative names are *Type 1 SS* for sequential SS and *Type 3 SS* for partial SS. *Type 2 SS* is an alternative partial SS that adjusts only for effects not containing the given effect, such as adjusting x_1 for x_2 but not for x_1x_2 when that interaction term is also in the model.

SSR is then $\text{SSR}(x_2) + \cdots + \text{SSR}(x_p)$. Also, with an intercept in the model, orthogonality of the parameters implies that pairs of explanatory variables are uncorrelated (Exercise 2.20).

When the explanatory variables in a linear model are uncorrelated, the sequential SS values do not depend on their order of entry into a model. They are then identical to the corresponding partial SS values, and the regression SS decomposes exactly in terms of them. We would not expect this in observational studies, but some balanced experimental designs have such simplicity.

For instance, consider the main effects model for the *two-way layout* with two binary qualitative factors and an equal sample size n in each cell,

$$E(y_{ijk}) = \beta_0 + \beta_i + \gamma_j,$$

for $i = 1, 2$, $j = 1, 2$, and $k = 1, \ldots, n$. With constraints $\beta_1 + \beta_2 = 0$ and $\gamma_1 + \gamma_2 = 0$ for identifiability and with $\boldsymbol{y}$ listing (i,j) in the order (1,1), (1,2), (2,1), (2,2), we can express the model as $E(\boldsymbol{y}) = \boldsymbol{X\beta}$ with

$$\boldsymbol{X\beta} = \begin{pmatrix} \mathbf{1}_n & \mathbf{1}_n & \mathbf{1}_n \\ \mathbf{1}_n & \mathbf{1}_n & -\mathbf{1}_n \\ \mathbf{1}_n & -\mathbf{1}_n & \mathbf{1}_n \\ \mathbf{1}_n & -\mathbf{1}_n & -\mathbf{1}_n \end{pmatrix} \begin{pmatrix} \beta_0 \\ \beta_1 \\ \gamma_1 \end{pmatrix}.$$

The scatterplot for the two indicator explanatory variables has n observations that occur at each of the points (−1,−1), (−1,1), (1,−1), and (1,1). Thus, those explanatory variables are uncorrelated (and orthogonal), and SSR decomposes into its separate parts for the row effects and for the column effects.

2.4.6 R-Squared and the Multiple Correlation

For a particular dataset and TSS value, the larger the value of SSR relative to SSE, the more effective the explanatory variables are in predicting the response variable. A summary of this predictive power is

$$R^2 = \frac{\text{SSR}}{\text{TSS}} = \frac{\text{TSS} - \text{SSE}}{\text{TSS}} = \frac{\sum_i (y_i - \bar{y})^2 - \sum_i (y_i - \hat{\mu}_i)^2}{\sum_i (y_i - \bar{y})^2}.$$

Here SSR = TSS−SSE measures the reduction in the sum of squared prediction errors after adding the explanatory variables to the model containing only an intercept. So, R^2 measures the *proportional reduction in error*, and it falls between 0 and 1. Sometimes called the *coefficient of determination*, it is usually merely referred to as "*R-squared*."

A related way to measure predictive power is with the sample correlation between the $\{y_i\}$ and $\{\hat{\mu}_i\}$ values. From (2.1), the normal equations solved to find the least squares estimates are $\sum_i y_i x_{ij} = \sum_i \hat{\mu}_i x_{ij}$, $j = 1, \ldots, p$. The equation corresponding to the intercept term is $\sum_i y_i = \sum_i \hat{\mu}_i$, so the sample mean of $\{\hat{\mu}_i\}$ equals $\bar{y}$. Therefore, the sample value of

$$\text{corr}(\boldsymbol{y}, \hat{\boldsymbol{\mu}}) = \frac{\sum_i (y_i - \bar{y})(\hat{\mu}_i - \bar{\hat{\mu}})}{\sqrt{\left[\sum_i (y_i - \bar{y})^2\right]\left[\sum_i (\hat{\mu}_i - \bar{\hat{\mu}})^2\right]}} = \frac{\sum_i (y_i - \hat{\mu}_i + \hat{\mu}_i - \bar{y})(\hat{\mu}_i - \bar{y})}{\sqrt{\left[\sum_i (y_i - \bar{y})^2\right]\left[\sum_i (\hat{\mu}_i - \bar{y})^2\right]}}.$$

The numerator simplifies to $\sum_i (\hat{\mu}_i - \bar{y})^2 = \text{SSR}$, since $\sum_i (y_i - \hat{\mu}_i)\hat{\mu}_i = 0$ by the orthogonality of $(\boldsymbol{y} - \hat{\boldsymbol{\mu}})$ and $\hat{\boldsymbol{\mu}}$, and the denominator equals $\sqrt{(\text{TSS})(\text{SSR})}$. So, $\text{corr}(\boldsymbol{y}, \hat{\boldsymbol{\mu}}) = \sqrt{\text{SSR/TSS}} = +\sqrt{R^2}$. This positive square root of R^2 is called the *multiple correlation*. Note that $0 \leq R \leq 1$. With a single explanatory variable, $R = |\text{corr}(\boldsymbol{x}, \boldsymbol{y})|$.

Out of all possible linear prediction equations $\tilde{\boldsymbol{\mu}} = \boldsymbol{X}\tilde{\boldsymbol{\beta}}$ that use the given model matrix, the least squares solution $\hat{\boldsymbol{\mu}}$ has the maximum correlation with $\boldsymbol{y}$. To ease notation as we show this, we suppose that all variables have been centered, which does not affect correlations. For an arbitrary $\tilde{\boldsymbol{\beta}}$ and constant c, for the least squares fit,

$$||\boldsymbol{y} - \hat{\boldsymbol{\mu}}||^2 \leq ||\boldsymbol{y} - c\tilde{\boldsymbol{\mu}}||^2.$$

Expanding both sides, subtracting the common term $||\boldsymbol{y}||^2$, and dividing by a common denominator yields

$$\frac{2\boldsymbol{y}^{\mathrm{T}}\hat{\boldsymbol{\mu}}}{||\boldsymbol{y}||\,||\hat{\boldsymbol{\mu}}||} - \frac{||\hat{\boldsymbol{\mu}}||^2}{||\boldsymbol{y}||\,||\hat{\boldsymbol{\mu}}||} \geq \frac{2c\boldsymbol{y}^{\mathrm{T}}\tilde{\boldsymbol{\mu}}}{||\boldsymbol{y}||\,||\hat{\boldsymbol{\mu}}||} - \frac{c^2||\tilde{\boldsymbol{\mu}}||^2}{||\boldsymbol{y}||\,||\hat{\boldsymbol{\mu}}||}.$$

Now, taking $c^2 = ||\hat{\boldsymbol{\mu}}||^2/||\tilde{\boldsymbol{\mu}}||^2$, we have

$$\frac{\boldsymbol{y}^{\mathrm{T}}\hat{\boldsymbol{\mu}}}{||\boldsymbol{y}||\,||\hat{\boldsymbol{\mu}}||} \geq \frac{\boldsymbol{y}^{\mathrm{T}}\tilde{\boldsymbol{\mu}}}{||\boldsymbol{y}||\,||\tilde{\boldsymbol{\mu}}||}.$$

But since the variables are centered, this says that $R = \text{corr}(\boldsymbol{y}, \hat{\boldsymbol{\mu}}) \geq \text{corr}(\boldsymbol{y}, \tilde{\boldsymbol{\mu}})$.

When explanatory variables are added to a model, since SSE cannot increase, R and R^2 are monotone increasing. For a model matrix $\boldsymbol{X}$, let $\boldsymbol{x}_{*j}$ denote column j for explanatory variable j. For the special case in which the sample $\text{corr}(\boldsymbol{x}_{*j}, \boldsymbol{x}_{*k}) = 0$ for each pair of the p explanatory variables, by the decomposition (2.8) of $\text{SSR}(x_1, \ldots, x_p)$,

$$R^2 = [\text{corr}(\boldsymbol{y}, \boldsymbol{x}_{*1})]^2 + [\text{corr}(\boldsymbol{y}, \boldsymbol{x}_{*2})]^2 + \cdots + [\text{corr}(\boldsymbol{y}, \boldsymbol{x}_{*p})]^2.$$

When n is small and a model has several explanatory variables, R^2 tends to overestimate the corresponding population value. An *adjusted R-squared* is designed to reduce this bias. It is defined to be the proportional reduction in variance based on the unbiased variance estimates, s_y^2 for the marginal distribution and s^2 for the conditional distributions; that is,

$$\text{adjusted } R^2 = \frac{s_y^2 - s^2}{s_y^2} = 1 - \frac{\text{SSE}/(n-p)}{\text{TSS}/(n-1)} = 1 - \frac{n-1}{n-p}(1 - R^2).$$

It is slightly smaller than ordinary R^2, and need not monotonically increase as we add explanatory variables to a model.

2.5 RESIDUALS, LEVERAGE, AND INFLUENCE

Since the residuals from the linear model fit are in the error space, orthogonal to the model space, they contain the information in the data that is not explained by the model. Thus, they are useful for investigating a model's lack of fit. This section takes a closer look at the residuals, including their moments and ways of plotting them to help check a model. We also present descriptions of the influence that each observation has on the least squares fit, using the residuals and "leverage" values from the hat matrix.

2.5.1 Residuals and Fitted Values Are Uncorrelated

From Section 2.4.6, the normal equation corresponding to the intercept term is $\sum_i y_i = \sum_i \hat{\mu}_i$. Thus, $\sum_i e_i = \sum_i (y_i - \hat{\mu}_i) = 0$, and the residuals have a sample mean of 0. Also,

$$E(\boldsymbol{e}) = E(\boldsymbol{y} - \hat{\boldsymbol{\mu}}) = \boldsymbol{X\beta} - \boldsymbol{X}E(\hat{\boldsymbol{\beta}}) = \boldsymbol{X\beta} - \boldsymbol{X\beta} = \boldsymbol{0}.$$

For linear models with an intercept, the sample correlation between the residuals $\boldsymbol{e}$ and fitted values $\hat{\boldsymbol{\mu}}$ has numerator $\sum_i e_i \hat{\mu}_i = \boldsymbol{e}^{\mathrm{T}} \hat{\boldsymbol{\mu}}$. So, the orthogonality of $\boldsymbol{e}$ and $\hat{\boldsymbol{\mu}}$ implies that $\text{corr}(\boldsymbol{e}, \hat{\boldsymbol{\mu}}) = 0$.

2.5.2 Plots of Residuals

Because $\text{corr}(\boldsymbol{e}, \hat{\boldsymbol{\mu}}) = 0$, the least squares line fitted to a scatterplot of the elements of $\boldsymbol{e} = (\boldsymbol{y} - \hat{\boldsymbol{\mu}})$ versus the corresponding elements of $\hat{\boldsymbol{\mu}}$ has slope 0. A scatterplot of the residuals against the fitted values helps to identify patterns of a model's lack of fit. Examples are nonconstant variance, sometimes referred to as *heteroscedasticity*, and nonlinearity. Likewise, since the residuals are also orthogonal to $C(\boldsymbol{X})$, they can be plotted against each explanatory variable to detect lack of fit.

Figure 2.8 shows how a plot of $\boldsymbol{e}$ against $\hat{\boldsymbol{\mu}}$ tends to look if (a) the linear model holds, (b) the variance is constant (homoscedasticity), but the mean of y is a quadratic rather than a linear function of the predictor, and (c) the linear trend predictor is correct, but the variance increases dramatically as the mean increases. In practice,

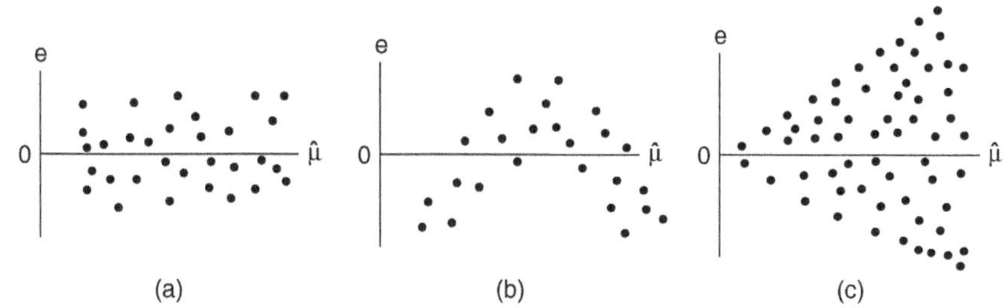

Figure 2.8 Residuals plotted against linear-model fitted values that reflect (a) model adequacy, (b) quadratic rather than linear relationship, and (c) nonconstant variance.

plots do not have such a neat appearance, but these illustrate how the plots can highlight model inadequacy. Section 2.6 shows an example.

For the normal linear model, the conditional distribution of $\boldsymbol{y}$, given the explanatory variables, is normal. This implies that the residuals, being linear in $\boldsymbol{y}$, also have normal distributions. A histogram of the residuals provides some information about the actual conditional distribution. Another check of the normality assumption is a plot of ordered residual values against expected values of order statistics from a $N(0, 1)$ distribution, called a *Q–Q plot*. We will discuss this type of plot in Section 3.4.2, in the chapter about the normal linear model.

2.5.3 Standardized and Studentized Residuals

For the ordinary linear model, the covariance matrix for the observations is $\boldsymbol{V} = \sigma^2\boldsymbol{I}$. In terms of the hat matrix $\boldsymbol{H} = \boldsymbol{X}(\boldsymbol{X}^{\mathrm{T}}\boldsymbol{X})^{-1}\boldsymbol{X}^{\mathrm{T}}$, this decomposes into

$$\boldsymbol{V} = \sigma^2\boldsymbol{I} = \sigma^2\boldsymbol{H} + \sigma^2(\boldsymbol{I} - \boldsymbol{H}).$$

Since $\hat{\boldsymbol{\mu}} = \boldsymbol{H}\boldsymbol{y}$ and since $\boldsymbol{H}$ is idempotent,

$$\mathrm{var}(\hat{\boldsymbol{\mu}}) = \sigma^2\boldsymbol{H}.$$

So, $\mathrm{var}(\hat{\mu}_i) = \sigma^2 h_{ii}$, where $\{h_{ii}\}$ denote the main diagonal elements of $\boldsymbol{H}$. Since variances are nonnegative, $h_{ii} \geq 0$. Likewise, since $(\boldsymbol{y} - \hat{\boldsymbol{\mu}}) = (\boldsymbol{I} - \boldsymbol{H})\boldsymbol{y}$ and since $(\boldsymbol{I} - \boldsymbol{H})$ is idempotent,

$$\mathrm{var}(\boldsymbol{y} - \hat{\boldsymbol{\mu}}) = \sigma^2(\boldsymbol{I} - \boldsymbol{H}).$$

So, the residuals are correlated, and their variance need not be constant, with

$$\mathrm{var}(e_i) = \mathrm{var}(y_i - \hat{\mu}_i) = \sigma^2(1 - h_{ii}).$$

Again since variances are nonnegative, $0 \leq h_{ii} \leq 1$. Also, $\text{var}(\hat{\mu}_i) = \sigma^2 h_{ii} \leq \sigma^2$ reveals a consequence of model parsimony: If the model holds (or nearly holds), $\hat{\mu}_i$ is better than y_i as an unbiased estimator of μ_i.

A standardized version of $e_i = (y_i - \hat{\mu}_i)$ that divides it by $\sigma\sqrt{1 - h_{ii}}$ has a standard deviation of 1. In practice, σ is unknown, so we replace it by the estimate s of σ derived in Section 2.4.1. The *standardized residual* is

$$r_i = \frac{y_i - \hat{\mu}_i}{s\sqrt{1 - h_{ii}}}. \tag{2.9}$$

This describes the number of estimated standard deviations that $(y_i - \hat{\mu}_i)$ departs from 0. If the normal linear model truly holds, these should nearly all fall between about -3 and $+3$. A slightly different residual, called[11] a *studentized residual*, estimates σ in the expression for $\text{var}(y_i - \hat{\mu}_i)$ based on the fit of the model to the $n - 1$ observations after excluding observation i. Then, that estimate is independent of observation i.

2.5.4 Leverages from Hat Matrix Measure Potential Influence

The element h_{ii} from $\boldsymbol{H}$, on which $\text{var}(e_i)$ depends, is called the *leverage* of observation i. Since $\text{var}(\hat{\mu}_i) = \sigma^2 h_{ii}$ with $0 \leq h_{ii} \leq 1$, the leverage determines the precision with which $\hat{\mu}_i$ estimates μ_i. For large h_{ii} close to 1, $\text{var}(\hat{\mu}_i) \approx \text{var}(y_i)$ and $\text{var}(e_i) \approx 0$. In this case, y_i may have a large influence on $\hat{\mu}_i$. In the extreme case $h_{ii} = 1$, $\text{var}(e_i) = 0$, and $\hat{\mu}_i = y_i$. By contrast, when h_{ii} is close to 0 and thus $\text{var}(\hat{\mu}_i)$ is relatively small, this suggests that $\hat{\mu}_i$ is based on contributions from many observations.

Here are two other ways to visualize how a relatively large leverage h_{ii} indicates that y_i may have a large influence on $\hat{\mu}_i$. First, since $\hat{\mu}_i = \sum_j h_{ij} y_j$, $\partial\hat{\mu}_i/\partial y_i = h_{ii}$. Second, since $\{y_i\}$ are uncorrelated[12],

$$\text{cov}(y_i, \hat{\mu}_i) = \text{cov}\left(y_i, \sum_{j=1}^{n} h_{ij} y_j\right) = \sum_{j=1}^{n} h_{ij}\text{cov}(y_i, y_j) = h_{ii}\text{cov}(y_i, y_i) = \sigma^2 h_{ii}.$$

Then, since $\text{var}(\hat{\mu}_i) = \sigma^2 h_{ii}$, it follows that the theoretical correlation,

$$\text{corr}(y_i, \hat{\mu}_i) = \frac{\sigma^2 h_{ii}}{\sqrt{\sigma^2 \cdot \sigma^2 h_{ii}}} = \sqrt{h_{ii}}.$$

When the leverage is relatively large, y_i is highly correlated with $\hat{\mu}_i$.

[11]Student is a pseudonym for W. S. Gosset, who discovered the t distribution in 1908. For the normal linear model, each studentized residual has a t distribution with $df = n - p$.
[12]Recall that for matrices of constants $\boldsymbol{A}$ and $\boldsymbol{B}$, $\text{cov}(\boldsymbol{Ax}, \boldsymbol{By}) = \boldsymbol{A}\text{cov}(\boldsymbol{x}, \boldsymbol{y})\boldsymbol{B}^{\text{T}}$.

So, what do the leverages look like? For the bivariate linear model $E(y_i) = \beta_0 + \beta x_i$, Section 2.1.3 showed the hat matrix. The leverage for observation i is

$$h_{ii} = \frac{1}{n} + \frac{(x_i - \bar{x})^2}{\sum_{k=1}^{n}(x_k - \bar{x})^2}.$$

The n leverages have a mean of $2/n$. They tend to be smaller with larger datasets. With multiple explanatory variables and values $\boldsymbol{x}_i$ for observation i and means $\bar{\boldsymbol{x}}$ (as row vectors), let $\tilde{\boldsymbol{X}}$ denote the model matrix using centered variables. Then, the leverage for observation i is

$$h_{ii} = \frac{1}{n} + (\boldsymbol{x}_i - \bar{\boldsymbol{x}})(\tilde{\boldsymbol{X}}^{\mathrm{T}}\tilde{\boldsymbol{X}})^{-1}(\boldsymbol{x}_i - \bar{\boldsymbol{x}})^{\mathrm{T}} \tag{2.10}$$

(Belsley et al. 1980, Appendix 2A). The leverage increases as $\boldsymbol{x}_i$ is farther from $\bar{\boldsymbol{x}}$. With p explanatory variables, including the intercept, the leverages have a mean of p/n. Observations with relatively large leverages, say exceeding about $3p/n$, may be influential in the fitting process.

2.5.5 Influential Points for Least Squares Fits

An observation having small leverage is not influential in its impact on $\{\hat{\mu}_i\}$ and $\{\hat{\beta}_j\}$, even if it is an outlier in the y direction. A point with extremely large leverage can be influential, but need not be so. It is influential when the observation is a "regression outlier," falling far from the least squares line that results using only the other $n - 1$ observations. See the first panel of Figure 2.9. By contrast, when the observation has a large leverage but is consistent with the trend shown by the other observations, it is not influential. See the second panel of Figure 2.9. To be influential, a point needs to have both a large leverage and a large standardized residual.

Summary measures that describe an observation's influence combine information from the leverages and the residuals. For any such measure of influence, larger values correspond to greater influence. *Cook's distance* (Cook 1977) is based on the change in $\hat{\boldsymbol{\beta}}$ when the observation is removed from the dataset. Let $\hat{\boldsymbol{\beta}}_{(i)}$ denote the least

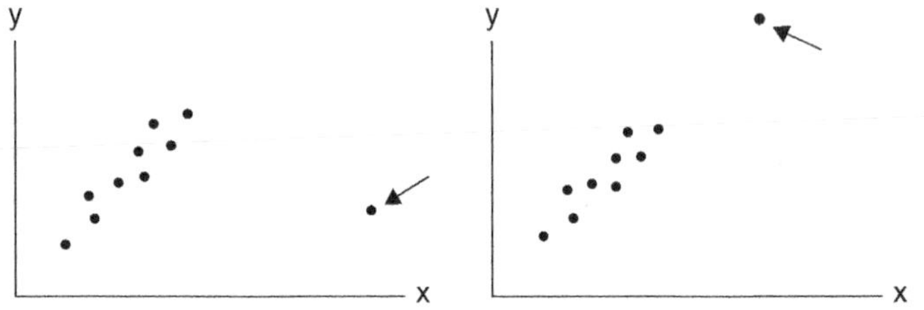

Figure 2.9 High leverage points in a linear model fit may be influential (first panel) or noninfluential (second panel).

squares estimate of $\boldsymbol{\beta}$ for the $n-1$ observations after excluding observation i. Then, Cook's distance for observation i is

$$D_i = \frac{(\hat{\boldsymbol{\beta}}_{(i)} - \hat{\boldsymbol{\beta}})^{\mathrm{T}}[\widehat{\mathrm{var}}(\hat{\boldsymbol{\beta}})]^{-1}(\hat{\boldsymbol{\beta}}_{(i)} - \hat{\boldsymbol{\beta}})}{p} = \frac{(\hat{\boldsymbol{\beta}}_{(i)} - \hat{\boldsymbol{\beta}})^{\mathrm{T}}(\boldsymbol{X}^{\mathrm{T}}\boldsymbol{X})(\hat{\boldsymbol{\beta}}_{(i)} - \hat{\boldsymbol{\beta}})}{ps^2}.$$

Incorporating the estimated variance of $\hat{\boldsymbol{\beta}}$ makes the measure free of the units of measurement and approximately free of the sample size. An equivalent expression uses the standardized residual r_i and the leverage h_{ii},

$$D_i = r_i^2 \left[\frac{h_{ii}}{p(1-h_{ii})}\right] = \frac{(y_i - \hat{\mu}_i)^2 h_{ii}}{ps^2(1-h_{ii})^2}. \tag{2.11}$$

A relatively large D_i, usually on the order of 1, occurs when both the standardized residual and the leverage are relatively large.

A measure with a similar purpose, *DFFIT*, describes the change in $\hat{\mu}_i$ due to deleting observation i. A standardized version (*DFFITS*) equals the studentized residual multiplied by the "leverage factor" $\sqrt{h_{ii}/(1-h_{ii})}$. A variable-specific measure, *DFBETA* (with standardized version *DFBETAS*), is based on the change in $\hat{\beta}_j$ alone when the observation is removed from the dataset. Each observation has a separate *DFBETA* for each $\hat{\beta}_j$.

2.5.6 Adjusting for Explanatory Variables by Regressing Residuals

Residuals are at the heart of what we mean by "adjusting for the other explanatory variables in the model," in describing the partial effect of an explanatory variable x_j. Suppose we use least squares to (1) regress y on the explanatory variables other than x_j, (2) regress x_j on those other explanatory variables. When we regress the residuals from (1) on the residuals from (2), Yule (1907) showed that the fit has slope that is identical to the partial effect of variable x_j in the multiple regression model. A scatterplot of these two sets of residuals is called a *partial regression plot*, also sometimes called an *added variable plot*. The residuals for the least squares line between these two sets of residuals are identical to the residuals in the multiple regression model that regresses y on all the explanatory variables.

To show Yule's result, we use his notation for linear model coefficients, introduced in Section 1.2.3. To ease formula complexity, we do this for the case of two explanatory variables, with all variables centered to eliminate intercept terms. Consider the models

$$E(y_i) = \beta_{y2}x_{i2}, \quad E(x_{i1}) = \beta_{12}x_{i2}, \quad E(y_i) = \beta_{y1\cdot 2}x_{i1} + \beta_{y2\cdot 1}x_{i2}.$$

The normal equations (2.1) for the bivariate models are

$$\sum_{i=1}^{n} x_{i2}(y_i - \beta_{y2}x_{i2}) = 0, \quad \sum_{i=1}^{n} x_{i2}(x_{i1} - \beta_{12}x_{i2}) = 0.$$

The normal equations for the multiple regression model are

$$\sum_{i=1}^{n} x_{i1}(y_i - \beta_{y1\cdot 2}x_{i1} - \beta_{y2\cdot 1}x_{i2}) = 0, \quad \sum_{i=1}^{n} x_{i2}(y_i - \beta_{y1\cdot 2}x_{i1} - \beta_{y2\cdot 1}x_{i2}) = 0.$$

From these two equations for the multiple regression model,

$$0 = \sum_{i=1}^{n}(y_i - \beta_{y1\cdot 2}x_{i1} - \beta_{y2\cdot 1}x_{i2})(x_{i1} - \beta_{12}x_{i2}).$$

Using this and the normal equation for the second bivariate model,

$$\begin{aligned} 0 &= \sum_{i=1}^{n} y_i(x_{i1} - \beta_{12}x_{i2}) - \beta_{y1\cdot 2}\sum_{i=1}^{n} x_{i1}(x_{i1} - \beta_{12}x_{i2}) \\ &= \sum_{i=1}^{n}(y_i - \beta_{y2}x_{i2})(x_{i1} - \beta_{12}x_{i2}) - \beta_{y1\cdot 2}\sum_{i=1}^{n}(x_{i1} - \beta_{12}x_{i2})^2. \end{aligned}$$

It follows that the estimated partial effect of x_1 on y, adjusting for x_2, is

$$\hat{\beta}_{y1\cdot 2} = \frac{\sum_{i=1}^{n}(y_i - \hat{\beta}_{y2}x_{i2})(x_{i1} - \hat{\beta}_{12}x_{i2})}{\sum_{i=1}^{n}(x_{i1} - \hat{\beta}_{12}x_{i2})^2}.$$

But from (2.5) this is precisely the result of regressing the residuals from the regression of y on x_2 on the residuals from the regression of x_1 on x_2.

This result has an interesting consequence. From the regression of residuals just mentioned, the fit for the full model satisfies

$$\hat{\mu}_i - \hat{\beta}_{y2}x_{i2} = \hat{\beta}_{y1\cdot 2}(x_{i1} - \hat{\beta}_{12}x_{i2})$$

so that

$$\hat{\mu}_i = \hat{\beta}_{y2}x_{i2} + \hat{\beta}_{y1\cdot 2}(x_{i1} - \hat{\beta}_{12}x_{i2}) = \hat{\beta}_{y1\cdot 2}x_{i1} + (\hat{\beta}_{y2} - \hat{\beta}_{y1\cdot 2}\hat{\beta}_{12})x_{i2}.$$

Therefore, the partial effect of x_2 on y, adjusting for x_1, has the expression

$$\hat{\beta}_{y2\cdot 1} = \hat{\beta}_{y2} - \hat{\beta}_{y1\cdot 2}\hat{\beta}_{12}.$$

In particular, $\hat{\beta}_{y2\cdot 1} = \hat{\beta}_{y2}$ if $\hat{\beta}_{12} = 0$, which is equivalent to $\text{corr}(\boldsymbol{x}_{*1}, \boldsymbol{x}_{*2}) = 0$. They are also equal if $\hat{\beta}_{y1\cdot 2} = 0$. Likewise, $\hat{\beta}_{y1\cdot 2} = \hat{\beta}_{y1} - \hat{\beta}_{y2\cdot 1}\hat{\beta}_{21}$. An implication is that if the primary interest in a study is the effect of x_1 while adjusting for x_2 but the model does not include x_2, then the difference between the effect of interest and the effect actually estimated is the *omitted variable bias*, $\beta_{y2\cdot 1}\beta_{21}$.

2.5.7 Partial Correlation

A *partial correlation* describes the association between two variables after adjusting for other variables. Yule (1907) also showed how to formalize this concept using residuals. For example, the partial correlation between y and x_1 while adjusting for x_2 and x_3 is obtained by (1) finding the residuals for predicting y using x_2 and x_3, (2) finding the residuals for predicting x_1 using x_2 and x_3, and then (3) finding the ordinary correlation between these two sets of residuals.

The squared partial correlation between y and a particular explanatory variable considers the variability unexplained without that variable and evaluates the proportional reduction in variability after adding it. That is, if R_0^2 is the proportional reduction in error without it, and R_1^2 is the value after adding it to the model, then the squared partial correlation between y and the variable, adjusting for the others, is $(R_1^2 - R_0^2)/(1 - R_0^2)$.

2.6 EXAMPLE: SUMMARIZING THE FIT OF A LINEAR MODEL

Each year the Scottish Hill Runners Association (`www.shr.uk.com`) publishes a list of hill races in Scotland for the year. Table 2.3 shows data on the record time for some of the races (in minutes). Explanatory variables listed are the distance of the race (in miles) and the cumulative climb (in thousands of feet).

Table 2.3 Record Time to Complete Race Course (in minutes), by Distance of Race (miles) and Climb (in thousands of feet)

Race	Distance	Climb	Record Time
Greenmantle New Year Dash	2.5	0.650	16.08
Craig Dunain Hill Race	6	0.900	33.65
Ben Rha Hill Race	7.5	0.800	45.60
Ben Lomond Hill Race	8	3.070	62.27
Bens of Jura Fell Race	16	7.500	204.62
Lairig Ghru Fun Run	28	2.100	192.67

Source: From Atkinson (1986), by permission of the Institute of Mathematical Statistics, with correction by Hoaglin[13] (2012). The complete data for 35 races are in the file `ScotsRaces.dat` at the text website, `www.stat.ufl.edu/~aa/glm/data`.

[13]Thanks to David Hoaglin for showing me his article and this data set.

We suggest that you download all 35 observations from the text website and view some summary statistics and graphics, such as follows:

```
-------------------------------------------------------------------------
> attach(ScotsRaces) # complete data at www.stat.ufl.edu/~aa/glm/data
> matrix(cbind(mean(time),sd(time),mean(climb),sd(climb),
+              mean(distance),sd(distance)),nrow=2)
         [,1]     [,2]     [,3] # e.g., time has mean = 56, std.dev.= 50
[1,] 56.0897  1.8153  7.5286
[2,] 50.3926  1.6192  5.5239
> pairs(~time+climb+distance) # scatterplot matrix for variable pairs
> cor(cbind(climb,distance,time)) # correlation matrix
          climb distance    time
climb    1.0000   0.6523  0.8327
distance 0.6523   1.0000  0.9431
time     0.8327   0.9431  1.0000
-------------------------------------------------------------------------
```

Figure 2.10 is a *scatterplot matrix*, showing a plot for each pair of variables. It seems natural that longer races would tend to have greater record times per mile, so we might expect the record time to be a convex increasing function of distance. However, the scatterplot relating these variables reveals a strong linear trend, apart from a single outlier. The scatterplot of record time by climb also shows linearity, apart from a rather severe outlier discussed below.

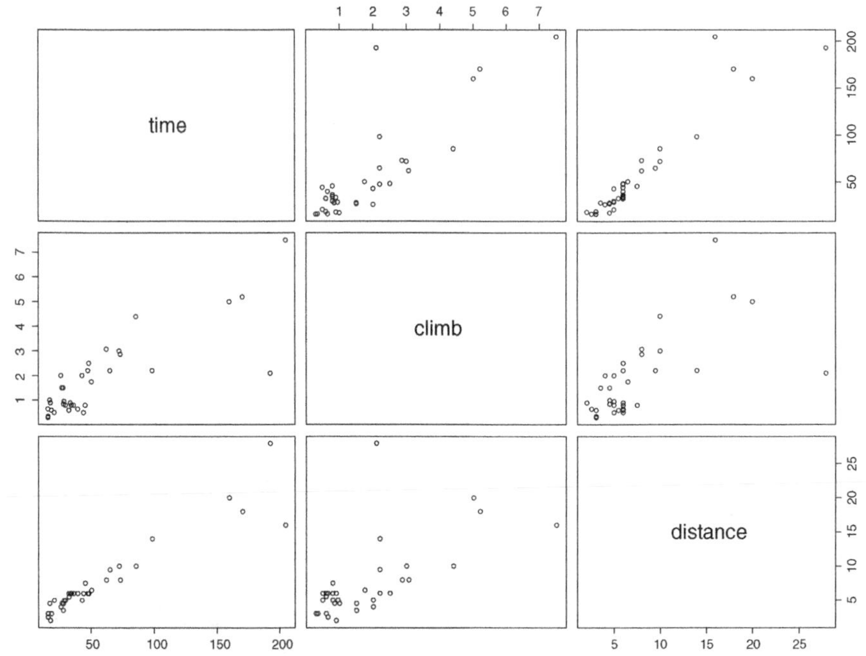

Figure 2.10 Scatterplot matrix for record time, climb, and distance, in Scottish hill races.

For the ordinary linear model that uses both explanatory variables, without interaction, here is basic R output, not showing inferential results that assume normality for y:

```
-------------------------------------------------------------------------
> fit.cd <- lm(time ~ climb + distance)
> summary(fit.cd)
Coefficients:
            Estimate  Std. Error
(Intercept) -13.1086      2.5608
climb        11.7801      1.2206
distance      6.3510      0.3578
---
Residual standard error: 8.734 on 32 degrees of freedom # This is s
Multiple R-squared: 0.9717,     Adjusted R-squared:  0.970
> cor(time, fitted(fit.cd)) # multiple correlation
[1] 0.9857611
-------------------------------------------------------------------------
```

The model fit indicates that, adjusted for climb, the predicted record time increases by 6.35 minutes for every additional mile of distance. The "Residual standard error" reported for the model fit is the estimated standard deviation of record times, at fixed values of climb and distance; that is, it is $s = 8.734$ minutes. From Section 2.4.1, the error variance estimate $s^2 = 76.29$ averages the variability of the residuals, with denominator $n - p$, which is here $df = 35 - 3 = 32$. The sample marginal variance for the record times is $s_y^2 = 2539.42$, considerably larger than s^2.

From the output, $R^2 = 0.972$ indicates a reduction of 97.2% in the sum of squared errors from using this prediction equation instead of $\bar{y}$ to predict the record times. The multiple correlation of $R = \sqrt{0.972} = 0.986$ equals the correlation between the 35 observed y_i and fitted $\hat{\mu}_i$ values. The output also reports adjusted $R^2 = 0.970$. We estimate that the conditional variance for record times is only 3% of the marginal variance.

The standardized residuals (`rstandard` in R) have an approximate mean of 0 and standard deviation of 1. A histogram (not shown here) of them or of the raw residuals exhibits some skew to the right. From Section 2.5, the residuals are orthogonal to the model fit, and we can check model assumptions by plots of them. Figure 2.11 plots the standardized residuals against the model-fitted values. We suggest you construct the plots against the explanatory variables. These plots do not suggest this model's lack of fit, but they and the histogram reveal an outlier. This is the record time of 204.62 minutes with fitted value of 176.86 for the Bens of Jura Fell Race, the race having the greatest climb. For this race, the standardized residual is 4.175 and Cook's distance is 4.215, the largest for the 32 observations and 13 times the next largest value. From Figure 2.10, the Lairig Ghru Fun Run is a severe outlier when record time is plotted against climb; yet when considered with both climb and distance predictors it has standardized residual of only 0.66 and Cook's distance of 0.32. Its record time of 192.67 minutes seems very large for a climb of 2.1 thousand feet, but not at all unusual when we take into account that it is the longest race (28 miles). Atkinson

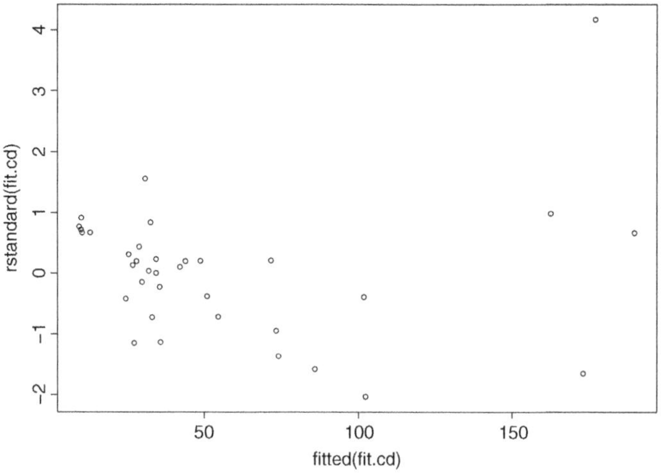

Figure 2.11 Plot of standardized residuals versus fitted values, for linear model predicting record time using climb and distance.

(1986) presented other diagnostic measures and plots for these data that are beyond the scope of this book.

```
--------------------------------------------------------------------
> hist(residuals(fit.cd)) # Histogram display of residuals
> quantile(rstandard(fit.cd), c(0,0.25,0.5,0.75,1))
        0%         25%         50%         75%        100%
-2.0343433 -0.5684549  0.1302666  0.6630338  4.1751367
> cor(fitted(fit.cd),residuals(fit.cd)) # correlation equals zero
[1] -7.070225e-17
> mean(rstandard(fit.cd)); sd(rstandard(fit.cd))
[1] 0.03068615 # Standardized residuals have mean approximately = 0
[1] 1.105608   # and standard deviation approximately = 1
> plot(distance, rstandard(fit.cd)) # scatterplot display
> plot(fitted(fit.cd), rstandard(fit.cd))
> cooks.distance(fit.cd)
> plot(cooks.distance(fit.cd))
--------------------------------------------------------------------
```

When we fit the model using the `glm` function in R, the output states:

```
--------------------------------------------------------------------
    Null deviance: 86340.1 on 34 degrees of freedom
Residual deviance:  2441.3 on 32 degrees of freedom
--------------------------------------------------------------------
```

We introduce the *deviance* in Chapter 4. For now, we mention that for the normal linear model, the null deviance is the corrected TSS and the residual deviance is the SSE. Thus, $R^2 = (86340.1 - 2441.3)/86340.1 = 0.972$. The difference $86340.1 - 2441.3 = 83898.8$ is the SSR for the model.

Next, we show ANOVA tables that provide SSE and the sequential SS for each explanatory variable in the order in which it enters the model, considering both possible sequences:

```
------------------------------------------------------------------
> anova(lm(time ~ climb + distance)) # climb entered, then distance
Analysis of Variance Table
           Df  Sum Sq  Mean Sq
climb       1   59861    59861
distance    1   24038    24038
Residuals  32    2441       76
---
> anova(lm(time ~ distance + climb)) # distance entered, then climb
Analysis of Variance Table
           Df  Sum Sq  Mean Sq
distance    1   76793    76793
climb       1    7106     7106
Residuals  32    2441       76
------------------------------------------------------------------
```

The sequential SS values differ substantially according to the order of entering the explanatory variables into the model, because the correlation is 0.652 between distance and climb. However, the SSE and SSR values for the full model, and hence R^2, do not depend on this. For each ANOVA table display, SSE = 2441 and SSR = 59861 + 24038 = 76793 + 7106 = 83,899.

One way this model containing only main effects fails is if the effect of distance is greater when the climb is greater, as seems plausible. To allow the effect of distance to depend on the climb, we add an interaction term:

```
------------------------------------------------------------------
> summary(lm(time ~ climb + distance + climb:distance))
Coefficients:
                Estimate  Std. Error
(Intercept)      -0.7672      3.9058
climb             3.7133      2.3647
distance          4.9623      0.4742
climb:distance    0.6598      0.1743
---
Residual standard error: 7.338 on 31 degrees of freedom
Multiple R-squared: 0.9807,      Adjusted R-squared: 0.9788
------------------------------------------------------------------
```

The effect on record time of a 1 mile increase in distance now changes from 4.962 + 0.660(0.3) = 5.16 minutes at the minimum climb of 0.3 thousand feet to 4.962 + 0.660(7.5) = 9.91 minutes at the maximum climb of 7.5 thousand feet. As R^2 has increased from 0.972 to 0.981 and adjusted R^2 from 0.970 to 0.979, this more informative summary explains about a third of the variability that had been unexplained by the main effects model. That is, the squared partial correlation, which summarizes the impact of adding the interaction term, is $(0.981 - 0.972)/(1 - 0.972) = 0.32$.

2.7 OPTIMALITY OF LEAST SQUARES AND GENERALIZED LEAST SQUARES

In this chapter, we have used least squares to estimate parameters in the ordinary linear model, which assumes independent observations with constant variance. We next show a criterion by which such estimators are optimal. We then generalize least squares to permit observations to be correlated and to have nonconstant variance.

2.7.1 The Gauss–Markov Theorem

For the ordinary linear model, least squares provides the best possible estimator of model parameters, in a certain restricted sense. Like most other results in this chapter, this one does not require an assumption (such as normality) about the distribution of the response variable. We express it here for linear combinations $\boldsymbol{a}^{\mathrm{T}}\boldsymbol{\beta}$ of the parameters, but then we apply it to the individual parameters.

> **Gauss–Markov theorem:** Suppose $E(\boldsymbol{y}) = \boldsymbol{X\beta}$, where $\boldsymbol{X}$ has full rank, and $\mathrm{var}(\boldsymbol{y}) = \sigma^2\boldsymbol{I}$. The least squares estimator $\hat{\boldsymbol{\beta}} = (\boldsymbol{X}^{\mathrm{T}}\boldsymbol{X})^{-1}\boldsymbol{X}^{\mathrm{T}}\boldsymbol{y}$ is the *best linear unbiased estimator* (BLUE) of $\boldsymbol{\beta}$, in this sense: For any $\boldsymbol{a}^{\mathrm{T}}\boldsymbol{\beta}$, of the estimators that are linear in $\boldsymbol{y}$ and unbiased, $\boldsymbol{a}^{\mathrm{T}}\hat{\boldsymbol{\beta}}$ has minimum variance.

To prove this, we express $\boldsymbol{a}^{\mathrm{T}}\hat{\boldsymbol{\beta}}$ in its linear form in $\boldsymbol{y}$ as

$$\boldsymbol{a}^{\mathrm{T}}\hat{\boldsymbol{\beta}} = \boldsymbol{a}^{\mathrm{T}}(\boldsymbol{X}^{\mathrm{T}}\boldsymbol{X})^{-1}\boldsymbol{X}^{\mathrm{T}}\boldsymbol{y} = \boldsymbol{c}^{\mathrm{T}}\boldsymbol{y},$$

where $\boldsymbol{c}^{\mathrm{T}} = \boldsymbol{a}^{\mathrm{T}}(\boldsymbol{X}^{\mathrm{T}}\boldsymbol{X})^{-1}\boldsymbol{X}^{\mathrm{T}}$. Suppose $\boldsymbol{b}^{\mathrm{T}}\boldsymbol{y}$ is an alternative linear estimator of $\boldsymbol{a}^{\mathrm{T}}\boldsymbol{\beta}$ that is unbiased. Then,

$$E(\boldsymbol{b}-\boldsymbol{c})^{\mathrm{T}}\boldsymbol{y} = E(\boldsymbol{b}^{\mathrm{T}}\boldsymbol{y}) - E(\boldsymbol{c}^{\mathrm{T}}\boldsymbol{y}) = \boldsymbol{a}^{\mathrm{T}}\boldsymbol{\beta} - \boldsymbol{a}^{\mathrm{T}}\boldsymbol{\beta} = 0$$

for all $\boldsymbol{\beta}$. But this also equals $(\boldsymbol{b}-\boldsymbol{c})^{\mathrm{T}}\boldsymbol{X\beta} = [\boldsymbol{\beta}^{\mathrm{T}}\boldsymbol{X}^{\mathrm{T}}(\boldsymbol{b}-\boldsymbol{c})]^{\mathrm{T}}$ for all $\boldsymbol{\beta}$. Therefore[14], $\boldsymbol{X}^{\mathrm{T}}(\boldsymbol{b}-\boldsymbol{c}) = \boldsymbol{0}$. So, $(\boldsymbol{b}-\boldsymbol{c})$ is in the error space $C(\boldsymbol{X})^{\perp} = N(\boldsymbol{X}^{\mathrm{T}})$ for the model. Now,

$$\mathrm{var}(\boldsymbol{b}^{\mathrm{T}}\boldsymbol{y}) = \mathrm{var}[\boldsymbol{c}^{\mathrm{T}}\boldsymbol{y} + (\boldsymbol{b}-\boldsymbol{c})^{\mathrm{T}}\boldsymbol{y}] = \mathrm{var}(\boldsymbol{c}^{\mathrm{T}}\boldsymbol{y}) + ||\boldsymbol{b}-\boldsymbol{c}||^2\sigma^2 + 2\mathrm{cov}[\boldsymbol{c}^{\mathrm{T}}\boldsymbol{y}, (\boldsymbol{b}-\boldsymbol{c})^{\mathrm{T}}\boldsymbol{y}].$$

But since $\boldsymbol{X}^{\mathrm{T}}(\boldsymbol{b}-\boldsymbol{c}) = \boldsymbol{0}$,

$$\mathrm{cov}[\boldsymbol{c}^{\mathrm{T}}\boldsymbol{y}, (\boldsymbol{b}-\boldsymbol{c})^{\mathrm{T}}\boldsymbol{y}] = \boldsymbol{c}^{\mathrm{T}}\mathrm{var}(\boldsymbol{y})(\boldsymbol{b}-\boldsymbol{c}) = \sigma^2\boldsymbol{a}^{\mathrm{T}}(\boldsymbol{X}^{\mathrm{T}}\boldsymbol{X})^{-1}\boldsymbol{X}^{\mathrm{T}}(\boldsymbol{b}-\boldsymbol{c}) = 0.$$

Thus, $\mathrm{var}(\boldsymbol{b}^{\mathrm{T}}\boldsymbol{y}) \geq \mathrm{var}(\boldsymbol{c}^{\mathrm{T}}\boldsymbol{y}) = \mathrm{var}(\boldsymbol{a}^{\mathrm{T}}\hat{\boldsymbol{\beta}})$, with equality if and only if $\boldsymbol{b} = \boldsymbol{c}$.

From the theorem's proof, any other linear unbiased estimator of $\boldsymbol{a}^{\mathrm{T}}\boldsymbol{\beta}$ can be expressed as $\boldsymbol{a}^{\mathrm{T}}\hat{\boldsymbol{\beta}} + \boldsymbol{d}^{\mathrm{T}}\boldsymbol{y}$ where $E(\boldsymbol{d}^{\mathrm{T}}\boldsymbol{y}) = 0$ and $\boldsymbol{d}^{\mathrm{T}}\boldsymbol{y}$ is uncorrelated with $\boldsymbol{a}^{\mathrm{T}}\hat{\boldsymbol{\beta}}$; that is,

[14]Recall that if $\boldsymbol{\beta}^{\mathrm{T}}\boldsymbol{L} = \boldsymbol{\beta}^{\mathrm{T}}\boldsymbol{M}$ for all $\boldsymbol{\beta}$, then $\boldsymbol{L} = \boldsymbol{M}$; here we identify $\boldsymbol{L} = \boldsymbol{X}^{\mathrm{T}}(\boldsymbol{b}-\boldsymbol{c})$ and $\boldsymbol{M} = \boldsymbol{0}$.

the variate added to $\boldsymbol{a}^{\mathrm{T}}\hat{\boldsymbol{\beta}}$ is like extra noise. The Gauss–Markov theorem extends to non-full-rank models. Using a generalized inverse of $(\boldsymbol{X}^{\mathrm{T}}\boldsymbol{X})$ in obtaining $\hat{\boldsymbol{\beta}}$, $\boldsymbol{a}^{\mathrm{T}}\hat{\boldsymbol{\beta}}$ is a BLUE of an estimable function $\boldsymbol{a}^{\mathrm{T}}\boldsymbol{\beta}$.

With the added assumption of normality for the distribution of $\boldsymbol{y}$, $\boldsymbol{a}^{\mathrm{T}}\hat{\boldsymbol{\beta}}$ is the minimum variance unbiased estimator (MVUE) of $\boldsymbol{a}^{\mathrm{T}}\boldsymbol{\beta}$. Here, the restriction is still unbiasedness, but not linearity in $\boldsymbol{y}$. This follows from the Lehmann–Scheffé theorem, which states that a function of a complete, sufficient statistic is the unique MVUE of its expectation.

Let $\boldsymbol{a}$ have 1 in position j and 0 elsewhere. Then the Gauss–Markov theorem implies that, for all j, $\mathrm{var}(\hat{\beta}_j)$ takes minimum value out of all linear unbiased estimators of β_j.

At first glance, the Gauss–Markov theorem is impressive, the least squares estimator being declared "best." However, the restriction to estimators that are both linear and unbiased is severe. In later chapters, maximum likelihood (ML) estimators for parameters in non-normal GLMs usually satisfy neither of these properties. Also, in some cases in Statistics, the best unbiased estimator is not sensible (e.g., see Exercise 2.41). In multivariate settings, Bayesian-like biased estimators often obtain a marked improvement in mean squared error by shrinking the ML estimate toward a prior mean[15].

2.7.2 Generalized Least Squares

The ordinary linear model, for which $E(\boldsymbol{y}) = \boldsymbol{X}\boldsymbol{\beta}$ with $\mathrm{var}(\boldsymbol{y}) = \sigma^2\boldsymbol{I}$, assumes that the response observations have identical variances and are uncorrelated. In practice, this is often not plausible. With count data, the variance is typically larger when the mean is larger. With time series data, observations close together in time are often highly correlated. With survey data, sampling designs are usually more complex than simple random sampling, and analysts weight observations so that they receive their appropriate influence.

A linear model with a more general structure for the covariance matrix is

$$E(\boldsymbol{y}) = \boldsymbol{X}\boldsymbol{\beta} \quad \text{with} \quad \mathrm{var}(\boldsymbol{y}) = \sigma^2\boldsymbol{V},$$

where $\boldsymbol{V}$ need not be the identity matrix. We next see that ordinary least squares is still relevant for a linear transformation of $\boldsymbol{y}$, and the method then corresponds to a weighted version of least squares on the original scale.

Suppose the model matrix $\boldsymbol{X}$ has full rank and $\boldsymbol{V}$ is a known positive definite matrix. Then, $\boldsymbol{V}$ can be expressed as $\boldsymbol{V} = \boldsymbol{B}\boldsymbol{B}^{\mathrm{T}}$ for a square matrix $\boldsymbol{B}$ that is denoted by $\boldsymbol{V}^{1/2}$. This results from using the *spectral decomposition* for a symmetric matrix as $\boldsymbol{V} = \boldsymbol{Q}\boldsymbol{\Lambda}\boldsymbol{Q}^{\mathrm{T}}$, where $\boldsymbol{\Lambda}$ is a diagonal matrix of the eigenvalues of $\boldsymbol{V}$ and $\boldsymbol{Q}$ is orthogonal[16] with columns that are its eigenvectors, from which $\boldsymbol{V}^{1/2} = \boldsymbol{Q}\boldsymbol{\Lambda}^{1/2}\boldsymbol{Q}^{\mathrm{T}}$

[15] A classic example is Charles Stein's famous result that, in estimating a vector of normal means, the sample mean vector is inadmissible. See Efron and Morris (1975).

[16] Recall that an orthogonal matrix $\boldsymbol{Q}$ is a square matrix having $\boldsymbol{Q}\boldsymbol{Q}^{\mathrm{T}} = \boldsymbol{Q}^{\mathrm{T}}\boldsymbol{Q} = \boldsymbol{I}$.

using the positive square roots of the eigenvalues. Then, $\boldsymbol{V}^{-1}$ exists, as does $\boldsymbol{V}^{-1/2} = \boldsymbol{Q}\boldsymbol{\Lambda}^{-1/2}\boldsymbol{Q}^{\mathrm{T}}$. Let

$$\boldsymbol{y}^* = \boldsymbol{V}^{-1/2}\boldsymbol{y}, \quad \boldsymbol{X}^* = \boldsymbol{V}^{-1/2}\boldsymbol{X}.$$

For these linearly transformed values,

$$E(\boldsymbol{y}^*) = \boldsymbol{V}^{-1/2}\boldsymbol{X}\boldsymbol{\beta} = \boldsymbol{X}^*\boldsymbol{\beta}, \quad \mathrm{var}(\boldsymbol{y}^*) = \sigma^2\boldsymbol{V}^{-1/2}\boldsymbol{V}(\boldsymbol{V}^{-1/2})^{\mathrm{T}} = \sigma^2\boldsymbol{I}.$$

So $\boldsymbol{y}^*$ satisfies the ordinary linear model, and we can apply least squares to the transformed values. The sum of squared errors comparing $\boldsymbol{y}^*$ and $\boldsymbol{X}^*\boldsymbol{\beta}$ that is minimized is

$$(\boldsymbol{y}^* - \boldsymbol{X}^*\boldsymbol{\beta})^{\mathrm{T}}(\boldsymbol{y}^* - \boldsymbol{X}^*\boldsymbol{\beta}) = (\boldsymbol{y} - \boldsymbol{X}\boldsymbol{\beta})^{\mathrm{T}}\boldsymbol{V}^{-1}(\boldsymbol{y} - \boldsymbol{X}\boldsymbol{\beta}).$$

The normal equations $[(\boldsymbol{X}^*)^{\mathrm{T}}\boldsymbol{X}^*]\boldsymbol{\beta} = (\boldsymbol{X}^*)^{\mathrm{T}}\boldsymbol{y}^*$ become

$$(\boldsymbol{X}^{\mathrm{T}}\boldsymbol{V}^{-1/2}\boldsymbol{V}^{-1/2}\boldsymbol{X})\boldsymbol{\beta} = \boldsymbol{X}^{\mathrm{T}}\boldsymbol{V}^{-1/2}\boldsymbol{V}^{-1/2}\boldsymbol{y}, \quad \text{or} \quad \boldsymbol{X}^{\mathrm{T}}\boldsymbol{V}^{-1}(\boldsymbol{y} - \boldsymbol{X}\boldsymbol{\beta}) = \boldsymbol{0}.$$

From (2.3), the least squares solution for the transformed values is

$$\hat{\boldsymbol{\beta}}_{GLS} = [(\boldsymbol{X}^*)^{\mathrm{T}}\boldsymbol{X}^*]^{-1}(\boldsymbol{X}^*)^{\mathrm{T}}\boldsymbol{y}^* = (\boldsymbol{X}^{\mathrm{T}}\boldsymbol{V}^{-1}\boldsymbol{X})^{-1}\boldsymbol{X}^{\mathrm{T}}\boldsymbol{V}^{-1}\boldsymbol{y}. \tag{2.12}$$

The estimator $\hat{\boldsymbol{\beta}}_{GLS}$ is called the *generalized least squares* estimator of $\boldsymbol{\beta}$. When $\boldsymbol{V}$ is diagonal and $\mathrm{var}(y_i) = \sigma^2/w_i$ for a known positive weight w_i, as in a survey design that gives more weight to some observations than others, $\hat{\boldsymbol{\beta}}_{GLS}$ is also referred to as a *weighted least squares* estimator. This form of estimator arises in fitting GLMs (Section 4.5.4).

The generalized least squares estimator has

$$E(\hat{\boldsymbol{\beta}}_{GLS}) = (\boldsymbol{X}^{\mathrm{T}}\boldsymbol{V}^{-1}\boldsymbol{X})^{-1}\boldsymbol{X}^{\mathrm{T}}\boldsymbol{V}^{-1}E(\boldsymbol{y}) = \boldsymbol{\beta}.$$

Like the OLS estimator, it is unbiased. The covariance matrix is

$$\begin{aligned}\mathrm{var}(\hat{\boldsymbol{\beta}}_{GLS}) &= (\boldsymbol{X}^{\mathrm{T}}\boldsymbol{V}^{-1}\boldsymbol{X})^{-1}\boldsymbol{X}^{\mathrm{T}}\boldsymbol{V}^{-1}(\sigma^2\boldsymbol{V})\boldsymbol{V}^{-1}\boldsymbol{X}(\boldsymbol{X}^{\mathrm{T}}\boldsymbol{V}^{-1}\boldsymbol{X})^{-1} \\ &= \sigma^2(\boldsymbol{X}^{\mathrm{T}}\boldsymbol{V}^{-1}\boldsymbol{X})^{-1}.\end{aligned}$$

It shares other properties of the ordinary least squares estimator, such as $\hat{\boldsymbol{\beta}}$ being the BLUE estimator of $\boldsymbol{\beta}$ and also the maximum likelihood estimator under the normality assumption.

The fitted values for this more general model are

$$\hat{\boldsymbol{\mu}} = \boldsymbol{X}\hat{\boldsymbol{\beta}}_{GLS} = \boldsymbol{X}(\boldsymbol{X}^{\mathrm{T}}\boldsymbol{V}^{-1}\boldsymbol{X})^{-1}\boldsymbol{X}^{\mathrm{T}}\boldsymbol{V}^{-1}\boldsymbol{y}.$$

Here, $\boldsymbol{H} = \boldsymbol{X}(\boldsymbol{X}^{\mathrm{T}}\boldsymbol{V}^{-1}\boldsymbol{X})^{-1}\boldsymbol{X}^{\mathrm{T}}\boldsymbol{V}^{-1}$ plays the role of a *hat matrix*. In this case, $\boldsymbol{H}$ is idempotent but need not be symmetric, so it is not a projection matrix as defined in Section 2.2. However, $\boldsymbol{H}$ is a projection matrix in a more general sense if we instead define the inner product to be $(\boldsymbol{w}, \boldsymbol{z}) = \boldsymbol{w}^{\mathrm{T}}\boldsymbol{V}^{-1}\boldsymbol{z}$, as motivated by the normal equations given above. Namely, if $\boldsymbol{w} \in C(\boldsymbol{X})$, say $\boldsymbol{w} = \boldsymbol{X}\boldsymbol{v}$, then

$$\begin{aligned}\boldsymbol{H}\boldsymbol{w} &= \boldsymbol{X}(\boldsymbol{X}^{\mathrm{T}}\boldsymbol{V}^{-1}\boldsymbol{X})^{-1}\boldsymbol{X}^{\mathrm{T}}\boldsymbol{V}^{-1}\boldsymbol{w}\\ &= \boldsymbol{X}(\boldsymbol{X}^{\mathrm{T}}\boldsymbol{V}^{-1}\boldsymbol{X})^{-1}\boldsymbol{X}^{\mathrm{T}}\boldsymbol{V}^{-1}\boldsymbol{X}\boldsymbol{v} = \boldsymbol{X}\boldsymbol{v} = \boldsymbol{w}.\end{aligned}$$

Also, if $\boldsymbol{w} \in C(\boldsymbol{X})^{\perp} = N(\boldsymbol{X}^{\mathrm{T}})$, then for all $\boldsymbol{v} \in C(\boldsymbol{X})$, $(\boldsymbol{w}, \boldsymbol{v}) = \boldsymbol{w}^{\mathrm{T}}\boldsymbol{V}^{-1}\boldsymbol{v} = 0$, so $\boldsymbol{H}\boldsymbol{w} = \boldsymbol{0}$.

The estimate of σ^2 in the generalized model with $\mathrm{var}(\boldsymbol{y}) = \sigma^2\boldsymbol{V}$ uses the usual unbiased estimator for the linearly transformed values. If $\mathrm{rank}(\boldsymbol{X}) = r$, the estimate is

$$s^2 = \frac{(\boldsymbol{y}^* - \boldsymbol{X}^*\hat{\boldsymbol{\beta}})^{\mathrm{T}}(\boldsymbol{y}^* - \boldsymbol{X}^*\hat{\boldsymbol{\beta}})}{n-r} = \frac{(\boldsymbol{y} - \hat{\boldsymbol{\mu}})^{\mathrm{T}}\boldsymbol{V}^{-1}(\boldsymbol{y} - \hat{\boldsymbol{\mu}})}{n-r}.$$

Statistical inference for the model parameters can be based directly on the regular inferences of the next chapter for the ordinary linear model but using the transformed variables.

2.7.3 Adjustment Using Estimated Heteroscedasticity

This generalization of the model seems straightforward, but we have neglected a crucial point: In applications, $\boldsymbol{V}$ itself is also often unknown and must be estimated. Once we have done that, we can use $\hat{\boldsymbol{\beta}}_{GLS}$ in (2.12) with $\boldsymbol{V}$ replaced by $\hat{\boldsymbol{V}}$. But this estimator is no longer unbiased nor has an exact formula for the covariance matrix, which also must be estimated.

Since $\hat{\boldsymbol{\beta}}_{GLS}$ is no longer optimal once we have substituted estimated variances, we could instead use the ordinary least squares estimator, which does not require estimating the variances and is still unbiased and consistent (i.e., converging in probability to $\boldsymbol{\beta}$ as $n \to \infty$). In doing so, however, we should adapt standard errors to adjust for the departure from the ordinary linear model assumptions. An important case (heteroscedasticity) is when $\boldsymbol{V}$ is diagonal. Let $\mathrm{var}(y_i) = \sigma_i^2$. Then, with $\boldsymbol{x}_i$ as row i of $\boldsymbol{X}$, $\hat{\boldsymbol{\beta}} = (\boldsymbol{X}^{\mathrm{T}}\boldsymbol{X})^{-1}\boldsymbol{X}^{\mathrm{T}}\boldsymbol{y} = \left(\boldsymbol{X}^{\mathrm{T}}\boldsymbol{X}\right)^{-1}\left(\sum_{i=1}^{n}\boldsymbol{x}_i^{\mathrm{T}}y_i\right)$, so

$$\mathrm{var}(\hat{\boldsymbol{\beta}}) = \left(\boldsymbol{X}^{\mathrm{T}}\boldsymbol{X}\right)^{-1}\left(\sum_{i=1}^{n}\sigma_i^2\boldsymbol{x}_i^{\mathrm{T}}\boldsymbol{x}_i\right)\left(\boldsymbol{X}^{\mathrm{T}}\boldsymbol{X}\right)^{-1}.$$

Since $\mathrm{var}(e_i) = \sigma_i^2(1 - h_{ii})$, we can estimate $\mathrm{var}(\hat{\boldsymbol{\beta}})$ by replacing σ_i^2 by $e_i^2/(1 - h_{ii})$, for each i.

CHAPTER NOTES

Section 2.1: Least Squares Model Fitting

2.1 **History:** Harter (1974), Plackett (1972), and Stigler (1981, 1986, Chapters 1 and 4) discussed the history of the least squares method.

2.2 **Computing $\hat{\boldsymbol{\beta}}$:** For details about how software inverts $\boldsymbol{X}^{\mathrm{T}}\boldsymbol{X}$ or uses another method to compute $\hat{\boldsymbol{\beta}}$, see McCullagh and Nelder (1989, pp. 81–89), Seber and Lee (2003, Chapter 11), Wood (2006, Chapter 1), or do an Internet search on the Gauss–Jordan elimination method (e.g., see the *Gaussian elimination* article in Wikipedia).

Section 2.2: Projections of Data onto Model Spaces

2.3 **Geometry:** For more on the geometry of least squares for linear models, see Rawlings et al. (1998, Chapter 6), Taylor (2013), and Wood (2006, Section 1.4).

Section 2.4: Summarizing Variability in a Linear Model

2.4 **Correlation measures:** The correlation is due to Galton (1888), but later received much more attention from Karl Pearson, such as in Pearson (1920). Yule (1897), in extending Galton's ideas about correlation and regression to multiple variables, introduced the multiple correlation and partial correlation. Wherry (1931) justified adjusted R^2 as a reduced-bias version of R^2.

Section 2.5: Residuals, Leverage, and Influence

2.5 **Influence:** For details about influence measures and related diagnostics, see Belsley et al. (1980), Cook (1977, 1986), Cook and Weisberg (1982), Davison and Tsai (1992), Fox (2008, Chapters 11–13), and Hoaglin and Welsch (1978).

Section 2.7: Optimality of Least Squares and Generalized Least Squares

2.6 **Gauss–Markov and GLS:** The Gauss–Markov theorem is named after results established in 1821 by Carl Friedrich Gauss and published in 1912 by the Russian probabilist Andrei A. Markov. Generalized least squares was introduced by the New Zealand mathematician/statistician A. C. Aitken (1935), and the model with general covariance structure is often called the *Aitken model.*

EXERCISES

2.1 For independent observations $y_1, \ldots, y_n$ from a probability distribution with mean μ, show that the least squares estimate of μ is $\bar{y}$.

2.2 In the linear model $\boldsymbol{y} = \boldsymbol{X}\boldsymbol{\beta} + \boldsymbol{\epsilon}$, suppose ϵ_i has the Laplace density, $f(\epsilon) = (1/2b)\exp(-|\epsilon|/b)$. Show that the ML estimate minimizes $\sum_i |y_i - \mu_i|$.

2.3 Consider the least squares fit of the linear model $E(y_i) = \beta_0 + \beta_1 x_i$.

a. Show that $\hat{\beta}_1 = [\sum_i (x_i - \bar{x})(y_i - \bar{y})]/[\sum_i (x_i - \bar{x})^2]$.

b. Derive var($\hat{\beta}_1$). State the estimated standard error of $\hat{\beta}_1$, and discuss how its magnitude is affected by (i) n, (ii) the variability around the fitted line, (iii) the sample variance of x. In experiments with control over setting values of x, what does (iii) suggest about the optimal way to do this?

2.4 In the linear model $E(y_i) = \beta_0 + \beta_1 x_i$, suppose that instead of observing x_i we observe $x_i^* = x_i + u_i$, where u_i is independent of x_i for all i and $\text{var}(u_i) = \sigma_u^2$. Analyze the expected impact of this *measurement error* on $\hat{\beta}_1$ and r.

2.5 In the linear model $E(y_i) = \beta_0 + \beta_1 x_i$, consider the fitted line that minimizes the sum of squared perpendicular distances from the points to the line. Is this fit invariant to the units of measurement of either variable? Show that such invariance is a property of the usual least squares fit.

2.6 For the model in Section 2.3.4 for the two-way layout, construct a full-rank model matrix. Show that the normal equations imply that the marginal row and column sample totals for $\boldsymbol{y}$ equal the row and column totals of the fitted values.

2.7 Refer to the analysis of covariance model $\mu_i = \beta_0 + \beta_1 x_{i1} + \beta_2 x_{i2}$ for quantitative x_1 and binary x_2 for two groups, with $x_{i2} = 0$ for group 1 and $x_{i2} = 1$ for group 2. Denote the sample means on x_1 and y by $(\bar{x}_1^{(1)}, \bar{y}^{(1)})$ for group 1 and $(\bar{x}_1^{(2)}, \bar{y}^{(2)})$ for group 2. Show that the least squares fit corresponds to parallel lines for the two groups, which pass through these points. (At the overall $\bar{x}_1$, the fitted values $\hat{\beta}_0 + \hat{\beta}_1 \bar{x}_1$ and $\hat{\beta}_0 + \hat{\beta}_1 \bar{x}_1 + \hat{\beta}_2$ are called *adjusted means* of y.)

2.8 By the *QR decomposition*, $\boldsymbol{X}$ can be decomposed as $\boldsymbol{X} = \boldsymbol{QR}$, where $\boldsymbol{Q}$ consists of the first p columns of a $n \times n$ orthogonal matrix and $\boldsymbol{R}$ is a $p \times p$ upper triangular matrix. Show that the least squares estimate $\hat{\boldsymbol{\beta}} = \boldsymbol{R}^{-1}\boldsymbol{Q}^{\mathrm{T}}\boldsymbol{y}$.

2.9 In an ordinary linear model with two explanatory variables x_1 and x_2 having sample corr$(\boldsymbol{x}_{*1}, \boldsymbol{x}_{*2}) > 0$, show that the estimated corr$(\hat{\beta}_1, \hat{\beta}_2) < 0$.

2.10 For a projection matrix $\boldsymbol{P}$, for any $\boldsymbol{y}$ in $\mathbb{R}^n$ show that $\boldsymbol{Py}$ and $\boldsymbol{y} - \boldsymbol{Py}$ are orthogonal vectors; that is, the projection is an *orthogonal projection*.

2.11 Prove that $\boldsymbol{I} - \frac{1}{n}\boldsymbol{1}_n\boldsymbol{1}_n^{\mathrm{T}}$ is symmetric and idempotent (i.e., a projection matrix), and identify the vector to which it projects an arbitrary $\boldsymbol{y}$.

2.12 For a full-rank model matrix $\boldsymbol{X}$, show that rank$(\boldsymbol{H})$ = rank$(\boldsymbol{X})$, where $\boldsymbol{H} = \boldsymbol{X}(\boldsymbol{X}^{\mathrm{T}}\boldsymbol{X})^{-1}\boldsymbol{X}^{\mathrm{T}}$.

2.13 From Exercise 1.17, if $\boldsymbol{A}$ is nonsingular and $\boldsymbol{X}^* = \boldsymbol{XA}$ (such as in using a different parameterization for a factor), then $C(\boldsymbol{X}^*) = C(\boldsymbol{X})$. Show that the linear models with the model matrices $\boldsymbol{X}$ and $\boldsymbol{X}^*$ have the same hat matrix and the same fitted values.

2.14 For a linear model with full rank $\boldsymbol{X}$ and projection matrix $\boldsymbol{P}_X$, show that $\boldsymbol{P}_X\boldsymbol{X} = \boldsymbol{X}$ and that $C(\boldsymbol{P}_X) = C(\boldsymbol{X})$.

2.15 Denote the hat matrix by $\boldsymbol{P}_0$ for the null model and $\boldsymbol{H}$ for any linear model that contains an intercept term. Explain why $\boldsymbol{P}_0\boldsymbol{H} = \boldsymbol{H}\boldsymbol{P}_0 = \boldsymbol{P}_0$. Show this implies that each row and each column of $\boldsymbol{H}$ sums to 1.

2.16 When $\boldsymbol{X}$ does not have full rank, let's see why $\boldsymbol{P}_X = \boldsymbol{X}(\boldsymbol{X}^{\mathrm{T}}\boldsymbol{X})^{-}\boldsymbol{X}^{\mathrm{T}}$ is invariant to the choice of generalized inverse. Let $\boldsymbol{G}$ and $\boldsymbol{H}$ be two generalized inverses of $\boldsymbol{X}^{\mathrm{T}}\boldsymbol{X}$. For an arbitrary $\boldsymbol{v} \in \mathbb{R}^n$, let $\boldsymbol{v} = \boldsymbol{v}_1 + \boldsymbol{v}_2$ with $\boldsymbol{v}_1 = \boldsymbol{X}\boldsymbol{b} \in C(\boldsymbol{X})$ for some $\boldsymbol{b}$.

a. Show that $\boldsymbol{v}^{\mathrm{T}}\boldsymbol{X}\boldsymbol{G}\boldsymbol{X}^{\mathrm{T}}\boldsymbol{X} = \boldsymbol{v}^{\mathrm{T}}\boldsymbol{X}$, so that $\boldsymbol{X}\boldsymbol{G}\boldsymbol{X}^{\mathrm{T}}\boldsymbol{X} = \boldsymbol{X}$ for any generalized inverse.

b. Show that $\boldsymbol{X}\boldsymbol{G}\boldsymbol{X}^{\mathrm{T}}\boldsymbol{v} = \boldsymbol{X}\boldsymbol{H}\boldsymbol{X}^{\mathrm{T}}\boldsymbol{v}$, and thus $\boldsymbol{X}\boldsymbol{G}\boldsymbol{X}^{\mathrm{T}}$ is invariant to the choice of generalized inverse.

2.17 When $\boldsymbol{X}$ has less than full rank and we use a generalized inverse to estimate $\boldsymbol{\beta}$, explain why the space of possible least squares solutions $\hat{\boldsymbol{\beta}}$ does not form a vector space. (For a solution, $\hat{\boldsymbol{\beta}}$, this space is the set of $\tilde{\boldsymbol{\beta}} = \hat{\boldsymbol{\beta}} + \boldsymbol{\gamma}$ for all $\boldsymbol{\gamma} \in N(\boldsymbol{X})$; such a shifted vector space is called an *affine space*.)

2.18 In $\mathbb{R}^3$, let $\boldsymbol{W}$ be the vector subspace spanned by (1, 0, 0), that is, the "x-axis" in three-dimensional space. Specify its orthogonal complement. For any $\boldsymbol{y}$ in $\mathbb{R}^3$, show its orthogonal decomposition $\boldsymbol{y} = \boldsymbol{y}_1 + \boldsymbol{y}_2$ with $\boldsymbol{y}_1 \in \boldsymbol{W}$ and $\boldsymbol{y}_2 \in \boldsymbol{W}^{\perp}$.

2.19 Two vectors that are orthogonal or that have zero correlation are linearly independent. However, orthogonal vectors need not be uncorrelated, and uncorrelated vectors need not be orthogonal.

a. Show this with two particular pairs of 4×1 vectors.

b. Suppose $\boldsymbol{u}$ and $\boldsymbol{v}$ have $\mathrm{corr}(\boldsymbol{u}, \boldsymbol{v}) = 0$. Explain why the centered versions $\boldsymbol{u}^* = (\boldsymbol{u} - \bar{\boldsymbol{u}})$ and $\boldsymbol{v}^* = (\boldsymbol{v} - \bar{\boldsymbol{v}})$ are orthogonal (where, e.g., $\bar{\boldsymbol{u}}$ denotes the vector having the mean of the elements of $\boldsymbol{u}$ in each component). Show that $\boldsymbol{u}$ and $\boldsymbol{v}$ themselves are orthogonal if and only if $\bar{\boldsymbol{u}} = \boldsymbol{0}$, $\bar{\boldsymbol{v}} = \boldsymbol{0}$, or both.

c. If $\boldsymbol{u}$ and $\boldsymbol{v}$ are orthogonal, then explain why they also have $\mathrm{corr}(\boldsymbol{u}, \boldsymbol{v}) = 0$ iff $\bar{\boldsymbol{u}} = \boldsymbol{0}$, $\bar{\boldsymbol{v}} = \boldsymbol{0}$, or both. (From (**b**) and (**c**), orthogonality and zero correlation are equivalent only when $\bar{\boldsymbol{u}} = \boldsymbol{0}$ and/or $\bar{\boldsymbol{v}} = \boldsymbol{0}$. Zero correlation means that the centered vectors are perpendicular. Centering typically changes the angle between the two vectors.)

2.20 Suppose that all the parameters in a linear model are orthogonal (Section 2.2.4).

a. When the model contains an intercept term, show that orthogonality implies that each column in $\boldsymbol{X}$ after the first (for the intercept) has mean 0; i.e., each explanatory variable is centered. Thus, based on the previous exercise, explain why each pair of explanatory variables is uncorrelated.

b. When the explanatory variables for the model are all centered, explain why the intercept estimate does not change as the variables are added to the linear predictor. Show that that estimate equals $\bar{y}$ in each case.

2.21 Using the normal equations for a linear model, show that SSE decomposes into

$$(\boldsymbol{y} - \boldsymbol{X}\hat{\boldsymbol{\beta}})^{\mathrm{T}}(\boldsymbol{y} - \boldsymbol{X}\hat{\boldsymbol{\beta}}) = \boldsymbol{y}^{\mathrm{T}}\boldsymbol{y} - \hat{\boldsymbol{\beta}}^{\mathrm{T}}\boldsymbol{X}^{\mathrm{T}}\boldsymbol{y}.$$

Thus, for nested M_1 and M_0, explain why

$$\mathrm{SSR}(M_1 \mid M_0) = \hat{\boldsymbol{\beta}}_1^{\mathrm{T}}\boldsymbol{X}_1^{\mathrm{T}}\boldsymbol{y} - \hat{\boldsymbol{\beta}}_0^{\mathrm{T}}\boldsymbol{X}_0^{\mathrm{T}}\boldsymbol{y}.$$

2.22 In Section 2.3.1 we showed the sum of squares decomposition for the null model $E(y_i) = \beta$, $i = 1, \ldots, n$. Suppose you have $n = 2$ observations.

a. Specify the model space $C(\boldsymbol{X})$ and its orthogonal complement, and find $\boldsymbol{P_X}$ and $(\boldsymbol{I} - \boldsymbol{P_X})$.

b. Suppose $y_1 = 5$ and $y_2 = 10$. Find $\hat{\beta}$ and $\hat{\boldsymbol{\mu}}$. Show the sum of squares decomposition, and find s. Sketch a graph that shows $\boldsymbol{y}$, $\hat{\boldsymbol{\mu}}$, $C(\boldsymbol{X})$, and the projection of $\boldsymbol{y}$ to $\hat{\boldsymbol{\mu}}$.

2.23 In complete contrast to the null model is the *saturated model*, $E(y_i) = \beta_i$, $i = 1, \ldots, n$, which has a separate parameter for each observation. For this model:

a. Specify $\boldsymbol{X}$, the model space $C(\boldsymbol{X})$, and its orthogonal complement, and find $\boldsymbol{P_X}$ and $(\boldsymbol{I} - \boldsymbol{P_X})$.

b. Find $\hat{\boldsymbol{\beta}}$ and $\hat{\boldsymbol{\mu}}$ in terms of $\boldsymbol{y}$. Find s, and explain why this model is not sensible for practice.

2.24 Verify that the $n \times n$ identity matrix $\boldsymbol{I}$ is a projection matrix, and describe the linear model to which it corresponds.

2.25 Section 1.4.2 stated "When $\boldsymbol{X}$ has full rank, $\boldsymbol{\beta}$ is identifiable, and then all linear combinations $\boldsymbol{\ell}^{\mathrm{T}}\boldsymbol{\beta}$ are estimable." Find $\boldsymbol{a}$ such that $E(\boldsymbol{a}^{\mathrm{T}}\boldsymbol{y}) = \boldsymbol{\ell}^{\mathrm{T}}\boldsymbol{\beta}$ for all $\boldsymbol{\beta}$.

2.26 For a linear model with p explanatory variables, explain why sample multiple correlation $R = 0$ is equivalent to sample $\mathrm{corr}(\boldsymbol{y}, \boldsymbol{x}_{*j}) = 0$ for $j = 1, \ldots, p$.

2.27 In Section 2.5.1 we noted that for linear models containing an intercept term, $\mathrm{corr}(\hat{\boldsymbol{\mu}}, \boldsymbol{e}) = 0$, and plotting $\boldsymbol{e}$ against $\hat{\boldsymbol{\mu}}$ helps detect violations of model assumptions. However, it is not helpful to plot $\boldsymbol{e}$ against $\boldsymbol{y}$. To see why not, using formula (2.5), show that (**a**) the regression of $\boldsymbol{y}$ on $\boldsymbol{e}$ has slope 1, (**b**) the regression of $\boldsymbol{e}$ on $\boldsymbol{y}$ has slope $1 - R^2$, (**c**) $\text{corr}(\boldsymbol{y}, \boldsymbol{e}) = \sqrt{1 - R^2}$.

2.28 Derive the hat matrix for the centered-data formulation of the linear model with a single explanatory variable. Explain what factors cause an observation to have a relatively large leverage.

2.29 Show that an observation in a one-way layout has the maximum possible leverage if it is the only observation for its group.

2.30 Consider the leverages for a linear model with full-rank model matrix and p parameters.

a. Prove that the leverages fall between 0 and 1 and have a mean of p/n.

b. Show how expression (2.10) for h_{ii} simplifies when each pair of explanatory variables is uncorrelated.

2.31 **a.** Give an example of actual variables y, x_1, x_2 for which you would expect $\beta_1 \neq 0$ in the model $E(y_i) = \beta_0 + \beta_1 x_{i1}$ but $\beta_1 \approx 0$ in the model $E(y_i) = \beta_0 + \beta_1 x_{i1} + \beta_2 x_{i2}$ (e.g., perhaps x_2 is a "lurking variable," such that the association of x_1 with y disappears when we adjust for x_2).

b. Let $r_1 = \text{corr}(\boldsymbol{y}, \boldsymbol{x}_{*1})$, $r_2 = \text{corr}(\boldsymbol{y}, \boldsymbol{x}_{*2})$, and let R be the multiple correlation with predictors x_1 and x_2. For the case described in (**a**), explain why you would expect R to be close to $|r_2|$.

c. For the case described in (**a**), which would you expect to be relatively near $SSR(x_1, x_2)$: $SSR(x_1)$ or $SSR(x_2)$? Why?

2.32 In studying the model for the one-way layout in Section 2.3.2, we found the projection matrices and sums of squares and constructed the ANOVA table.

a. We did the analysis for a non-full-rank model matrix $\boldsymbol{X}$. Show that the simple form for $(\boldsymbol{X}^{\mathrm{T}}\boldsymbol{X})^{-}$ stated there is in fact a generalized inverse.

b. Verify the corresponding projection matrix $\boldsymbol{P}_X$ specified there.

c. Verify that $\boldsymbol{y}^{\mathrm{T}}(\boldsymbol{I} - \boldsymbol{P}_X)\boldsymbol{y}$ is the within-groups sum of squares stated there.

2.33 Refer to the previous exercise. Conduct a similar analysis, but making parameters identifiable by setting $\beta_0 = 0$. Specify $\boldsymbol{X}$ and find $\boldsymbol{P}_X$ and $\boldsymbol{y}^{\mathrm{T}}(\boldsymbol{I} - \boldsymbol{P}_X)\boldsymbol{y}$.

2.34 From the previous exercise, setting $\beta_0 = 0$ results in $\{\hat{\beta}_i = \bar{y}_i\}$. Explain why imposing only this constraint is inadequate for models with multiple factors, and a constraint such as $\beta_1 = 0$ is more generalizable. Illustrate for the two-way layout.

2.35 Consider the main-effects linear model for the two-way layout with one observation per cell. Section 2.3.4 stated the projection matrix $\boldsymbol{P}_r$ that generates the treatment means. Find the projection matrix $\boldsymbol{P}_c$ that generates the block means.

2.36 For the two-way $r \times c$ layout with one observation per cell, find the hat matrix.

2.37 In the model for the balanced one-way layout, $E(y_{ij}) = \beta_0 + \beta_i$ with identical n_i, show that $\{\beta_i\}$ are orthogonal with β_0 if we impose the constraint $\sum_i \beta_i = 0$.

2.38 Section 2.4.5 considered the "main effects" model for a balanced 2×2 layout, showing there is orthogonality between each pair of parameters when we constrain $\sum_i \beta_i = \sum_j \gamma_j = 0$.

a. If you instead constrain $\beta_1 = \gamma_1 = 0$, show that pairs of columns of $\boldsymbol{X}$ are uncorrelated but not orthogonal.

b. Explain why β_2 for the coding $\beta_1 = 0$ in (**a**) is identical to $2\beta_2$ for the coding $\beta_1 + \beta_2 = 0$.

c. Explain how the results about constraints and orthogonality generalize if the model also contains a term δ_{ij} to permit interaction between A and B in their effects on y.

2.39 Extend results in Section 2.3.4 to the $r \times c$ factorial with n observations per cell.

a. Express the orthogonal decomposition of y_{ijk} to include main effects, interaction, and residual error.

b. Show how $\boldsymbol{P_r}$ generalizes from the matrix given in Section 2.3.4.

c. Show the relevant sum of squares decomposition in an ANOVA table that also shows the *df* values. (It may help you to refer to (**b**) and (**c**) in Exercise 3.13.)

2.40 A genetic association study considers a large number of explanatory variables, with nearly all expected to have no effect or a very minor effect on the response. An alternative to the least squares estimator $\hat{\boldsymbol{\beta}}$ for the linear model incorporating those explanatory variables is the null model and its estimator, $\tilde{\boldsymbol{\beta}} = \boldsymbol{0}$ except for the intercept. Is $\tilde{\boldsymbol{\beta}}$ unbiased? How does $\text{var}(\tilde{\beta}_j)$ compare to $\text{var}(\hat{\beta}_j)$? Explain why $\sum_j E(\tilde{\beta}_j - \beta_j)^2 < \sum_j E(\hat{\beta}_j - \beta_j)^2$ unless n is extremely large.

2.41 The Gauss–Markov theorem shows the best way to form a linear unbiased estimator in a linear model. Are unbiased estimators always sensible? Consider a sequence of independent Bernoulli trials with parameter π.

a. Let y be the number of failures before the first success. Show that the only unbiased estimator (and thus the best unbiased estimator) of π is $T(y) = 1$ if $y = 0$ and $T(y) = 0$ if $y > 0$. Show that the ML estimator of π is $\hat{\pi} = 1/(1 + y)$. Although biased, is this a more efficient estimator? Why?

b. For n trials, show there is no unbiased estimator of the logit, $\log[\pi/(1 - \pi)]$.

2.42 In some applications, such as regressing annual income on the number of years of education, the variance of y tends to be larger at higher values of x. Consider the model $E(y_i) = \beta x_i$, assuming $\text{var}(y_i) = x_i\sigma^2$ for unknown σ^2.

a. Show that the generalized least squares estimator minimizes $\sum_i (y_i - \beta x_i)^2/x_i$ (i.e., giving more weight to observations with smaller x_i) and has $\hat{\beta}_{GLS} = \bar{y}/\bar{x}$, with $\text{var}(\hat{\beta}_{GLS}) = \sigma^2/(\sum_i x_i)$.

b. Show that the ordinary least squares estimator is $\hat{\beta} = (\sum_i x_i y_i)/(\sum_i x_i^2)$ and has $\text{var}(\hat{\beta}) = \sigma^2(\sum_i x_i^3)/(\sum_i x_i^2)^2$.

c. Show that $\text{var}(\hat{\beta}) \geq \text{var}(\hat{\beta}_{GLS})$.

2.43 Write a simple program to simulate data so that when you plot residuals against x after fitting the bivariate linear model $E(y_i) = \beta_0 + \beta_1 x_i$, the plot shows inadequacy of (**a**) the linear predictor, (**b**) the constant variance assumption.

2.44 Exercise 1.21 concerned a study comparing forced expiratory volume (y = *fev*1 in the data file `FEV.dat` at the text website) for three drugs, adjusting for a baseline measurement. For the R output shown, using notation you define, state the model that was fitted, and interpret all results shown.

```
-------------------------------------------------------------
> summary(lm(fev1 ~ base + factor(drug)))
                  Estimate  Std. Error
(Intercept)         1.1139      0.2999
base                0.8900      0.1063
factor(drug)b       0.2181      0.1375
factor(drug)p      -0.6448      0.1376
---
Residual standard error: 0.4764 on 68 degrees of freedom
Multiple R-squared: 0.6266,     Adjusted R-squared:  0.6101
> anova(lm(fev1 ~ base + factor(drug)))
Analysis of Variance Table
              Df    Sum Sq  Mean Sq
base           1   16.2343  16.2343
factor(drug)   2    9.6629   4.8315
Residuals     68   15.4323   0.2269
> quantile(rstandard(lm(fev1 ~ base + factor(drug))))
     0%       25%       50%      75%     100%
-2.0139   -0.7312   -0.1870   0.6341   2.4772
-------------------------------------------------------------
```

2.45 A data set shown partly in Table 2.4 and fully available in the `Optics.dat` file at the text website is taken from a math education graduate student research project. For the optics module in a high school freshman physical science class, the randomized study compared two instruction methods (1 = model building inquiry, 0 = traditional scientific). The response variable was an optics post-test score. Other explanatory variables were an optics pre-test score, gender (1 = female, 0 = male), OAA (Ohio Achievement Assessment) reading score, OAA science score, attendance for optics module (number of days), and individualized education program (IEP) for student with disabilities (1 = yes, 0 = no).

a. Fit the linear model with instruction type, pre-test score, and attendance as explanatory variables. Summarize and interpret the software output.

b. Find and interpret diagnostics, including residual plots and measures of influence, for this model.

Table 2.4 Partial Optics Instruction Data[a] for Exercise 2.45

ID	Post	Inst	Pre	Gender	Reading	Science	Attend	IEP
1	50	1	50	0	368	339	14	0
2	67	1	50	0	372	389	11	0
...								
37	55	0	42	1	385	373	7	0

Source: Thanks to Harry Khamis, Wright State University, Statistical Consulting Center, for these data, provided with client permission. Complete data (n = 37) are in the file `Optics.dat` at `www.stat.ufl.edu/~aa/glm/data`.

2.46 Download from the text website the data file `Crabs.dat` introduced in Section 1.5.1. Fit the linear model with both weight and color as explanatory variables for the number of satellites for each crab, without interaction, treating color as qualitative. Summarize and interpret the software output, including the prediction equation, error variance, R^2, adjusted R^2, and multiple correlation. Plot the residuals against the fitted values for the model, and interpret. What explains the lower nearly straight-line boundary? By contrast, what residual pattern would you expect if the response variable is normal and the linear model holds with constant variance?

2.47 The horseshoe crab dataset[17] `Crabs3.dat` at `www.stat.ufl.edu/~aa/glm/data` collects several variables for female horseshoe crabs that have males attached during mating, over several years at Seahorse Key, Florida. Use linear modeling to describe the relation between y = attached male's carapace width (AMCW) and x_1 = female's carapace width (FCW), x_2 = female's color (Fcolor, where 1 = light, 3 = medium, 5 = dark), and x_3 = female's surface condition (Fsurf, where lower scores represent better condition). Summarize and interpret the output, including the prediction equation, error variance, R^2, adjusted R^2, multiple correlation, and model diagnostics.

2.48 Refer to the anorexia study in Exercise 1.24. For the model fitted there, interpret the output relating to predictive power, and check the model using residuals and influence measures. Summarize your findings.

[17]Thanks to Jane Brockmann for making these data available.

2.49 In later chapters, we use functions in the useful R package, `VGAM`. In that package, the `venice` data set contains annual data between 1931 and 1981 on the annual maximum sea level (variable $r1$) in Venice. Analyze the relation between year and maximum sea level. Summarize results in a two-page report, with software output as an appendix. (An alternative to least squares uses ML with a distribution suitable for modeling extremes, as in Davison (2003, p. 475).)

CHAPTER 3

Normal Linear Models: Statistical Inference

Chapter 2 introduced least squares fitting of ordinary linear models. For n independent observations $\boldsymbol{y} = (y_1, \dots, y_n)^{\mathrm{T}}$, with $\boldsymbol{\mu} = (\mu_1, \dots, \mu_n)^{\mathrm{T}}$ for $\mu_i = E(y_i)$ and a model matrix $\boldsymbol{X}$ and parameter vector $\boldsymbol{\beta}$, this model states that

$$\boldsymbol{\mu} = \boldsymbol{X}\boldsymbol{\beta} \quad \text{with} \quad \boldsymbol{V} = \mathrm{var}(\boldsymbol{y}) = \sigma^2 \boldsymbol{I}.$$

We now add to this model the assumption that $\{y_i\}$ have normal distributions. The model is then the *normal linear model*. This chapter presents the foundations of statistical inference about the parameters of the normal linear model.

We begin this chapter by reviewing relevant distribution theory for normal linear models. Quadratic forms incorporating normally distributed response variables and projection matrices generate chi-squared distributions. One such result, *Cochran's theorem*, is the basis of significance tests about $\boldsymbol{\beta}$ in the normal linear model. Section 3.2 shows how the tests use the chi-squared quadratic forms to construct test statistics having F distributions. A useful general result about comparing two nested models is also derived as a likelihood-ratio test. Section 3.3 presents confidence intervals for elements of $\boldsymbol{\beta}$ and expected responses as well as prediction intervals for future observations. Following an example in Section 3.4, Section 3.5 presents methods for making multiple inferences with a fixed overall error rate, such as multiple comparison methods for constructing simultaneous confidence intervals for differences between all pairs of a set of means. Without the normality assumption, the exact inference methods of this chapter apply to the ordinary linear model in an approximate manner for large n.

Foundations of Linear and Generalized Linear Models, First Edition. Alan Agresti.

3.1 DISTRIBUTION THEORY FOR NORMAL VARIATES

Statistical inference for normal linear models uses sampling distributions derived from quadratic forms with multivariate normal random variables. We now review the multivariate normal distribution and related sampling distributions.

3.1.1 Multivariate Normal Distribution

Let $N(\boldsymbol{\mu}, \boldsymbol{V})$ denote the multivariate normal distribution with mean $\boldsymbol{\mu}$ and covariance matrix $\boldsymbol{V}$. If $\boldsymbol{y} = (y_1, \ldots, y_n)^{\mathrm{T}}$ has this distribution and $\boldsymbol{V}$ is positive definite, then the probability density function (pdf) is

$$f(\boldsymbol{y}) = (2\pi)^{-\frac{n}{2}} |\boldsymbol{V}|^{-\frac{1}{2}} \exp\left[-\frac{1}{2}(\boldsymbol{y} - \boldsymbol{\mu})^{\mathrm{T}}\boldsymbol{V}^{-1}(\boldsymbol{y} - \boldsymbol{\mu})\right],$$

where $|\boldsymbol{V}|$ denotes the determinant of $\boldsymbol{V}$. Here are a few properties, when $\boldsymbol{y} \sim N(\boldsymbol{\mu}, \boldsymbol{V})$.

- If $\boldsymbol{x} = \boldsymbol{A}\boldsymbol{y} + \boldsymbol{b}$, then $\boldsymbol{x} \sim N(\boldsymbol{A}\boldsymbol{\mu} + \boldsymbol{b}, \boldsymbol{A}\boldsymbol{V}\boldsymbol{A}^{\mathrm{T}})$.
- Suppose that $\boldsymbol{y}$ partitions as

$$\boldsymbol{y} = \begin{pmatrix} \boldsymbol{y}_1 \\ \boldsymbol{y}_2 \end{pmatrix}, \text{ with } \boldsymbol{\mu} = \begin{pmatrix} \boldsymbol{\mu}_1 \\ \boldsymbol{\mu}_2 \end{pmatrix} \text{ and } \boldsymbol{V} = \begin{pmatrix} \boldsymbol{V}_{11} & \boldsymbol{V}_{12} \\ \boldsymbol{V}_{21} & \boldsymbol{V}_{22} \end{pmatrix}.$$

  The marginal distribution of $\boldsymbol{y}_a$ is $N(\boldsymbol{\mu}_a, \boldsymbol{V}_{aa})$, $a = 1, 2$. The conditional distribution

$$(\boldsymbol{y}_1 \mid \boldsymbol{y}_2) \sim N\left[\boldsymbol{\mu}_1 + \boldsymbol{V}_{12}\boldsymbol{V}_{22}^{-1}(\boldsymbol{y}_2 - \boldsymbol{\mu}_2),\ \boldsymbol{V}_{11} - \boldsymbol{V}_{12}\boldsymbol{V}_{22}^{-1}\boldsymbol{V}_{21}\right].$$

  In addition, $\boldsymbol{y}_1$ and $\boldsymbol{y}_2$ are independent if and only if $\boldsymbol{V}_{12} = \boldsymbol{0}$.
- From the previous property, if $\boldsymbol{V} = \sigma^2\boldsymbol{I}$, then $y_i \sim N(\mu_i, \sigma^2)$ and $\{y_i\}$ are independent.

The normal linear model assumes that $\boldsymbol{y} \sim N(\boldsymbol{\mu}, \boldsymbol{V})$ with $\boldsymbol{V} = \sigma^2\boldsymbol{I}$. The least squares estimator $\hat{\boldsymbol{\beta}}$ and the residuals $\boldsymbol{e}$ also have multivariate normal distributions, since they are linear functions of $\boldsymbol{y}$, but their elements are typically correlated. This estimator $\hat{\boldsymbol{\beta}}$ is also the maximum likelihood (ML) estimator under the normality assumption (as we showed in Section 2.1).

3.1.2 Chi-Squared, *F*, and *t* Distributions

Let χ^2_p denote a chi-squared distribution with p degrees of freedom (*df*). A chi-squared random variable is nonnegative with mean = *df* and variance = 2(*df*). Its distribution[1] is skewed to the right but becomes more bell-shaped as *df* increases.

[1]The pdf is the special case of the gamma distribution pdf (4.29) with shape parameter $k = df/2$.

Recall that when $y_1, \dots, y_p$ are independent standard normal random variables, $\sum_{i=1}^{p} y_i^2 \sim \chi_p^2$. In particular, if $y \sim N(0, 1)$, then $y^2 \sim \chi_1^2$. More generally

- If a p-dimensional random variable $\boldsymbol{y} \sim N(\boldsymbol{\mu}, \boldsymbol{V})$ with $\boldsymbol{V}$ nonsingular of rank p, then

$$x = (\boldsymbol{y} - \boldsymbol{\mu})^{\mathrm{T}} \boldsymbol{V}^{-1} (\boldsymbol{y} - \boldsymbol{\mu}) \sim \chi_p^2.$$

  Exercise 3.1 outlines a proof.

- If $z \sim N(0, 1)$ and $x \sim \chi_p^2$, with x and z independent, then

$$\frac{z}{\sqrt{x/p}} \sim t_p,$$

  the t distribution with $df = p$.

The t distribution is symmetric around 0 with variance $= df/(df - 2)$ when $df > 2$. The term x/p in the denominator is a mean of p independent squared $N(0, 1)$ random variables, so as $p \to \infty$ it converges in probability to their expected value of 1. Therefore, the t distribution converges to a $N(0, 1)$ distribution as df increases.

Here is a classic way the t distribution occurs for independent responses $y_1, \dots, y_n$ from a $N(\mu, \sigma^2)$ distribution with sample mean $\bar{y}$ and sample variance s^2: For testing H_0: $\mu = \mu_0$, the test statistic $z = \sqrt{n}(\bar{y} - \mu_0)/\sigma$ has the $N(0, 1)$ null distribution. Also, s^2/σ^2 is a χ_{n-1}^2 variate $x = (n-1)s^2/\sigma^2$ divided by its df. Since $\bar{y}$ and s^2 are independent for independent observations from a normal distribution, under H_0

$$t = \frac{z}{\sqrt{x/(n-1)}} = \frac{\bar{y} - \mu_0}{s/\sqrt{n}} \sim t_{n-1}.$$

Larger values of $|t|$ provide stronger evidence against H_0.

- If $x \sim \chi_p^2$ and $y \sim \chi_q^2$, with x and y independent, then

$$\frac{x/p}{y/q} \sim F_{p,q},$$

  the F distribution with $df_1 = p$ and $df_2 = q$.

An F random variable takes nonnegative values. When $df_2 > 2$, it has mean $= df_2/(df_2 - 2)$, approximately 1 for large df_2. We shall use this distribution for testing hypotheses in ANOVA and regression by taking a ratio of independent mean squares. For a t random variable, t^2 has the F distribution with $df_1 = 1$ and df_2 equal to the df for that t.

3.1.3 Noncentral Distributions

In significance testing, to analyze the behavior of test statistics when null hypotheses are false, we use *noncentral* sampling distributions that occur under parameter values from the alternative hypothesis. Such distributions determine the power of a test (i.e., the probability of rejecting H_0), which can be analyzed as a function of the actual parameter value. When observations have a multivariate normal distribution, sampling distributions in such non-null cases contain the ones just summarized as special cases.

Let $\chi^2_{p,\lambda}$ denote a noncentral chi-squared distribution with $df = p$ and with noncentrality parameter λ. This is the distribution of $x = \sum_{i=1}^{p} y_i^2$ in which $\{y_i\}$ are independent with $y_i \sim N(\mu_i, 1)$ and $\lambda = \sum_{i=1}^{p} \mu_i^2$. For this distribution[2], $E(x) = p + \lambda$ and $\text{var}(x) = 2(p + 2\lambda)$. The ordinary (central) chi-squared distribution is the special case with $\lambda = 0$.

- If a p-dimensional random variable $\boldsymbol{y} \sim N(\boldsymbol{\mu}, \boldsymbol{V})$ with $\boldsymbol{V}$ nonsingular of rank p, then

$$x = \boldsymbol{y}^{\mathrm{T}}\boldsymbol{V}^{-1}\boldsymbol{y} \sim \chi^2_{p,\lambda} \quad \text{with} \quad \lambda = \boldsymbol{\mu}^{\mathrm{T}}\boldsymbol{V}^{-1}\boldsymbol{\mu}.$$

The construction of the noncentral chi-squared from a sum of squared independent $N(\mu_i, 1)$ random variables results when $\boldsymbol{V} = \boldsymbol{I}$.

- If $z \sim N(\mu, 1)$ and $x \sim \chi^2_p$, with x and z independent, then

$$t = \frac{z}{\sqrt{x/p}} \sim t_{p,\mu},$$

  the noncentral t distribution with $df = p$ and noncentrality μ.

The noncentral t distribution is unimodal, but skewed in the direction of the sign of $\mu = E(z)$. When $p > 1$ and $\mu \neq 0$, its mean $E(t) \approx [1 - 3/(4p - 1)]^{-1}\mu$, which is near μ but slightly larger in absolute value. For large p, the distribution of t is approximately the $N(\mu, 1)$ distribution.

- If $x \sim \chi^2_{p,\lambda}$ and $y \sim \chi^2_q$, with x and y independent, then

$$\frac{x/p}{y/q} \sim F_{p,q,\lambda},$$

  the noncentral F distribution with $df_1 = p$, $df_2 = q$, and noncentrality λ.

[2]Here is an alternative way to define noncentrality: Let $z \sim \text{Poisson}(\phi)$ and $(x \mid z) \sim \chi^2_{p+2z}$. Then unconditionally $x \sim \chi^2_{p,\phi}$. This noncentrality ϕ relates to the noncentrality λ we defined by $\phi = \lambda/2$.

For large df_2, the noncentral F has mean approximately $1 + \lambda/df_1$, which increases in λ from the approximate mean of 1 for the central case.

As reality deviates farther from a particular null hypothesis, the noncentrality λ increases. The noncentral chi-squared and noncentral F distributions are stochastically increasing in λ. That is, evaluated at any positive value, the cumulative distribution function (cdf) decreases as λ increases, so values of the statistic tend to be larger.

3.1.4 Normal Quadratic Forms with Projection Matrices Are Chi-Squared

Two results about quadratic forms involving normal random variables are especially useful for statistical inference with normal linear models. The first generalizes the above quadratic form result for the noncentral chi-squared, which follows with $\boldsymbol{A} = \boldsymbol{V}^{-1}$.

- Suppose $\boldsymbol{y} \sim N(\boldsymbol{\mu}, \boldsymbol{V})$ and $\boldsymbol{A}$ is a symmetric matrix. Then,

$$\boldsymbol{y}^{\mathrm{T}}\boldsymbol{A}\boldsymbol{y} \sim \chi^2_{r,\boldsymbol{\mu}^{\mathrm{T}}\boldsymbol{A}\boldsymbol{\mu}} \quad \Leftrightarrow \quad \boldsymbol{A}\boldsymbol{V} \text{ is idempotent of rank } r.$$

For the normal linear model, the n independent observations $\boldsymbol{y} \sim N(\boldsymbol{\mu}, \sigma^2\boldsymbol{I})$ with $\boldsymbol{\mu} = \boldsymbol{X}\boldsymbol{\beta}$, and so $\boldsymbol{y}/\sigma \sim N(\boldsymbol{\mu}/\sigma, \boldsymbol{I})$. By this result, if $\boldsymbol{P}$ is a projection matrix (which is symmetric and idempotent) with rank r, then $\boldsymbol{y}^{\mathrm{T}}\boldsymbol{P}\boldsymbol{y}/\sigma^2 \sim \chi^2_{r,\boldsymbol{\mu}^{\mathrm{T}}\boldsymbol{P}\boldsymbol{\mu}/\sigma^2}$. Applying the result with the standardized normal variables $(\boldsymbol{y} - \boldsymbol{\mu})/\sigma \sim N(\boldsymbol{0}, \boldsymbol{I})$, we have

Normal quadratic form with projection matrix and chi-squared: Suppose $\boldsymbol{y} \sim N(\boldsymbol{\mu}, \sigma^2\boldsymbol{I})$ and $\boldsymbol{P}$ is symmetric. Then,

$$\frac{1}{\sigma^2}(\boldsymbol{y} - \boldsymbol{\mu})^{\mathrm{T}}\boldsymbol{P}(\boldsymbol{y} - \boldsymbol{\mu}) \sim \chi^2_r \quad \Leftrightarrow \quad \boldsymbol{P} \text{ is a projection matrix of rank } r.$$

Cochran (1934) showed[3] this result, which also provides an interpretation for degrees of freedom.

- Since the *df* for the chi-squared distribution of a quadratic form with a normal linear model equals the rank of $\boldsymbol{P}$, *degrees of freedom* represent the dimension of the vector subspace to which $\boldsymbol{P}$ projects.

The following key result also follows from Cochran (1934), building on the first result.

[3]From Cochran's result I, since a symmetric matrix whose eigenvalues are 0 and 1 is idempotent.

Cochran's theorem: Suppose n observations $\boldsymbol{y} \sim N(\boldsymbol{\mu}, \sigma^2\boldsymbol{I})$ and $\boldsymbol{P}_1, \dots, \boldsymbol{P}_k$ are projection matrices having $\sum_i \boldsymbol{P}_i = \boldsymbol{I}$. Then, $\{\boldsymbol{y}^{\mathrm{T}}\boldsymbol{P}_i\boldsymbol{y}\}$ are independent and $(\frac{1}{\sigma^2})\boldsymbol{y}^{\mathrm{T}}\boldsymbol{P}_i\boldsymbol{y} \sim \chi^2_{r_i,\lambda_i}$ where $r_i = \text{rank}(\boldsymbol{P}_i)$ and $\lambda_i = \frac{1}{\sigma^2}\boldsymbol{\mu}^{\mathrm{T}}\boldsymbol{P}_i\boldsymbol{\mu}$, $i = 1, \dots, k$, with $\sum_i r_i = n$.

If we replace $\boldsymbol{y}$ by $(\boldsymbol{y} - \boldsymbol{\mu})$ in the quadratic forms, we obtain central chi-squared distributions ($\lambda_i = 0$). This result is the basis of significance tests for parameters in normal linear models. The proof of the independence result shows that all pairs of projection matrices in this decomposition satisfy $\boldsymbol{P}_i\boldsymbol{P}_j = 0$.

3.1.5 Proof of Cochran's Theorem

We next show a proof[4] of Cochran's theorem. You may wish to skip these technical details for now and go to the next section, which uses this result to construct significance tests for the normal linear model.

We first show that if $\boldsymbol{y} \sim N(\boldsymbol{\mu}, \sigma^2\boldsymbol{I})$ and $\boldsymbol{P}$ is a projection matrix having rank r, then $(\frac{1}{\sigma^2})\boldsymbol{y}^{\mathrm{T}}\boldsymbol{P}\boldsymbol{y} \sim \chi^2_{r,\lambda}$ with $\lambda = \frac{1}{\sigma^2}\boldsymbol{\mu}^{\mathrm{T}}\boldsymbol{P}\boldsymbol{\mu}$. Since $\boldsymbol{P}$ is symmetric and idempotent with rank r, its eigenvalues are 1 (r times) and 0 ($n - r$ times). By the spectral decomposition of a symmetric matrix, we can express $\boldsymbol{P} = \boldsymbol{Q}\boldsymbol{\Lambda}\boldsymbol{Q}^{\mathrm{T}}$, where $\boldsymbol{\Lambda}$ is a diagonal matrix of (1, 1, ..., 1, 0, ..., 0), the eigenvalues of $\boldsymbol{P}$, and $\boldsymbol{Q}$ is an orthogonal matrix with columns that are the eigenvectors of $\boldsymbol{P}$. Let $\boldsymbol{z} = \boldsymbol{Q}^{\mathrm{T}}\boldsymbol{y}/\sigma$. Then, $\boldsymbol{z} \sim N(\boldsymbol{Q}^{\mathrm{T}}\boldsymbol{\mu}/\sigma, \boldsymbol{I})$, and $(\frac{1}{\sigma^2})\boldsymbol{y}^{\mathrm{T}}\boldsymbol{P}\boldsymbol{y} = \boldsymbol{z}^{\mathrm{T}}\boldsymbol{\Lambda}\boldsymbol{z} = \sum_{i=1}^{r} z_i^2$. Since each z_i is normal with standard deviation 1, $\sum_{i=1}^{r} z_i^2$ has a noncentral chi-squared distribution with $df = r$ and noncentrality parameter

$$\sum_{i=1}^{r}[E(z_i)]^2 = [E(\boldsymbol{\Lambda}\boldsymbol{z})]^{\mathrm{T}}[E(\boldsymbol{\Lambda}\boldsymbol{z})] = \left(\frac{1}{\sigma^2}\right)[\boldsymbol{\Lambda}\boldsymbol{Q}^{\mathrm{T}}\boldsymbol{\mu}]^{\mathrm{T}}[\boldsymbol{\Lambda}\boldsymbol{Q}^{\mathrm{T}}\boldsymbol{\mu}]$$
$$= \left(\frac{1}{\sigma^2}\right)\boldsymbol{\mu}^{\mathrm{T}}\boldsymbol{Q}\boldsymbol{\Lambda}\boldsymbol{Q}^{\mathrm{T}}\boldsymbol{\mu} = \left(\frac{1}{\sigma^2}\right)\boldsymbol{\mu}^{\mathrm{T}}\boldsymbol{P}\boldsymbol{\mu}.$$

Now we consider k quadratic forms with k projection matrices that are a decomposition of $\boldsymbol{I}$, the $n \times n$ identity matrix. The rank of a projection matrix is its trace, so $\sum_i r_i = \sum_i \text{trace}(\boldsymbol{P}_i) = \text{trace}(\sum_i \boldsymbol{P}_i) = \text{trace}(\boldsymbol{I}) = n$. We apply the spectral decomposition to each projection matrix, with $\boldsymbol{P}_i = \boldsymbol{Q}_i\boldsymbol{\Lambda}_i\boldsymbol{Q}_i^{\mathrm{T}}$, where $\boldsymbol{\Lambda}_i$ is a diagonal matrix of $(1, 1, \dots, 1, 0, \dots, 0)$ with r_i entries that are 1. By the form of $\boldsymbol{\Lambda}_i$, this is identical to $\boldsymbol{P}_i = \tilde{\boldsymbol{Q}}_i\boldsymbol{I}_{r_i}\tilde{\boldsymbol{Q}}_i^{\mathrm{T}} = \tilde{\boldsymbol{Q}}_i\tilde{\boldsymbol{Q}}_i^{\mathrm{T}}$, where $\tilde{\boldsymbol{Q}}_i$ is a $n \times r_i$ matrix of the first r_i columns of $\boldsymbol{Q}_i$. Note that $\tilde{\boldsymbol{Q}}_i^{\mathrm{T}}\tilde{\boldsymbol{Q}}_i = \boldsymbol{I}_{r_i}$. We stack the $\{\tilde{\boldsymbol{Q}}_i\}$ together as

$$\boldsymbol{Q} = [\tilde{\boldsymbol{Q}}_1 \;:\; \tilde{\boldsymbol{Q}}_2 \;:\; \cdots \;:\; \tilde{\boldsymbol{Q}}_k],$$

[4]This proof is based on one in Monahan (2008, pp. 113–114).

for which

$$QQ^{\mathrm{T}} = \tilde{Q}_1\tilde{Q}_1^{\mathrm{T}} + \cdots + \tilde{Q}_k\tilde{Q}_k^{\mathrm{T}} = P_1 + \cdots + P_k = I_n.$$

Thus, Q is an orthogonal $n \times n$ matrix and also $Q^{\mathrm{T}}Q = I_n$ and $\tilde{Q}_i^{\mathrm{T}}\tilde{Q}_j = \mathbf{0}$ for $i \neq j$. So $Q^{\mathrm{T}}y \sim N(Q^{\mathrm{T}}\mu, \sigma^2 I)$, and its components $\{\tilde{Q}_i^{\mathrm{T}}y\}$ are independent, as are $\{\|\tilde{Q}_i^{\mathrm{T}}y\|^2 = y^{\mathrm{T}}\tilde{Q}_i\tilde{Q}_i^{\mathrm{T}}y = y^{\mathrm{T}}P_iy\}$. Note[5] also that for $i \neq j$, $P_iP_j = \tilde{Q}_i\tilde{Q}_i^{\mathrm{T}}\tilde{Q}_j\tilde{Q}_j^{\mathrm{T}} = \mathbf{0}$.

3.2 SIGNIFICANCE TESTS FOR NORMAL LINEAR MODELS

We now use Cochran's theorem to derive fundamental significance tests for the normal linear model. We first revisit the one-way layout and then present inference for the more general context of comparing two nested normal linear models.

3.2.1 Example: ANOVA for the One-Way Layout

For the one-way layout (introduced in Sections 1.3.3 and 2.3.2), let y_{ij} denote observation j in group i, for $i = 1, \ldots, c$ and $j = 1, \ldots, n_i$, with $n = \sum_i n_i$. The observations are assumed to be independent. The linear predictor for $\mu_i = E(y_{ij})$ is

$$E(y_{ij}) = \beta_0 + \beta_i,$$

with a constraint such as $\beta_1 = 0$. We construct a significance test of H_0: $\mu_1 = \cdots = \mu_c$, assuming that $\{y_{ij} \sim N(\mu_i, \sigma^2)\}$. Under H_0, which is equivalently H_0: $\beta_1 = \cdots = \beta_c$, the model simplifies to the null model, $E(y_{ij}) = \beta_0$ for all i and j.

The projection matrix P_X for this model is a block-diagonal matrix with components $\frac{1}{n_i}\mathbf{1}_{n_i}\mathbf{1}_{n_i}^{\mathrm{T}}$, shown in Equation 2.6 of Section 2.3.2. Let $P_0 = \frac{1}{n}\mathbf{1}_n\mathbf{1}_n^{\mathrm{T}}$ denote the projection matrix for the null model. We use the decomposition

$$I = P_0 + (P_X - P_0) + (I - P_X).$$

Each of the three components is a projection matrix, so we can apply Cochran's theorem with $P_1 = P_0, P_2 = P_X - P_0$, and $P_3 = I - P_X$. The ranks of the components, which equal their traces, are 1, $c - 1$, and $n - c$.

From Section 2.3.3, the corrected total sum of squares (TSS) decomposes into two parts,

$$y^{\mathrm{T}}(P_X - P_0)y = \sum_{i=1}^{c} n_i(\bar{y}_i - \bar{y})^2, \quad y^{\mathrm{T}}(I - P_X)y = \sum_{i=1}^{c}\sum_{j=1}^{n_i}(y_{ij} - \bar{y}_i)^2,$$

[5]The result that $P_iP_j = \mathbf{0}$ is also a special case of the stronger result about the decomposition of projection matrices stated at the end of Section 2.1.1.

the "between-groups" and "within-groups" sums of squares. By Cochran's theorem,

$$\frac{1}{\sigma^2}\sum_{i=1}^{c} n_i(\bar{y}_i - \bar{y})^2 \sim \chi^2_{c-1,\lambda}, \quad \text{with} \quad \lambda = \frac{1}{\sigma^2}\boldsymbol{\mu}^{\mathrm{T}}(\boldsymbol{P}_X - \boldsymbol{P}_0)\boldsymbol{\mu},$$

$$\frac{1}{\sigma^2}\left[\sum_{i=1}^{c}\sum_{j=1}^{n_i}(y_{ij} - \bar{y}_i)^2\right] \sim \chi^2_{n-c},$$

and the quadratic forms are independent. The second one has noncentrality 0 because $\boldsymbol{\mu}^{\mathrm{T}}(\boldsymbol{I} - \boldsymbol{P}_X)\boldsymbol{\mu} = \boldsymbol{\mu}^{\mathrm{T}}(\boldsymbol{\mu} - \boldsymbol{P}_X\boldsymbol{\mu}) = \boldsymbol{\mu}^{\mathrm{T}}\boldsymbol{0} = 0$. As a consequence, the test statistic

$$F = \frac{\sum_{i=1}^{c} n_i(\bar{y}_i - \bar{y})^2/(c-1)}{\sum_{i=1}^{c}\sum_{j=1}^{n_i}(y_{ij} - \bar{y}_i)^2/(n-c)} \sim F_{c-1,n-c,\lambda}.$$

Using the expressions for $\boldsymbol{P}_0$ and $\boldsymbol{P}_X$, you can verify that $\boldsymbol{\mu}^{\mathrm{T}}\boldsymbol{P}_X\boldsymbol{\mu} = \sum_{i=1}^{c} n_i\mu_i^2$ and $\boldsymbol{\mu}^{\mathrm{T}}\boldsymbol{P}_0\boldsymbol{\mu} = n\bar{\mu}^2$, where $\bar{\mu} = \sum_i n_i\mu_i/n$. Thus, the noncentrality simplifies to $\lambda = \frac{1}{\sigma^2}\sum_{i=1}^{c} n_i(\mu_i - \bar{\mu})^2$. Under H_0, $\lambda = 0$, and the F test statistic has an F distribution with $df_1 = c - 1$ and $df_2 = n - c$. Larger F values are more contradictory to H_0, so the P-value is the right-tail probability from that distribution above the observed test statistic value, F_{obs}. When H_0 is false, λ and the power of the test increase as $\{n_i\}$ increase and as the variability in $\{\mu_i\}$ increases.

This significance test for the one-way layout is known as *(one-way) analysis of variance*, due to R. A. Fisher (1925). To complete the ANOVA table shown in Table 2.1, we include mean squares, which are ratios of the two SS values to their *df* values, and the F statistic as the ratio of those mean squares. The table has the form shown in Table 3.1, and would also include the P-value, $P_{H_0}(F > F_{\text{obs}})$. The first line refers to the null model, which specifies a common mean for all groups. Often, the ANOVA table does not show this line, essentially assuming the intercept is in the model. The table then shows the total sum of squares after subtracting $n\bar{y}^2$, giving the corrected total sum of squares, TSS $= \sum_i\sum_j(y_{ij} - \bar{y})^2$ based on $df = n - 1$.

Table 3.1 Complete ANOVA Table for the Normal Linear Model for the One-Way Layout

Source	*df*	Sum of Squares	Mean Square	F_{obs}
Mean	1	$n\bar{y}^2$		
Group	$c-1$	$\sum_i n_i(\bar{y}_i - \bar{y})^2$	$\frac{\sum_i n_i(\bar{y}_i-\bar{y})^2}{c-1}$	$\frac{\sum_i n_i(\bar{y}_i-\bar{y})^2/(c-1)}{\sum_i\sum_j(y_{ij}-\bar{y}_i)^2/(n-c)}$
Error	$n-c$	$\sum_i\sum_j(y_{ij} - \bar{y}_i)^2$	$\frac{\sum_i\sum_j(y_{ij}-\bar{y}_i)^2}{n-c}$	
Total	n	$\sum_{i=1}^{c}\sum_{j=1}^{n_i} y_{ij}^2$		

3.2.2 Comparing Two Nested Normal Linear Models

The model-building process often deals with comparing a model to a more complex one that has additional parameters or to a simpler one that has fewer parameters. An example of the first type is analyzing whether to add interaction terms to a model containing only main effects. An example of the second type is testing whether sufficiently strong evidence exists to keep a term in the model. Denote the simpler model by M_0 and the more complex model by M_1. Denote the numbers of parameters by p_0 for M_0 and p_1 for M_1, when both model matrices have full rank. We now construct a test of the null hypothesis that M_0 holds against the alternative hypothesis that M_1 holds.

Denote the projection matrices for the two models by $\boldsymbol{P}_0$ and $\boldsymbol{P}_1$. The decomposition using projection matrices

$$\boldsymbol{I} = \boldsymbol{P}_0 + (\boldsymbol{P}_1 - \boldsymbol{P}_0) + (\boldsymbol{I} - \boldsymbol{P}_1)$$

corresponds to the orthogonal decomposition of the data as

$$\boldsymbol{y} = \boldsymbol{P}_0\boldsymbol{y} + (\boldsymbol{P}_1 - \boldsymbol{P}_0)\boldsymbol{y} + (\boldsymbol{I} - \boldsymbol{P}_1)\boldsymbol{y}.$$

Here $\boldsymbol{P}_0\boldsymbol{y} = \hat{\boldsymbol{\mu}}_0$ and $\boldsymbol{P}_1\boldsymbol{y} = \hat{\boldsymbol{\mu}}_1$ are the fitted values for the two models. The corresponding sum-of-squares decomposition is

$$\boldsymbol{y}^{\mathrm{T}}\boldsymbol{y} = \boldsymbol{y}^{\mathrm{T}}\boldsymbol{P}_0\boldsymbol{y} + \boldsymbol{y}^{\mathrm{T}}(\boldsymbol{P}_1 - \boldsymbol{P}_0)\boldsymbol{y} + \boldsymbol{y}^{\mathrm{T}}(\boldsymbol{I} - \boldsymbol{P}_1)\boldsymbol{y}.$$

From Sections 2.4.1 and 2.4.2, $\boldsymbol{y}^{\mathrm{T}}(\boldsymbol{I} - \boldsymbol{P}_1)\boldsymbol{y} = \boldsymbol{y}^{\mathrm{T}}(\boldsymbol{I} - \boldsymbol{P}_1)^{\mathrm{T}}(\boldsymbol{I} - \boldsymbol{P}_1)\boldsymbol{y} = \sum_i (y_i - \hat{\mu}_{i1})^2$ is the residual sum of squares for M_1, which we denote by SSE_1. Likewise,

$$\begin{aligned}\boldsymbol{y}^{\mathrm{T}}(\boldsymbol{P}_1 - \boldsymbol{P}_0)\boldsymbol{y} &= \boldsymbol{y}^{\mathrm{T}}(\boldsymbol{I} - \boldsymbol{P}_0)\boldsymbol{y} - \boldsymbol{y}^{\mathrm{T}}(\boldsymbol{I} - \boldsymbol{P}_1)\boldsymbol{y} \\ &= \sum_i (y_i - \hat{\mu}_{i0})^2 - \sum_i (y_i - \hat{\mu}_{i1})^2 = \text{SSE}_0 - \text{SSE}_1.\end{aligned}$$

Since $(\boldsymbol{P}_1 - \boldsymbol{P}_0)$ is a projection matrix, this difference also equals

$$\boldsymbol{y}^{\mathrm{T}}(\boldsymbol{P}_1 - \boldsymbol{P}_0)\boldsymbol{y} = \boldsymbol{y}^{\mathrm{T}}(\boldsymbol{P}_1 - \boldsymbol{P}_0)^{\mathrm{T}}(\boldsymbol{P}_1 - \boldsymbol{P}_0)\boldsymbol{y} = (\hat{\boldsymbol{\mu}}_1 - \hat{\boldsymbol{\mu}}_0)^{\mathrm{T}}(\hat{\boldsymbol{\mu}}_1 - \hat{\boldsymbol{\mu}}_0).$$

So $\text{SSE}_0 - \text{SSE}_1 = \sum_i (\hat{\mu}_{i1} - \hat{\mu}_{i0})^2 = \text{SSR}(M_1 \mid M_0)$, the difference between the regression SS values for M_1 and M_0.

Now $\boldsymbol{I} - \boldsymbol{P}_1$ has rank $n - p_1$, since trace$(\boldsymbol{I} - \boldsymbol{P}_1)$ = trace$(\boldsymbol{I})$ − trace$(\boldsymbol{P}_1)$ and $\boldsymbol{P}_1$ has full rank p_1. Likewise, $\boldsymbol{P}_1 - \boldsymbol{P}_0$ has rank $p_1 - p_0$. Under H_0, by Cochran's theorem,

$$\frac{\text{SSE}_0 - \text{SSE}_1}{\sigma^2} \sim \chi^2_{p_1 - p_0} \quad \text{and} \quad \frac{\text{SSE}_1}{\sigma^2} \sim \chi^2_{n - p_1},$$

and these are independent. Here, under H_0, the noncentralities of the two chi-squared variates are

$$\boldsymbol{\mu}^{\mathrm{T}}(\boldsymbol{P}_1 - \boldsymbol{P}_0)\boldsymbol{\mu} = 0, \quad \boldsymbol{\mu}^{\mathrm{T}}(\boldsymbol{I} - \boldsymbol{P}_1)\boldsymbol{\mu} = 0$$

since for $\boldsymbol{\mu}$ satisfying M_0, $\boldsymbol{P}_1\boldsymbol{\mu} = \boldsymbol{P}_0\boldsymbol{\mu} = \boldsymbol{\mu}$. It follows that, under H_0, the test statistic

$$F = \frac{(\mathrm{SSE}_0 - \mathrm{SSE}_1)/(p_1 - p_0)}{\mathrm{SSE}_1/(n - p_1)} \tag{3.1}$$

has an F distribution with $df_1 = p_1 - p_0$ and $df_2 = n - p_1$. The denominator $\mathrm{SSE}_1/(n - p_1)$ is the error mean square, which is the s^2 estimator of σ^2 for M_1. Larger differences in SSE values, and larger values of the F test statistic, provide stronger evidence against H_0. The P-value is $P_{H_0}(F > F_{\mathrm{obs}})$.

3.2.3 Likelihood-Ratio Test Comparing Models

The test comparing two nested normal linear models can also be derived as a likelihood-ratio test[6]. For the normal linear model with model matrix $\boldsymbol{X}$, the likelihood function is

$$\ell(\boldsymbol{\beta}, \sigma) = \left(\frac{1}{\sigma\sqrt{2\pi}}\right)^n \exp\left[-(1/2\sigma^2)(\boldsymbol{y} - \boldsymbol{X}\boldsymbol{\beta})^{\mathrm{T}}(\boldsymbol{y} - \boldsymbol{X}\boldsymbol{\beta})\right].$$

The log-likelihood function is

$$L(\boldsymbol{\beta}, \sigma) = -(n/2)\log(2\pi) - n\log(\sigma) - (\boldsymbol{y} - \boldsymbol{X}\boldsymbol{\beta})^{\mathrm{T}}(\boldsymbol{y} - \boldsymbol{X}\boldsymbol{\beta})/2\sigma^2.$$

From Section 2.1.1, differentiating with respect to $\boldsymbol{\beta}$ yields the normal equations and the least squares estimate, $\hat{\boldsymbol{\beta}}$. Differentiating with respect to σ yields

$$\partial L(\boldsymbol{\beta}, \sigma)/\partial\sigma = -\frac{n}{\sigma} + \frac{(\boldsymbol{y} - \boldsymbol{X}\boldsymbol{\beta})^{\mathrm{T}}(\boldsymbol{y} - \boldsymbol{X}\boldsymbol{\beta})}{\sigma^3}.$$

Setting this equal to 0 and solving yields the ML estimator

$$\hat{\sigma}^2 = \frac{(\boldsymbol{y} - \boldsymbol{X}\hat{\boldsymbol{\beta}})^{\mathrm{T}}(\boldsymbol{y} - \boldsymbol{X}\hat{\boldsymbol{\beta}})}{n} = \frac{\mathrm{SSE}}{n}.$$

This estimator is the multiple $(n - p)/n$ of the unbiased estimator, which is $s^2 = \left[\sum_i (y_i - \hat{\mu}_i)^2\right]/(n - p)$. The maximized likelihood function simplifies to

$$\ell(\hat{\boldsymbol{\beta}}, \hat{\sigma}) = \left(\frac{1}{\hat{\sigma}\sqrt{2\pi}}\right)^n e^{-n/2}.$$

[6]The likelihood-ratio test is introduced in a more general context, for GLMs, in Section 4.3.1.

Now, for testing M_0 against M_1, let $\hat{\sigma}_0^2$ and $\hat{\sigma}_1^2$ denote the two ML variance estimates. The ratio of the maximized likelihood functions is

$$\frac{\sup_{M_0} \ell(\boldsymbol{\beta}, \sigma)}{\sup_{M_1} \ell(\boldsymbol{\beta}, \sigma)} = \left(\frac{\hat{\sigma}_1^2}{\hat{\sigma}_0^2}\right)^{n/2}$$

$$= \left(\frac{\text{SSE}_1}{\text{SSE}_0}\right)^{n/2} = \left(1 + \frac{\text{SSE}_0 - \text{SSE}_1}{\text{SSE}_1}\right)^{-n/2} = \left(1 + \frac{p_1 - p_0}{n - p_1} F\right)^{-n/2}$$

for the F test statistic (3.1) derived above. A small value of the likelihood ratio, and thus strong evidence against H_0, corresponds to a large value of the F statistic.

3.2.4 Example: Test That All Effects in a Normal Linear Model Equal Zero

In an important special case of the test comparing two nested normal linear models, the simpler model M_0 is the null model, $E(y_i) = \beta_0$, and M_1 has a set of explanatory variables[7],

$$E(y_i) = \beta_0 + \beta_1 x_{i1} + \cdots + \beta_{p-1} x_{i,p-1}.$$

Comparing the models corresponds to testing the global null hypothesis H_0: $\beta_1 = \cdots = \beta_{p-1} = 0$.

The projection matrix for M_0 is $\boldsymbol{P}_0 = \frac{1}{n}\mathbf{1}\mathbf{1}^{\mathrm{T}}$. The sum-of-squares decomposition corresponding to the orthogonal decomposition

$$\boldsymbol{y} = \boldsymbol{P}_0\boldsymbol{y} + (\boldsymbol{P}_1 - \boldsymbol{P}_0)\boldsymbol{y} + (\boldsymbol{I} - \boldsymbol{P}_1)\boldsymbol{y}$$

yields the ANOVA table shown in Table 3.2, where $\{\hat{\mu}_i\}$ are the fitted values for the full model. The F test statistic, which is the ratio of the mean squares, has $df_1 = p - 1$ and $df_2 = n - p$.

Table 3.2 ANOVA Table for Testing That All Effects in a Normal Linear Model Equal Zero

Source	Projection Matrix	df	Sum of Squares	Mean Square
Intercept	$\boldsymbol{P}_0 = \frac{1}{n}\mathbf{1}\mathbf{1}^{\mathrm{T}}$	1	$\boldsymbol{y}^{\mathrm{T}}\boldsymbol{P}_0\boldsymbol{y} = n\bar{y}^2$	
Regression	$\boldsymbol{P}_1 - \boldsymbol{P}_0$	$p - 1$	$\boldsymbol{y}^{\mathrm{T}}(\boldsymbol{P}_1 - \boldsymbol{P}_0)\boldsymbol{y} = \sum_i(\hat{\mu}_i - \bar{y})^2$	$\frac{\sum_i(\hat{\mu}_i - \bar{y})^2}{p-1}$
Error	$\boldsymbol{I} - \boldsymbol{P}_1$	$n - p$	$\boldsymbol{y}^{\mathrm{T}}(\boldsymbol{I} - \boldsymbol{P}_1)\boldsymbol{y} = \sum_i(y_i - \hat{\mu}_i)^2$	$\frac{\sum_i(y_i - \hat{\mu}_i)^2}{n-p}$
Total	$\boldsymbol{I}$	n	$\sum_{i=1}^n y_i^2$	

[7]We use $p - 1$ for the highest index, so p is, as usual, the number of model parameters.

The one-way ANOVA test for c means constructed in Section 3.2.1 results when $p = c$ and the explanatory variables are indicator variables for $c - 1$ of the c groups. Testing H_0: $\beta_1 = \cdots = \beta_{c-1} = 0$ is then equivalent to testing H_0: $\mu_1 = \cdots = \mu_c$. The fitted value $\hat{\mu}_{ij}$ is then $\bar{y}_i$.

3.2.5 Non-null Behavior of F Statistic Comparing Nested Models

The numerator of the F test statistic for comparing two models summarizes the sample information about how much better M_1 fits than M_0. A relatively large value for $\text{SSE}_0 - \text{SSE}_1 = \|\hat{\boldsymbol{\mu}}_1 - \hat{\boldsymbol{\mu}}_0\|^2$ yields a large F value. If M_1 holds but M_0 does not, how large can we expect $\|\hat{\boldsymbol{\mu}}_1 - \hat{\boldsymbol{\mu}}_0\|^2$ and the F test statistic to be?

When M_1 holds, $E(\boldsymbol{y}) = \boldsymbol{\mu}_1$. Since $(\boldsymbol{P}_1 - \boldsymbol{P}_0)$ is symmetric and idempotent,

$$E\|\hat{\boldsymbol{\mu}}_1 - \hat{\boldsymbol{\mu}}_0\|^2 = E\|(\boldsymbol{P}_1 - \boldsymbol{P}_0)\boldsymbol{y}\|^2 = E[\boldsymbol{y}^{\mathrm{T}}(\boldsymbol{P}_1 - \boldsymbol{P}_0)\boldsymbol{y}].$$

Using the result (2.7) shown in Section 2.4.1 for $\boldsymbol{V} = \text{var}(\boldsymbol{y})$ and a matrix $\boldsymbol{A}$ that $E(\boldsymbol{y}^{\mathrm{T}}\boldsymbol{A}\boldsymbol{y}) = \text{trace}(\boldsymbol{A}\boldsymbol{V}) + \boldsymbol{\mu}^{\mathrm{T}}\boldsymbol{A}\boldsymbol{\mu}$, we have (with $\boldsymbol{V} = \sigma^2\boldsymbol{I}$)

$$\begin{aligned} E[\boldsymbol{y}^{\mathrm{T}}(\boldsymbol{P}_1 - \boldsymbol{P}_0)\boldsymbol{y}] &= \text{trace}[(\boldsymbol{P}_1 - \boldsymbol{P}_0)\sigma^2\boldsymbol{I}] + \boldsymbol{\mu}_1^{\mathrm{T}}(\boldsymbol{P}_1 - \boldsymbol{P}_0)\boldsymbol{\mu}_1 \\ &= \sigma^2[\text{rank}(\boldsymbol{P}_1) - \text{rank}(\boldsymbol{P}_0)] + \boldsymbol{\mu}_1^{\mathrm{T}}(\boldsymbol{P}_1 - \boldsymbol{P}_0)^{\mathrm{T}}(\boldsymbol{P}_1 - \boldsymbol{P}_0)\boldsymbol{\mu}_1. \end{aligned}$$

Let $\boldsymbol{\mu}_0 = \boldsymbol{P}_0\boldsymbol{\mu}_1$ denote the projection of the true mean vector onto the model space for M_0. Then, with full-rank model matrices, the numerator of the F test statistic has expected value

$$E\left[\frac{\|\hat{\boldsymbol{\mu}}_1 - \hat{\boldsymbol{\mu}}_0\|^2}{p_1 - p_0}\right] = \sigma^2 + \frac{\|\boldsymbol{\mu}_1 - \boldsymbol{\mu}_0\|^2}{p_1 - p_0}.$$

The chi-squared component of the numerator of the F statistic is

$$\frac{\|\hat{\boldsymbol{\mu}}_1 - \hat{\boldsymbol{\mu}}_0\|^2}{\sigma^2} \sim \chi^2_{p_1 - p_0, \lambda}$$

with noncentrality $\lambda = \|\boldsymbol{\mu}_1 - \boldsymbol{\mu}_0\|^2/\sigma^2$.

Next, for this non-null case, consider the denominator of the F statistic, which is the estimate of the error variance σ^2 for model M_1. Since

$$E\|\boldsymbol{y} - \hat{\boldsymbol{\mu}}_1\|^2 = E[\boldsymbol{y}^{\mathrm{T}}(\boldsymbol{I} - \boldsymbol{P}_1)\boldsymbol{y}] = \text{trace}[(\boldsymbol{I} - \boldsymbol{P}_1)\sigma^2\boldsymbol{I}] + \boldsymbol{\mu}_1^{\mathrm{T}}(\boldsymbol{I} - \boldsymbol{P}_1)\boldsymbol{\mu}_1$$

and since $(\boldsymbol{I} - \boldsymbol{P}_1)\boldsymbol{\mu}_1 = \boldsymbol{0}$, this expected sum of squares equals $(n - p_1)\sigma^2$. Thus, regardless of whether H_0 is true, the F denominator has

$$E\left[\frac{\|\boldsymbol{y} - \hat{\boldsymbol{\mu}}_1\|^2}{n - p_1}\right] = \sigma^2.$$

Under H_0 for testing M_0 against M_1, $\boldsymbol{\mu}_1 = \boldsymbol{\mu}_0$ and the expected value of the numerator mean square is also σ^2. Then the F test statistic is a ratio of two unbiased estimators of σ^2. The ratio of expectations equals 1, and when $n - p_1$ (and hence df_2) is large, this is also the approximate expected value of the F test statistic itself. That is, under H_0 we expect to observe F values near 1, within limits of sampling variability. Under the alternative, the ratio of the expected value of the numerator to the expected value of the denominator is $1 + \|\boldsymbol{\mu}_1 - \boldsymbol{\mu}_0\|^2/(p_1 - p_0)\sigma^2$. The noncentrality of the F test is the noncentrality of the numerator chi-squared, $\lambda = \|\boldsymbol{\mu}_1 - \boldsymbol{\mu}_0\|^2/\sigma^2$. The power of the F test increases as n increases, since then $\boldsymbol{\mu}_0$ and $\boldsymbol{\mu}_1$ contain more elements that contribute to the numerator sum of squares in λ.

3.2.6 Expected Mean Squares and Power for One-Way ANOVA

To illustrate expected non-null behavior, consider the one-way ANOVA F test for c groups, derived in Section 3.2.1. For it, the expected value of the numerator mean square is

$$E\left[\frac{\|\hat{\boldsymbol{\mu}}_1 - \hat{\boldsymbol{\mu}}_0\|^2}{p_1 - p_0}\right] = E\left[\frac{\sum_{i=1}^{c} n_i(\bar{y}_i - \bar{y})^2}{c-1}\right] = \sigma^2 + \frac{\sum_{i=1}^{c} n_i(\mu_i - \bar{\mu})^2}{c-1}.$$

Suppose the ANOVA compares $c = 3$ groups with $n_i = 10$ observations per group. The F test statistic for H_0: $\mu_1 = \mu_2 = \mu_3$ has $df_1 = 2$ and $df_2 = n - 3 = 27$. Let $F_{q,a,b}$ denote the q quantile of the central F distribution with $df_1 = a$ and $df_2 = b$. Consider the relatively large effects $\mu_1 - \mu_2 = \mu_2 - \mu_3 = \sigma$. The noncentrality (derived in Section 3.2.1) of $\lambda = \frac{1}{\sigma^2}\sum_i n_i(\mu_i - \bar{\mu})^2$ then equals 20. The power of the F test with size $\alpha = 0.05$ is the probability that a noncentral F random variable with $df_1 = 2$, $df_2 = 27$, and $\lambda = 20$ exceeds $F_{0.95,2,27}$. Using R, we find that the power is quite high, 0.973:

```
-----------------------------------------------------------------------
> qf(0.95, 2, 27) # 0.95 quantile of F dist. with df1 = 2, df2 = 27
[1] 3.354131
> 1 - pf(3.354131, 2, 27, 20) # right-tail prob. for noncentral F
[1] 0.9732551
-----------------------------------------------------------------------
```

In planning a study, it is sensible to find the power for various n for a variety of plausible effect sizes.

3.2.7 Testing a General Linear Hypothesis

In practice, nearly all hypotheses tested about effects in linear models can be expressed as H_0: $\boldsymbol{\Lambda\beta} = \mathbf{0}$ for a $\ell \times p$ matrix of constants $\boldsymbol{\Lambda}$ and a vector of estimable quantities $\boldsymbol{\Lambda\beta}$. A special case is the example just considered of H_0: $\beta_1 = \cdots = \beta_{p-1} = 0$ for comparing a full model to the null model. Another example is a test for a contrast or set of contrasts, such as H_0: $\beta_j - \beta_k = 0$ for comparing means j and k in a one-way layout (see Section 3.4.5). The form H_0: $\boldsymbol{\Lambda\beta} = \mathbf{0}$ is called the *general linear hypothesis*.

Suppose $\boldsymbol{X}$ and $\boldsymbol{\Lambda}$ are full rank, so the hypotheses contain no redundancies. That is, H_0 imposes ℓ independent constraints on an identifiable $\boldsymbol{\beta}$. The estimator $\boldsymbol{\Lambda}\hat{\boldsymbol{\beta}}$ of $\boldsymbol{\Lambda\beta}$ is the BLUE, and it is maximum likelihood under the assumption of normality for y. As a vector of ℓ linear transformations of $\hat{\boldsymbol{\beta}}$, $\boldsymbol{\Lambda}\hat{\boldsymbol{\beta}}$ has a $N[\boldsymbol{\Lambda\beta}, \boldsymbol{\Lambda}(\boldsymbol{X}^{\mathrm{T}}\boldsymbol{X})^{-1}\boldsymbol{\Lambda}^{\mathrm{T}}\sigma^2]$ distribution. The quadratic form

$$(\boldsymbol{\Lambda}\hat{\boldsymbol{\beta}} - \mathbf{0})^{\mathrm{T}} \left[\boldsymbol{\Lambda}(\boldsymbol{X}^{\mathrm{T}}\boldsymbol{X})^{-1}\boldsymbol{\Lambda}^{\mathrm{T}}\sigma^2\right]^{-1} (\boldsymbol{\Lambda}\hat{\boldsymbol{\beta}} - \mathbf{0})$$

compares the estimate $\boldsymbol{\Lambda}\hat{\boldsymbol{\beta}}$ of $\boldsymbol{\Lambda\beta}$ to its H_0 value of $\mathbf{0}$, relative to the inverse covariance matrix of $\boldsymbol{\Lambda}\hat{\boldsymbol{\beta}}$. Under H_0, it has a chi-squared distribution with $df = \ell$. By the orthogonality of the model space and the error space, we can form an F test statistic (with $df_1 = \ell$ and $df_2 = n - p$) from the ratio of chi-squared variates divided by their df values,

$$F = \frac{(\boldsymbol{\Lambda}\hat{\boldsymbol{\beta}})^{\mathrm{T}} \left[\boldsymbol{\Lambda}(\boldsymbol{X}^{\mathrm{T}}\boldsymbol{X})^{-1}\boldsymbol{\Lambda}^{\mathrm{T}}\right]^{-1} \boldsymbol{\Lambda}\hat{\boldsymbol{\beta}}/\ell}{\mathrm{SSE}/(n-p)},$$

where σ^2 has canceled from the numerator and denominator.

The restriction $\boldsymbol{\Lambda\beta} = \mathbf{0}$ implies a new model that is a special case M_0 of the original model. In fact, the F statistic just derived is identical to the F statistic (3.1) for comparing the full model to the special case M_0. So, how can we express the original model and the constraints $\boldsymbol{\Lambda\beta} = \mathbf{0}$ as an equivalent model M_0? It is the model having model matrix $\boldsymbol{X}_0$ found as follows. Let $\boldsymbol{U}$ be a matrix such that $C(\boldsymbol{U})$ is the orthogonal complement of $C(\boldsymbol{\Lambda}^T)$. That is, $\boldsymbol{\beta}$ is such that $\boldsymbol{\Lambda\beta} = \mathbf{0}$ if and only if $\boldsymbol{\beta} \in C(\boldsymbol{U})$. Then $\boldsymbol{\beta} = \boldsymbol{U\gamma}$ for some vector $\boldsymbol{\gamma}$. But under this restriction the original model $E(\boldsymbol{y}) = \boldsymbol{X\beta}$ simplifies to $E(\boldsymbol{y}) = \boldsymbol{XU\gamma} = \boldsymbol{X}_0\boldsymbol{\gamma}$ for $\boldsymbol{X}_0 = \boldsymbol{XU}$. Also, M_0 is a simpler model than the original model, with $C(\boldsymbol{X}_0)$ contained in $C(\boldsymbol{X})$, since any vector that is a linear combination of columns of $\boldsymbol{X}_0$ (e.g., $\boldsymbol{X}_0\boldsymbol{\gamma}$) is also a linear combination of columns of $\boldsymbol{X}$ (e.g., $\boldsymbol{X\beta}$ with $\boldsymbol{\beta} = \boldsymbol{U\gamma}$).

In the F statistic for comparing the two models, it can be shown[8] that

$$\mathrm{SSE}_0 - \mathrm{SSE}_1 = (\boldsymbol{\Lambda}\hat{\boldsymbol{\beta}})^{\mathrm{T}} \left[\boldsymbol{\Lambda}(\boldsymbol{X}^{\mathrm{T}}\boldsymbol{X})^{-1}\boldsymbol{\Lambda}^{\mathrm{T}}\right]^{-1} \boldsymbol{\Lambda}\hat{\boldsymbol{\beta}}.$$

When we developed the F test for comparing nested models in Section 3.2.2, we observed that $\mathrm{SSE}_0 - \mathrm{SSE}_1$ was merely $\boldsymbol{y}^{\mathrm{T}}(\boldsymbol{P}_1 - \boldsymbol{P}_0)\boldsymbol{y}$ based on the projection matrices for the two models. For the general linear hypothesis, what is the difference $(\boldsymbol{P}_1 - \boldsymbol{P}_0)$ projection matrix? Using the least squares solution for $\hat{\boldsymbol{\beta}}$,

$$\begin{aligned}\mathrm{SSE}_0 - \mathrm{SSE}_1 &= (\boldsymbol{\Lambda}\hat{\boldsymbol{\beta}})^{\mathrm{T}} \left[\boldsymbol{\Lambda}(\boldsymbol{X}^{\mathrm{T}}\boldsymbol{X})^{-1}\boldsymbol{\Lambda}^{\mathrm{T}}\right]^{-1} \boldsymbol{\Lambda}\hat{\boldsymbol{\beta}} \\ &= \boldsymbol{y}^{\mathrm{T}}\boldsymbol{X}(\boldsymbol{X}^{\mathrm{T}}\boldsymbol{X})^{-1}\boldsymbol{\Lambda}^{\mathrm{T}} \left[\boldsymbol{\Lambda}(\boldsymbol{X}^{\mathrm{T}}\boldsymbol{X})^{-1}\boldsymbol{\Lambda}^{\mathrm{T}}\right]^{-1} \boldsymbol{\Lambda}(\boldsymbol{X}^{\mathrm{T}}\boldsymbol{X})^{-1}\boldsymbol{X}^{\mathrm{T}}\boldsymbol{y} \\ &= \boldsymbol{y}^{\mathrm{T}}\boldsymbol{A}(\boldsymbol{A}^{\mathrm{T}}\boldsymbol{A})^{-1}\boldsymbol{A}^{\mathrm{T}}\boldsymbol{y},\end{aligned}$$

[8]See Christensen (2011, Section 3.3) or Monahan (2008, Section 6.3–6.5).

where $A = X(X^TX)^{-1}\Lambda^T$ (Doss 2010). The projection matrix $(P_1 - P_0)$ is $A(A^TA)^{-1}A^T$ for A as just defined.

A yet more general form of the general linear hypothesis is H_0: $\Lambda\beta = c$ for constants c. In the F test statistic, we then merely replace $(\Lambda\hat{\beta} - 0)$ by $(\Lambda\hat{\beta} - c)$. This more general H_0 is useful for inverting significance tests to construct confidence regions (Exercise 3.18). Another useful application is *noninferiority testing* in drug research, which analyzes whether the effect of a new drug falls within some acceptable margin c of the effect for an established drug.

3.2.8 Example: Testing That a Single Model Parameter Equals Zero

A common inference in linear modeling is testing H_0: $\beta_j = 0$ that a single explanatory variable in the model can be dropped. This is the special case of H_0: $\Lambda\beta = 0$ that substitutes for Λ a row vector λ with a multiple 1 of β_j and 0 elsewhere. Since the denominator of the F test statistic for comparing two nested models is s^2 (the error mean square) for the full model, the F test statistic then simplifies to

$$F = \frac{(\text{SSE}_0 - \text{SSE}_1)/1}{\text{SSE}_1/(n-p)} = \frac{(\lambda\hat{\beta})^T \left[\lambda(X^TX)^{-1}\lambda^T\right]^{-1} \lambda\hat{\beta}}{s^2} = \frac{\hat{\beta}_j^2}{(SE_j)^2},$$

where SE_j denotes the standard error of $\hat{\beta}_j$, the square of which is s^2 times the element from the corresponding row and column of $(X^TX)^{-1}$. This test statistic has $df_1 = 1$ and $df_2 = n - p$.

In the first ratio in this expression, $(\text{SSE}_0 - \text{SSE}_1)$ is the partial sum of squares explained by adding term j to the linear predictor, once the other terms are already there. The last ratio is $F = t^2$, where $t = \hat{\beta}_j/(SE_j)$. The null distribution of this t statistic is the t distribution with $df = n - p$.

3.2.9 Testing Terms in an Unbalanced Factorial ANOVA

In Section 3.2.1 (Table 3.1) we showed sum-of-squares formulas for the sources in the one-way layout. Analogous relatively simple formulas occur in factorial ANOVA with two or more factors, in the balanced case of equal sample sizes in the cells (e.g., Exercise 3.13). Unbalanced cases do not yield such formulas.

Consider, for example, the two-way layout in which y_{ijk} is observation k in the cell for level i of factor A and level j of factor B, for $i = 1, \ldots, r, j = 1, \ldots, c, k = 1, \ldots, n_{ij}$, where n_{ij} varies with i and j. The model with linear predictor

$$E(y_{ijk}) = \beta_0 + \beta_i + \gamma_j + \delta_{ij}$$

permits interaction between A and B in their effects on y. To achieve identifiability, we can express this as a linear model in which $r - 1$ of $\{\beta_i\}$ are coefficients of indicator variables for all except one level of A, $c - 1$ of $\{\gamma_j\}$ are coefficients of indicator variables for all except one level of B, and $(r-1)(c-1)$ of $\{\delta_{ij}\}$ are coefficients of products of the $r-1$ indicator variables for A with the $c-1$ indicator variables for B. With unbalanced data, a simple formula no longer occurs for the partial sum of squares explained by the interaction terms, or when those terms are not in the model, by the main effects. However, it is straightforward to fit the full model, fit a reduced model such as with $\{\delta_{ij}=0\}$, and then conduct the F test to compare these two nested models.

More complex models have several factors as well as higher-order interactions. Moreover, some combinations of the factors may have no observations, or the levels of some factors may be nested in levels of other factors, and the model may also contain quantitative explanatory variables. It may not even be obvious how to constrain parameters to achieve identifiability. Good software properly determines this, when we enter the terms as predictors in the linear model. Then we can test whether we need high-order terms in the model by fitting the model with and without those terms and using the F test for nested models to evaluate whether the partial SS explained by those terms is statistically significant. That test is a very general and useful one.

3.3 CONFIDENCE INTERVALS AND PREDICTION INTERVALS FOR NORMAL LINEAR MODELS

We learn more from constructing confidence intervals for parameter values than from significance testing. A confidence interval shows us the entire range of plausible values for a parameter, rather than focusing merely on whether a particular value is plausible.

3.3.1 Confidence Interval for a Parameter of a Normal Linear Model

To construct a confidence interval for a parameter β_j in a normal linear model, we construct and then invert a t test of H_0: $\beta_j = \beta_{j0}$ about potential values for β_j. The test statistic is

$$t = \frac{\hat{\beta}_j - \beta_{j0}}{SE_j},$$

the number of standard errors that $\hat{\beta}_j$ falls from β_{j0}. Recall that SE_j is the square root of the element in row j and column j of the estimated covariance matrix $s^2(\boldsymbol{X}^{\mathrm{T}}\boldsymbol{X})^{-1}$ of $\hat{\boldsymbol{\beta}}$, where s^2 is the error mean square. Just as the residuals are orthogonal to the model space, the residuals are uncorrelated with $\hat{\boldsymbol{\beta}}$. Specifically, the $p \times n$ covariance matrix

$$\mathrm{cov}(\hat{\boldsymbol{\beta}}, \boldsymbol{y} - \hat{\boldsymbol{\mu}}) = \mathrm{cov}\left[(\boldsymbol{X}^{\mathrm{T}}\boldsymbol{X})^{-1}\boldsymbol{X}^{\mathrm{T}}\boldsymbol{y}, (\boldsymbol{I}-\boldsymbol{H})\boldsymbol{y}\right] = (\boldsymbol{X}^{\mathrm{T}}\boldsymbol{X})^{-1}\boldsymbol{X}^{\mathrm{T}}\sigma^2\boldsymbol{I}(\boldsymbol{I}-\boldsymbol{H})^{\mathrm{T}},$$

and this is $\boldsymbol{0}$ because $\boldsymbol{H}\boldsymbol{X} = \boldsymbol{X}(\boldsymbol{X}^{\mathrm{T}}\boldsymbol{X})^{-1}\boldsymbol{X}^{\mathrm{T}}\boldsymbol{X} = \boldsymbol{X}$. Being linear functions of $\boldsymbol{y}$, $\hat{\boldsymbol{\beta}}$ and $(\boldsymbol{y} - \hat{\boldsymbol{\mu}})$ are jointly normally distributed, so uncorrelatedness implies independence.

Since s^2 is a function of the residuals, $\hat{\boldsymbol{\beta}}$ and s^2 are independent, and so are the numerator and denominator of the t statistic, as is required to obtain a t distribution.

The $100(1-\alpha)\%$ confidence interval for β_j is the set of all β_{j0} values for which the test has P-value $> \alpha$, that is, for which $|t| < t_{\alpha/2,n-p}$, the $1-\alpha/2$ quantile of the t distribution having $df = n-p$. For example, the 95% confidence interval is

$$\hat{\beta}_j \pm t_{0.025,n-p}(SE_j).$$

3.3.2 Confidence Interval for $E(y) = \boldsymbol{x}_0\boldsymbol{\beta}$

At a fixed setting $\boldsymbol{x}_0$ (a row vector) for the explanatory variables, we can construct a confidence interval for $E(y) = \boldsymbol{x}_0\boldsymbol{\beta}$. We do this by constructing and then inverting a t test about values for that linear predictor.

Let $\hat{\mu} = \boldsymbol{x}_0\hat{\boldsymbol{\beta}}$. Now

$$\text{var}(\hat{\mu}) = \text{var}(\boldsymbol{x}_0\hat{\boldsymbol{\beta}}) = \boldsymbol{x}_0\text{var}(\hat{\boldsymbol{\beta}})\boldsymbol{x}_0^{\text{T}} = \sigma^2\boldsymbol{x}_0(\boldsymbol{X}^{\text{T}}\boldsymbol{X})^{-1}\boldsymbol{x}_0^{\text{T}}.$$

Since $\boldsymbol{x}_0\hat{\boldsymbol{\beta}}$ is a linear function of $\boldsymbol{y}$, it has a normal distribution. Thus,

$$z = \frac{\boldsymbol{x}_0\hat{\boldsymbol{\beta}} - \boldsymbol{x}_0\boldsymbol{\beta}}{\sigma\sqrt{\boldsymbol{x}_0(\boldsymbol{X}^{\text{T}}\boldsymbol{X})^{-1}\boldsymbol{x}_0^{\text{T}}}} \sim N(0,1),$$

and

$$t = \frac{\boldsymbol{x}_0\hat{\boldsymbol{\beta}} - \boldsymbol{x}_0\boldsymbol{\beta}}{s\sqrt{\boldsymbol{x}_0(\boldsymbol{X}^{\text{T}}\boldsymbol{X})^{-1}\boldsymbol{x}_0^{\text{T}}}} = \frac{\boldsymbol{x}_0\hat{\boldsymbol{\beta}} - \boldsymbol{x}_0\boldsymbol{\beta}}{\sigma\sqrt{\boldsymbol{x}_0(\boldsymbol{X}^{\text{T}}\boldsymbol{X})^{-1}\boldsymbol{x}_0^{\text{T}}}} \Big/ \sqrt{\frac{s^2}{\sigma^2}} \sim t_{n-p}.$$

This last result follows because $(n-p)s^2/\sigma^2$ has a χ^2_{n-p} distribution for a normal linear model, by Cochran's theorem, so the t statistic is a $N(0,1)$ variate divided by the square root of the ratio of a χ^2_{n-p} variate to its df value. Also, since s^2 and $\hat{\boldsymbol{\beta}}$ are independent, so are the numerator and denominator of the t statistic. It follows that a $100(1-\alpha)\%$ confidence interval for $E(y) = \boldsymbol{x}_0\boldsymbol{\beta}$ is

$$\boldsymbol{x}_0\hat{\boldsymbol{\beta}} \pm t_{\alpha/2,n-p}s\sqrt{\boldsymbol{x}_0(\boldsymbol{X}^{\text{T}}\boldsymbol{X})^{-1}\boldsymbol{x}_0^{\text{T}}}. \tag{3.2}$$

When $\boldsymbol{x}_0$ is the explanatory variable value $\boldsymbol{x}_i$ for a particular observation, the term under the square root is the leverage h_{ii} from the model's hat matrix.

The construction for this interval extends directly to confidence intervals for linear combinations $\boldsymbol{\ell\beta}$. An example is a contrast of the parameters, such as $\beta_j - \beta_k$ for a pair of levels of a factor.

3.3.3 Prediction Interval for a Future y

At a particular value $\boldsymbol{x}_0$, how can we form an interval that is very likely to contain a future observation y at that value? This is more challenging than forming a confidence interval for the expected response. With lots of data, we can make precise inference about the mean but not precise prediction about a single future observation.

The normal linear model states that a future value y satisfies

$$y = \boldsymbol{x}_0\boldsymbol{\beta} + \epsilon, \quad \text{where} \quad \epsilon \sim N(0, \sigma^2).$$

From the fit of the model, the prediction of the future y value is $\hat{\mu} = \boldsymbol{x}_0\hat{\boldsymbol{\beta}}$. Now the future y also satisfies

$$y = \boldsymbol{x}_0\hat{\boldsymbol{\beta}} + e, \quad \text{where} \quad e = y - \hat{\mu}$$

is the residual for that observation. Since the future y is independent of the observations $y_1, \ldots, y_n$ used to determine $\hat{\boldsymbol{\beta}}$ and then $\hat{\mu}$,

$$\text{var}(e) = \text{var}(y - \hat{\mu}) = \text{var}(y) + \text{var}(\hat{\mu}) = \sigma^2[1 + \boldsymbol{x}_0(\boldsymbol{X}^{\text{T}}\boldsymbol{X})^{-1}\boldsymbol{x}_0^{\text{T}}].$$

It follows that

$$\frac{y - \hat{\mu}}{\sigma\sqrt{1 + \boldsymbol{x}_0(\boldsymbol{X}^{\text{T}}\boldsymbol{X})^{-1}\boldsymbol{x}_0^{\text{T}}}} \sim N(0, 1) \quad \text{and} \quad \frac{y - \hat{\mu}}{s\sqrt{1 + \boldsymbol{x}_0(\boldsymbol{X}^{\text{T}}\boldsymbol{X})^{-1}\boldsymbol{x}_0^{\text{T}}}} \sim t_{n-p}.$$

Inverting this yields a $100(1 - \alpha)\%$ *prediction interval* for the future y observation,

$$\hat{\mu} \pm t_{\alpha/2,n-p}s\sqrt{1 + \boldsymbol{x}_0(\boldsymbol{X}^{\text{T}}\boldsymbol{X})^{-1}\boldsymbol{x}_0^{\text{T}}}. \tag{3.3}$$

3.3.4 Example: Confidence Interval and Prediction Interval for Simple Linear Regression

We illustrate the confidence interval for the mean and the prediction interval for a future observation with the bivariate linear model,

$$E(y_i) = \beta_0 + \beta_1 x_i.$$

It is simpler to use the explanatory variable in centered form $x_i^* = x_i - \bar{x}$, which (from Section 2.1.3) results in uncorrelated $\hat{\beta}_0$ and $\hat{\beta}_1$. For the centered predictor values, $\hat{\beta}_0$ changes value to $\bar{y}$, but $\hat{\beta}_1$ and $\text{var}(\hat{\beta}_1) = \sigma^2/[\sum_i(x_i - \bar{x})^2]$ do not change. So, at a particular value x_0 for x,

$$\begin{aligned}\text{var}(\hat{\mu}) &= \text{var}[\hat{\beta}_0 + \hat{\beta}_1(x_0 - \bar{x})] \\ &= \text{var}(\bar{y}) + (x_0 - \bar{x})^2\text{var}(\hat{\beta}_1) = \sigma^2\left[\frac{1}{n} + \frac{(x_0 - \bar{x})^2}{\sum_{i=1}^n (x_i - \bar{x})^2}\right].\end{aligned}$$

For a future observation y and its independent prediction $\hat{\mu}$,

$$\text{var}(y - \hat{\mu}) = \sigma^2\left[1 + \frac{1}{n} + \frac{(x_0 - \bar{x})^2}{\sum_{i=1}^n (x_i - \bar{x})^2}\right].$$

The variances are smallest at $x_0 = \bar{x}$ and increase in a symmetric quadratic manner as x_0 moves away from $\bar{x}$. At $x_0 = \bar{x}$, we see that $\text{var}(\hat{\mu}) = \text{var}(\bar{y}) = \sigma^2/n$, whereas $\text{var}(y - \hat{\mu}) = \sigma^2(1 + 1/n)$. As n increases, $\text{var}(\hat{\mu})$ decreases toward 0, but $\text{var}(y - \hat{\mu})$ has σ^2 as its lower bound. Even if we can estimate nearly perfectly the regression line, we are limited in how accurately we can predict any future observation.

Figure 3.1 sketches the confidence interval and prediction interval, as a function of x_0. As n increases, the width of a confidence interval for the mean at any x_0 decreases toward 0, but the width of the 95% prediction interval decreases toward $2(1.96)\sigma$.

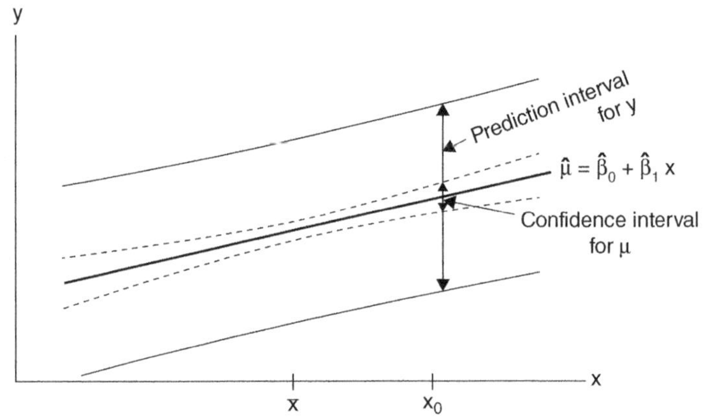

Figure 3.1 Portrayal of confidence intervals for the mean, $E(y) = \beta_0 + \beta_1 x_0$, and prediction intervals for a future observation y, at various x_0 values.

3.3.5 Interpretation and Limitations of Prediction Intervals

Interpreting a prediction interval is awkward. With $\alpha = 0.05$, we would like to say that conditional on the observed data and the model fit, we have 95% confidence that the future y will fall in the interval; that is, close to 95% of a large number of future observations would fall in the interval. However, the probability distributions in the derivation of Section 3.3.3 treat $\hat{\mu}$ as well as the future y as random, whereas in practice we use the interval after observing the data and hence $\hat{\mu}$. The conditional probability that the prediction interval captures a future y, given $\hat{\mu}$, is not 0.95. From the reasoning that led to Equation 3.3, before collecting any data, for the $\hat{\mu}$ (and s) to be found and then the future y,

$$P\left[|y - \hat{\mu}|/s\sqrt{1 + \boldsymbol{x}_0(\boldsymbol{X}^{\mathrm{T}}\boldsymbol{X})^{-1}\boldsymbol{x}_0^{\mathrm{T}}} \le t_{0.025,n-p}\right] = 0.95.$$

Once we observe the data and find $\hat{\mu}$ and s, this probability (with y as the only random part) does not equal 0.95. It depends on where $\hat{\mu}$ happened to fall. It need not be close to 0.95 unless var($\hat{\mu}$) is negligible compared to var(y). The 95% confidence for a prediction interval means the following: If we repeatedly used this method with many such datasets of independent observations satisfying the model (i.e., to construct both the fitted equation and this interval) and each time made a future observation, in the long run 95% of the time the interval formed would contain the future observation.

To this interpretation, we add the vital qualifier, *if the model truly holds*. In practice, we should have considerable faith in the model before forming prediction intervals. Even if we do not truly believe the model (the usual situation in practice), a confidence interval for $E(y) = \boldsymbol{x}_0\boldsymbol{\beta}$ at various $\boldsymbol{x}_0$ values is useful for describing the fit of the model in the population of interest. However, if the model fails, either in its description of the population mean as a function of the explanatory variables or in its assumptions of normality with constant variance, then the actual percentage of many future observations that fall within the limits of 95% prediction intervals may be quite different from 95%.

3.4 EXAMPLE: NORMAL LINEAR MODEL INFERENCE

What affects the selling price of a house? Table 3.3 shows observations on recent home sales in Gainesville, Florida. This table shows data for 8 houses from a data file for 100 home sales at the text website. Variables listed are selling price (in thousands of dollars), size of house (in square feet), annual property tax bill (in dollars), number of bedrooms, number of bathrooms, and whether the house is new. Since these 100 observations are from one city alone, we cannot use them to make inferences about the relationships in general. But for illustrative purposes, we treat them as a random sample of a conceptual population of home sales in this market and analyze how selling price seems to relate to these characteristics. We suggest that you download the data from the text website, so you can construct graphics not shown here and fit various models that seem sensible.

Table 3.3 Selling Prices and Related Characteristics for a Sample of Home Sales in Gainesville, Florida

Home	Selling Price	Size	Taxes	Bedrooms	Bathrooms	New
1	279.9	2048	3104	4	2	No
2	146.5	912	1173	2	1	No
3	237.7	1654	3076	4	2	No
4	200.0	2068	1608	3	2	No
5	159.9	1477	1454	3	3	No
6	499.9	3153	2997	3	2	Yes
7	265.5	1355	4054	3	2	No
8	289.9	2075	3002	3	2	Yes

Complete file for 100 homes is file `Houses.dat` at `www.stat.ufl.edu/~aa/glm/data`.

3.4.1 Inference for Modeling House Selling Prices

For modeling, we take y = selling price. Section 4.6 discusses issues in selecting explanatory variables for a model. For now, for simplicity we use only x_1 = size of house and x_2 = whether the house is new (1 = yes, 0 = no). We refer to these as "size" and "new." To begin, let us look at the data.

```
------------------------------------------------------------------------
> Houses # complete data at www.stat.ufl.edu/~aa/glm/data
   case  taxes  beds  baths  new  price  size
1     1   3104     4      2    0  279.9  2048
2     2   1173     2      1    0  146.5   912
...
> cbind(mean(price), sd(price), mean(size), sd(size))
          [,1]       [,2]        [,3]       [,4]
[1,]    155.33     101.26     1629.28     666.94
> table(new)
new
 0  1
89 11
> pch.list <- rep(0, 100)
> pch.list[new=="0"] <- 1; pch.list[new=="1"] <- 4 # pick symbols
> plot(size, price, pch=(pch.list)) # plot with symbols for new=0,1
------------------------------------------------------------------------
```

Figure 3.2 shows roughly an increasing linear trend for selling price as a function of size. An exception is a relatively low selling price for a very large dwelling that was not new (observation 64 in the data file). Only 11 houses in the sample were new, so the impact of that variable is rather unclear.

We next fit the model $E(y_i) = \beta_0 + \beta_1 x_{i1} + \beta_2 x_{i2}$, having additive effects of these explanatory variables. The least squares fit is $\hat{\mu}_i = -40.231 + 0.116x_{i1} + 57.736x_{i2}$.

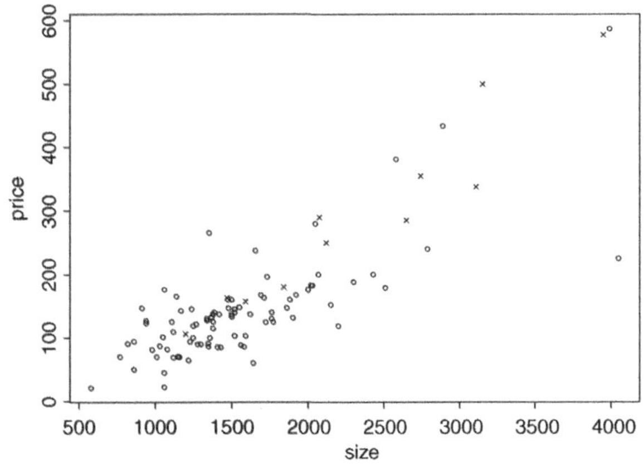

Figure 3.2 Scatterplot of selling price (in thousands of dollars) versus size of house (in square feet), labeled by whether new (× symbol for "yes" and o symbol for "no").

Adjusting for house size, the estimated mean selling price is \$57,736 higher for new homes. Because only 11 houses in the sample were new, this estimate is imprecise. For new or older houses, the estimated mean selling price increases by \$116 for each additional square foot of size. The sample R^2 value is large (0.72).

```
-----------------------------------------------------------------------
> fit <- lm(price ~ size + new)
> summary(fit)
Coefficients:
              Estimate   Std. Error   t value    Pr(>|t|)
(Intercept)   -40.2309      14.6961    -2.738     0.00737
size            0.1161       0.0088    13.204     < 2e-16
new            57.7363      18.6530     3.095     0.00257
---
Residual standard error: 53.88 on 97 degrees of freedom # This is s
Multiple R-squared: 0.7226, Adjusted R-squared: 0.7169
F-statistic: 126.3 on 2 and 97 DF, p-value: < 2.2e-16
> plot(fit)
-----------------------------------------------------------------------
```

Consider H_0: $\beta_1 = \beta_2 = 0$, stating that neither size nor new has an effect on selling price. The global F test statistic equals 126.3, with $df_1 = 2$ (since there are two effect parameters) and $df_2 = 100 - 3 = 97$. The P-value is 0 to many decimal places. This is no surprise. With this global test, H_0 states that *none* of the explanatory variables are truly correlated with the response. We usually expect a small P-value, and of greater interest is whether each explanatory variable has an effect, adjusting for the other explanatory variables in the model. The t statistic for testing the effect of whether the house is new, adjusting for size, is $t = 3.095$ ($df = 97$), highly significant ($P = 0.003$). Likewise, size has a highly significant partial effect, which again is no surprise.

Next we find a 95% confidence interval for the mean selling price of new homes at the mean size of the new homes, 2354.73 square feet. If the model truly holds, Equation 3.2 implies 95% confidence that the conceptual population mean selling price falls between \$258,721 and \$323,207. Equation 3.3 predicts that a selling price for another new house of that size will fall between \$179,270 and \$402,658.

```
-----------------------------------------------------------------------
> predict(fit,data.frame(size=2354.73, new=1), interval="confidence")
      fit        lwr        upr # 95% confidence is default
1 290.964   258.7207 323.2072
> predict(fit,data.frame(size=2354.73, new=1), interval="prediction")
      fit        lwr        upr
1 290.964   179.2701 402.6579
-----------------------------------------------------------------------
```

3.4.2 Model Checking

We next check the adequacy of the normal linear model and highlight influential observations. When the model holds, the standardized residuals have approximately a $N(0, 1)$ distribution. Let us look at a histogram and a *Q–Q plot*. The latter plots the standardized residual values against expected values of order statistics from a $N(0, 1)$ distribution (so-called *normal scores*). When a normal linear model holds, the points should lie roughly on a line through the origin with slope 1. Severe departures from that line indicate substantial non-normality in the conditional distribution of y. However, be cautious in interpreting such plots when n is not large, as they are affected by ordinary sampling variability.

```
-----------------------------------------------------------------
> hist(rstandard(fit)) # use rstudent instead for Studentized residuals
> qqnorm(rstandard(fit)) # Q-Q plot of standardized residuals
-----------------------------------------------------------------
```

For these data, the histogram in Figure 3.3 suggests that the conditional distribution of y is mound shaped, but possibly skewed to the right. Also, observation 64 has a relatively large negative standardized residual of −4.2. The Q–Q plot also shows evidence of skew to the right, because large positive theoretical quantiles have sample quantiles that are larger in absolute value whereas large negative theoretical quantiles have sample quantiles that are smaller in absolute value (except for the outlier). However, it is difficult to judge shape well unless n is quite large, and the actual error rate for two-sided statistical inference about β_j parameters in the linear model is *robust* to violations of the normality assumption. Inadequacy of statistical inference and consequent substantive conclusions are usually affected more by an inappropriate linear predictor (e.g., lacking an important interaction) and by practical sampling problems (e.g., missing data, errors of measurement) than by non-normality of the response. With clearly non-normal residuals, one can transform y to improve the normality. But the linear predictor may then more poorly describe the relationship, and effects on $E[g(y)]$ are of less interest than effects on $E(y)$. So, we recommend

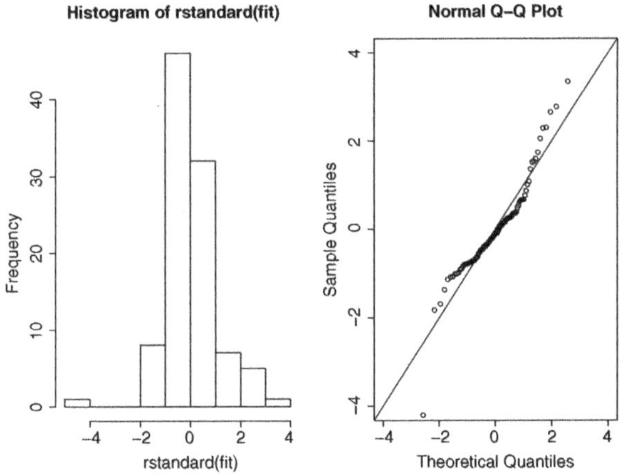

Figure 3.3 Histogram and Q–Q plot of standardized residuals, for normal linear model predicting selling price using size and new as explanatory variables.

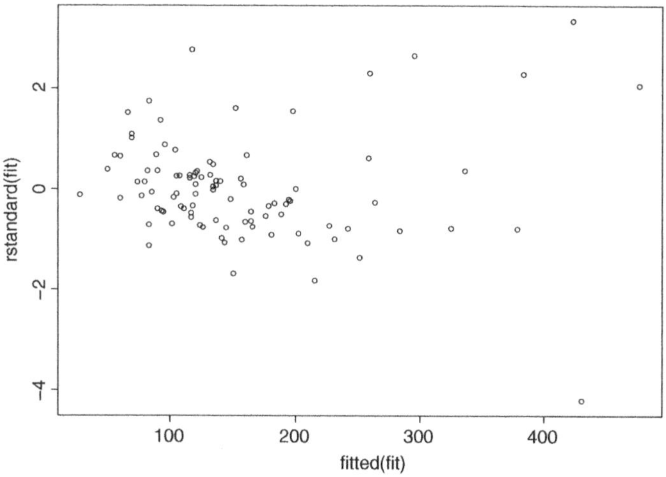

Figure 3.4 Plot of standardized residuals versus fitted values, for linear model predicting selling price using size and new as explanatory variables.

such plots mainly to help detect unusual observations that could influence substantive conclusions.

To investigate the adequacy of the linear predictor, we plot the residuals against the fitted values (Figure 3.4) and against size.

```
-----------------------------------------------------------------
> plot(fitted(fit), rstandard(fit))
> plot(size, rstandard(fit))
-----------------------------------------------------------------
```

If the normal linear model holds, a plot of the residuals against fitted values or values of explanatory variables should show a random pattern about 0 with relatively constant variability (Section 2.5.2). Figure 3.4 also highlights the unusual observation 64, but generally does not indicate lack of fit. There is a suggestion that residuals may tend to be larger in absolute value at higher values of the response. Rather than constant variance, it seems plausible that the variance may be larger at higher mean selling prices. We address this when we revisit the data in the next chapter.

The next table shows some standardized residuals and values of Cook's distance, including results for observation 64, which has the only Cook's distance exceeding 1.

```
------------------------------------------------------------------------
> cooks.distance(fit)
> plot(cooks.distance(fit))
> cbind(case,size,new,price,fitted(fit),rstandard(fit),cooks.distance(fit))
    case size new price
1      1 2048   0 279.9 197.607   1.541 1.462e-02
```

```
2        2  912    0 146.5   65.681    1.517 1.703e-02
...
64      64 4050    0 225.0 430.102   -4.202 1.284e+00
...
---------------------------------------------------------------------------
```

To check whether observation 64 is influential, we refit the model without it.

```
---------------------------------------------------------------------------
> fit2 <- lm(price ~ size + new, subset(Houses, case != 64))
> summary(fit2)
Coefficients:
             Estimate  Std. Error  t value  Pr(>|t|)
(Intercept)  -63.1545     14.2519   -4.431  2.49e-05
size           0.1328      0.0088   15.138   < 2e-16
new           41.3062     17.3269    2.384    0.0191
---
Residual standard error: 48.99 on 96 degrees of freedom
Multiple R-squared: 0.772,      Adjusted R-squared: 0.7672
---------------------------------------------------------------------------
```

The effect of a house being new has diminished from \$57,736 to \$41,306, the effect of size has increased some, and R^2 has increased considerably. This observation clearly is influential. We will see that it is not influential or even unusual when we consider an alternative model in Section 4.7.3 that allows the variability to grow with the mean.

There is no assurance that the effects of these two explanatory variables are truly additive. Perhaps the effect of size is different for new houses than for others. We can check by adding an interaction term, which we do for the dataset without the highly influential observation 64:

```
---------------------------------------------------------------------------
> fit3 <- lm(price ~ size + new + size:new, subset(Houses, case != 64))
> summary(fit3)
Coefficients:
             Estimate  Std. Error  t value  Pr(>|t|)
(Intercept)  -48.2431     15.6864   -3.075   0.00274
size           0.1230      0.0098   12.536   < 2e-16
new          -52.5122     47.6303   -1.102   0.27303
size:new       0.0434      0.0206    2.109   0.03757
---
Residual standard error: 48.13 on 95 degrees of freedom
Multiple R-squared: 0.7822,     Adjusted R-squared: 0.7753
---------------------------------------------------------------------------
```

Adjusted R^2 increases only from 0.767 to 0.775. The SSE values, not reported here (but available in R by requesting `deviance(fit2)` and `deviance(fit3)`), are 230,358 and 220,055. The F test comparing the two models has test statistic $F = t^2 = 2.109^2 = 4.45$ with $df_1 = 1$ and $df_2 = 95$, giving a P-value = 0.038. This model estimates that the effect of size is 0.123 for older houses and 0.123 + 0.043 = 0.166 for newer houses. The statistically significant improved fit at the 0.05 level must be weighed against a practically insignificant increase in R^2 and a relatively wide confidence interval for the true difference in size effects for new and older houses.

3.4.3 Conditional versus Marginal Effects: Simpson's Paradox

Alternatively, we could continue with the complete dataset of 100 observations and check whether an improved fit occurs from fitting other models. We might expect that the number of bedrooms is an important predictor of selling price, yet it was not included in the above model. Does it help to include "beds" in the model?

```
------------------------------------------------------------------
> cor(beds, price)
[1] 0.3940
> summary(lm(price ~ beds))
Coefficients:
            Estimate  Std. Error  t value   Pr(>|t|)
(Intercept)   -28.41       44.30   -0.641      0.523
beds           61.25       14.43    4.243   5.01e-05
---
> fit4 <- lm(price ~ size + new + beds)
> summary(fit4)
Coefficients:
            Estimate  Std. Error  t value   Pr(>|t|)
(Intercept) -25.1998     25.6022   -0.984    0.32745
size          0.1205      0.0107   11.229    < 2e-16
new          54.8996     19.1128    2.872    0.00501
beds         -7.2927     10.1588   -0.718    0.47458
---
Residual standard error: 54.02 on 96 degrees of freedom
Multiple R-squared: 0.7241,     Adjusted R-squared: 0.7155
------------------------------------------------------------------
```

Although the number of bedrooms has correlation 0.394 with selling price and is highly significant on its own, it has a P-value of 0.47 for its partial effect. Moreover, the adjusted $R^2 = 0.7155$ is smaller than the value 0.7169 without beds in the model. Apparently once size and new are explanatory variables in the model, it does not help to add beds.

Although the marginal effect of beds is positive, as described by the moderate positive correlation, the estimated partial effect of beds is negative! This illustrates *Simpson's paradox*[9]: An effect of a variable can change direction after adjusting for other variables. Figure 3.5 is a simplistic illustration of how this can happen.

[9]The name refers to Simpson (1951), but the result had been shown by Yule (1903).

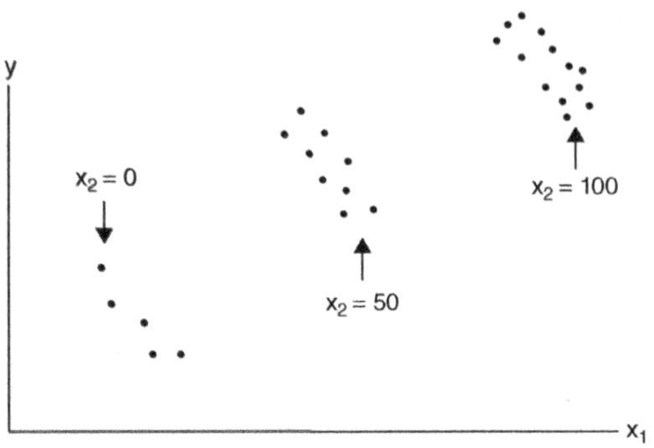

Figure 3.5 Portrayal of Simpson's paradox: The effect of x_1 on y is positive marginally but negative after adjusting for x_2.

3.4.4 Partial Correlation

The partial correlation between selling price and beds while adjusting for size and new is obtained by (1) finding the residuals for predicting selling price using size and new, (2) finding the residuals for predicting beds using size and new, and then (3) finding the ordinary correlation between these two sets of residuals:

```
--------------------------------------------------------------------
> cor(resid(lm(price ~ size + new)), resid(lm(beds ~ size + new)))
[1] -0.07307201 # partial correlation between selling price and beds
> summary(lm(resid(lm(price ~ size+new)) ~ resid(lm(beds ~ size+new))))
Coefficients: # this yields partial effect of beds on selling price
                              Estimate  Std. Error  t value  Pr(>|t|)
(Intercept)                  1.019e-14   5.346e+00    0.000      1.00
resid(lm(beds ~ size + new)) -7.293e+00  1.005e+01   -0.725      0.47
> plot(resid(lm(beds ~ size + new)), resid(lm(price ~ size + new)))
--------------------------------------------------------------------
```

The partial correlation value of −0.073 is weak. When a true partial correlation is 0, the standard error of a sample partial correlation r for a normal linear model with p parameters is $\sqrt{(1-r^2)/(n-p)}$, about 0.1 in this case.

Using the fact that the multiple correlation $R = \text{corr}(\boldsymbol{y}, \hat{\boldsymbol{\mu}})$, we use the formula at the end of Section 2.5.7 to find the squared partial correlation:

```
--------------------------------------------------------------------
> (cor(price,fitted(fit4))^2 - cor(price,fitted(fit))^2)/
+                        (1 - cor(price,fitted(fit))^2)
[1] 0.005339518
--------------------------------------------------------------------
```

The proportion of the variation in selling price unexplained by size and new that is explained by adding beds to the model is only $(-0.073)^2 = 0.0053$. Again, you can check that effects change substantially if you refit the model without observation 64 (e.g., the partial correlation changes to −0.240).

3.4.5 Testing Contrasts as a General Linear Hypothesis

For a factor in a model, we can test whether particular parameters are equal by expressing the null hypothesis as a set of contrasts. Such a hypothesis has the form of the general linear hypothesis H_0: $\boldsymbol{\Lambda\beta} = \mathbf{0}$. To illustrate, the analysis that suggested a lack of effect for beds, adjusting for size and new, investigated the linear effect. We could instead treat beds as a factor, with levels (2,3,4,5), to allow a nonlinear impact. Testing whether 3, 4, and 5 bedrooms have the same effect has a null hypothesis consisting of two contrasts and yields a F statistic with $df_1 = 2$ and $df_2 = 94$. The following code shows the contrasts expressed by equating the parameters for 3 bedrooms and 5 bedrooms and equating the parameters for 4 bedrooms and 5 bedrooms, for R constraints that set the parameter for 2 bedrooms equal to 0.

```
----------------------------------------------------------------------------
> fit5 <-lm(price ~ size + new + factor(beds))
> Lambda <- matrix(c(0,0,0,0,0,0,1,0,0,1,-1,-1), nrow=2)
> Lambda
      [,1] [,2] [,3] [,4] [,5] [,6]
[1,]     0    0    0    1    0   -1 # betas for intercept, size, new,
[2,]     0    0    0    0    1   -1 #           beds=3, beds=4, beds=5
> library(car)
> linearHypothesis(fit5, Lambda, test=c("F"))
Hypothesis:
factor(beds)3 - factor(beds)5 = 0
factor(beds)4 - factor(beds)5 = 0
  Res.Df     RSS  Df  Sum of Sq        F  Pr(>F)
1     96  275849
2     94  273722   2     2127.4   0.3653    0.695
----------------------------------------------------------------------------
```

3.4.6 Selecting or Building a Model

This chapter has presented inferences for normal linear models but has not discussed how to select a model or build a model from a set of potential explanatory variables. These issues are relevant for all generalized linear models (GLMs), and we discuss them in the next chapter (Section 4.6).

3.5 MULTIPLE COMPARISONS: BONFERRONI, TUKEY, AND FDR METHODS

Using a model to compare many groups or to evaluate the significance of many potential explanatory variables in a model can entail a very large number of inferences. For example, in a one-way layout, comparing each pair of c groups involves $c(c-1)/2$

inferences, which is considerable when c itself is large. Even if each inference has a small error probability, the probability may be substantial that at least one inference is in error. In such cases, we can construct the inferences so that the error probability applies to the entire family of inferences rather than to each individual one. For example, in constructing confidence intervals for pairwise comparisons of means, we can provide 95% *family-wise* confidence that the entire set of intervals simultaneously contains the true differences.

3.5.1 Bonferroni Method for Multiple Inferences

A popular way to conduct multiple inferences while controlling the overall error rate is based on a simple inequality shown by the British mathematician George Boole (1854), in an impressive treatise of which several chapters presented laws of probability.

Boole's inequality: Let $E_1, E_2, \ldots, E_t$ be t events in a sample space. Then, the probability that at least one of these events occurs has the upper bound

$$P(\cup_j E_j) \leq \sum_{j=1}^{t} P(E_j).$$

The proof of this is simple. We suggest that you construct a Venn diagram to illustrate. Let

$$B_1 = E_1,\ B_2 = E_1^c \cap E_2,\ B_3 = E_1^c \cap E_2^c \cap E_3, \ldots.$$

Then, $\cup_j B_j = \cup_j E_j$ and $B_j \subset E_j$, but the $\{B_j\}$ are disjoint and so $P(\cup_j B_j) = \sum_j P(B_j)$. Thus,

$$P(\cup_j E_j) = P(\cup_j B_j) = \sum_{j=1}^{t} P(B_j) \leq \sum_{j=1}^{t} P(E_j).$$

In the context of multiple confidence intervals, let E_j (for $j = 1, \ldots, t$) denote the event that interval j is in error, not containing the relevant parameter value. If each interval has confidence coefficient $(1 - \alpha/t)$, then the (a priori) probability that at least one of the t intervals is in error is bounded above by $t(\alpha/t) = \alpha$. So, the family-wise confidence coefficient for the set of the t intervals is bounded below by $1 - \alpha$. For example, for the one-way layout with $c = 5$ means, if we use confidence level 99% for each of the 10 pairwise comparisons, the overall confidence level is at least 90%. This method for constructing simultaneous confidence intervals is called the *Bonferroni method.* It relies merely on Boole's inequality, but the name refers to the Italian probabilist/mathematician Carlo Bonferroni, who in 1936 extended Boole's inequality in various ways.

An advantage of the Bonferroni method is its generality. It applies for any probability-based inferences for any distribution, not just confidence intervals for a normal linear model. A disadvantage is that the method is *conservative*: If we want overall 90% confidence (say), the method ensures that the actual confidence level is *at least* that high. As a consequence, the intervals are wider than ones that would produce *exactly* that confidence level. The next method discussed is more limited, being designed specifically for comparing means in balanced normal linear models, but it does not have this disadvantage.

3.5.2 Tukey Method of Multiple Comparisons

In 1953 the great statistician John Tukey proposed a method for simultaneously comparing means of several normal distributions. Using a probability distribution for the range of observations from a normal distribution, it applies to balanced designs such as one-way and two-way layouts with equal sample sizes.

Definition. Suppose $\{y_i\}$ are independent, with $y_i \sim N(\mu, \sigma^2)$, $i = 1, \ldots, c$. Let s^2 be an independent estimate of σ^2 with $\nu s^2/\sigma^2 \sim \chi^2_\nu$. Then,

$$Q = \frac{max_i\, y_i - min_i\, y_i}{s}$$

has the *Studentized range distribution* with parameters c and ν. We denote the distribution by $Q_{c,\nu}$ and its $1 - \alpha$ quantile by $Q_{1-\alpha,c,\nu}$.

To illustrate how Tukey's method uses the Studentized range distribution, we consider the balanced one-way layout for the normal linear model. The sample means $\bar{y}_1, \ldots, \bar{y}_c$ each have sample size $n_i = n$. Let $N = \sum_i n_i = cn$. Let $s^2 = \sum_{i=1}^{c} \sum_{j=1}^{n} (y_{ij} - \bar{y}_i)^2/(N - c)$ denote the pooled variance estimate from the one-way ANOVA (i.e., the error mean square in Section 3.2.1). Then each $\sqrt{n}(\bar{y}_i - \mu_i)$ has a $N(0, \sigma^2)$ distribution, and so

$$\sqrt{n}[\max_i(\bar{y}_i - \mu_i) - \min_i(\bar{y}_i - \mu_i)]/s \sim Q_{c,N-c}.$$

A priori, the probability is $(1 - \alpha)$ that this statistic is less than $Q_{1-\alpha,c,N-c}$. When the statistic *is* bounded above by $Q_{1-\alpha,c,N-c}$, then

$$\text{all } \ |(\bar{y}_i - \mu_i) - (\bar{y}_j - \mu_j)| < Q_{1-\alpha,c,N-c}(s/\sqrt{n})$$

and thus $(\mu_i - \mu_j)$ falls within $Q_{1-\alpha,c,N-c}(s/\sqrt{n})$ of $(\bar{y}_i - \bar{y}_j)$ for *all* pairs. So, we can construct family-wise confidence intervals for the pairs $\{\mu_i - \mu_j\}$ using simultaneously for all i and j,

$$(\bar{y}_i - \bar{y}_j) \pm Q_{1-\alpha,c,N-c}\left(\frac{s}{\sqrt{n}}\right).$$

The confidence coefficient for the family of all $t = c(c-1)/2$ such comparisons equals $1-\alpha$. A difference $|\bar{y}_i - \bar{y}_j|$ that exceeds $Q_{1-\alpha,c,N-c}(s/\sqrt{n})$ is considered statistically significant, as the interval for $(\mu_i - \mu_j)$ does not contain 0. The corresponding margin of error using the Bonferroni method is $t_{\alpha/c(c-1),N-c}s\sqrt{2/n}$.

To illustrate, suppose we plan to construct family-wise 95% confidence intervals for the 45 pairs of means for $c = 10$ groups, and we have $n = 20$ observations from each group and a standard deviation estimate of $s = 15$. The margin of error for each comparison is $Q_{0.95,10,190}(15/\sqrt{20}) = 15.19$ for the Tukey method and $t_{0.05/2(45),190}(15\sqrt{2/20}) = 15.71$ for the Bonferroni method. The Q and t quantiles used here are easily obtained with software:

```
-------------------------------------------------------------------
> qtukey(0.95, 10, 190); qt(1 - 0.05/(2*45), 190)
[1] 4.527912
[1] 3.311379
-------------------------------------------------------------------
```

The Tukey method applies exactly to this balanced case, for which the sample means have equal variances. A generalized version applies in a slightly conservative manner for unbalanced cases (see Note 3.5).

3.5.3 Controlling the False Discovery Rate

As the number of inferences (t) increases in multiple comparison methods designed to have fixed family-wise error rate α, the margin of error for each inference increases. When t is enormous, as in detecting differential expression in thousands of genes, there may be very low power for establishing significance with any individual inference. It can be difficult to discover any effects that truly exist, especially if those effects are weak. But, in the absence of a multiplicity adjustment, most significant results found could be Type I errors, especially when the number of true non-null effects is small. Some multiple inference methods attempt to address this issue. Especially popular are methods that control the *false discovery rate* (FDR). In the context of significance testing, this is the expected proportion of the rejected null hypotheses ("discoveries") that are erroneously rejected (i.e., that are actually true—"false discoveries").

Benjamini and Hochberg (1995) proposed a simple algorithm for ensuring FDR $\leq \alpha$ for a desired α. It applies with t independent[10] tests. Let $P_{(1)} \leq P_{(2)} \leq \cdots \leq P_{(t)}$ denote the ordered P-values for the t tests. We reject hypotheses (1), …, (j^*), where j^* is the maximum j for which $P_{(j)} \leq j\alpha/t$. The actual FDR for this method is bounded above by α times the proportion of rejected hypotheses that are actually true. This bound is α when the null hypothesis is always true.

Here is intuition for comparing $P_{(j)}$ to $j\alpha/t$ in this method: Suppose t_0 of the t hypotheses tested are actually true. Since P-values based on continuous test statistics

[10] Benjamini and Yekutieli (2001) showed that the method also works with tests that are positively dependent in a certain sense.

have a uniform distribution when H_0 is true, conditional on $P_{(j)}$ being the cutoff for rejection, a priori we expect to reject about $t_0 P_{(j)}$ of the t_0 true hypotheses. Of the j observed tests actually having P-value $\leq P_{(j)}$, this is a proportion of expected false rejections of $t_0 P_{(j)}/j$. In practice t_0 is unknown, but since $t_0 \leq t$, if $tP_{(j)}/j \leq \alpha$ then this ensures $t_0 P_{(j)}/j \leq \alpha$. Therefore, rejecting H_0 whenever $P_{(j)} \leq j\alpha/t$ ensures this.

With this method, the most significant test compares $P_{(1)}$ to α/t and has the same decision as in the ordinary Bonferroni method, but then the other tests have less conservative requirements. When some hypotheses are false, the FDR method tends to reject more of them than the Bonferroni method, which focuses solely on controlling the family-wise error rate. Benjamini and Hochberg illustrated the FDR for a study about myocardial infarction. For the 15 hypotheses tested, the ordered P-values were

$$0.0001, 0.0004, 0.0019, 0.0095, 0.020, 0.028, 0.030,$$
$$0.034, 0.046, 0.32, 0.43, 0.57, 0.65, 0.76, 1.00.$$

With $\alpha = 0.05$, these are compared with $j(0.05)/15$, starting with $j = 15$. The maximum j for which $P_{(j)} \leq j(0.0033)$ is $j = 4$, for which $P_{(4)} = 0.0095 < 4(0.0033)$. So the hypotheses with the four smallest P-values are rejected. By contrast, the Bonferroni approach with family-wise error rate 0.05 compares each P-value to $0.05/15 = 0.0033$ and rejects only three of these hypotheses.

CHAPTER NOTES

Section 3.1: Distribution Theory for Normal Variates

3.1 **Cochran's theorem**: Results on quadratic forms in normal variates were shown by the Scottish statistician William Cochran in 1934 when he was a 24-year old graduate student at the University of Cambridge, studying under the supervision of John Wishart. He left Cambridge without completing his Ph.D. degree to work at Rothamsted Experimental Station, recruited by Frank Yates after R. A. Fisher left to take a professorship at University College, London. In the 1934 article, Cochran showed that if $x_1, \ldots, x_n$ are iid $N(0, 1)$ and $\sum_i x_i^2 = Q_1 + \cdots + Q_k$ for quadratic forms having ranks $r_1, \ldots, r_k$, then $Q_1, \ldots, Q_k$ are independent chi-squared with *df* values $r_1, \ldots, r_k$ if and only if $r_1 + \cdots + r_k = n$.

3.2 **Independent normal quadratic forms**: The Cochran's theorem implication that $\{\boldsymbol{y}^{\mathrm{T}} \boldsymbol{P}_i \boldsymbol{y}\}$ are independent when $\boldsymbol{P}_i \boldsymbol{P}_j = \boldsymbol{0}$ extends to this result (Searle 1997, Chapter 2): When $\boldsymbol{y} \sim N(\boldsymbol{\mu}, \boldsymbol{V})$, $\boldsymbol{y}^{\mathrm{T}} \boldsymbol{A} \boldsymbol{y}$ and $\boldsymbol{y}^{\mathrm{T}} \boldsymbol{B} \boldsymbol{y}$ are independent if and only if $\boldsymbol{AVB} = \boldsymbol{0}$.

Section 3.2: Significance Tests for Normal Linear Models

3.3 **Fisher and ANOVA**: Application of ANOVA was stimulated by the 1925 publication of R. A. Fisher's classic text, *Statistical Methods for Research Workers*. Later contributions include Scheffé (1959) and Hoaglin et al. (1991).

3.4 **General linear hypothesis**: For further details about tests for the general linear hypothesis and in particular for one-way and two-way layouts, see Lehmann and Romano (2005, Chapter 7) and Scheffé (1959, Chapters 2–4).

Section 3.5: Multiple Comparisons: Bonferroni, Tukey, FDR Methods

3.5 **Boole, Bonferroni, Tukey, Scheffé**: Seneta (1992) surveyed probability inequalities presented by Boole and Bonferroni and related results of Fréchet. For an overview of Tukey's contributions to multiple comparisons, see Benjamini and Braun (2002) and Tukey (1994). With unbalanced data, Kramer (1956) suggested replacing $s/\sqrt{n}$ in the Tukey interval by $s\sqrt{\frac{1}{2}\left[(1/n_a)+(1/n_b)\right]}$ for groups a and b. Hayter (1984) showed this is slightly conservative. For the normal linear model, Scheffé (1959) proposed a method that applies simultaneously to all contrasts of c means. For estimating a contrast $\sum_i a_i\mu_i$ in the one-way layout (possibly unbalanced), it multiplies the usual estimated standard error $s\sqrt{\sum_i(a_i^2/n_i)}$ for $\sum_i a_i\bar{y}_i$ by $\sqrt{(c-1)F_{1-\alpha,c-1,n-c}}$ to obtain the margin of error. For simple differences between means, these are wider than the Tukey intervals, because they apply to a much larger family of contrasts. Hochberg and Tamhane (1987) and Hsu (1996) surveyed multiple comparison methods.

3.6 **False discovery rate**: For surveys of FDR methods and issues in large-scale multiple hypothesis testing, see Benjamini (2010), Dudoit et al. (2003), and Farcomeni (2008).

EXERCISES

3.1 Suppose $\boldsymbol{y} \sim N(\boldsymbol{\mu}, \boldsymbol{V})$ with $\boldsymbol{V}$ nonsingular of rank p. Show that $(\boldsymbol{y}-\boldsymbol{\mu})^{\mathrm{T}}\boldsymbol{V}^{-1}(\boldsymbol{y}-\boldsymbol{\mu}) \sim \chi_p^2$ by letting $\boldsymbol{z} = \boldsymbol{V}^{-1/2}(\boldsymbol{y}-\boldsymbol{\mu})$ and finding the distribution of $\boldsymbol{z}$ and $\boldsymbol{z}^{\mathrm{T}}\boldsymbol{z}$.

3.2 If T has a t distribution with $df = p$, then using the construction of t and F random variables, explain why T^2 has the F distribution with $df_1 = 1$ and $df_2 = p$.

3.3 Suppose $z = x + y$ where $z \sim \chi_p^2$ and $x \sim \chi_q^2$. Show how to find the distribution of y.

3.4 Applying the SS decomposition with the projection matrix for the null model (Section 2.3.1), use Cochran's theorem to show that for $y_1, \ldots, y_n$ independent from $N(\mu, \sigma^2)$, $\bar{y}$ and s^2 are independent (Cochran 1934).

3.5 For $y_1, \ldots, y_n$ independent from $N(\mu, \sigma^2)$, apply Cochran's theorem to construct a F test of H_0: $\mu = \mu_0$ against H_1: $\mu \neq \mu_0$ by applying the SS decomposition with the projection matrix for the null model shown in Section 2.3.1 to the adjusted observations $\{y_i - \mu_0\}$. State the null and alternative distributions of the test statistic. Show how to construct an equivalent t test.

3.6 Consider the normal linear model for the one-way layout (Section 3.2.1).

a. Explain why the F statistic used to test H_0: $\mu_1 = \cdots = \mu_c$ has, under H_0, an F distribution.

b. Why is the test is called analysis of *variance* when H_0 deals with *means*? (*Hint*: See Section 3.2.5.)

3.7 A one-way ANOVA uses n_i observations from group i, $i = 1, \dots, c$.

a. Verify the noncentrality parameter for the scaled between-groups sum of squares.

b. Suppose $c = 3$, with $\mu_1 - \mu_2 = \mu_2 - \mu_3 = \sigma/2$. Evaluate the noncentrality, and use it to find the power of a F test with size $\alpha = 0.05$ for a common sample size n, when (i) $n = 10$, (ii) $n = 30$, (iii) $n = 50$.

c. Now suppose $\mu_1 - \mu_2 = \mu_2 - \mu_3 = \Delta\sigma$. Evaluate the noncentrality when each $n_i = 10$, and use it to find the power of a F test with size $\alpha = 0.05$ when $\Delta = 0, 0.5, 1.0$.

3.8 Based on the formula $s^2(\boldsymbol{X}^{\mathrm{T}}\boldsymbol{X})^{-1}$ for the estimated var$(\hat{\boldsymbol{\beta}})$, explain why the standard errors of $\{\hat{\beta}_j\}$ tend to decrease as n increases.

3.9 Using principles from this chapter, inferentially compare μ_1 and μ_2 from $N(\mu_1, \sigma^2)$ and $N(\mu_2, \sigma^2)$ populations, based on independent random samples of sizes n_1 and n_2.

a. Put the analysis in a normal linear model context, showing a model matrix and explaining how to interpret the model parameters.

b. Find the projection matrix for the model space, and find SSR and SSE.

c. Construct a F test statistic for testing H_0: $\mu_1 = \mu_2$ against H_a: $\mu_1 \neq \mu_2$. Using Cochran's theorem, specify a null distribution for this statistic.

d. Relate the F test statistic in (**c**) to the t statistic for this test,

$$t = \frac{\bar{y}_1 - \bar{y}_2}{s\sqrt{\frac{1}{n_1} + \frac{1}{n_2}}}$$

where s^2 is the pooled variance estimate from the two samples.

3.10 Refer to the previous exercise. Based on inverting significance tests with nonzero null values, show how to construct a confidence interval for $\mu_1 - \mu_2$.

3.11 Section 2.3.4 considered the projection matrices and ANOVA table for the two-way layout with one observation per cell. For testing each main effect in that model, show how to construct test statistics and explain how to obtain their null distributions, based on theory in this chapter.

3.12 For the balanced two-way $r \times c$ layout with n observations $\{y_{ijk}\}$ in each cell, denote the sample means by $\{\bar{y}_{ij.}\}$ in the cells, $\bar{y}_{i..}$ in level i of A, $\bar{y}_{.j.}$ in level j of B, and $\bar{y}$ overall for all $N = nrc$ observations. Consider the model that assumes a lack of interaction.

a. Construct the ANOVA table, including SS and df values, showing how to construct F statistics for testing the main effects.

b. Show that the expected value of the numerator mean square for the test of the A factor effect is $\sigma^2 + \left(\frac{cn}{r-1}\right) \sum_{i=1}^{r}(\mu_{i..} - \bar{\mu})^2$.

3.13 Refer to the previous exercise. Now consider the model permitting interaction. Table 3.4 shows the resulting ANOVA table.

a. Argue intuitively and in analogy with results for one-way ANOVA that the SS values for factor A, factor B, and residual are as shown in the ANOVA table.

b. Based on the results in (**a**) and what you know about the total of the SS values, show that the SS for interaction is as shown in the ANOVA table.

c. In the ANOVA table, show the *df* values for each source. Show the mean squares, and show how to construct test statistics for testing no interaction and for testing each main effect. Specify the null distribution for each test statistic.

Table 3.4 ANOVA Table for Normal Linear Model with Two-Way Layout

Source	*df*	Sum of Squares	Mean Square	F_{obs}
Mean	1	$N\bar{y}^2$		
A (rows)	—	$cn\sum_i(\bar{y}_{i..} - \bar{y})^2$	—	—
B (columns)	—	$rn\sum_j(\bar{y}_{.j.} - \bar{y})^2$	—	—
Interaction	—	$n\sum_i\sum_j(\bar{y}_{ij.} - \bar{y}_{i..} - \bar{y}_{.j.} + \bar{y})^2$	—	—
Residual (error)	—	$\sum_i\sum_j\sum_k(y_{ijk} - \bar{y}_{ij.})^2$	—	
Total	N	$\sum_{i=1}^r\sum_{j=1}^c\sum_{k=1}^n y_{ijk}^2$		

3.14 **a.** Show that the F statistic in Section 3.2.4 for testing that all effects equal 0 has expression in terms of the R^2 value as

$$F = \frac{R^2/(p-1)}{(1-R^2)/(n-p)}$$

b. Show that the F statistic (3.1) for comparing nested models has expression in terms of the R^2 values for the models as

$$F = \frac{(R_1^2 - R_0^2)/(p_1 - p_0)}{(1-R_1^2)/(n-p_1)}.$$

3.15 Using the F formula for comparing models in the previous exercise, show that adjusted R^2 being larger for the more complex model is equivalent to $F > 1$.

3.16 For the linear model $E(y_{ij}) = \beta_0 + \beta_i$ for the one-way layout, explain how H_0: $\beta_1 = \cdots = \beta_c$ is a special case of the general linear hypothesis.

3.17 For a normal linear model with p parameters and n observations, explain how to test H_0: $\beta_j = \beta_k$ in the context of the (**a**) general linear hypothesis and (**b**) F test comparing two nested linear models.

3.18 Explain how to use the F test for the general linear hypothesis H_0: $\boldsymbol{\Lambda\beta} = \mathbf{c}$ to invert a test of H_0: $\boldsymbol{\beta} = \boldsymbol{\beta}_0$ to form a *confidence ellipsoid* for $\boldsymbol{\beta}$. For $p = 2$, describe how this could give you information beyond what you would learn from separate intervals for β_1 and β_2.

3.19 Suppose a one-way layout has ordered levels for the c groups, such as dose levels in a dose–response assessment. The model $E(y_{ij}) = \beta_0 + \beta_i$ treats the groups as a qualitative factor. The model $E(y_{ij}) = \beta_0 + \beta x_i$ has a quantitative predictor that assumes monotone group scores $\{x_i\}$.

a. Explain why the quantitative-predictor model is a special case of the qualitative-predictor model. Given the qualitative-predictor model, show how the null hypothesis that the quantitative-predictor model is adequate is a special case of the general linear hypothesis. Illustrate by showing $\boldsymbol{\Lambda}$ for the case $c = 5$ with $\{x_i = i\}$.

b. Explain how to use an F test to compare the models, specifying the *df* values.

c. Describe an advantage and disadvantage of each way of handling ordered categories.

3.20 Mimicking the derivation in Section 3.3.2, derive a confidence interval for the linear combination $\boldsymbol{\ell\beta}$. Explain how it simplifies for the case $\beta_j - \beta_k$.

3.21 When there are no explanatory variables, show how the confidence interval in Section 3.3.2 simplifies to a confidence interval for the marginal $E(y)$.

3.22 Consider the null model, for simplicity with known σ^2. After estimating $\mu = E(y)$ by $\bar{y}$, you plan to predict a future y from the $N(\mu, \sigma^2)$ distribution. State the formula for a 95% prediction interval for this model. Suppose, unknown to you, $\bar{y} = \mu + z_o\sigma/\sqrt{n}$ for some particular z_o value. Find an expression for the actual probability, conditional on $\bar{y}$, that the prediction interval contains the future y. Explain why this is not equal to 0.95 (e.g., what happens if $z_o = 0$?) but converges to it as $n \to \infty$.

3.23 Based on the expression for a squared partial correlation in Section 3.4.4, show how it relates to a partial SS for the full model and SSE for the model without that predictor.

3.24 For the normal linear model for the $r \times c$ two-way layout with n observations per cell, explain how to use the Tukey method for family-wise comparisons of all pairs of the r row means with confidence level 95%.

3.25 An analyst plans to construct family-wise confidence intervals for normal linear model parameters $\{\beta^{(1)}, \ldots, \beta^{(g)}\}$ in estimating an effect as part of a meta-analysis with g independent studies. Explain why constructing each interval with confidence level $(1-\alpha)^{1/g}$ provides exactly the family-wise confidence level $(1-\alpha)$. Prove that such intervals are narrower than Bonferroni intervals.

3.26 In the one-way layout with c groups and a fixed common sample size n, consider simultaneous confidence intervals for pairwise comparisons of means, using family-wise error probability $\alpha = 0.05$. Using software such as R, analyze how the ratio of margins of error for the Tukey method to the Bonferroni method behaves as c increases for fixed n and as n increases for fixed c. Show that this ratio converges to 1 as α approaches 0 (i.e., the Bonferroni method is only very slightly conservative when applied with very small α).

3.27 *Selection bias*: Suppose the normal linear model $\mu_i = \beta_0 + \beta_1 x_i$ holds with $\beta_1 > 0$, but the responses are *truncated* and we observe y_i only when $y_i > L$ (or perhaps only when $y_i < L$) for some threshold L.

a. Describe a practical scenario for which this could happen. How would you expect the truncation to affect $\hat{\beta}_1$ and s? Illustrate by sketching a graph. (You could check this with data, such as by fitting the model in Section 3.4.1 only to house sales having $y_i > 150$.)

b. Construct a likelihood function with the conditional distribution of y, to enable consistent estimation of $\boldsymbol{\beta}$. (See Amemiya (1984) for a survey of modeling with truncated or censored data. In R, see the `truncreg` package.)

3.28 In the previous exercise, suppose truncation instead occurs on x. Would you expect this to affect (**a**) $E(\hat{\beta}_1)$? (**b**) inference about β_1? Why?

3.29 Construct a Q–Q plot for the model for the house selling prices that uses size, new, and their interaction as the predictors, and interpret. To get a sense of how such a plot with a finite sample size may differ from its expected pattern when the model holds, randomly generate 100 standard normal variates a few times and form a Q-Q plot each time.

3.30 Suppose the relationship between y = college GPA and x = high school GPA satisfies $y_i \sim N(1.80 + 0.40x_i, 0.30^2)$. Simulate and construct a scatterplot for $n = 1000$ independent observations taken from this model when x_i has a uniform distribution (**a**) over (2.0, 4.0), (**b**) over (3.5, 4.0). In each case, find R^2. How do R^2 and corr$(\boldsymbol{x}, \boldsymbol{y})$ depend on the range sampled for $\{x_i\}$? Use the formula for R^2 to explain why this happens.

3.31 Refer to Exercise 1.21 on a study comparing forced expiratory volume (y = *fev*1 in the data file) for three drugs (x_2), adjusting for a baseline measurement (x_1).

a. Fit the normal linear model using both x_1 and x_2 and their interaction. Interpret model parameter estimates.

b. Test to see whether the interaction terms are needed. Interpret using confidence intervals for parameters in your chosen model.

3.32 For the horseshoe crab dataset `Crabs.dat` at the text website, analyze inferentially the effect of color on the mean number of satellites, treating the data as a random sample from a conceptual population of female crabs. Fit the normal one-way ANOVA model using color as a qualitative factor. Report results of the significance test for the color effect, and interpret. Provide evidence that the inferential assumption of a normal response with constant variance is badly violated. (Section 7.5 considers more appropriate models.)

3.33 Refer to Exercise 2.47 on carapace width of attached male horseshoe crabs. Extend your analysis of that exercise by conducting statistical inference, and interpret.

3.34 Section 3.4.1 used x_1 = size of house and x_2 = whether new to predict y = selling price. Suppose we instead use a GLM, $\log(\mu_i) = \beta_0 + \beta_1 \log(x_{i1}) + \beta_2 x_{i2}$.

a. For this GLM, interpret β_1 and β_2. (*Hint*: Adjusting for the other variable, find multiplicative effects on μ_i of (i) changing x_{i2} from 0 to 1, (ii) increasing x_{i1} by 1%.)

b. Fit the GLM, assuming normality for $\{y_i\}$, and interpret. Compare the predictive power of this model with the linear model of Section 3.4.1 by finding $R = \text{corr}(\boldsymbol{y}, \hat{\boldsymbol{\mu}})$ for each model.

c. For this GLM or the corresponding LM for $E[\log(y_i)]$, refit the model without the most influential observation and summarize. Also, determine whether the fit improves significantly by permitting interaction between $\log(x_{i1})$ and x_{i2}.

3.35 For the house selling price data of Section 3.4, when we include size, new, and taxes as explanatory variables, we obtain

```
-----------------------------------------------------------------
> summary(lm(price ~ size + new + taxes))
              Estimate  Std. Error  t value   Pr(>|t|)
(Intercept)   -21.3538     13.3115   -1.604    0.11196
size            0.0617      0.0125    4.937   3.35e-06
new            46.3737     16.4590    2.818    0.00588
taxes           0.0372      0.0067    5.528   2.78e-07
---
Residual standard error: 47.17 on 96 degrees of freedom
Multiple R-squared: 0.7896,   Adjusted R-squared: 0.783
F-statistic: 120.1 on 3 and 96 DF, p-value: < 2.2e-16
> anova(lm(price ~ size + new + taxes)) # sequential SS, size first
Analysis of Variance Table
Response: price
          Df  Sum Sq  Mean Sq  F value      Pr(>F)
```

```
size         1  705729   705729  317.165  < 2.2e-16
new          1   27814    27814   12.500  0.0006283
taxes        1   67995    67995   30.558  2.782e-07
Residuals 96   213611     2225
----------------------------------------------------------------
```

a. Report and interpret results of the global test of the hypothesis that none of the explanatory variables has an effect.

b. Report and interpret significance tests for the individual partial effects, adjusting for the other variables in the model.

c. What is the conceptual difference between the test of the size effect in the coefficients table and in the ANOVA table?

3.36 Using the house selling price data at the text website, describe the predictive power of various models by finding adjusted R^2 when (i) size is the sole predictor, (ii) size and new are main-effect predictors, (iii) size, new, and taxes are main-effect predictors, (iv) case (iii) with the addition of the three two-way interaction terms. Of these four, which is the simplest model that seems adequate? Why?

3.37 For the house selling price data, fit the model with size of home as the sole explanatory variable. Find a 95% confidence interval for $E(y)$ and a 95% prediction interval for y, at the sample mean size. Interpret.

3.38 In a study[11] at Iowa State University, a large field was partitioned into 20 equal-size plots. Each plot was planted with the same amount of seed corn, using a fixed spacing pattern between the seeds. The goal was to study how the yield of corn later harvested from the plots depended on the levels of use of nitrogen-based fertilizer (low = 45 kg per hectare, high = 135 kg per hectare) and manure (low = 84 kg per hectare, high = 168 kg per hectare). The corn yields (in metric tons) for this completely randomized two-factor study are shown in the table:

Fertilizer	Manure	Observations, by Plot				
High	High	13.7	15.8	13.9	16.6	15.5
High	Low	16.4	12.5	14.1	14.4	12.2
Low	High	15.0	15.1	12.0	15.7	12.2
Low	Low	12.4	10.6	13.7	8.7	10.9

a. Conduct a two-way ANOVA, assuming a lack of interaction between fertilizer level and manure level in their effects on crop yield. Report the ANOVA table. Summarize the main effect tests, and interpret the P-values.

[11] Thanks to Dan Nettleton, Iowa State University, for data on which this exercise is based.

b. If yield were instead measured in some other units, such as pounds or tons, then in your ANOVA table, what will change and what will stay the same?

c. Follow up the main-effect tests in (**a**) by forming 95% Bonferroni confidence intervals for the two main-effect comparisons of means. Interpret.

d. Now allow for interaction, and show results of the *F* test of the hypothesis of a lack of interaction. Interpret.

3.39 Refer to the study for comparing instruction methods mentioned in Exercise 2.45. Write a short report summarizing inference for the model fitted there, interpreting results and attaching edited software output as an appendix.

3.40 For the `Student survey.dat` data file at the text website, model how political ideology relates to number of times per week of newspaper reading and religiosity. Prepare a report, posing a research question, and then summarizing your graphical analyses, models and interpretations, inferences, checks of assumptions, and overall summary of the relationships.

3.41 For the anorexia study of Exercise 1.24, write a report in which you pose a research question and then summarize your analyses, including graphical description, interpretation of a model fit and its inferences, and checks of assumptions.

CHAPTER 4

Generalized Linear Models: Model Fitting and Inference

We now extend our scope from the linear model to the *generalized linear model* (GLM). This extension encompasses (1) non-normal response distributions and (2) link functions of the mean equated to the linear predictor. Section 1.1.5 introduced examples of GLMs: *Loglinear models* using the log-link function for a Poisson (count) response and *logistic models* using the logit-link function for a binomial (binary) response.

Section 4.1 provides more details about exponential family distributions for the random component of a GLM. In Section 4.2 we derive likelihood equations for the maximum likelihood (ML) estimators of model parameters and show their large-sample normal distribution. Section 4.3 summarizes the likelihood ratio, score, and Wald inference methods for the model parameters. Then in Section 4.4 we introduce the *deviance*, a generalization of the residual sum of squares used in inference, such as to compare nested GLMs. That section also presents residuals for GLMs and ways of checking the model. Section 4.5 presents two standard methods, *Newton–Raphson* and *Fisher scoring*, for solving the likelihood equations to fit GLMs. Section 4.6 discusses the selection of explanatory variables for a model, followed by an example. A chapter appendix shows that fundamental results for linear models about orthogonality of fitted values and residuals do not hold exactly for GLMs, but analogs hold for an adjusted, weighted version of the response variable that satisfies a linear model with approximately constant variance.

4.1 EXPONENTIAL DISPERSION FAMILY DISTRIBUTIONS FOR A GLM

In Section 1.1 we introduced the three components of a GLM: (1) random component, (2) linear predictor, (3) link function. We now take a closer look at the random

Foundations of Linear and Generalized Linear Models, First Edition. Alan Agresti.

component, showing an exponential family form that encompasses standard distributions such as the normal, Poisson, and binomial and that has general expressions for moments and for likelihood equations.

4.1.1 Exponential Dispersion Family for a Random Component

The *random component* of a GLM consists of a response variable y with independent observations $(y_1, \dots, y_n)$ from a distribution having probability density or mass function for y_i of the form

$$f(y_i; \theta_i, \phi) = \exp\{[y_i\theta_i - b(\theta_i)]/a(\phi) + c(y_i, \phi)\}. \tag{4.1}$$

This is called the *exponential dispersion family*. The parameter θ_i is called the *natural parameter*, and ϕ is called the *dispersion parameter*. Often $a(\phi) = 1$ and $c(y_i, \phi) = c(y_i)$, giving the *natural exponential family* of the form $f(y_i; \theta_i) = h(y_i)\exp[y_i\theta_i - b(\theta_i)]$. Otherwise, usually $a(\phi)$ has the form $a(\phi) = \phi$ or $a(\phi) = \phi/\omega_i$ for $\phi > 0$ and a known weight ω_i. For instance, when y_i is a mean of n_i independent readings, $\omega_i = n_i$. Various choices for the functions $b(\cdot)$ and $a(\cdot)$ give rise to different distributions.

Expressions for $E(y_i)$ and $\text{var}(y_i)$ use quantities in (4.1). Let $L_i = \log f(y_i; \theta_i, \phi)$ denote the contribution of y_i to the log-likelihood function, $L = \sum_i L_i$. Since

$$L_i = [y_i\theta_i - b(\theta_i)]/a(\phi) + c(y_i, \phi), \tag{4.2}$$

$$\partial L_i/\partial\theta_i = \left[y_i - b'(\theta_i)\right]/a(\phi), \quad \partial^2 L_i/\partial\theta_i^2 = -b''(\theta_i)/a(\phi),$$

where $b'(\theta_i)$ and $b''(\theta_i)$ denote the first two derivatives of $b(\cdot)$ evaluated at θ_i. We now apply the general likelihood results

$$E\left(\frac{\partial L}{\partial\theta}\right) = 0 \quad \text{and} \quad -E\left(\frac{\partial^2 L}{\partial\theta^2}\right) = E\left(\frac{\partial L}{\partial\theta}\right)^2,$$

which hold under regularity conditions satisfied by the exponential dispersion family. From the first formula applied with a single observation,

$$E\left[y_i - b'(\theta_i)\right]/a(\phi) = 0, \quad \text{so that} \quad \mu_i = E(y_i) = b'(\theta_i). \tag{4.3}$$

From the second formula,

$$b''(\theta_i)/a(\phi) = E\left[(y_i - b'(\theta_i))/a(\phi)\right]^2 = \text{var}(y_i)/[a(\phi)]^2,$$

so that

$$\text{var}(y_i) = b''(\theta_i)a(\phi). \tag{4.4}$$

In summary, the function $b(\cdot)$ in (4.1) determines moments of y_i. This function is called the *cumulant function*, because when $a(\phi) = 1$ its derivatives yield the cumulants[1] of the distribution.

4.1.2 Poisson, Binomial, and Normal in Exponential Dispersion Family

We illustrate the exponential dispersion family by showing its representations for Poisson, binomial, and normal distributions. We then evaluate the mean and variance expressions for these cases.

When y_i has a Poisson distribution, the probability mass function is

$$\begin{aligned} f(y_i; \mu_i) &= \frac{e^{-\mu_i}\mu_i^{y_i}}{y_i!} = \exp[y_i \log \mu_i - \mu_i - \log(y_i!)] \\ &= \exp[y_i\theta_i - \exp(\theta_i) - \log(y_i!)], \quad y_i = 0, 1, 2, \ldots, \end{aligned} \tag{4.5}$$

where the natural parameter $\theta_i = \log \mu_i$. This has exponential dispersion form (4.1) with $b(\theta_i) = \exp(\theta_i)$, $a(\phi) = 1$, and $c(y_i, \phi) = -\log(y_i!)$. By (4.3) and (4.4),

$$\begin{aligned} E(y_i) &= b'(\theta_i) = \exp(\theta_i) = \mu_i, \\ \text{var}(y_i) &= b''(\theta_i) = \exp(\theta_i) = \mu_i. \end{aligned}$$

Next, suppose that $n_i y_i$ has a bin(n_i, π_i) distribution; that is, here y_i is the sample *proportion* (rather than *number*) of successes, so $E(y_i) = \pi_i$ does not depend on n_i. Let $\theta_i = \log[\pi_i/(1 - \pi_i)]$. Then $\pi_i = \exp(\theta_i)/[1 + \exp(\theta_i)]$ and $\log(1 - \pi_i) = -\log[1 + \exp(\theta_i)]$. We can express

$$\begin{aligned} f(y_i; \pi_i, n_i) &= \binom{n_i}{n_i y_i} \pi_i^{n_i y_i}(1 - \pi_i)^{n_i - n_i y_i}, \quad y_i = 0, \frac{1}{n_i}, \frac{2}{n_i}, \ldots, 1, \\ &= \exp\left[\frac{y_i\theta_i - \log[1 + \exp(\theta_i)]}{1/n_i} + \log\binom{n_i}{n_i y_i}\right]. \end{aligned} \tag{4.6}$$

This has exponential dispersion form (4.1) with $b(\theta_i) = \log[1 + \exp(\theta_i)]$, $a(\phi) = 1/n_i$, and $c(y_i, \phi) = \log\binom{n_i}{n_i y_i}$. The natural parameter is $\theta_i = \log[\pi_i/(1 - \pi_i)]$, the *logit*. By (4.3) and (4.4),

$$\begin{aligned} E(y_i) &= b'(\theta_i) = \exp(\theta_i)/[1 + \exp(\theta_i)] = \pi_i, \\ \text{var}(y_i) &= b''(\theta_i)a(\phi) = \exp(\theta_i)/\{[1 + \exp(\theta_i)]^2 n_i\} = \pi_i(1 - \pi_i)/n_i. \end{aligned}$$

[1]Recall that cumulants $\{\kappa_n\}$ are coefficients in a power series expansion of the log mgf, $\log[E(e^{ty})] = \sum_{n=1}^{\infty} \kappa_n t^n/n!$. The moments determine the cumulants, and vice versa.

For the normal distribution, observation i has probability density function

$$f(y_i; \mu, \sigma^2) = \frac{1}{\sqrt{2\pi\sigma}} \exp\left[-\frac{(y_i - \mu_i)^2}{2\sigma^2}\right]$$

$$= \exp\left[\frac{y_i\mu_i - \frac{1}{2}\mu_i^2}{\sigma^2} - \frac{1}{2}\log(2\pi\sigma^2) - \frac{y_i^2}{2\sigma^2}\right].$$

This satisfies the exponential dispersion family (4.1) with natural parameter $\theta_i = \mu_i$ and

$$b(\theta_i) = \frac{1}{2}\mu_i^2 = \frac{1}{2}\theta_i^2, \quad a(\phi) = \sigma^2, \quad c(y_i; \phi) = -\frac{1}{2}\log(2\pi\sigma^2) - \frac{y_i^2}{2\sigma^2}.$$

Then

$$E(y_i) = b'(\theta_i) = \theta_i = \mu_i \quad \text{and} \quad \text{var}\left(y_i\right) = b''(\theta_i)a(\phi) = \sigma^2.$$

4.1.3 The Canonical Link Function of a Generalized Linear Model

The *link function* of a GLM connects the random component and the linear predictor. That is, a GLM states that a linear predictor $\eta_i = \sum_{j=1}^{p} \beta_j x_{ij}$ relates to μ_i by $\eta_i = g(\mu_i)$, for a link function g. Equivalently, the *response function* g^{-1} maps linear predictor values to the mean.

The link function g that transforms the mean μ_i to the natural parameter θ_i in (4.1) is called the *canonical link*. For it, the direct relationship

$$\theta_i = \sum_{j=1}^{p} \beta_j x_{ij}$$

equates the natural parameter to the linear predictor. From the exponential dispersion family expressions just derived, the canonical link functions are the log link for the Poisson distribution, the logit link for the binomial distribution, and the identity link for the normal distribution. Section 4.5.5 shows special results that apply for GLMs that use the canonical link function.

4.2 LIKELIHOOD AND ASYMPTOTIC DISTRIBUTIONS FOR GLMS

We next obtain general expressions for likelihood equations and asymptotic distributions of ML parameter estimators for GLMs. For n independent observations, from (4.2) the log likelihood is

$$L(\boldsymbol{\beta}) = \sum_{i=1}^{n} L_i = \sum_{i=1}^{n} \log f(y_i; \theta_i, \phi) = \sum_{i=1}^{n} \frac{y_i\theta_i - b(\theta_i)}{a(\phi)} + \sum_{i=1}^{n} c(y_i, \phi). \quad (4.7)$$

The notation $L(\boldsymbol{\beta})$ reflects the dependence of $\boldsymbol{\theta}$ on the model parameters $\boldsymbol{\beta}$. For the canonical link function, $\theta_i = \sum_j \beta_j x_{ij}$, so when $a(\phi)$ is a fixed constant, the part of the log likelihood involving both the data and the model parameters is

$$\sum_{i=1}^{n} y_i \left(\sum_{j=1}^{p} \beta_j x_{ij} \right) = \sum_{j=1}^{p} \beta_j \left(\sum_{i=1}^{n} y_i x_{ij} \right).$$

Then the sufficient statistics for $\{\beta_j\}$ are $\{\sum_{i=1}^{n} y_i x_{ij},\ j = 1, \ldots, p\}$.

4.2.1 Likelihood Equations for a GLM

For a GLM $\eta_i = \sum_j \beta_j x_{ij} = g(\mu_i)$ with link function g, the likelihood equations are

$$\partial L(\boldsymbol{\beta})/\partial \beta_j = \sum_{i=1}^{n} \partial L_i/\partial \beta_j = 0, \quad \text{for all } j.$$

To differentiate the log likelihood (4.7), we use the chain rule,

$$\frac{\partial L_i}{\partial \beta_j} = \frac{\partial L_i}{\partial \theta_i} \frac{\partial \theta_i}{\partial \mu_i} \frac{\partial \mu_i}{\partial \eta_i} \frac{\partial \eta_i}{\partial \beta_j}. \tag{4.8}$$

Since $\partial L_i/\partial \theta_i = [y_i - b'(\theta_i)]/a(\phi)$, and since $\mu_i = b'(\theta_i)$ and $\text{var}(y_i) = b''(\theta_i)a(\phi)$ from (4.3) and (4.4),

$$\partial L_i/\partial \theta_i = (y_i - \mu_i)/a(\phi), \quad \partial \mu_i/\partial \theta_i = b''(\theta_i) = \text{var}(y_i)/a(\phi).$$

Also, since $\eta_i = \sum_{j=1}^{p} \beta_j x_{ij}$, $\partial \eta_i/\partial \beta_j = x_{ij}$. Finally, since $\eta_i = g(\mu_i)$, $\partial \mu_i/\partial \eta_i$ depends on the link function for the model. In summary, substituting into (4.8) gives us

$$\begin{aligned}\frac{\partial L_i}{\partial \beta_j} &= \frac{\partial L_i}{\partial \theta_i} \frac{\partial \theta_i}{\partial \mu_i} \frac{\partial \mu_i}{\partial \eta_i} \frac{\partial \eta_i}{\partial \beta_j} \\ &= \frac{(y_i - \mu_i)}{a(\phi)} \frac{a(\phi)}{\text{var}(y_i)} \frac{\partial \mu_i}{\partial \eta_i} x_{ij} = \frac{(y_i - \mu_i)x_{ij}}{\text{var}(y_i)} \frac{\partial \mu_i}{\partial \eta_i}.\end{aligned} \tag{4.9}$$

Summing over the n observations yields the likelihood equations.

Likelihood equations for a GLM:

$$\frac{\partial L(\boldsymbol{\beta})}{\partial \beta_j} = \sum_{i=1}^{n} \frac{(y_i - \mu_i)x_{ij}}{\text{var}(y_i)} \frac{\partial \mu_i}{\partial \eta_i} = 0, \quad j = 1, 2, \ldots, p, \tag{4.10}$$

where $\eta_i = \sum_{j=1}^{p} \beta_j x_{ij} = g(\mu_i)$ for link function g.

Let $\boldsymbol{V}$ denote the diagonal matrix of variances of the observations, and let $\boldsymbol{D}$ denote the diagonal matrix with elements $\partial\mu_i/\partial\eta_i$. For the GLM expression $\boldsymbol{\eta} = \boldsymbol{X}\boldsymbol{\beta}$ with a model matrix $\boldsymbol{X}$, these likelihood equations have the form

$$\boldsymbol{X}^{\mathrm{T}}\boldsymbol{D}\boldsymbol{V}^{-1}(\boldsymbol{y} - \boldsymbol{\mu}) = \boldsymbol{0}. \tag{4.11}$$

Although $\boldsymbol{\beta}$ does not appear in these equations, it is there implicitly through $\boldsymbol{\mu}$, since $\mu_i = g^{-1}\left(\sum_{j=1}^{p} \beta_j x_{ij}\right)$. Different link functions yield different sets of equations. The likelihood equations are nonlinear functions of $\boldsymbol{\beta}$ that must be solved iteratively. We defer details to Section 4.5.

4.2.2 Likelihood Equations for Poisson Loglinear Model

For count data, one possible GLM assumes a Poisson random component and uses the log-link function. The *Poisson loglinear model* is $\log(\mu_i) = \sum_{j=1}^{p} \beta_j x_{ij}$. For the log link, $\eta_i = \log \mu_i$, so $\mu_i = \exp(\eta_i)$ and $\partial\mu_i/\partial\eta_i = \exp(\eta_i) = \mu_i$. Since $\mathrm{var}(y_i) = \mu_i$, the likelihood equations (4.10) simplify to

$$\sum_{i=1}^{n}(y_i - \mu_i)x_{ij} = 0, \quad j = 1, 2, \ldots, p. \tag{4.12}$$

These equate the sufficient statistics $\{\sum_i y_i x_{ij}\}$ for $\boldsymbol{\beta}$ to their expected values. Section 4.5.5 shows that these equations occur for GLMs that use the canonical link function.

4.2.3 The Key Role of the Mean–Variance Relation

Interestingly, the likelihood equations (4.10) depend on the distribution of y_i only through μ_i and $\mathrm{var}(y_i)$. The variance itself depends on the mean through a functional form[2]

$$\mathrm{var}(y_i) = v(\mu_i),$$

for some function v. For example, $v(\mu_i) = \mu_i$ for the Poisson, $v(\mu_i) = \mu_i(1 - \mu_i)/n_i$ for the binomial proportion, and $v(\mu_i) = \sigma^2$ (i.e., constant) for the normal.

When the distribution of y_i is in the exponential dispersion family, the relation between the mean and the variance characterizes[3] the distribution. For instance, if y_i has distribution in the exponential dispersion family and if $v(\mu_i) = \mu_i$, then necessarily y_i has the Poisson distribution.

4.2.4 Large-Sample Normal Distribution of Model Parameter Estimators

From a fundamental property of maximum likelihood, under standard regularity conditions[4], for large n the ML estimator $\hat{\boldsymbol{\beta}}$ of $\boldsymbol{\beta}$ for a GLM is efficient and has an

[2]We express the variance of y as $v(\mu)$ to emphasize that it is a function of the mean.

[3]See Jørgensen (1987), Tweedie (1947), and Wedderburn (1974).

[4]See Cox and Hinkley (1974, p. 281). Mainly, $\boldsymbol{\beta}$ falls in the interior of the parameter space and p is fixed as n increases.

approximate normal distribution. We next use the log-likelihood function for a GLM to find the covariance matrix of that distribution. The covariance matrix is the inverse of the information matrix $\mathcal{J}$, which has elements $E[-\partial^2 L(\boldsymbol{\beta})/\partial\beta_h\,\partial\beta_j]$. The estimator $\hat{\boldsymbol{\beta}}$ is more precise when the log-likelihood function has greater curvature at $\boldsymbol{\beta}$. To find the covariance matrix, for the contribution L_i to the log likelihood we use the helpful result

$$E\left(\frac{-\partial^2 L_i}{\partial\beta_h\,\partial\beta_j}\right) = E\left[\left(\frac{\partial L_i}{\partial\beta_h}\right)\left(\frac{\partial L_i}{\partial\beta_j}\right)\right],$$

which holds for distributions in the exponential dispersion family. Thus, using (4.9),

$$\begin{aligned} E\left(\frac{-\partial^2 L_i}{\partial\beta_h\,\partial\beta_j}\right) &= E\left[\frac{(y_i-\mu_i)x_{ih}}{\mathrm{var}(y_i)}\frac{\partial\mu_i}{\partial\eta_i}\frac{(y_i-\mu_i)x_{ij}}{\mathrm{var}(y_i)}\frac{\partial\mu_i}{\partial\eta_i}\right] \\ &= \frac{x_{ih}x_{ij}}{\mathrm{var}(y_i)}\left(\frac{\partial\mu_i}{\partial\eta_i}\right)^2. \end{aligned}$$

Since $L(\boldsymbol{\beta}) = \sum_{i=1}^n L_i$,

$$E\left(-\frac{\partial^2 L(\boldsymbol{\beta})}{\partial\beta_h\,\partial\beta_j}\right) = \sum_{i=1}^n \frac{x_{ih}x_{ij}}{\mathrm{var}(y_i)}\left(\frac{\partial\mu_i}{\partial\eta_i}\right)^2.$$

Let $\boldsymbol{W}$ be the diagonal matrix with main-diagonal elements

$$w_i = \frac{(\partial\mu_i/\partial\eta_i)^2}{\mathrm{var}(y_i)}.$$

Then, generalizing from the typical element of the information matrix to the entire matrix, with the model matrix $\boldsymbol{X}$,

$$\mathcal{J} = \boldsymbol{X}^{\mathrm{T}}\boldsymbol{W}\boldsymbol{X}. \tag{4.13}$$

The form of $\boldsymbol{W}$, and hence $\mathcal{J}$, depends on the link function g, since $\partial\eta_i/\partial\mu_i = g'(\mu_i)$. In summary,

Asymptotic distribution of $\hat{\boldsymbol{\beta}}$ for GLM $\boldsymbol{\eta} = \boldsymbol{X}\boldsymbol{\beta}$:

$$\hat{\boldsymbol{\beta}} \text{ has an approximate } N[\boldsymbol{\beta}, (\boldsymbol{X}^{\mathrm{T}}\boldsymbol{W}\boldsymbol{X})^{-1}] \text{ distribution,} \tag{4.14}$$

where $\boldsymbol{W}$ is the diagonal matrix with elements $w_i = (\partial\mu_i/\partial\eta_i)^2/\mathrm{var}(y_i)$.

The asymptotic covariance matrix is estimated by $\widehat{\text{var}}(\hat{\boldsymbol{\beta}}) = (\boldsymbol{X}^{\text{T}}\widehat{\boldsymbol{W}}\boldsymbol{X})^{-1}$, where $\widehat{\boldsymbol{W}}$ is $\boldsymbol{W}$ evaluated at $\hat{\boldsymbol{\beta}}$.

For example, the Poisson loglinear model has the GLM form

$$\log \boldsymbol{\mu} = \boldsymbol{X\beta}.$$

For this case, $\eta_i = \log(\mu_i)$, so $\partial\eta_i/\partial\mu_i = 1/\mu_i$. Thus,$w_i = (\partial\mu_i/\partial\eta_i)^2/\text{var}(y_i) = \mu_i$, and in the asymptotic covariance matrix (4.14) of $\hat{\boldsymbol{\beta}}$, $\boldsymbol{W}$ is the diagonal matrix with the elements of $\boldsymbol{\mu}$ on the main diagonal.

For some GLMs, the parameter vector partitions into the parameters $\boldsymbol{\beta}$ for the linear predictor and other parameters $\boldsymbol{\phi}$ (such as a dispersion parameter) needed to specify the model completely. Sometimes[5], $E(\partial^2 L/\partial\beta_j\partial\phi_k) = 0$ for each j and k. Similarly, the inverse of the expected information matrix has 0 elements connecting each β_j with each ϕ_k. Because this inverse is the asymptotic covariance matrix, $\hat{\boldsymbol{\beta}}$ and $\hat{\boldsymbol{\phi}}$ are then asymptotically independent. The parameters $\boldsymbol{\beta}$ and $\boldsymbol{\phi}$ are said to be *orthogonal*. This is the generalization to GLMs of the notion of orthogonal parameters for linear models (Cox and Reid 1987). For the exponential dispersion family (4.1), θ and ϕ are orthogonal parameters.

4.2.5 Delta Method Yields Covariance Matrix for Fitted Values

The estimated linear predictor relates to $\hat{\boldsymbol{\beta}}$ by $\hat{\boldsymbol{\eta}} = \boldsymbol{X}\hat{\boldsymbol{\beta}}$. Thus, for large samples, its covariance matrix

$$\text{var}(\hat{\boldsymbol{\eta}}) = \boldsymbol{X}\text{var}(\hat{\boldsymbol{\beta}})\boldsymbol{X}^{\text{T}} \approx \boldsymbol{X}(\boldsymbol{X}^{\text{T}}\boldsymbol{W}\boldsymbol{X})^{-1}\boldsymbol{X}^{\text{T}}.$$

We can obtain the asymptotic $\text{var}(\hat{\boldsymbol{\mu}})$ from $\text{var}(\hat{\boldsymbol{\eta}})$ by the *delta method*, which gives approximate variances using linearizations from a Taylor-series expansion. For example, in the univariate case with a smooth function h, the linearization $h(y) - h(\mu) \approx (y - \mu)h'(\mu)$, which holds for y near μ, implies that $\text{var}[h(y)] \approx [h'(\mu)]^2\text{var}(y)$ when $\text{var}(y)$ is small. For a vector $\boldsymbol{y}$ with covariance matrix $\boldsymbol{V}$ and a vector $\boldsymbol{h}(\boldsymbol{y}) = (h_1(\boldsymbol{y}), \dots, h_n(\boldsymbol{y}))^{\text{T}}$, let $(\partial\boldsymbol{h}/\partial\boldsymbol{\mu})$ denote the Jacobian matrix with entry in row i and column j equal to $\partial h_i(\boldsymbol{y})/\partial y_j$ evaluated at $\boldsymbol{y} = \boldsymbol{\mu}$. Then the delta method yields $\text{var}[\boldsymbol{h}(\boldsymbol{y})] \approx (\partial\boldsymbol{h}/\partial\boldsymbol{\mu})\boldsymbol{V}(\partial\boldsymbol{h}/\partial\boldsymbol{\mu})^{\text{T}}$. So, by the delta method, using the diagonal matrix $\boldsymbol{D}$ with elements $\partial\mu_i/\partial\eta_i$, for large samples the covariance matrix of the fitted values

$$\text{var}(\hat{\boldsymbol{\mu}}) \approx \boldsymbol{D}\text{var}(\hat{\boldsymbol{\eta}})\boldsymbol{D} \approx \boldsymbol{D}\boldsymbol{X}(\boldsymbol{X}^{\text{T}}\boldsymbol{W}\boldsymbol{X})^{-1}\boldsymbol{X}^{\text{T}}\boldsymbol{D}.$$

However, to obtain a confidence interval for μ_i when g is not the identity link, it is preferable to construct one for η_i and then apply the response function g^{-1} to the endpoints, thus avoiding the further delta method approximation.

[5]An example is the negative binomial GLM for counts in Section 7.3.3.

These results for $\hat{\eta}$ and $\hat{\mu}$ are based on those for $\hat{\boldsymbol{\beta}}$, for which the asymptotics refer to $n \to \infty$. However, $\hat{\eta}$ and $\hat{\mu}$ have length n. Asymptotics make more sense for them when n is fixed and each component is based on an increasing number of subunits, such that the observations themselves become approximately normal. One such example is a fixed number of binomial observations, in which the asymptotics refer to each binomial sample size $n_i \to \infty$. In another example, each observation is a Poisson cell count in a contingency table with fixed dimensions, and the asymptotics refer to each expected cell count growing. Such cases can be expressed as exponential dispersion families in which the dispersion parameter $a(\phi) = \phi/\omega_i$ has weight ω_i growing. This component-specific large-sample theory is called *small-dispersion asymptotics* (Jørgensen 1987). The covariance matrix formulas are also used in an approximate sense in the more standard asymptotic cases with large n.

4.2.6 Model Misspecification: Robustness of GLMs with Correct Mean

Like other ML estimators of a fixed-length parameter vector, $\hat{\boldsymbol{\beta}}$ is consistent (i.e., $\hat{\boldsymbol{\beta}} \xrightarrow{p} \boldsymbol{\beta}$ as $n \to \infty$). As n increases, $\boldsymbol{X}$ has more rows, the diagonal elements of the asymptotic covariance matrix $(\boldsymbol{X}^{\mathrm{T}}\boldsymbol{W}\boldsymbol{X})^{-1}$ of $\hat{\boldsymbol{\beta}}$ tend to be smaller, and $\hat{\boldsymbol{\beta}}$ tends to fall closer to $\boldsymbol{\beta}$.

But what if we have misspecified the probability distribution for $\boldsymbol{y}$? Models, such as GLMs, that assume a response distribution from an exponential family have a certain robustness property. If the model for the mean is correct, that is, if we have specified the link function and linear predictor correctly, then $\hat{\boldsymbol{\beta}}$ is still consistent[6] for $\boldsymbol{\beta}$. However, if the assumed variance function is incorrect (which is likely when the assumed distribution for $\boldsymbol{y}$ is incorrect), then so is the formula for var$(\hat{\boldsymbol{\beta}})$. Moreover, not knowing the actual distribution for $\boldsymbol{y}$, we would not know the correct expression for var$(\hat{\boldsymbol{\beta}})$. Section 8.3 discusses model misspecification issues and ways of dealing with it, including using the sample variability to help obtain a consistent estimator of the appropriate covariance matrix.

4.3 LIKELIHOOD-RATIO/WALD/SCORE METHODS OF INFERENCE FOR GLM PARAMETERS

Inference about GLMs has three standard ways to use the likelihood function. For a generic scalar model parameter β, we focus on tests[7] of H_0: $\beta = \beta_0$ against H_1: $\beta \neq \beta_0$. We then explain how to construct confidence intervals using those tests.

4.3.1 Likelihood-Ratio Tests

A general purpose significance test method uses the likelihood function through the ratio of (1) its value ℓ_0 at β_0, and (2) its maximum ℓ_1 over β values permitting H_0

[6]Gourieroux et al. (1984) proved this and showed the key role of the natural exponential family and a generalization that includes the exponential dispersion family.

[7]Here, β_0 denotes a particular null value, typically 0, not the intercept parameter.

or H_1 to be true. The ratio $\Lambda = \ell_0/\ell_1 \leq 1$, since ℓ_0 results from maximizing at a restricted β value. The *likelihood-ratio test statistic* is[8]

$$-2\log\Lambda = -2\log(\ell_0/\ell_1) = -2(L_0 - L_1),$$

where L_0 and L_1 denote the maximized log-likelihood functions. Under regularity conditions, it has a limiting null chi-squared distribution as $n \to \infty$, with $df = 1$. The P-value is the chi-squared probability above the observed test statistic value.

This test extends directly to multiple parameters. For instance, for $\boldsymbol{\beta} = (\boldsymbol{\beta}_0, \boldsymbol{\beta}_1)$, consider H_0: $\boldsymbol{\beta}_0 = \mathbf{0}$. Then ℓ_1 is the likelihood function calculated at the $\boldsymbol{\beta}$ value for which the data would have been most likely, and ℓ_0 is the likelihood function calculated at the $\boldsymbol{\beta}_1$ value for which the data would have been most likely when $\boldsymbol{\beta}_0 = \mathbf{0}$. The chi-squared df equal the difference in the dimensions of the parameter spaces under $H_0 \cup H_1$ and under H_0, which is $\dim(\boldsymbol{\beta}_0)$ when the model is parameterized to achieve identifiability. The test also extends to the general linear hypothesis H_0: $\boldsymbol{\Lambda\beta} = \mathbf{0}$, since the linear constraints imply a new model that is a special case of the original one.

4.3.2 Wald Tests

Standard errors obtained from the inverse of the information matrix depend on the unknown parameter values. When we substitute the unrestricted ML estimates (i.e., not assuming the null hypothesis), we obtain an *estimated* standard error (SE) of $\hat{\beta}$. For H_0: $\beta = \beta_0$, the test statistic using this non-null estimated standard error,

$$z = (\hat{\beta} - \beta_0)/SE,$$

is called[9] a *Wald statistic*. It has an approximate standard normal distribution when $\beta = \beta_0$, and z^2 has an approximate chi-squared distribution with $df = 1$.

For multiple parameters $\boldsymbol{\beta} = (\boldsymbol{\beta}_0, \boldsymbol{\beta}_1)$, to test H_0: $\boldsymbol{\beta}_0 = \mathbf{0}$, the Wald chi-squared statistic is

$$\hat{\boldsymbol{\beta}}_0^{\mathrm{T}}[\widehat{\mathrm{var}}(\hat{\boldsymbol{\beta}}_0)]^{-1}\hat{\boldsymbol{\beta}}_0,$$

where $\hat{\boldsymbol{\beta}}_0$ is the unrestricted ML estimate of $\boldsymbol{\beta}_0$ and $\widehat{\mathrm{var}}(\hat{\boldsymbol{\beta}}_0)$ is a block of the unrestricted estimated covariance matrix of $\hat{\boldsymbol{\beta}}$.

4.3.3 Score Tests

A third inference method uses the *score statistic*. The score test, referred to in some literature as the *Lagrange multiplier test*, uses the slope (i.e., the *score function*) and

[8]The general form was proposed by Samuel S. Wilks in 1938; see Cox and Hinkley (1974, pp. 313, 314, 322, 323) for a derivation of the chi-squared limit.
[9]The general form was proposed by Abraham Wald in 1943.

expected curvature of the log-likelihood function, evaluated at the null value β_0. The chi-squared form[10] of the score statistic is

$$\frac{[\partial L(\beta)/\partial \beta_0]^2}{-E[\partial^2 L(\beta)/\partial \beta_0^2]},$$

where the notation reflects derivatives with respect to β that are evaluated at β_0. In the multiparameter case, the score statistic is a quadratic form based on the vector of partial derivatives of the log likelihood and the inverse information matrix, both evaluated at the H_0 estimates.

4.3.4 Illustrating the Likelihood-Ratio, Wald, and Score Tests

Figure 4.1 plots a generic log-likelihood function $L(\beta)$ and illustrates the three tests of H_0: $\beta = \beta_0$, at $\beta_0 = 0$. The Wald test uses $L(\beta)$ at the ML estimate $\hat{\beta}$, having chi-squared form $(\hat{\beta}/SE)^2$ with SE of $\hat{\beta}$ based on the curvature of $L(\beta)$ at $\hat{\beta}$. The score test uses the slope and curvature of $L(\beta)$ at $\beta_0 = 0$. The likelihood-ratio test combines information about $L(\beta)$ at $\hat{\beta}$ and at $\beta_0 = 0$. In Figure 4.1, this statistic is twice the vertical distance between values of $L(\beta)$ at $\beta = \hat{\beta}$ and at $\beta = 0$.

To illustrate, consider a binomial parameter π and testing H_0: $\pi = \pi_0$. With sample proportion $\hat{\pi} = y$ for n observations, you can show that the chi-squared forms of the

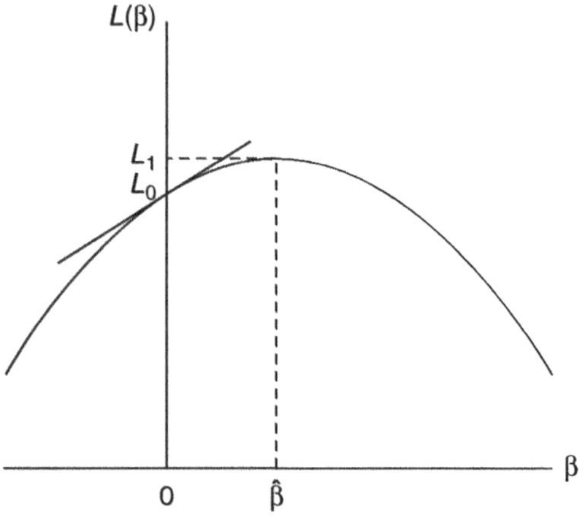

Figure 4.1 Log-likelihood function and information used in likelihood-ratio, score, and Wald tests of H_0: $\beta = 0$.

[10]The general form was proposed by C. R. Rao in 1948.

test statistics are

$$\text{Likelihood-ratio:} \quad -2(L_0 - L_1) = -2\log\left[\frac{\pi_0^{ny}(1-\pi_0)^{n(1-y)}}{y^{ny}(1-y)^{n(1-y)}}\right];$$

$$\text{Wald:} \quad z^2 = \frac{(y-\pi_0)^2}{[y(1-y)/n]};$$

$$\text{Score:} \quad z^2 = \frac{(y-\pi_0)^2}{[\pi_0(1-\pi_0)/n]}.$$

As $n \to \infty$, the three tests have certain asymptotic equivalences[11]. For the best-known GLM, the normal linear model, the three types of inference provide identical results. Unlike the other methods, though, we show in Section 5.3.3 that the results of the Wald test depend on the scale for the parameterization. Also, Wald inference is useless when an estimate or H_0 value is on the boundary of the parameter space. Examples are $\hat{\pi} = 0$ for a binomial and $\hat{\beta} = \infty$ in a GLM (not unusual in logistic regression).

4.3.5 Constructing Confidence Intervals by Inverting Tests

For any of the three test methods, we can construct a confidence interval by inverting the test. For instance, in the single-parameter case a 95% confidence interval for β is the set of β_0 for which the test of H_0: $\beta = \beta_0$ has P-value exceeding 0.05.

Let z_a denote the $(1-a)$ quantile of the standard normal distribution. A $100(1-\alpha)\%$ confidence interval based on asymptotic normality uses $z_{\alpha/2}$, for instance, $z_{0.025} = 1.96$ for 95% confidence. The Wald confidence interval is the set of β_0 for which $|\hat{\beta} - \beta_0|/SE < z_{\alpha/2}$. This gives the interval $\hat{\beta} \pm z_{\alpha/2}(SE)$. The score-test-based confidence interval often simplifies to the set of β_0 for which $|\hat{\beta} - \beta_0|/SE_0 < z_{\alpha/2}$, where SE_0 is the standard error estimated under the restriction that $\beta = \beta_0$. Let $\chi^2_d(a)$ denote the $(1-a)$ quantile of the chi-squared distribution with $df = d$. The likelihood-ratio-based confidence interval is the set of β_0 for which $-2[L(\beta_0) - L(\hat{\beta})] < \chi^2_1(\alpha)$. [Note that $\chi^2_1(\alpha) = z^2_{\alpha/2}$.]

When $\hat{\beta}$ has a normal distribution, the log-likelihood function is a second-degree polynomial and thus has a parabolic shape. For small samples of highly non-normal data or when β falls near the boundary of the parameter space, $\hat{\beta}$ may have distribution far from normality, and the log-likelihood function can be far from a symmetric, parabolic curve. A marked divergence in the results of Wald and likelihood-ratio inference indicates that the distribution of $\hat{\beta}$ may not be close to normality. It is then preferable to use the likelihood-ratio inference or higher order asymptotic methods[12].

[11]See, for example, Cox and Hinkley (1974, Section 9.3).

[12]For an introduction to higher-order asymptotics, see Brazzale et al. (2007).

4.3.6 Profile Likelihood Confidence Intervals

For confidence intervals for multiparameter models, especially useful is the *profile likelihood* approach. It is based on inverting likelihood-ratio tests for the various possible null values of β, regarding the other parameters $\boldsymbol{\psi}$ in the model as *nuisance parameters*. In inverting a likelihood-ratio test of H_0: $\beta = \beta_0$ to check whether β_0 belongs in the confidence interval, the ML estimate $\hat{\boldsymbol{\psi}}(\beta_0)$ of $\boldsymbol{\psi}$ that maximizes the likelihood under the null varies as β_0 does. The *profile log-likelihood function* is $L(\beta_0, \hat{\boldsymbol{\psi}}(\beta_0))$, viewed as a function of β_0. For each β_0 this function gives the maximum of the ordinary log-likelihood subject to the constraint $\beta = \beta_0$. Evaluated at $\beta_0 = \hat{\beta}$, this is the maximized log likelihood $L(\hat{\beta}, \hat{\boldsymbol{\psi}})$, which occurs at the unrestricted ML estimates. The *profile likelihood confidence interval* for β is the set of β_0 for which

$$-2[L(\beta_0, \hat{\boldsymbol{\psi}}(\beta_0)) - L(\hat{\beta}, \hat{\boldsymbol{\psi}})] < \chi_1^2(\alpha).$$

The interval contains all β_0 not rejected in likelihood-ratio tests of nominal size α. The profile likelihood interval is more complex to calculate than the Wald interval, but it is available in software[13].

4.4 DEVIANCE OF A GLM, MODEL COMPARISON, AND MODEL CHECKING

For a particular GLM with observations $\boldsymbol{y} = (y_1, \ldots, y_n)$, let $L(\boldsymbol{\mu}; \boldsymbol{y})$ denote the log-likelihood function expressed in terms of the means $\boldsymbol{\mu} = (\mu_1, \ldots, \mu_n)$. Let $L(\hat{\boldsymbol{\mu}}; \boldsymbol{y})$ denote the maximum of the log likelihood for the model. Considered for all possible models, the maximum achievable log likelihood is $L(\boldsymbol{y}; \boldsymbol{y})$. This occurs for the most general model, having a separate parameter for each observation and the perfect fit $\hat{\boldsymbol{\mu}} = \boldsymbol{y}$. This model is called the *saturated model*. It explains all variation by the linear predictor of the model. A perfect fit sounds good, but the saturated model is not a helpful one. It does not smooth the data or have the advantages that a simpler model has because of its parsimony, such as better estimation of the true relation. However, it often serves as a baseline for comparison with other model fits, such as for checking goodness of fit.

4.4.1 Deviance Compares Chosen Model with Saturated Model

For a chosen model, for all i denote the ML estimate of the natural parameter θ_i by $\hat{\theta}_i$, corresponding to the estimated mean $\hat{\mu}_i$. Let $\tilde{\theta}_i$ denote the estimate of θ_i for the saturated model, with corresponding $\tilde{\mu}_i = y_i$. For maximized log likelihoods $L(\hat{\boldsymbol{\mu}}; \boldsymbol{y})$ for the chosen model and $L(\boldsymbol{y}; \boldsymbol{y})$ for the saturated model,

$$-2 \log \left[\frac{\text{maximum likelihood for model}}{\text{maximum likelihood for saturated model}} \right] = -2[L(\hat{\boldsymbol{\mu}}; \boldsymbol{y}) - L(\boldsymbol{y}; \boldsymbol{y})]$$

[13]Examples are the `confint` function and `ProfileLikelihood` and `cond` packages in R.

is the likelihood-ratio statistic for testing H_0 that the model holds against H_1 that a more general model holds. It describes lack of fit. From (4.7),

$$-2[L(\hat{\boldsymbol{\mu}};\mathbf{y}) - L(\mathbf{y};\mathbf{y})]$$
$$= 2\sum_i [y_i\tilde{\theta}_i - b(\tilde{\theta}_i)]/a(\phi) - 2\sum_i [y_i\hat{\theta}_i - b(\hat{\theta}_i)]/a(\phi).$$

Usually $a(\phi) = \phi/\omega_i$, in which case this difference equals

$$2\sum_i \omega_i[y_i(\tilde{\theta}_i - \hat{\theta}_i) - b(\tilde{\theta}_i) + b(\hat{\theta}_i)]/\phi = D(\mathbf{y}; \hat{\boldsymbol{\mu}})/\phi, \tag{4.15}$$

called the *scaled deviance*. The statistic $D(\mathbf{y}; \hat{\boldsymbol{\mu}})$ is called the *deviance*.

Since $L(\hat{\boldsymbol{\mu}};\mathbf{y}) \leq L(\mathbf{y};\mathbf{y})$, $D(\mathbf{y}; \hat{\boldsymbol{\mu}}) \geq 0$. The greater the deviance, the poorer the fit. For some GLMs, such as binomial and Poisson GLMs under small-dispersion asymptotics in which the number of observations n is fixed and the individual observations converge to normality, the scaled deviance has an approximate chi-squared distribution. The *df* equal the difference between the numbers of parameters in the saturated model and in the chosen model. When ϕ is known, we use the scaled deviance for model checking. The main use of the deviance is for inferential comparisons of models (Section 4.4.3).

4.4.2 The Deviance for Poisson GLMs and Normal GLMs

For Poisson GLMs, from Section 4.1.2, $\hat{\theta}_i = \log\, \hat{\mu}_i$ and $b(\hat{\theta}_i) = \exp(\hat{\theta}_i) = \hat{\mu}_i$. Similarly, $\tilde{\theta}_i = \log\, y_i$ and $b(\tilde{\theta}_i) = y_i$ for the saturated model. Also $a(\phi) = 1$, so the deviance and scaled deviance (4.15) equal

$$D(\mathbf{y}; \hat{\boldsymbol{\mu}}) = 2\sum_i [y_i \log(y_i/\hat{\mu}_i) - y_i + \hat{\mu}_i].$$

When a model with log link contains an intercept term, the likelihood equation (4.12) implied by that parameter is $\sum_i y_i = \sum_i \hat{\mu}_i$. Then the deviance simplifies to

$$D(\mathbf{y}; \hat{\boldsymbol{\mu}}) = 2\sum_i y_i \log(y_i/\hat{\mu}_i). \tag{4.16}$$

For some applications with Poisson GLMs, such as modeling cell counts in contingency tables, the number n of counts is fixed. With p model parameters, as the expected counts grow the deviance converges in distribution to chi-squared with $df = n - p$. Chapter 7 shows that the deviance then provides a test of model fit.

For normal GLMs, by Section 4.1.2, $\hat{\theta}_i = \hat{\mu}_i$ and $b(\hat{\theta}_i) = \hat{\theta}_i^2/2$. Similarly, $\tilde{\theta}_i = y_i$ and $b(\tilde{\theta}_i) = y_i^2/2$ for the saturated model. So the deviance equals

$$D(\mathbf{y}; \hat{\boldsymbol{\mu}}) = 2\sum_i \left[y_i(y_i - \hat{\mu}_i) - \frac{y_i^2}{2} + \frac{\hat{\mu}_i^2}{2} \right] = \sum_i (y_i - \hat{\mu}_i)^2.$$

For linear models, this is the residual sum of squares, which we have denoted by SSE. Also $\phi = \sigma^2$, so the scaled deviance is $[\sum_i (y_i - \hat{\mu}_i)^2]/\sigma^2$. When the model holds, we have seen (Section 3.2.2, by Cochran's theorem) that this has a χ^2_{n-p} distribution.

For a particular GLM, *maximizing the likelihood corresponds to minimizing the deviance*. Using least squares to minimize SSE for a linear model generalizes to using ML to minimize a deviance for a GLM.

4.4.3 Likelihood-Ratio Model Comparison Uses Deviance Difference

Methods for comparing deviances generalize methods for normal linear models that compare residual sums of squares. When $\phi = 1$, such as for a Poisson or binomial model, the deviance (4.15) equals

$$D(\boldsymbol{y}; \hat{\boldsymbol{\mu}}) = -2[L(\hat{\boldsymbol{\mu}}; \boldsymbol{y}) - L(\boldsymbol{y}; \boldsymbol{y})].$$

Consider two nested models, M_0 with p_0 parameters and fitted values $\hat{\boldsymbol{\mu}}_0$ and M_1 with p_1 parameters and fitted values $\hat{\boldsymbol{\mu}}_1$, with M_0 a special case of M_1. Section 3.2.2 showed how to compare nested linear models. Since the parameter space for M_0 is contained in that for M_1, $L(\hat{\boldsymbol{\mu}}_0; \boldsymbol{y}) \leq L(\hat{\boldsymbol{\mu}}_1; \boldsymbol{y})$. Since $L(\boldsymbol{y}; \boldsymbol{y})$ is identical for each model,

$$D(\boldsymbol{y}; \hat{\boldsymbol{\mu}}_1) \leq D(\boldsymbol{y}; \hat{\boldsymbol{\mu}}_0).$$

Simpler models have larger deviances.

Assuming that model M_1 holds, the likelihood-ratio test of the hypothesis that M_0 holds uses the test statistic

$$\begin{aligned} -2[L(\hat{\boldsymbol{\mu}}_0; \boldsymbol{y}) - L(\hat{\boldsymbol{\mu}}_1; \boldsymbol{y})] &= -2[L(\hat{\boldsymbol{\mu}}_0; \boldsymbol{y}) - L(\boldsymbol{y}; \boldsymbol{y})] - \{-2[L(\hat{\boldsymbol{\mu}}_1; \boldsymbol{y}) - L(\boldsymbol{y}; \boldsymbol{y})]\} \\ &= D(\boldsymbol{y}; \hat{\boldsymbol{\mu}}_0) - D(\boldsymbol{y}; \hat{\boldsymbol{\mu}}_1), \end{aligned}$$

when $\phi = 1$. This statistic is large when M_0 fits poorly compared with M_1. In expression (4.15) for the deviance, since the terms involving the saturated model cancel,

$$D(\boldsymbol{y}; \hat{\boldsymbol{\mu}}_0) - D(\boldsymbol{y}; \hat{\boldsymbol{\mu}}_1) = 2\sum_i \omega_i[y_i(\hat{\theta}_{1i} - \hat{\theta}_{0i}) - b(\hat{\theta}_{1i}) + b(\hat{\theta}_{0i})].$$

This also has the form of the deviance. Under standard regularity conditions for which likelihood-ratio statistics have large-sample chi-squared distributions, this difference has approximately a chi-squared null distribution with $df = p_1 - p_0$.

For example, for a Poisson loglinear model with an intercept term, from expression (4.16) for the deviance, the difference in deviances uses the observed counts and the two sets of fitted values in the form

$$D(\boldsymbol{y}; \hat{\boldsymbol{\mu}}_0) - D(\boldsymbol{y}; \hat{\boldsymbol{\mu}}_1) = 2\sum_i y_i \log(\hat{\mu}_{1i}/\hat{\mu}_{0i}).$$

We denote the likelihood-ratio statistic for comparing nested models by $G^2(M_0 \mid M_1)$.

4.4.4 Score Tests and Pearson Statistics for Model Comparison

For GLMs having variance function $\text{var}(y_i) = v(\mu_i)$ with $\phi = 1$, the score statistic for comparing a chosen model with the saturated model is[14]

$$X^2 = \sum_i \frac{(y_i - \hat{\mu}_i)^2}{v(\hat{\mu}_i)}. \tag{4.17}$$

For Poisson y_i, for which $v(\hat{\mu}_i) = \hat{\mu}_i$, this has the form

$$\sum (\text{observed} - \text{fitted})^2/\text{fitted}.$$

This is known as the *Pearson chi-squared statistic*, because Karl Pearson introduced it in 1900 for testing various hypotheses using the chi-squared distribution, such as the hypothesis of independence in a two-way contingency table (Section 7.2.2). The generalized Pearson statistic (4.17) is an alternative to the deviance for testing the fit of certain GLMs.

For two nested models, a generalized Pearson statistic for comparing nested models is

$$X^2(M_0 \mid M_1) = \sum_i (\hat{\mu}_{1i} - \hat{\mu}_{0i})^2/v(\hat{\mu}_{0i}). \tag{4.18}$$

This is a quadratic approximation for $G^2(M_0 \mid M_1)$, with the same null asymptotic behavior. However, this is not the score statistic for comparing the models, which is more complex. See Note 4.4.

4.4.5 Residuals and Fitted Values Asymptotically Uncorrelated

Examining residuals helps us find where the fit of a GLM is poor or where unusual observations occur. As in ordinary linear models, we would like to exploit the decomposition

$$\boldsymbol{y} = \hat{\boldsymbol{\mu}} + (\boldsymbol{y} - \hat{\boldsymbol{\mu}}) \quad \text{(i.e., data = fit + residuals)}.$$

With GLMs, however, $\hat{\boldsymbol{\mu}}$ and $(\boldsymbol{y} - \hat{\boldsymbol{\mu}})$ are not orthogonal when we leave the simple linear model case of identity link with constant variance. Pythagoras's theorem does not apply, because maximizing the likelihood does not correspond to minimizing $\|\boldsymbol{y} - \hat{\boldsymbol{\mu}}\|$. With a nonlinear link function, although the space of linear predictor values $\boldsymbol{\eta}$ that satisfy a particular GLM is a linear vector space, the corresponding set of $\boldsymbol{\mu} = g^{-1}(\boldsymbol{\eta})$ values is not. Fundamental results for ordinary linear models about projections and orthogonality of fitted values and residuals do not hold exactly for GLMs.

[14]See Lovison (2005, 2014), Pregibon (1982), and Smyth (2003).

We next obtain an asymptotic covariance matrix for the residuals. From Section 4.2.4, $\boldsymbol{W} = \text{diag}\{(\partial\mu_i/\partial\eta_i)^2/\text{var}(y_i)\}$ and $\boldsymbol{D} = \text{diag}\{\partial\mu_i/\partial\eta_i\}$, so we can express the diagonal matrix $\boldsymbol{V} = \text{var}(\boldsymbol{y})$ as $\boldsymbol{V} = \boldsymbol{D}\boldsymbol{W}^{-1}\boldsymbol{D}$. For large n, if $\hat{\boldsymbol{\mu}}$ is approximately uncorrelated with $(\boldsymbol{y} - \hat{\boldsymbol{\mu}})$, then $\boldsymbol{V} \approx \text{var}(\hat{\boldsymbol{\mu}}) + \text{var}(\boldsymbol{y} - \hat{\boldsymbol{\mu}})$. Then, using the approximate expression for $\text{var}(\hat{\boldsymbol{\mu}})$ from Section 4.2.5 and $\boldsymbol{V}^{1/2} = \boldsymbol{D}\boldsymbol{W}^{-1/2}$,

$$\begin{aligned}\text{var}(\boldsymbol{y} - \hat{\boldsymbol{\mu}}) &\approx \boldsymbol{V} - \text{var}(\hat{\boldsymbol{\mu}}) \approx \boldsymbol{D}\boldsymbol{W}^{-1}\boldsymbol{D} - \boldsymbol{D}\boldsymbol{X}(\boldsymbol{X}^{\mathrm{T}}\boldsymbol{W}\boldsymbol{X})^{-1}\boldsymbol{X}^{\mathrm{T}}\boldsymbol{D}\\ &= \boldsymbol{D}\boldsymbol{W}^{-1/2}[\boldsymbol{I} - \boldsymbol{W}^{1/2}\boldsymbol{X}(\boldsymbol{X}^{\mathrm{T}}\boldsymbol{W}\boldsymbol{X})^{-1}\boldsymbol{X}^{\mathrm{T}}\boldsymbol{W}^{1/2}]\boldsymbol{W}^{-1/2}\boldsymbol{D}.\end{aligned}$$

This has the form $\boldsymbol{V}^{1/2}[\boldsymbol{I} - \boldsymbol{H}_W]\boldsymbol{V}^{1/2}$, where $\boldsymbol{I}$ is the identity matrix and

$$\boldsymbol{H}_W = \boldsymbol{W}^{1/2}\boldsymbol{X}(\boldsymbol{X}^{\mathrm{T}}\boldsymbol{W}\boldsymbol{X})^{-1}\boldsymbol{X}^{\mathrm{T}}\boldsymbol{W}^{1/2}. \tag{4.19}$$

You can verify that $\boldsymbol{H}_W$ is a projection matrix by showing it is symmetric and idempotent. McCullagh and Nelder (1989, p. 397) noted that it is approximately a hat matrix for standardized units of $\boldsymbol{y}$, with

$$\boldsymbol{H}_W\boldsymbol{V}^{-1/2}(\boldsymbol{y} - \boldsymbol{\mu}) \approx \boldsymbol{V}^{-1/2}(\hat{\boldsymbol{\mu}} - \boldsymbol{\mu}).$$

The chapter appendix shows that the estimate of $\boldsymbol{H}_W$ is also a type of hat matrix, applying to weighted versions of the response and the linear predictor.

So why is $(\boldsymbol{y} - \hat{\boldsymbol{\mu}})$ asymptotically uncorrelated with $\hat{\boldsymbol{\mu}}$, thus generalizing the exact orthogonal decomposition for linear models? Lovison (2014) gave an argument that seems relevant for small-dispersion asymptotic cases in which "large samples" refer to the individual components, such as binomial indices. If $(\boldsymbol{y} - \hat{\boldsymbol{\mu}})$ and $\hat{\boldsymbol{\mu}}$ were not approximately uncorrelated, one could construct an asymptotically unbiased estimator of $\boldsymbol{\mu}$ that is asymptotically more efficient than $\hat{\boldsymbol{\mu}}$ using $\hat{\boldsymbol{\mu}}^* = [\hat{\boldsymbol{\mu}} + \boldsymbol{L}(\boldsymbol{y} - \hat{\boldsymbol{\mu}})]$ for a matrix of constants $\boldsymbol{L}$. But this would contradict the ML estimator $\hat{\boldsymbol{\mu}}$ being asymptotically efficient. Such an argument is an asymptotic version for ML estimators of the one in the Gauss–Markov theorem (Section 2.7.1) that unbiased estimators other than the least squares estimator have difference from that estimator that is uncorrelated with it. The small-dispersion asymptotic setting applies for the discrete-data models we will present in the next three chapters for situations in which residuals are mainly useful, in which individual y_i have approximate normal distributions. Then $(\boldsymbol{y} - \boldsymbol{\mu})$ and $(\hat{\boldsymbol{\mu}} - \boldsymbol{\mu})$ jointly have an approximate normal distribution, as does their difference.

4.4.6 Pearson, Deviance, and Standardized Residuals for GLMs

For a particular model with variance function $v(\mu)$, the *Pearson residual* for observation y_i and its fitted value $\hat{\mu}_i$ is

$$\text{Pearson residual:} \quad e_i = \frac{y_i - \hat{\mu}_i}{\sqrt{v(\hat{\mu}_i)}}. \tag{4.20}$$

Their squared values sum to the generalized Pearson statistic (4.17). For instance, consider a Poisson GLM. The Pearson residual is

$$e_i = (y_i - \hat{\mu}_i)/\sqrt{\hat{\mu}_i},$$

and when $\{\mu_i\}$ are large and the model holds, e_i has an approximate normal distribution and $X^2 = \sum_i e_i^2$ has an approximate chi-squared distribution (Chapter 7). For a binomial GLM in which $n_i y_i$ has a bin(n_i, π_i) distribution, the Pearson residual is

$$e_i = (y_i - \hat{\pi}_i)/\sqrt{\hat{\pi}_i(1 - \hat{\pi}_i)/n_i},$$

and when $\{n_i\}$ are large, $X^2 = \sum_i e_i^2$ also has an approximate chi-squared distribution (Chapter 5). In these cases, such statistics are used in model goodness-of-fit tests.

In expression (4.15) for the deviance, let $D(\boldsymbol{y}; \hat{\boldsymbol{\mu}}) = \sum_i d_i$, where

$$d_i = 2\omega_i[y_i(\tilde{\theta}_i - \hat{\theta}_i) - b(\tilde{\theta}_i) + b(\hat{\theta}_i)].$$

The *deviance residual* is

$$\text{Deviance residual:} \quad \sqrt{d_i} \times \text{sign}(y_i - \hat{\mu}_i). \tag{4.21}$$

The sum of squares of these residuals equals the deviance.

To judge when a residual is "large" it is helpful to have residual values that, when the model holds, have means of 0 and variances of 1. However, Pearson and deviance residuals tend to have variance less than 1 because they compare y_i with the fitted mean $\hat{\mu}_i$ rather than the true mean μ_i. For example, the denominator of the Pearson residual estimates $[v(\mu_i)]^{1/2} = [\text{var}(y_i - \mu_i)]^{1/2}$ rather than $[\text{var}(y_i - \hat{\mu}_i)]^{1/2}$. The *standardized residual* divides each raw residual $(y_i - \hat{\mu}_i)$ by its standard error. From Section 4.4.5, $\text{var}(y_i - \hat{\mu}_i) \approx v(\mu_i)(1 - h_{ii})$, where h_{ii} is the diagonal element of the generalized hat matrix $\boldsymbol{H}_W$ for observation i, its *leverage*. Let $\hat{h}_{ii}$ denote the estimate of h_{ii}. Then, standardizing by dividing $y_i - \hat{\mu}_i$ by its estimated SE yields

$$\text{Standardized residual:} \quad r_i = \frac{y_i - \hat{\mu}_i}{\sqrt{v(\hat{\mu}_i)(1 - \hat{h}_{ii})}} = \frac{e_i}{\sqrt{1 - \hat{h}_{ii}}}. \tag{4.22}$$

For Poisson GLMs, for instance, $r_i = (y_i - \hat{\mu}_i)/\sqrt{\hat{\mu}_i(1 - \hat{h}_{ii})}$. Likewise, deviance residuals have standardized versions. They are most useful for small-dispersion asymptotic cases, such as for relatively large Poisson means and relatively large binomial indices. In such cases their model-based distribution is approximately standard normal.

To detect a model's lack of fit, any particular type of residual can be plotted against the component fitted values in $\hat{\boldsymbol{\mu}}$ and against each explanatory variable. As with the linear model, the fit could be quite different when we delete an observation that has a large standardized residual and a large leverage. The estimated leverages fall between 0 and 1 and sum to p. Unlike in ordinary linear models, the generalized hat matrix depends on the fit as well as on the model matrix, and points that have extreme values for the explanatory variables need not have high estimated leverage. To gauge influence, an analog of Cook's distance (2.11) uses both the standardized residuals and the estimated leverages, by $r_i^2[\hat{h}_{ii}/p(1-\hat{h}_{ii})]$.

4.5 FITTING GENERALIZED LINEAR MODELS

How do we find the ML estimator $\hat{\boldsymbol{\beta}}$ of GLM parameters? The likelihood equations (4.10) are usually nonlinear in $\hat{\boldsymbol{\beta}}$. We next describe a general purpose iterative method for solving nonlinear equations and apply it in two ways to determine the maximum of the likelihood function.

4.5.1 Newton–Raphson Method

The *Newton–Raphson method* iteratively solves nonlinear equations, for example, to determine the point at which a function takes its maximum. It begins with an initial approximation for the solution. It obtains a second approximation by approximating the function in a neighborhood of the initial approximation by a second-degree polynomial and then finding the location of that polynomial's maximum value. It then repeats this step to generate a sequence of approximations. These converge to the location of the maximum when the function is suitable and/or the initial approximation is good.

Mathematically, here is how the Newton–Raphson method determines the value $\hat{\boldsymbol{\beta}}$ at which a function $L(\boldsymbol{\beta})$ is maximized. Let

$$\boldsymbol{u} = \left(\frac{\partial L(\boldsymbol{\beta})}{\partial \beta_1}, \frac{\partial L(\boldsymbol{\beta})}{\partial \beta_2}, \ldots, \frac{\partial L(\boldsymbol{\beta})}{\partial \beta_p}\right)^{\mathrm{T}}.$$

Let $\boldsymbol{H}$ denote[15] the matrix having entries $h_{ab} = \partial^2 L(\boldsymbol{\beta})/\partial\beta_a\partial\beta_b$, called the *Hessian matrix*. Let $\boldsymbol{u}^{(t)}$ and $\boldsymbol{H}^{(t)}$ be $\boldsymbol{u}$ and $\boldsymbol{H}$ evaluated at $\boldsymbol{\beta}^{(t)}$, approximation t for $\hat{\boldsymbol{\beta}}$. Step t in the iterative process ($t = 0, 1, 2, \ldots$) approximates $L(\boldsymbol{\beta})$ near $\boldsymbol{\beta}^{(t)}$ by the terms up to the second order in its Taylor series expansion,

$$L(\boldsymbol{\beta}) \approx L(\boldsymbol{\beta}^{(t)}) + \boldsymbol{u}^{(t)T}(\boldsymbol{\beta} - \boldsymbol{\beta}^{(t)}) + \left(\tfrac{1}{2}\right)(\boldsymbol{\beta} - \boldsymbol{\beta}^{(t)})^{\mathrm{T}}\boldsymbol{H}^{(t)}(\boldsymbol{\beta} - \boldsymbol{\beta}^{(t)}).$$

Solving $\partial L(\boldsymbol{\beta})/\partial\boldsymbol{\beta} \approx \boldsymbol{u}^{(t)} + \boldsymbol{H}^{(t)}(\boldsymbol{\beta} - \boldsymbol{\beta}^{(t)}) = \boldsymbol{0}$ for $\boldsymbol{\beta}$ yields the next approximation,

$$\boldsymbol{\beta}^{(t+1)} = \boldsymbol{\beta}^{(t)} - (\boldsymbol{H}^{(t)})^{-1}\boldsymbol{u}^{(t)}, \tag{4.23}$$

assuming that $\boldsymbol{H}^{(t)}$ is nonsingular.

[15]Here, $\boldsymbol{H}$ is *not* the hat matrix; it is conventional to use $\boldsymbol{H}$ for a Hessian matrix.

Iterations proceed until changes in $L(\boldsymbol{\beta}^{(t)})$ between successive cycles are sufficiently small. The ML estimator is the limit of $\boldsymbol{\beta}^{(t)}$ as $t \to \infty$; however, this need not happen if $L(\boldsymbol{\beta})$ has other local maxima at which $\boldsymbol{u}(\boldsymbol{\beta}) = \boldsymbol{0}$. In that case, a good initial approximation is crucial. Figure 4.2 illustrates a cycle of the method, showing the parabolic (second-order) approximation at a given step.

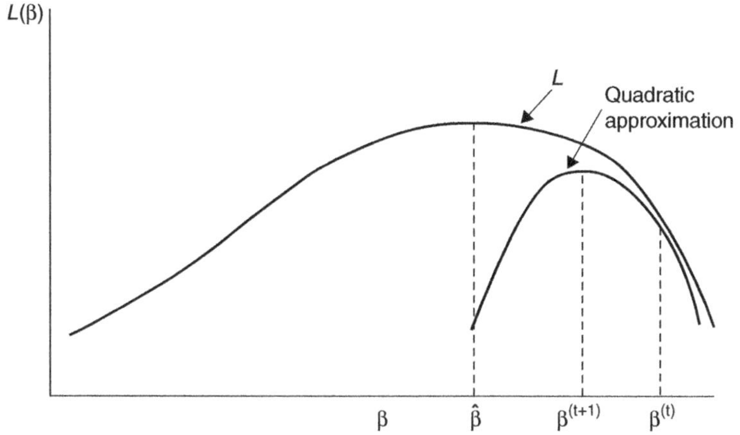

Figure 4.2 Illustration of a cycle of the Newton–Raphson method.

For many GLMs, including Poisson loglinear models and binomial logistic models, with full-rank model matrix the Hessian is negative definite, and the log likelihood is a strictly concave function. Then ML estimates of model parameters exist and are unique under quite general conditions[16]. The convergence of $\boldsymbol{\beta}^{(t)}$ to $\hat{\boldsymbol{\beta}}$ in the neighborhood of $\hat{\boldsymbol{\beta}}$ is then usually fast.

4.5.2 Fisher Scoring Method

Fisher scoring is an alternative iterative method for solving likelihood equations. The difference from Newton–Raphson is in the way it uses the Hessian matrix. Fisher scoring uses the *expected value* of this matrix, called the *expected information*, whereas Newton–Raphson uses the Hessian matrix itself, called the *observed information*.

Let $\boldsymbol{\mathcal{J}}^{(t)}$ denote approximation t for the ML estimate of the expected information matrix; that is, $\boldsymbol{\mathcal{J}}^{(t)}$ has elements $-E\,(\partial^2 L(\boldsymbol{\beta})/\partial\beta_a\,\partial\beta_b)$, evaluated at $\boldsymbol{\beta}^{(t)}$. The formula for Fisher scoring is

$$\boldsymbol{\beta}^{(t+1)} = \boldsymbol{\beta}^{(t)} + (\boldsymbol{\mathcal{J}}^{(t)})^{-1}\boldsymbol{u}^{(t)}, \quad \text{or} \quad \boldsymbol{\mathcal{J}}^{(t)}\boldsymbol{\beta}^{(t+1)} = \boldsymbol{\mathcal{J}}^{(t)}\boldsymbol{\beta}^{(t)} + \boldsymbol{u}^{(t)}. \tag{4.24}$$

Formula (4.13) showed that $\boldsymbol{\mathcal{J}} = \boldsymbol{X}^{\mathrm{T}}\boldsymbol{W}\boldsymbol{X}$, where $\boldsymbol{W}$ is diagonal with elements $w_i = (\partial\mu_i/\partial\eta_i)^2/\mathrm{var}(y_i)$. Similarly, $\boldsymbol{\mathcal{J}}^{(t)} = \boldsymbol{X}^{\mathrm{T}}\boldsymbol{W}^{(t)}\boldsymbol{X}$, where $\boldsymbol{W}^{(t)}$ is $\boldsymbol{W}$ evaluated at $\boldsymbol{\beta}^{(t)}$. The estimated asymptotic covariance matrix $\boldsymbol{\mathcal{J}}^{-1}$ of $\hat{\boldsymbol{\beta}}$ [see (4.14)] occurs as

[16]See, for example, Wedderburn (1976).

a by-product of this algorithm as $(\boldsymbol{\mathcal{J}}^{(t)})^{-1}$ for t at which convergence is adequate. For GLMs with a canonical link function, Section 4.5.5 shows that the observed and expected information are the same.

A simple way to begin either iterative process takes the initial estimate of $\boldsymbol{\mu}$ to be the data $\boldsymbol{y}$, smoothed to avoid boundary values. This determines the initial estimate of the weight matrix $\boldsymbol{W}$ and hence the initial approximation for $\hat{\boldsymbol{\beta}}$.

4.5.3 Newton–Raphson and Fisher Scoring for a Binomial Parameter

In the next three chapters we use the Newton–Raphson and Fisher scoring methods for models for categorical data and count data. We illustrate them here with a simpler problem for which we know the answer, maximizing the log likelihood with a sample proportion y from a bin(n, π) distribution. The log likelihood to be maximized is then $L(\pi) = \log[\pi^{ny}(1-\pi)^{n-ny}] = ny\log\pi + (n-ny)\log(1-\pi)$.

The first two derivatives of $L(\pi)$ are

$$u = (ny - n\pi)/\pi(1-\pi), \quad H = -[ny/\pi^2 + (n-ny)/(1-\pi)^2].$$

Each Newton–Raphson step has the form

$$\pi^{(t+1)} = \pi^{(t)} + \left[\frac{ny}{(\pi^{(t)})^2} + \frac{n-ny}{(1-\pi^{(t)})^2}\right]^{-1} \frac{ny - n\pi^{(t)}}{\pi^{(t)}(1-\pi^{(t)})}.$$

This adjusts $\pi^{(t)}$ up if $y > \pi^{(t)}$ and down if $y < \pi^{(t)}$. For instance, with $\pi^{(0)} = \frac{1}{2}$, you can check that $\pi^{(1)} = y$. When $\pi^{(t)} = y$, no adjustment occurs and $\pi^{(t+1)} = y$, which is the correct answer for $\hat{\pi}$. From the expectation of H above, the information is $n/[\pi(1-\pi)]$. A step of Fisher scoring gives

$$\begin{aligned}\pi^{(t+1)} &= \pi^{(t)} + \left[\frac{n}{\pi^{(t)}(1-\pi^{(t)})}\right]^{-1} \frac{ny - n\pi^{(t)}}{\pi^{(t)}(1-\pi^{(t)})} \\ &= \pi^{(t)} + (y - \pi^{(t)}) = y.\end{aligned}$$

This gives the correct answer for $\hat{\pi}$ after a single iteration and stays at that value for successive iterations.

4.5.4 ML as Iteratively Reweighted Least Squares

A relation exists between using Fisher scoring to find ML estimates and *weighted least squares estimation*. We refer here to the general linear model

$$z = \boldsymbol{X\beta} + \boldsymbol{\epsilon}.$$

When the covariance matrix of ϵ is V, from Section 2.7.2 the generalized least squares estimator of β is

$$(X^{\mathrm{T}}V^{-1}X)^{-1}X^{\mathrm{T}}V^{-1}z.$$

When V is diagonal, this is referred to as a *weighted least squares* estimator.

From (4.11), the score vector for a GLM is $X^{\mathrm{T}}DV^{-1}(y-\mu)$. Since $D = \mathrm{diag}\{\partial\mu_i/\partial\eta_i\}$ and $W = \mathrm{diag}\{(\partial\mu_i/\partial\eta_i)^2/\mathrm{var}(y_i)\}$, we have $DV^{-1} = WD^{-1}$ and we can express the score function as

$$u = X^{\mathrm{T}}WD^{-1}(y-\mu).$$

Since $\mathcal{J} = X^{\mathrm{T}}WX$, it follows that in the Fisher scoring formula (4.24),

$$\begin{aligned}\mathcal{J}^{(t)}\beta^{(t)} + u^{(t)} &= (X^{\mathrm{T}}W^{(t)}X)\beta^{(t)} + X^{\mathrm{T}}W^{(t)}(D^{(t)})^{-1}(y-\mu^{(t)})\\ &= X^{\mathrm{T}}W^{(t)}[X\beta^{(t)} + (D^{(t)})^{-1}(y-\mu^{(t)})] = X^{\mathrm{T}}W^{(t)}z^{(t)},\end{aligned}$$

where $z^{(t)}$ has elements

$$z_i^{(t)} = \sum_j x_{ij}\beta_j^{(t)} + \left(y_i - \mu_i^{(t)}\right)\frac{\partial\eta_i^{(t)}}{\partial\mu_i^{(t)}} = \eta_i^{(t)} + \left(y_i - \mu_i^{(t)}\right)\frac{\partial\eta_i^{(t)}}{\partial\mu_i^{(t)}}.$$

The Fisher scoring equations then have the form

$$(X^{\mathrm{T}}W^{(t)}X)\beta^{(t+1)} = X^{\mathrm{T}}W^{(t)}z^{(t)}.$$

These are the normal equations for using weighted least squares to fit a linear model for a response variable $z^{(t)}$, when the model matrix is X and the inverse of the covariance matrix is $W^{(t)}$. The equations have the solution

$$\beta^{(t+1)} = (X^{\mathrm{T}}W^{(t)}X)^{-1}X^{\mathrm{T}}W^{(t)}z^{(t)}.$$

The vector $z^{(t)}$ in this formulation is an estimated linearized form of the link function g, evaluated at y,

$$g(y_i) \approx g\left(\mu_i^{(t)}\right) + \left(y_i - \mu_i^{(t)}\right)g'\left(\mu_i^{(t)}\right) = \eta_i^{(t)} + \left(y_i - \mu_i^{(t)}\right)\frac{\partial\eta_i^{(t)}}{\partial\mu_i^{(t)}} = z_i^{(t)}. \qquad (4.25)$$

The *adjusted response variable* z has element i approximated by $z_i^{(t)}$ for cycle t of the iterative scheme. That cycle regresses $z^{(t)}$ on X with weight (i.e., inverse covariance) $W^{(t)}$ to obtain a new approximation $\beta^{(t+1)}$. This estimate yields a new linear predictor value $\eta^{(t+1)} = X\beta^{(t+1)}$ and a new approximation $z^{(t+1)}$ for the adjusted response for the next cycle. The ML estimator results from iterative use of weighted least squares, in which the weight matrix changes at each cycle. The process is called *iteratively reweighted least squares* (IRLS). The weight matrix $\boldsymbol{W}$ used in $\text{var}(\hat{\boldsymbol{\beta}}) \approx (\boldsymbol{X}^{\text{T}}\boldsymbol{W}\boldsymbol{X})^{-1}$, in the generalized hat matrix (4.19), and in Fisher scoring is the inverse covariance matrix of the linearized form $\boldsymbol{z} = \boldsymbol{X}\boldsymbol{\beta} + \boldsymbol{D}^{-1}(\boldsymbol{y} - \boldsymbol{\mu})$ of $g(\boldsymbol{y})$. At convergence,

$$\hat{\boldsymbol{\beta}} = (\boldsymbol{X}^{\text{T}}\widehat{\boldsymbol{W}}\boldsymbol{X})^{-1}\boldsymbol{X}^{\text{T}}\widehat{\boldsymbol{W}}\hat{\boldsymbol{z}},$$

for the estimated adjusted response $\hat{\boldsymbol{z}} = \boldsymbol{X}\hat{\boldsymbol{\beta}} + \widehat{\boldsymbol{D}}^{-1}(\boldsymbol{y} - \hat{\boldsymbol{\mu}})$.

4.5.5 Simplifications for Canonical Link Functions

Certain simplifications result for GLMs that use the canonical link function. For that link,

$$\eta_i = \theta_i = \sum_{j=1}^{p} \beta_j x_{ij},$$

and

$$\partial\mu_i/\partial\eta_i = \partial\mu_i/\partial\theta_i = \partial b'(\theta_i)/\partial\theta_i = b''(\theta_i).$$

Since $\text{var}(y_i) = b''(\theta_i)a(\phi)$, the contribution (4.9) to the likelihood equation for β_j simplifies to

$$\frac{\partial L_i}{\partial \beta_j} = \frac{(y_i - \mu_i)}{\text{var}(y_i)} b''(\theta_i)x_{ij} = \frac{(y_i - \mu_i)x_{ij}}{a(\phi)}. \tag{4.26}$$

Often $a(\phi)$ is identical for all observations, such as for Poisson GLMs $[a(\phi) = 1]$ and for binomial GLMs with each $n_i = 1$ [for which $a(\phi) = 1$]. Then, the likelihood equations are

$$\sum_{i=1}^{n} x_{ij}y_i = \sum_{i=1}^{n} x_{ij}\mu_i, \quad j = 1, 2, \ldots, p. \tag{4.27}$$

We noted at the beginning of Section 4.2 that $\{\sum_{i=1}^{n} x_{ij}y_i\}$ are the sufficient statistics for $\{\beta_j\}$. So equation (4.27) illustrates a fundamental result:

- For GLMs with canonical link function, the likelihood equations equate the sufficient statistics for the model parameters to their expected values.

For a normal distribution with identity link, these are the *normal equations*. We obtained them for Poisson loglinear models in (4.12).

From expression (4.26) for $\partial L_i/\partial \beta_j$, with the canonical link function the second partial derivatives of the log likelihood are

$$\frac{\partial^2 L_i}{\partial \beta_h \partial \beta_j} = -\frac{x_{ij}}{a(\phi)}\left(\frac{\partial \mu_i}{\partial \beta_h}\right).$$

This does not depend on y_i, so

$$\partial^2 L(\boldsymbol{\beta})/\partial \beta_h \partial \beta_j = E[\partial^2 L(\boldsymbol{\beta})/\partial \beta_h \partial \beta_j].$$

That is, $\boldsymbol{H} = -\boldsymbol{\mathcal{J}}$, so the Newton–Raphson and Fisher scoring algorithms are identical for GLMs that use the canonical link function (Nelder and Wedderburn 1972).

Finally, in the canonical link case the log likelihood is necessarily a concave function, because the log likelihood for an exponential family distribution is concave in the natural parameter. In using iterative methods to find the ML estimates, we do not need to worry about the possibility of multiple maxima for the log likelihood.

4.6 SELECTING EXPLANATORY VARIABLES FOR A GLM

Model selection for GLMs faces the same issues as for ordinary linear models. The selection process becomes more difficult as the number of explanatory variables increases, because of the rapid increase in possible effects and interactions. The selection process has two competing goals. The model should be complex enough to fit the data well. On the other hand, it should smooth rather than overfit the data and ideally be relatively simple to interpret.

Most research studies are designed to answer certain questions. Those questions guide the choice of model terms. Confirmatory analyses then use a restricted set of models. For instance, a study hypothesis about an effect may be tested by comparing models with and without that effect. For studies that are exploratory rather than confirmatory, a search among possible models may provide clues about the structure of effects and raise questions for future research. In either case, it is helpful first to study the marginal effect of each predictor by itself with descriptive statistics and a scatterplot matrix, to get a feel for those effects.

This section discusses some model-selection procedures and issues that affect the selection process. Section 4.7 presents an example and illustrates that the variables selected, and the influence of individual observations, can be highly sensitive to the assumed distribution for y.

4.6.1 Stepwise Procedures: Forward Selection and Backward Elimination

With p explanatory variables, the number of potential models is 2^p, as each variable either is or is not in the chosen model. The *best subset selection* identifies the model that performs best according to a criterion such as maximizing the adjusted R^2 value. This is computationally intensive when p is large. Alternative algorithmic methods can search among the models. In exploratory studies, such methods can be informative if we use the results cautiously.

Forward selection adds terms sequentially. At each stage it selects the term giving the greatest improvement in fit. A point of diminishing returns occurs in adding explanatory variables when new ones added are themselves so well predicted by ones already used that they do not provide a substantive improvement in R^2. The process stops when further additions do not improve the fit, according to statistical significance or a criterion for judging the model fit (such as the AIC, introduced below in Section 4.6.3). A stepwise variation of this procedure rechecks, at each stage, whether terms added at previous stages are still needed. *Backward elimination* begins with a complex model and sequentially removes terms. At each stage, it selects the term whose removal has the least damaging effect on the model, such as the largest P-value in a test of its significance or the least deterioration in a criterion for judging the model fit. The process stops when any further deletion leads to a poorer fit.

With either approach, an interaction term should not be in a model without its component main effects. Also, for qualitative predictors with more than two categories, the process should consider the entire variable at any stage rather than just individual indicator variables. Add or drop the entire variable rather than only one of its indicators. Otherwise, the result depends on the choice of reference category for the indicator coding.

Some statisticians prefer backward elimination over forward selection, feeling it safer to delete terms from an overly complex model than to add terms to an overly simple one. Forward selection based on significance testing can stop prematurely because a particular test in the sequence has low power. It also has the theoretical disadvantage that in early stages both models being compared are likely to be inadequate, making the basis for a significance test dubious. Neither strategy necessarily yields a meaningful model. When you evaluate many terms, some that are not truly important may seem so merely because of chance. For instance, when all the true effects are weak, the largest sample effect is likely to overestimate substantially its true effect. Also, the use of standard significance tests in the process lacks theoretical justification, because the distribution of the minimum or maximum P-value evaluated over a set of explanatory variables is not the same as that of a P-value for a preselected variable. Use variable-selection algorithms in an informal manner and with caution. Backward and forward selection procedures yielding quite different models is an indication that such results are of dubious value.

For any method, since statistical significance is not the same as practical significance, a significance test should not be the sole criterion for including a term in a model. It is sensible to include a variable that is central to the purposes of the study and report its estimated effect even if it is not statistically significant. Keeping it in the model may make it possible to compare results with other studies where the effect is significant, perhaps because of a larger sample size. If the variable is a potential confounder, including it in the model may help to reduce bias in estimating relevant effects of key explanatory variables. But also a variable should not be kept merely because it is statistically significant. For example, if a selection method results in a model having adjusted $R^2 = 0.39$ but a simpler model without the interaction terms has adjusted $R^2 = 0.38$, for ease of interpretation it may be preferable to drop the interaction terms. Algorithmic selection procedures are no substitute for careful thought in guiding the formulation of models.

Some variable-selection methods adapt stepwise procedures to take such issues into account. For example, Hosmer et al. (2013, Chapter 4) recommended a *purposeful selection* model-building process that also pays attention to potential confounding variables. In outline, they suggest constructing an initial main-effects model by (1) choosing a set of explanatory variables that include the known clinically important variables and others that show *any* evidence of being relevant predictors in a univariable analysis (e.g., having P-value <0.25), (2) conducting backward elimination with the full set from (1), keeping a variable if it is either significant at a somewhat more stringent level or shows evidence of being a relevant confounder, in the sense that the estimated effect of a key variable changes by at least 20% when it is removed, (3) checking whether any variables not included in (1) are significant when adjusting for the variables in the model after Step (2). One then checks for plausible interactions among variables in the model after Step (3), using significance tests at conventional levels such as 0.05, followed by the usual diagnostic investigations presented in Section 4.4.

4.6.2 Model Selection: The Bias–Variance Tradeoff

In selecting a model from a set of candidates, we are mistaken if we think that there is a "correct" one. Any model is a simplification of reality. For instance, an explanatory variable will not have exactly a linear effect, no matter which link function we use. And it is not always a good idea to choose a more complex model in order to obtain a better fit. A simple model that fits adequately has the advantages of model parsimony, including a tendency to provide more accurate estimates of the quantities of interest. The choice of how complex a model to use is at the heart of the basic statistical tradeoff between the variance of an estimator and its bias. Here, bias occurs when the true $\{E(y_i)\}$ values differ from the values $\{\mu_{Mi}\}$ corresponding to fitting model M to the population. Using a simpler model has the disadvantage of increasing the bias; that is, the differences $\{|\mu_{Mi} - E(y_i)|\}$ between the model-based means and the true means tend to be larger. But a simpler model has the advantage that the decrease in the number of model parameters results in decreased variance in the estimators. This can result in overall lower mean squared error[17] in estimating characteristics such as the true $\{E(y_i)\}$ values.

In practice, many models can be consistent with the data. If not one of them is "correct," it is logically inconsistent to choose one model based on its fitting the data well and then make subsequent inferences as if the model had been chosen before seeing the data. Although this is common practice, it results in a tendency to underestimate uncertainty and to exaggerate significance. Keep in mind the selection uncertainty in making inferences based on a model, because those inferences use the same data that helped you to select the model. Although selection procedures are

[17] Recall that MSE = variance + $(\text{bias})^2$.

helpful tools, results of an exploratory study are highly tentative and useful mainly for suggesting effects and hypotheses to analyze in future studies. The model-building process should also be based on theory and common sense.

Other criteria besides significance tests comparing models can help you to select a sensible model. We next introduce the best known of such criteria.

4.6.3 AIC: Minimizing Distance of the Fit from the Truth

The *Akaike information criterion* (AIC) judges a model by how close we can expect its sample fit to be to the true model fit. In the population of interest, even though a simple model is farther from the true relationship than is a more complex model, for a sample it may tend to provide a closer fit because of the advantages of model parsimony. In a set of potential models, the optimal model is the one that tends to have sample fit closest to the true model fit.

Here "closeness" is defined in terms of the *Kullback–Leibler divergence* of a model M from the unknown true model. Let $p(\mathbf{y})$ denote the density (or probability, in the discrete case) of the data under the true model, and let $p_M(\mathbf{y};\boldsymbol{\beta}_M)$ be the density under model M with parameters $\boldsymbol{\beta}_M$. For a given value of the ML estimator $\hat{\boldsymbol{\beta}}_M$ of $\boldsymbol{\beta}_M$ and for a future sample $\mathbf{y}^*$ from $p(\cdot)$, the Kullback–Leibler divergence between the true and fitted distributions is

$$KL[p, p_M(\hat{\boldsymbol{\beta}}_M)] = E\left[\log \frac{p(\mathbf{y}^*)}{p_M(\mathbf{y}^*;\hat{\boldsymbol{\beta}}_M)}\right],$$

where the expectation is taken relative to the true distribution $p(\cdot)$. The goal of AIC is to choose the model to minimize $E[KL(p, p_M(\hat{\boldsymbol{\beta}}_M))]$ for a set of potential models, where this expectation also is taken relative to $p(\cdot)$, now with $\hat{\boldsymbol{\beta}}_M$ as the random variable for another sample. To do this, it is sufficient to minimize $E\{-E\log[p_M(\mathbf{y}^*;\hat{\boldsymbol{\beta}}_M)]\}$ over the set of models. The true distribution $p(\cdot)$ needed to evaluate this expectation is unknown, but the expectation can be estimated consistently. Akaike (1973) showed that when M is reasonably close to the true model, the maximized log likelihood $L(\hat{\boldsymbol{\beta}}_M)$ for M is a biased estimator of $E\{E\log[p_M(\mathbf{y}^*;\hat{\boldsymbol{\beta}}_M)]\}$, and for large sample sizes the bias is reduced by subtracting the number of parameters in M. This implies that out of a set of reasonably fitting models, the optimal model minimizes[18]

$$\text{AIC} = -2\left[L(\hat{\boldsymbol{\beta}}_M) - \text{number of parameters in } M\right].$$

Although the role of subtracting the number of parameters in M is to adjust for bias, the AIC essentially penalizes a model for having many parameters. With many potential explanatory variables, using AIC can aid in variable selection. Out of a set of candidate models, we identify the one with smallest AIC or identify parsimonious

[18]Akaike introduced the multiple of 2 merely for convenience, to link the AIC formula with likelihood-ratio chi-squared statistics.

models that have AIC near the minimum value. The candidate models need not be nested or even based on the same family of distributions for the random component.

An alternative to AIC, a *Bayesian information criterion* (BIC), penalizes more severely for the number of model parameters. It replaces 2 by $\log(n)$ as its multiple. Compared with AIC, BIC gravitates less quickly toward more complex models as n increases. It is based on a Bayesian argument for determining which of a set of models has highest posterior probability (Schwarz 1978). Because of selection bias, however, model-selection criteria such as minimizing AIC or minimizing BIC can result in inclusion of irrelevant variables (George 2000). This can be especially problematic when p is large and few variables truly have an effect[19].

4.6.4 Summarizing Predictive Power: R-Squared and Other Measures

In ordinary linear models, R^2 and the multiple correlation R describe how well the explanatory variables predict the sample response values, with $R = 1$ for perfect prediction. For any GLM, the correlation between the fitted values $\{\hat{\mu}_i\}$ and the observed responses $\{y_i\}$ measures predictive power. It is also useful for comparing fits of different models for the same data. For the ordinary linear model, $\text{corr}(\boldsymbol{y}, \hat{\boldsymbol{\mu}})$ is the multiple correlation. An advantage of the correlation, relative to its square, is the appeal of working on the original scale and its approximate proportionality to effect size: For a small effect with a single explanatory variable, doubling the slope corresponds approximately to doubling the correlation. For GLMs, unlike linear models, $\text{corr}(\boldsymbol{y}, \hat{\boldsymbol{\mu}})$ need not be nondecreasing as the model gets more complex, although it usually is.

Other measures of predictive power directly use the likelihood function. Denote the maximized log likelihood by L_M for a given model, L_S for the saturated model, and L_0 for the null model containing only an intercept term. Then, $L_0 \le L_M \le L_S$, and

$$\frac{L_M - L_0}{L_S - L_0} \tag{4.28}$$

falls between 0 and 1. It equals 0 when the model provides no improvement in fit over the null model, and it equals 1 when the model fits as well as the saturated model. A weakness is that the scale for the log likelihood may not be as easy to interpret as the scale for the response variable itself. The measure is mainly useful for comparing models.

With any such measure, with many explanatory variables, the sample estimators can be biased upward in estimating the true population value. It can be misleading to compare sample values for models with quite different numbers of parameters. Bias corrections are possible, for example, by using cross-validation (Stone 1974) or the jackknife (Zheng and Agresti 2000).

[19]For example, when no variables truly have an effect, for t tests of the individual partial effects, $E(t^2_{\max}) \approx 2\log p$ (George 2000).

4.6.5 Effects of Collinearity

In an observational study with many explanatory variables, relations among them may suggest that not one variable is important when all the others are in the model. A variable may have little partial effect because it is predicted well by the others. Deleting a nearly redundant predictor can be helpful, for instance, to reduce standard errors of other estimated effects.

In a linear model, the variance of $\hat{\beta}_j$ is

$$\mathrm{var}(\hat{\beta}_j) = \frac{1}{1 - R_j^2}\left[\frac{\sigma^2}{\sum_i (x_{ij} - \bar{x}_j)^2}\right],$$

where R_j^2 denotes the value of R^2 for predicting x_j as a response using the other explanatory variables in the model. One can derive this formula from an expression of $\hat{\beta}_j$ for a regression using two sets of residuals, as in Section 2.5.6 (e.g., see Greene 2011, p. 90). The ratio $VIF_j = 1/(1 - R_j^2)$ is called the *variance inflation factor* for predictor x_j. It is the multiple by which the variance increases because the other predictors are correlated with x_j. As R_j^2 increases, $\mathrm{var}(\hat{\beta}_j)$ increases. If $R_j^2 = 1$, there is extrinsic aliasing (Section 1.3.2): The model matrix has less than full rank, and there are infinitely many solutions for $\hat{\boldsymbol{\beta}}$. When R_j^2 is near 1, $\hat{\beta}_j$ can be unstable. When $R_j^2 = 0$, $\hat{\beta}_j$ and its variance are identical to their values when x_j is the sole explanatory variable in the model.

To illustrate, for the horseshoe crab data (Section 1.5.1), the width of the carapace shell is highly statistically significant as a predictor of a female crab's number of satellites. What happens if we add the crab's weight as a predictor? Here is the result of fitting Poisson loglinear models:

```
-------------------------------------------------------------------
> attach(Crabs) # y is number of satellites
> summary(glm(y ~ width, family=poisson(link=log)))
            Estimate Std. Error z value Pr(>|z|)
(Intercept) -3.30476    0.54224  -6.095  1.1e-09
width        0.16405    0.01997   8.216  < 2e-16
----
> summary(glm(y ~ weight + width, family=poisson(link=log)))
            Estimate Std. Error z value Pr(>|z|)
(Intercept) -1.29521    0.89890  -1.441  0.14962
weight       0.44697    0.15862   2.818  0.00483
width        0.04608    0.04675   0.986  0.32433
----
> cor(weight, width)
[1] 0.8868715
-------------------------------------------------------------------
```

Width loses its significance. The loss also happens with normal linear models and with a more appropriate two-parameter distribution for count data that Chapter 7 uses. The dramatic reduction in the significance of the crab's shell width when its weight is added to the model reflects the correlation of 0.887 between weight and width. The variance inflation factor for the effect of either predictor in a linear model is $1/[1 - (0.887)^2] = 4.685$. The *SE* for the effect of width more than doubles when weight is added to the model, and the estimate itself is much smaller, reflecting also the strong correlation.

This example illustrates a general phenomenon in modeling. When an explanatory variable x_j is highly correlated with a linear combination of other explanatory variables in the model, the relation is said to exhibit[20] *collinearity* (also referred to as *multicollinearity*).

When collinearity exists, one approach chooses a subset of the explanatory variables, removing those variables that explain a small portion of the remaining unexplained variation in y. When several predictors are highly correlated and are indicators of a common feature, another approach constructs a summary index by combining responses on those variables. Also, methods such as *principal components analysis* create artificial variables from the original ones in such a way that the new variables are uncorrelated. In most applications, though, it is more advisable from an interpretive standpoint to use a subset of the variables or create some new variables directly. The effect of interaction terms on collinearity is diminished if we center the explanatory variables before entering them in the model. Section 11.1.2 introduces alternative methods, such as *ridge regression*, that produce estimates that are biased but less severely affected by collinearity.

Collinearity does not adversely affect all aspects of regression. Although collinearity makes it difficult to assess partial effects of explanatory variables, it does not hinder the assessment of their joint effects. If newly added explanatory variables overlap substantially with ones already in the model, R^2 will not increase much, but the presence of collinearity has little effect on the global test of significance.

4.7 EXAMPLE: BUILDING A GLM

Section 3.4 introduced a dataset on home selling prices. The response variable is selling *price* in thousands of dollars. The explanatory variables are *size* of the home in square feet, whether it is *new* (1 = yes, 0 = no), annual *tax* bill in dollars, number of *bedrooms*, and number of *bathrooms*. A scatterplot matrix has limited use for highly discrete variables such as new, beds, and baths, but Figure 4.3 does reveal the strong positive correlation for each pair of price, size, and taxes.

```
-----------------------------------------------------------------------
> attach(Houses) # data at www.stat.ufl.edu/~aa/glm/data
> pairs(cbind(price,size,taxes)) # scatterplot matrix for pairs of var's
```

[20]Technically, collinearity refers to an *exact* linear dependence, but the term is used in practice when there is a *near* dependence.

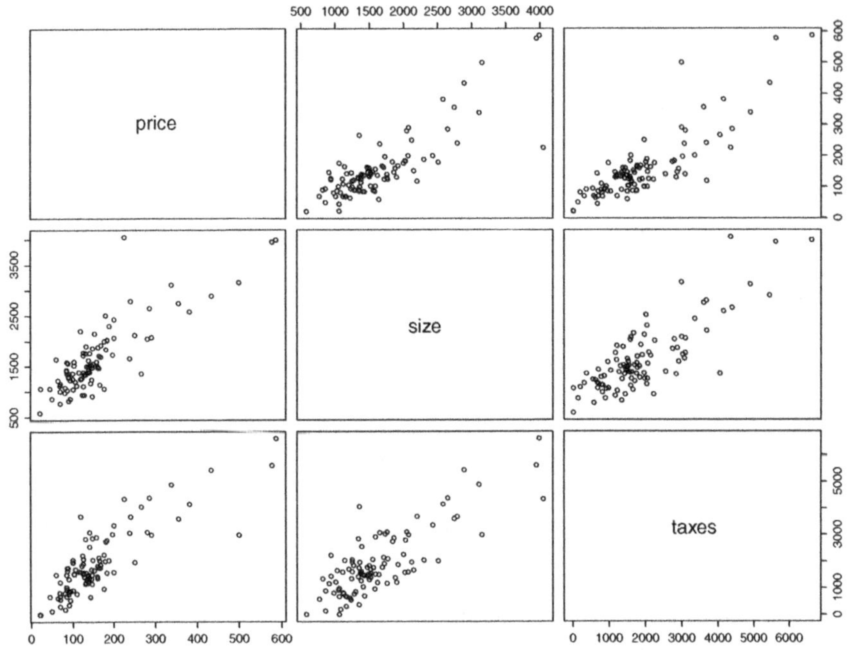

Figure 4.3 Scatterplot matrix for price, size, and taxes in dataset on house selling prices.

```
> cor(cbind(price,size,taxes,beds,baths)) # correlation matrix
        price    size   taxes    beds   baths
price  1.0000  0.8338  0.8420  0.3940  0.5583
size   0.8338  1.0000  0.8188  0.5448  0.6582
taxes  0.8420  0.8188  1.0000  0.4739  0.5949
beds   0.3940  0.5448  0.4739  1.0000  0.4922
baths  0.5583  0.6582  0.5949  0.4922  1.0000
-------------------------------------------------------------------------
```

4.7.1 Backward Elimination with House Selling Price Data

We illustrate a backward elimination process for selecting a model, using all the variables except taxes. (A chapter exercise uses all the variables.) Rather than relying solely on significance tests, we combine a backward process with judgments about practical significance.

To gauge how complex a model may be needed, we begin by comparing models containing the main effects only, also the second-order interactions, and also the third-order interactions. The anova function in R executes the F test comparing nested normal linear models (Section 3.2.2).

```
---------------------------------------------------------------------
> fit1 <- lm(price ~ size + new + baths + beds)
> fit2 <- lm(price ~ (size + new + baths + beds)^2)
```

```
> fit3 <- lm(price ~ (size + new + baths + beds)^3)
> anova(fit1, fit2)
Analysis of Variance Table
Model 1: price ~ size + new + baths + beds
Model 2: price ~ (size + new + baths + beds)^2
  Res.Df     RSS  Df  Sum of Sq       F      Pr(>F)
1     95  279624
2     89  217916   6      61708  4.2004  0.0009128
-------------------------------------------------------------------
```

A statistically significant improvement results from adding six pairwise interactions to the main effects model, with a drop in SSE of 61,708. A similar analysis (not shown here) indicates that we do not need three-way interactions. The R^2 values for the three models are 0.724, 0.785, and 0.804. In this process we compare models with quite different numbers of parameters, so we instead focus on the adjusted R^2 values: 0.713, 0.761, and 0.771. So we search for a model that fits adequately but is simpler than the model with all the two-way interactions.

In *fit2* (not shown), the least significant two-way interaction is baths × beds. Removing that interaction yields *fit4* with adjusted $R^2 = 0.764$. Then the least significant remaining two-way interaction is size × baths. With *fit5* we remove it, obtaining adjusted $R^2 = 0.766$. At that stage, the new × beds interaction is least significant, and we remove it, yielding adjusted $R^2 = 0.769$. The result is *fit6*:

```
-------------------------------------------------------------------
> summary(fit6)
              Estimate  Std. Error  t value  Pr(>|t|)
(Intercept)   135.6459     54.1902    2.503    0.0141
size           -0.0032      0.0323   -0.098    0.9219
new            90.7242     77.5413    1.170    0.2450
baths          12.2813     12.1814    1.008    0.3160
beds          -55.0541     17.6201   -3.125    0.0024
size:new        0.1040      0.0286    3.630    0.0005
size:beds       0.0309      0.0091    3.406    0.0010
new:baths    -111.5444     45.3086   -2.462    0.0157
---
Multiple R-squared: 0.7851,    Adjusted R-squared: 0.7688
-------------------------------------------------------------------
```

The three remaining two-way interactions are statistically significant at the 0.02 level. However, the P-values are only rough guidelines, and dropping the new × baths interaction (*fit7*, not shown) has only a slight effect, adjusted R^2 dropping to 0.756. At this stage we could drop baths from the model, as it is not in the remaining interaction terms and its $t = 0.40$.

```
-------------------------------------------------------------------
> fit8 <- update(fit7, .~. - baths)
> summary(fit8)
```

```
              Estimate  Std. Error  t value  Pr(>|t|)
(Intercept)  143.47098     54.1412    2.650    0.0094
size           0.00684      0.0326    0.210    0.8345
new          -56.68578     49.3006   -1.150    0.2531
beds         -53.63734     17.9848   -2.982    0.0036
size:new       0.05441      0.0210    2.588    0.0112
size:beds      0.03002      0.0092    3.254    0.0016
---
Multiple R-squared: 0.7706,     Adjusted R-squared: 0.7584
---
> plot(fit8)
-------------------------------------------------------------------
```

Both interactions are highly statistically significant, and adjusted R^2 drops to 0.716 if we drop them both. Viewing this as a provisional model, let us interpret the effects in *fit8*:

- For an older two-bedroom home, the effect on the predicted selling price of a 100 square foot increase in size is 100[0.00684 + 2(0.03002), or \$6688. For an older three-bedroom home, it is 100[0.00684 + 3(0.03002)], or \$9690, and for an older four-bedroom home, it is 100[0.00684 + 4(0.03002)], or \$12,692. For a new home, \$5441 is added to each of these three effects.
- Adjusted for the number of bedrooms, the effect on the predicted selling price of a home's being new (instead of older) is −56.686 + 1000(0.0544), or −\$2277, for a 1000-square-foot home, −56.686 + 2000(0.0544), or \$52,132, for a 2000-square-foot home, and −56.686 + 3000(0.0544), or \$106,541 for a 3000-square-foot home.
- Adjusted for whether a house is new, the effect on the predicted selling price of an extra bedroom is $-53.637 + 1000(0.0300)$, or $-\$23,616$, for a 1000-square-foot home, $-53.637 + 2000(0.0300)$, or \$6405, for a 2000-square-foot home, and $-53.637 + 3000(0.0300)$, or \$36,426, for a 3000-square-foot home.

For many purposes in an exploratory study, a simple model is adequate. We obtain a reasonably effective fit by removing the beds effects from *fit8*, yielding adjusted R^2 = 0.736 and very simple interpretations from the fit $\hat{\mu} = -22.228 + 0.1044(\text{size}) - 78.5275(\text{new}) + 0.0619(\text{size} \times \text{new})$. For example, the estimated effect of a 100 square-foot increase in size is \$10,440 for an older home and \$16,630 for a new home. In fact, this is the model having minimum BIC. The model having minimum AIC is[21] slightly more complex, the same as *fit6* above.

```
-------------------------------------------------------------------
> step(lm(price ~ (size + new + beds + baths)^2))
Start:  AIC=790.67 # AIC for initial model with two-factor interactions
...
```

[21]The AIC value reported by the `step` and `extractAIC` functions in R ignores certain constants, which the `AIC` function in R includes.

```
Step:  AIC=784.78 # lowest AIC for special cases of starting model
price ~ size + new + beds + baths + size:new + size:beds + new:baths
> AIC(lm(price ~ size+new+beds+baths+size:new+size:beds+new:baths))
[1] 1070.565 # correct value using AIC formula for normal linear model
> BIC(lm(price ~ size+new+size:new)) # this is model with lowest BIC
[1] 1092.973
-----------------------------------------------------------------------
```

4.7.2 Gamma GLM Has Standard Deviation Proportional to Mean

We ignored an important detail in the above model selection process. Section 3.4.2 noted that observation 64 in the dataset is an outlier that is highly influential in least squares fitting. Repeating the backward elimination process without it yields a different final model. This makes any conclusions even more highly tentative.

Section 3.4 noted some evidence of greater variability when mean selling prices are greater. This seems plausible and often happens for positive-valued response variables. At settings of explanatory variables for which $E(y)$ is low, we would not expect much variability in y (partly because y cannot be < 0), whereas when $E(y)$ is high, we would expect considerable variability. In each case, we would also expect some skew to the right in the response distribution, which could partly account for relatively large values. For such data, ordinary least squares is not optimal. One approach instead uses weighted least squares, by weighting observations according to how the variance depends on the mean. An alternative GLM approach assumes a distribution for y for which the variance increases as the mean increases. The family of *gamma distributions* has this property.

The two-parameter gamma probability density function for y, parameterized in terms of its mean μ and the shape parameter $k > 0$, is

$$f(y; k, \mu) = \frac{(k/\mu)^k}{\Gamma(k)} e^{-ky/\mu} y^{k-1}, \quad y \geq 0, \tag{4.29}$$

$$\text{for which } E(y) = \mu, \quad \text{var}(y) = \mu^2/k.$$

Gamma GLMs usually assume k to be constant but unknown, like σ^2 in ordinary linear models. Then the coefficient of variation, $\sqrt{\text{var}(y)}/\mu = 1/\sqrt{k}$, is constant as μ varies, and the standard deviation increases proportionally with the mean. The density is skewed to the right, but the degree of skewness (which equals $2/\sqrt{k}$) decreases as k increases. The mode is 0 when $k \leq 1$ and $\mu(k-1)/k$ when $k > 1$, with $k = 1$ giving the exponential distribution. The chi-squared distribution is the special case with $\mu = df$ and $k = df/2$.

The gamma distribution is in the exponential dispersion family with natural parameter $\theta = -1/\mu$, $b(\theta) = -\log(-\theta)$, and dispersion parameter $\phi = 1/k$. The scaled deviance for a gamma GLM has approximately a chi-squared distribution. However, the dispersion parameter is usually treated as unknown. We can mimic how we eliminate it in ordinary linear models by constructing an F statistic. For example, consider testing M_0 against M_1 for nested GLMs M_0 and M_1 with $p_0 < p_1$ parameters. Using the model deviances, the test statistic

$$\frac{[D(M_0) - D(M_1)]/(p_1 - p_0)}{D(M_1)/(n - p_1)},$$

has an approximate $F_{p_1-p_0,n-p_1}$ distribution, if the numerator and denominator are approximately independent[22]. Or, we can explicitly estimate ϕ for the more complex model and use the approximation

$$\frac{[D(M_0) - D(M_1)]/(p_1 - p_0)}{\hat{\phi}} \sim F_{p_1-p_0,n-p_1}.$$

Some software (e.g., SAS) uses ML to estimate ϕ. However, the ML estimator is inconsistent if the variance function is correct but the distribution is not truly the assumed one (McCullagh and Nelder 1989, p. 295). Other software (e.g., R) uses[23] the scaling $\hat{\phi} = X^2/(n - p)$ of the Pearson statistic (4.17), which is based on equating the average squared Pearson residual to 1, adjusted by using the dimension of the error space $n - p$ instead of n in the denominator (Wedderburn 1974). It is consistent when $\hat{\boldsymbol{\beta}}$ is. In the gamma context, this estimate is

$$\hat{\phi} = \frac{1}{n - p} \sum_{i=1}^{n} \frac{(y_i - \hat{\mu}_i)^2}{\hat{\mu}_i^2}.$$

When k is large, a gamma variate y has distribution close to normal. However, the gamma GLM fit is more appropriate than the least squares fit because the standard deviation increases as the mean does. Sometimes the identity link function is inadequate, because y must be nonnegative. It is then more common to use the log link. With that link, results are similar to least squares with a log-normal assumption for the response, that is, applying least squares to a linear model expressed in terms of $\log(y)$ (Exercise 4.27).

4.7.3 Gamma GLMs for House Selling Price Data

For the house selling price data, perhaps observation 64 is *not* especially unusual if we assume a gamma distribution for price. Using the same linear predictor as in the model (with *fit8*) interpreted in Section 4.7.1, we obtain:

```
-----------------------------------------------------------------
> fit.gamma <- glm(price ~ size + new + beds + size:new + size:beds,
                   family = Gamma(link = identity))
> summary(fit.gamma)$coef
                Estimate  Std. Error  t value   Pr(>|t|)
(Intercept)      44.3759     48.5978   0.9131     0.3635
```

[22]This holds when the dispersion parameter is small, so the gamma distribution is approximately normal. See Jørgensen (1987) for the general case using the F.

[23]But ML is available in R with the `gamma.dispersion` function in the MASS package.

```
size              0.0740       0.0400     1.8495     0.0675
new             -60.0290      65.7655    -0.9128     0.3637
beds            -22.7131      17.6312    -1.2882     0.2008
size:new          0.0538       0.0376     1.4325     0.1553
size:beds         0.0100       0.0126     0.7962     0.4279
---------------------------------------------------------------------
```

Now, neither interaction is significant! This also happens if we fit the model without observation 64. Including that observation, its standardized residual is now only −1.63, not at all unusual, because this model expects more variability in the data when the mean is larger. In fact, we may not need any interaction terms:

```
--------------------------------------------------------------------------
> fit.g1 <- glm(price ~ size+new+baths+beds, family=Gamma(link=identity))
> fit.g2 <- glm(price~(size+new+baths+beds)^2,family=Gamma(link=identity))
> anova(fit.g1, fit.g2, test="F")
Analysis of Deviance Table
  Resid. Df  Resid. Dev  Df  Deviance        F  Pr(>F)
1         95    10.4417
2         89     9.8728   6    0.5689  0.8438  0.5396
--------------------------------------------------------------------------
```

Further investigation using various model-building strategies reveals that according to AIC the model with size alone does well (AIC = 1050.7), as does the model with size and beds (AIC = 1048.3) and the model with size and new (AIC = 1049.5), with a slight improvement from adding the size × new interaction (AIC = 1047.9). Here is the output for the latter gamma model and for the corresponding normal linear model that we summarized near the end of Section 4.7.1:

```
--------------------------------------------------------------------------
> summary(glm(price ~ size+new+size:new, family=Gamma(link=identity)))
             Estimate  Std. Error  t value  Pr(>|t|)
(Intercept)   -7.4522     12.9738   -0.574    0.5670
size           0.0945      0.0100    9.396  2.95e-15
new          -77.9033     64.5827   -1.206    0.2307
size:new       0.0649      0.0367    1.769    0.0801 .
---
(Dispersion parameter for Gamma family taken to be 0.11021)
Residual deviance: 10.563  on 96  degrees of freedom
AIC: 1047.9
> plot(glm(price ~ size + new + size:new, family=Gamma(link=identity)))
> summary(lm(price ~ size + new + size:new))
             Estimate  Std. Error  t value  Pr(>|t|)
(Intercept)  -22.2278     15.5211   -1.432    0.1554
size           0.1044      0.0094   11.082   < 2e-16
new          -78.5275     51.0076   -1.540    0.1270
size:new       0.0619      0.0217    2.855    0.0053
---
Residual standard error: 52 on 96 degrees of freedom
Multiple R-squared:  0.7443,    Adjusted R-squared:  0.7363
-----------------------------------------------------------------------
```

Effects are similar, but the interaction term in the gamma model has larger *SE*. For this gamma model, $\hat{\phi} = 0.11021$, so the estimated shape parameter is $\hat{k} = 1/\hat{\phi} = 9.07$, which corresponds to a bell shape with some skew to the right. The estimated standard deviation $\hat{\sigma}$ of the conditional distribution of y relates to the estimated mean $\hat{\mu}$ by

$$\hat{\sigma} = \sqrt{\hat{\phi}\hat{\mu}} = \hat{\mu}/\sqrt{\hat{k}} = 0.33197\hat{\mu}.$$

For example, at predictor values having estimated mean selling price $\hat{\mu} = \$100{,}000$, the estimated standard deviation is \$33,197, whereas at $\hat{\mu} = \$400{,}000$, $\hat{\sigma}$ is four times as large.

The reported AIC value of 1047.9 for this gamma model is much better than the AIC for the normal linear model with the same explanatory variables, or for the normal linear model (*fit6*) in Section 4.7.1 that minimized AIC, of the models with main effects and two-way interactions.

```
--------------------------------------------------------------------------
> AIC(lm(price ~ size + new + size:new))
[1] 1079.9
> AIC(lm(price ~ size +new +beds +baths +size:new +size:beds +new:baths))
[1] 1070.6
--------------------------------------------------------------------------
```

We learn an important lesson from this example:

- In modeling, it is not sufficient to focus on how $E(y_i)$ depends on $\boldsymbol{x}_i$ for all i. The assumption about how var(y_i) depends on $E(y_i)$ can have a significant impact on conclusions about the effects.

Other approaches, such as using the log link instead of the identity link, yield other plausible models. Analyses that are beyond our scope here (such as Q–Q plots) indicate that selling prices may have a somewhat longer right tail than gamma and log-normal models permit. An alternative response distribution having this property is the *inverse Gaussian*, which has variance proportional to μ^3 (Seshadri 1994).

APPENDIX: GLM ANALOGS OF ORTHOGONALITY RESULTS FOR LINEAR MODELS

This appendix presents approximate analogs of linear model orthogonality results. Lovison (2014) showed that a weighted version of the estimated adjusted responses that has approximately constant variance has the same orthogonality of fitted values and residuals as occurs in ordinary linear models.

Recall that $\boldsymbol{D} = \text{diag}\{\partial\mu_i/\partial\eta_i\}$ and $\boldsymbol{W} = \text{diag}\{(\partial\mu_i/\partial\eta_i)^2/\text{var}(y_i)\}$. From Section 4.5.4, the IRLS fitting process is naturally expressed in terms of the estimate $\hat{z} = \boldsymbol{X}\hat{\boldsymbol{\beta}} + \widehat{\boldsymbol{D}}^{-1}(\boldsymbol{y} - \hat{\boldsymbol{\mu}})$ of an *adjusted* response variable $z = \boldsymbol{X}\boldsymbol{\beta} + \boldsymbol{D}^{-1}(\boldsymbol{y} - \boldsymbol{\mu})$. Since

$$\hat{\boldsymbol{\eta}} = \boldsymbol{X}\hat{\boldsymbol{\beta}} = \boldsymbol{X}(\boldsymbol{X}^{\mathrm{T}}\widehat{\boldsymbol{W}}\boldsymbol{X})^{-1}\boldsymbol{X}^{\mathrm{T}}\widehat{\boldsymbol{W}}\hat{z}$$

for the fitted linear predictor values, $\boldsymbol{X}(\boldsymbol{X}^{\mathrm{T}}\widehat{\boldsymbol{W}}\boldsymbol{X})^{-1}\boldsymbol{X}^{\mathrm{T}}\widehat{\boldsymbol{W}} = \widehat{\boldsymbol{W}}^{-1/2}\widehat{\boldsymbol{H}}_W\widehat{\boldsymbol{W}}^{1/2}$ is a sort of asymmetric projection adaptation of the estimate of the generalized hat matrix (4.19), namely,

$$\widehat{\boldsymbol{H}}_W = \widehat{\boldsymbol{W}}^{1/2}\boldsymbol{X}(\boldsymbol{X}^{\mathrm{T}}\widehat{\boldsymbol{W}}\boldsymbol{X})^{-1}\boldsymbol{X}^{\mathrm{T}}\widehat{\boldsymbol{W}}^{1/2}.$$

Consider the weighted adjusted responses and linear predictor, $z_0 = \boldsymbol{W}^{1/2}z$ and $\boldsymbol{\eta}_0 = \boldsymbol{W}^{1/2}\boldsymbol{\eta}$. For $\boldsymbol{V} = \text{var}(\boldsymbol{y})$, $\boldsymbol{W} = \boldsymbol{D}\boldsymbol{V}^{-1}\boldsymbol{D}$ and $\boldsymbol{W}^{-1} = \boldsymbol{D}^{-1}\boldsymbol{V}\boldsymbol{D}^{-1}$. Since $\text{var}(z) = \boldsymbol{D}^{-1}\boldsymbol{V}\boldsymbol{D}^{-1} = \boldsymbol{W}^{-1}$, it follows that $\text{var}(z_0) = \boldsymbol{I}$. Likewise, let $\hat{z}_0 = \widehat{\boldsymbol{W}}^{1/2}\hat{z}$ and $\hat{\boldsymbol{\eta}}_0 = \widehat{\boldsymbol{W}}^{1/2}\hat{\boldsymbol{\eta}}$. Then

$$\hat{\boldsymbol{\eta}}_0 = \widehat{\boldsymbol{W}}^{1/2}\boldsymbol{X}\hat{\boldsymbol{\beta}} = \widehat{\boldsymbol{W}}^{1/2}\boldsymbol{X}(\boldsymbol{X}^{\mathrm{T}}\widehat{\boldsymbol{W}}\boldsymbol{X})^{-1}\boldsymbol{X}^{\mathrm{T}}\widehat{\boldsymbol{W}}\hat{z} = \widehat{\boldsymbol{H}}_W\hat{z}_0.$$

So the weighted fitted linear predictor values are the orthogonal projection of the estimated weighted adjusted response variable onto the vector space spanned by the columns of the weighted model matrix $\widehat{\boldsymbol{W}}^{1/2}\boldsymbol{X}$. The estimated generalized hat matrix $\widehat{\boldsymbol{H}}_W$ equals $\boldsymbol{X}_0(\boldsymbol{X}_0^{\mathrm{T}}\boldsymbol{X}_0)^{-1}\boldsymbol{X}_0^{\mathrm{T}}$ for the weighted model matrix $\boldsymbol{X}_0 = \widehat{\boldsymbol{W}}^{1/2}\boldsymbol{X}$.

For the estimated weighted adjusted response, the raw residual is

$$\boldsymbol{e}_0 = \hat{z}_0 - \hat{\boldsymbol{\eta}}_0 = (\boldsymbol{I} - \widehat{\boldsymbol{H}}_W)\hat{z}_0,$$

so these residuals are orthogonal to the weighted fitted linear predictor values. Also, these residuals equal

$$\boldsymbol{e}_0 = \widehat{\boldsymbol{W}}^{1/2}(\hat{z} - \hat{\boldsymbol{\eta}}) = \widehat{\boldsymbol{W}}^{1/2}\widehat{\boldsymbol{D}}^{-1}(\boldsymbol{y} - \hat{\boldsymbol{\mu}}) = \widehat{\boldsymbol{V}}^{-1/2}(\boldsymbol{y} - \hat{\boldsymbol{\mu}}),$$

which are the Pearson residuals defined in (4.20).

A corresponding approximate version of Pythagoras's theorem states that

$$\|\hat{z}_0 - \boldsymbol{\eta}_0\|^2 \approx \|\hat{z}_0 - \hat{\boldsymbol{\eta}}_0\|^2 + \|\hat{\boldsymbol{\eta}}_0 - \boldsymbol{\eta}_0\|^2 = \|\boldsymbol{e}_0\|^2 + \|\hat{\boldsymbol{\eta}}_0 - \boldsymbol{\eta}_0\|^2.$$

The relation is not exact, because $\boldsymbol{\eta}_0 = \boldsymbol{W}^{1/2}\boldsymbol{X}\boldsymbol{\beta}$ lies in $C(\boldsymbol{W}^{1/2}\boldsymbol{X})$, not $C(\widehat{\boldsymbol{W}}^{1/2}\boldsymbol{X})$. Likewise, other decompositions for linear models occur only in an approximate manner for GLMs. For example, Firth (1991) noted that orthogonality of columns of $\boldsymbol{X}$ does not imply orthogonality of corresponding model parameters, except when the link function is such that $\boldsymbol{W}$ is a constant multiple of the identity matrix.

CHAPTER NOTES

Section 4.1: Exponential Dispersion Family Distributions for a GLM

4.1 Exponential dispersion: Jørgensen (1987, 1997) developed properties of the exponential dispersion family, including showing a convolution result and approximate normality for small values of the dispersion parameter. Davison (2003, Section 5.2), Morris (1982, 1983a), and Pace and Salvan (1997, Chapters 5 and 6) surveyed properties of exponential family models and their extensions.

4.2 GLMs: For more on GLMs, see Davison (2003), Fahrmeir and Tutz (2001), Faraway (2006), Firth (1991), Hastie and Pregibon (1991), Lee et al. (2006), Lovison (2014), Madsen and Thyregod (2011), McCullagh and Nelder (1989), McCulloch et al. (2008), and Nelder and Wedderburn (1972). For asymptotic theory, including conditions for consistency of $\hat{\boldsymbol{\beta}}$, see Fahrmeir and Kaufmann (1985).

Section 4.4: Deviance of a GLM, Model Comparison, and Model Checking

4.3 Diagnostics: Cox and Snell (1968) generalized residuals from ordinary linear models, including standardizations. Haberman (1974, Chapter 4) proposed standardized residuals for Poisson models, and Gilchrist (1981) proposed them for GLMs. For other justification for them, see Davison and Snell (1991). Pierce and Schafer (1986) and Williams (1984) evaluated residuals and presented standardized deviance residuals. Lovison (2014) proposed other adjusted residuals and showed their relations with test statistics for comparing nested models. See also Fahrmeir and Tutz (2001, pp. 147–148) and Tutz (2011, Section 3.10). Atkinson and Riani (2000), Davison and Tsai (1992), and Williams (1987) proposed other diagnostic measures for GLMs. Since residuals have limited usefulness for assessing GLMs, Cook and Weisberg (1997) proposed marginal model plots that compare nonparametric smoothings of the data to the model fit, both plotted as a function of characteristics such as individual predictors and the linear predictor values.

4.4 Score statistics: For comparing nested models M_0 and M_1, let $\boldsymbol{X}$ be the model matrix for M_1 and let $\boldsymbol{V}(\hat{\boldsymbol{\mu}}_0)$ be the estimated variances of $\boldsymbol{y}$ under M_0. With the canonical link, Lovison (2005) showed that the score statistic is

$$(\hat{\boldsymbol{\mu}}_1 - \hat{\boldsymbol{\mu}}_0)^{\mathrm{T}}\boldsymbol{X}[\boldsymbol{X}^{\mathrm{T}}\boldsymbol{V}(\hat{\boldsymbol{\mu}}_0)\boldsymbol{X}]^{-1}\boldsymbol{X}^{\mathrm{T}}(\hat{\boldsymbol{\mu}}_1 - \hat{\boldsymbol{\mu}}_0)$$

and this statistic bounds below the $X^2(M_0 \mid M_1)$ statistic in (4.18). Pregibon (1982) showed that the score statistic equals $X^2(M_0) - X^2(M_1)$ when $X^2(M_1)$ uses a one-step approximation to $\hat{\boldsymbol{\mu}}_1$. Pregibon (1982) and Williams (1984) showed that the squared standardized residual is a score statistic for testing whether the observation is an outlier.

Section 4.5: Fitting Generalized Linear Models

4.5 IRLS: For more on iteratively reweighted least squares and ML, see Bradley (1973), Green (1984), and Jørgensen (1983). Wood (2006, Chapter 2) illustrated the geometry of GLMs and IRLS.

4.6 Observed versus expected information: Fisher scoring has the advantages that it produces the asymptotic covariance matrix as a by-product, the expected information is necessarily nonnegative-definite, and the method relates to weighted least squares for ordinary linear models. For complex models, the observed information is often simpler to calculate. Efron and Hinkley (1978) argued that observed information has variance estimates that better approximate a relevant conditional variance (conditional on ancillary statistics not relevant to the parameter being estimated), it is "close to the data" rather than averaged over data that could have occurred but did not, and it tends to agree more closely with variances from Bayesian analyses.

Section 4.6: Selecting Explanatory Variables for a GLM

4.7 **Bias–variance tradeoff**: See Davison (2003, p. 405) and James et al. (2013, Section 2.2) for informative discussions of the bias–variance tradeoff.

4.8 **AIC and BIC**: Burnham and Anderson (2010) and Davison (2003, Sections 4.7 and 8.7) justified and illustrated the use of AIC for model comparison and suggested adjustments when n/p is not large. Raftery (1995) showed that differences between BIC values for two models relate to a Bayes factor comparing them. George (2000) presented a brief survey of variable selection methods and cautioned against using a criterion such as minimizing AIC or BIC to select a model.

4.9 **Collinearity**: Other measures besides *VIF* summarize the severity of collinearity and detect the variables involved. A *condition number* is the ratio of largest to smallest eigenvalues of $\boldsymbol{X}$, with large values (e.g., above 30) being problematic. See Belsley et al. (1980) and Rawlings et al. (1998, Chapter 13) for details.

EXERCISES

4.1 Suppose that y_i has a $N(\mu_i, \sigma^2)$ distribution, $i = 1, \dots, n$. Formulate the normal linear model as a GLM, specifying the random component, linear predictor, and link function.

4.2 Show the exponential dispersion family representation for the gamma distribution (4.29). When do you expect it to be a useful distribution for GLMs?

4.3 Show that the t distribution is not in the exponential dispersion family. (Although GLM theory works out neatly for family (4.1), in practice it is sometimes useful to use other distributions, such as the Cauchy special case of the t.)

4.4 Show that an alternative expression for the GLM likelihood equations is

$$\sum_{i=1}^{n} \frac{(y_i - \mu_i)}{\mathrm{var}(y_i)} \frac{\partial \mu_i}{\partial \beta_j} = 0, \quad j = 1, 2, \dots, p.$$

Show that these equations result from the generalized least squares problem of minimizing $\sum_i [(y_i - \mu_i)^2/\mathrm{var}(y_i)]$, treating the variances as known constants.

4.5 For a GLM with canonical link function, explain how the likelihood equations imply that the residual vector $\boldsymbol{e} = (\boldsymbol{y} - \hat{\boldsymbol{\mu}})$ is orthogonal with $C(\boldsymbol{X})$.

4.6 Suppose y_i has a Poisson distribution with $g(\mu_i) = \beta_0 + \beta_1 x_i$, where $x_i = 1$ for $i = 1, \dots, n_A$ from group A and $x_i = 0$ for $i = n_A + 1, ..., n_A + n_B$ from group B, and with all observations being independent. Show that for the log-link function, the GLM likelihood equations imply that the fitted means $\hat{\mu}_A$ and $\hat{\mu}_B$ equal the sample means.

4.7 Refer to the previous exercise. Using the likelihood equations, show that the same result holds for (**a**) *any* link function for this Poisson model, (**b**) any GLM of the form $g(\mu_i) = \beta_0 + \beta_1 x_i$ with a binary indicator predictor.

4.8 For the two-way layout with one observation per cell, consider the model whereby $y_{ij} \sim N(\mu_{ij}, \sigma^2)$ with

$$\mu_{ij} = \beta_0 + \beta_i + \gamma_j + \lambda \beta_i \gamma_j.$$

For independent observations, is this a GLM? Why or why not? (Tukey (1949) proposed a test of H_0: $\lambda = 0$ as a way of testing for interaction; in this setting, after we form the usual interaction SS, the residual SS is 0, so the ordinary test that applies with multiple observations degenerates.)

4.9 Consider the expression for the weight matrix $\boldsymbol{W}$ in $\text{var}(\hat{\boldsymbol{\beta}}) = (\boldsymbol{X}^{\text{T}} \boldsymbol{W} \boldsymbol{X})^{-1}$ for a GLM. Find $\boldsymbol{W}$ for the ordinary normal linear model, and show how $\text{var}(\hat{\boldsymbol{\beta}})$ follows from the GLM formula.

4.10 For the normal bivariate linear model, the asymptotic variance of the correlation r is $(1 - \rho^2)^2/n$. Using the delta method, show that the transform $\frac{1}{2}\log[(1 + r)/(1 - r)]$ is variance stabilizing. (Fisher (1921) noted this, showing that $1/(n - 3)$ is an improved variance for the transform.) Explain how to use this result to construct a confidence interval for ρ.

4.11 For a binomial random variable ny with parameter π, consider the null model.

a. Explain how to invert the Wald, likelihood-ratio, and score tests of H_0: $\pi = \pi_0$ against H_1: $\pi \neq \pi_0$ to obtain 95% confidence intervals for π.

b. In teaching an introductory statistics class, one year I collected data from the students to use for lecture examples. One question in the survey asked whether the student was a vegetarian. Of 25 students, 0 said "yes." Treating this as a random sample from some population, find the 95% confidence interval for π using each method in (**a**).

c. Do you trust the Wald interval in (**b**)? (Your answer may depend on whether you regard the standard error estimate for the interval to be credible.) Explain why the Wald method may behave poorly when a parameter takes value near the parameter space boundary.

4.12 For the normal linear model, Section 3.3.2 showed how to construct a confidence interval for $E(y)$ at a fixed $\boldsymbol{x}_0$. Explain how to do this for a GLM.

4.13 For a GLM assuming $y_i \sim N(\mu_i, \sigma^2)$, show that the Pearson chi-squared statistic is the same as the deviance. Find the form of the difference between the deviances for nested models M_0 and M_1.

4.14 In a GLM that uses a noncanonical link function, explain why it need not be true that $\sum_i \hat{\mu}_i = \sum_i y_i$. Hence, the residuals need not have a mean of 0. Explain why a canonical link GLM needs an intercept term in order to ensure that this happens.

4.15 For a binomial GLM, explain why the Pearson residual for observation i, $e_i = (y_i - \hat{\pi}_i)/\sqrt{\hat{\pi}_i(1-\hat{\pi}_i)/n_i}$, does not have an approximate standard normal distribution, even for a large n_i.

4.16 Find the form of the deviance residual (4.21) for an observation in (**a**) a binomial GLM, (**b**) a Poisson GLM.

4.17 Suppose x is uniformly distributed between 0 and 100, and y is binary with $\log[\pi_i/(1-\pi_i)] = -2.0 + 0.04x_i$. Randomly generate $n = 25$ independent observations from this model. Fit the model, and find corr$(\mathbf{y} - \hat{\boldsymbol{\mu}}, \hat{\boldsymbol{\mu}})$. Do the same for $n = 100$, $n = 1000$, and $n = 10,000$, and summarize how the correlation seems to depend on n.

4.18 Derive the formula $\text{var}(\hat{\beta}_j) = \sigma^2/\{(1-R_j^2)[\sum_i (x_{ij} - \bar{x}_j)^2]\}$.

4.19 Consider the value $\hat{\beta}$ that maximizes a function $L(\beta)$. This exercise motivates the Newton–Raphson method by focusing on the single-parameter case.

a. Using $L'(\hat{\beta}) = L'(\beta^{(0)}) + (\hat{\beta} - \beta^{(0)})L''(\beta^{(0)}) + \cdots$, argue that for an initial approximation $\beta^{(0)}$ close to $\hat{\beta}$, approximately $0 = L'(\beta^{(0)}) + (\hat{\beta} - \beta^{(0)})L''(\beta^{(0)})$. Solve this equation to obtain an approximation $\beta^{(1)}$ for $\hat{\beta}$.

b. Let $\beta^{(t)}$ denote approximation t for $\hat{\beta}$, $t = 0, 1, 2, \ldots$. Justify that the next approximation is

$$\beta^{(t+1)} = \beta^{(t)} - L'(\beta^{(t)})/L''(\beta^{(t)}).$$

4.20 For n independent observations from a Poisson distribution with parameter μ, show that Fisher scoring gives $\mu^{(t+1)} = \bar{y}$ for all $t > 0$. By contrast, what happens with the Newton–Raphson method?

4.21 For an observation y from a Poisson distribution, write a short computer program to use the Newton–Raphson method to maximize the likelihood. With $y = 0$, summarize the effects of the starting value on speed of convergence.

4.22 For noncanonical link functions in a GLM, show that the observed information matrix may depend on the data and hence differs from the expected information matrix. Thus, the Newton–Raphson method and Fisher scoring may provide different standard errors.

4.23 The bias–variance tradeoff: Before an election, a polling agency randomly samples $n = 100$ people to estimate π = population proportion who prefer candidate A over candidate B. You estimate π by the sample proportion $\hat{\pi}$. I estimate it by $\frac{1}{2}\hat{\pi} + \frac{1}{2}(0.50)$. Which estimator is biased? Which estimator has smaller variance? For what range of π values does my estimator have smaller mean squared error?

4.24 In selecting explanatory variables for a linear model, what is inadequate about the strategy of selecting the model with largest R^2 value?

4.25 For discrete probability distributions of $\{p_j\}$ for the "true" model and $\{p_{Mj}\}$ for a model M, prove that the Kullback–Leibler divergence $E\{\log[p(y)/p_M(y)]\} \geq 0$.

4.26 For a normal linear model M_1 with $p + 1$ parameters, namely, $\{\beta_j\}$ and σ^2, which has ML estimator $\hat{\sigma}^2 = [\sum_{i=1}^{n}(y_i - \hat{\mu}_i)^2]/n$, show that

$$\text{AIC} = n[\log(2\pi\hat{\sigma}^2) + 1] + 2(p + 1).$$

Using this, when M_2 has q additional terms, show that M_2 has smaller AIC value if $\text{SSE}_2/\text{SSE}_1 < e^{-2q/n}$.

4.27 Section 4.7.2 mentioned that using a gamma GLM with log-link function gives similar results to applying a normal linear model to $\log(y)$.

a. Use the delta method to show that when y has standard deviation σ proportional to μ (as does the gamma GLM), $\log(y)$ has approximately constant variance for small σ.

b. The gamma GLM with log link refers to $\log[E(y_i)]$, whereas the ordinary linear model for the transformed response refers to $E[\log(y_i)]$. Show that if $\log(y_i) \sim N(\mu_i, \sigma^2)$, then $\log[E(y_i)] = E[\log(y_i)] + \sigma^2/2$.

c. For the lognormal fitted mean L_i for the linear model for $\log(y_i)$, explain why $\exp(L_i)$ is the fitted median for the conditional distribution of y_i. Explain why the fitted median would often be more relevant than the fitted mean of that distribution.

4.28 Download the `Houses.dat` data file from `www.stat.ufl.edu/~aa/glm/data`. Summarize the data with descriptive statistics and plots. Using a forward selection procedure with all five predictors together with judgments about practical significance, select and interpret a linear model for selling price. Check whether results depend on any influential observations.

4.29 Refer to the previous exercise. Use backward elimination to select a model.

a. Use an initial model containing the two-factor interactions. When you reach the stage at which all terms are statistically significant, adjusted R^2 should still be about 0.87. See whether you can simplify further without serious loss of practical significance. Interpret your final model.

b. A simple model for these data has only main effects for size, new, and taxes. Compare your model with this model in terms of adjusted R^2, AIC, and the summaries of effects.

c. If any observations seem to be influential, redo the analyses to analyze their impact.

4.30 Refer to the previous two exercises. Conduct a model-selection process assuming a gamma distribution for y, using (**a**) identity link, (**b**) log link. For each, interpret the final model.

4.31 For the Scottish races data of Section 2.6, the Bens of Jura Fell Race was an outlier for an ordinary linear model with main effects of climb and distance in predicting record times. Alternatively the residual plots might merely suggest increasing variability at higher record times. Fit this model and the corresponding interaction model, assuming a gamma response instead of normal. Interpret results. According to AIC, what is your preferred model for these data?

4.32 Exercise 1.21 presented a study comparing forced expiratory volume after 1 hour of treatment for three drugs (a, b, and p = placebo), adjusting for a baseline measurement x_1. Table 4.1 shows the results of fitting some normal GLMs (with identity link, except one with log link) and a GLM assuming a gamma response. Interpret results.

Table 4.1 Results of Fitting GLMs for Exercise 4.32

Explanatory Variables	R^2	AIC	Fitted Linear Predictor
base	0.393	134.4	$0.95 + .90x_1$
drug	0.242	152.4	$3.49 + .20b - .67p$
base + drug	0.627	103.4	$1.11 + .89x_1 + .22b - .64p$
base + drug (gamma)	0.626	106.2	$0.93 + .97x_1 + .20b - .66p$
base + drug (log link)	0.609	106.8	$0.55 + .25x_1 + .06b - .20p$
base + drug + base:drug	0.628	107.1	$1.33 + .81x_1 - .17b - .91p + .15x_1b + .10x_1p$

4.33 Refer to Exercise 2.45 and the study for comparing instruction methods. Write a report summarizing a model-building process. Include instruction type in the chosen model, because of the study goals and the small n, which results in little power for finding significance for that effect. Check and interpret the final model.

4.34 The horseshoe crab dataset `Crabs2.dat` at the text website comes from a study of factors that affect sperm traits of males. One response variable is ejaculate size, measured as the log of the amount of ejaculate (microliters) measured after 10 seconds of stimulation. Explanatory variables are the location of the observation, carapace width (centimeters), mass (grams), color (1 = dark, 2 = medium, 3 = light), the operational sex ratio (OSR, the number of males per females on the beach), and a subjective condition number that takes into account mucus, pitting on the prosoma, and eye condition (the higher the better). Prepare a report (maximum 4 pages) describing a model-building process for these data. Attach edited software output as an appendix to your report.

4.35 The MASS package of R contains the `Boston` data file, which has several predictors of the median value of owner-occupied homes, for 506 neighborhoods in the suburbs near Boston. Describe a model-building process for these data, using the first 253 observations. Fit your chosen model to the other 253 observations. Compare how well the model fits in the two cases. Attach edited software output in your report.

4.36 For x between 0 and 100, suppose the normal linear model holds with

$$E(y) = 45 + 0.1x + 0.0005x^2 + 0.0000005x^3 + 0.0000000005x^4$$
$$+ 0.0000000000005x^5$$

and $\sigma = 10.0$. Randomly generate 25 observations from the model, with x having a uniform distribution between 0 and 100. Fit the simple model $E(y) = \beta_0 + \beta_1 x$ and the "correct" model $E(y) = \beta_0 + \beta_1 x + \cdots + \beta_5 x^5$. Construct plots, showing the data, the true relationship, and the model fits. For each model, summarize the quality of the fit by the mean of $|\hat{\mu}_i - \mu_i|$. Summarize, and explain what this exercise illustrates about model parsimony.

4.37 What does the fit of the "correct" model in the previous exercise illustrate about collinearity?

4.38 Randomly generate 100 observations (x_i, y_i) that are independent uniform random variables over [0, 100]. Fit a sequence of successively more complex polynomial models for using x to predict y, of degree 1, 2, 3, ... In principle, even though the true model is $E(y) = 50$ with population $R^2 = 0$, you should be able to fit a polynomial of degree 99 to the data and achieve $R^2 = 1$. Note that when you get to $p \approx 15$, $(\boldsymbol{X}^{\mathrm{T}}\boldsymbol{X})$ is effectively singular and effects of collinearity appear. As p increases, monitor R^2, adjusted R^2, and the P-value for testing significance of the intercept term. Summarize your results.

CHAPTER 5

Models for Binary Data

For binary responses, analysts usually assume a binomial distribution for the random component of a generalized linear model (GLM). From its exponential dispersion representation (4.6) in Section 4.1.2, the binomial natural parameter is the log odds, the so-called *logit*. The canonical link function for binomial GLMs is the logit, for which the model itself is referred to as *logistic regression*. This is the most important model for binary response data and has been used for a wide variety of applications. Early uses were in biomedical studies, for instance to model the effects of smoking, cholesterol, and blood pressure on the presence or absence of heart disease. The past 25 years have seen of substantial use in social science research for modeling opinions (e.g., favor or oppose legalization of same-sex marriage) and behaviors, in marketing applications for modeling consumer decisions (e.g., a choice between two products), and in finance for modeling credit-related outcomes (e.g., whether a credit card bill is paid on time).

In this chapter we focus on logistic regression and other models for binary response data. Section 5.1 presents some link functions and a latent variable model that motivates particular cases. Section 5.2 shows properties of logistic regression models and interprets its parameters. In Section 5.3 we apply GLM methods to specify likelihood equations and then conduct inference based on the logistic regression model. Section 5.4 covers model fitting. In Section 5.5 we find the deviance for binomial GLMs and discuss ways of checking the model fit. In Section 5.6 we present alternatives to logistic regression, such as the model using the *probit* link. Section 5.7 illustrates the models with two examples.

5.1 LINK FUNCTIONS FOR BINARY DATA

In this chapter, we distinguish between two sample size measures: a measure n_i for the number of Bernoulli trials that constitute a particular binomial observation, and a measure N for the number of binomial observations. We assume that $y_1, \dots, y_N$ are

Foundations of Linear and Generalized Linear Models, First Edition. Alan Agresti.

independent binomial proportions, with $n_i y_i \sim \text{bin}(n_i, \pi_i)$. That is, y_i is the *proportion* of "successes" out of n_i independent Bernoulli trials, and $E(y_i) = \pi_i$ does not depend on n_i. Let $\boldsymbol{n} = (n_1, \ldots, n_N)$ denote the binomial sample sizes. The overall number of binary observations is $n = \sum_{i=1}^{N} n_i$.

5.1.1 Ungrouped versus Grouped Binary Data

Data files for binary data have two possible formats. For *ungrouped data*, $\boldsymbol{n} = (1, \ldots, 1)$. The data file takes this form when each observation y_i results from a single Bernoulli trial, and thus equals 0 or 1. Large-sample methods for statistical inference then apply as $N \rightarrow \infty$.

For *grouped data*, sets of observations have the same value for each explanatory variable. Most commonly this happens when all explanatory variables are categorical. Then, n_i refers to the number of observations at setting i of the explanatory variables, $i = 1, \ldots, N$. For example, in a dose–response study of the effect of various dosages of a drug on the probability of an adverse outcome, $\{n_i\}$ record the number of observations at the various dosages. For grouped data, the number N of combinations of the categorical predictors is fixed, and large-sample methods for inference and model checking apply as each $n_i \rightarrow \infty$. Under such *small-dispersion asymptotics*, as we obtain more data, the variance for each binomial observation decreases.

A grouped-data file for binary data can be converted to ungrouped form. The same maximum likelihood (ML) estimates $\hat{\boldsymbol{\beta}}$ and standard errors occur, with the same large-sample normal distributions; however, other summary measures of fit, such as the deviance, change. We will see that the grouped-data format is useful for checking model fit. An ungrouped-data file can be converted to grouped-data form only when multiple subjects share the same values for explanatory variables.

5.1.2 Latent Variable Threshold Model for Binary GLMs

A latent variable model called a *threshold model* provides motivation for families of GLMs. We express this model in terms of ungrouped data. The model assumes (1) there is an unobserved continuous response y_i^* for subject i satisfying $y_i^* = \sum_j \beta_j x_{ij} + \epsilon_i$, where $\{\epsilon_i\}$ are independent from a distribution with mean 0 and having cdf F, and (2) there is a threshold τ such that we observe $y_i = 0$ if $y_i^* \leq \tau$ and $y_i = 1$ if $y_i^* > \tau$. See Figure 5.1. Then

$$P(y_i = 1) = P(y_i^* > \tau) = P\left(\sum_{j=1}^{p} \beta_j x_{ij} + \epsilon_i > \tau\right)$$

$$= 1 - P\left(\epsilon_i \leq \tau - \sum_{j=1}^{p} \beta_j x_{ij}\right) = 1 - F\left(\tau - \sum_{j=1}^{p} \beta_j x_{ij}\right).$$

The data contain no information about τ, so without loss of generality we take $\tau = 0$. Likewise, an equivalent model results if we multiply all parameters by any positive

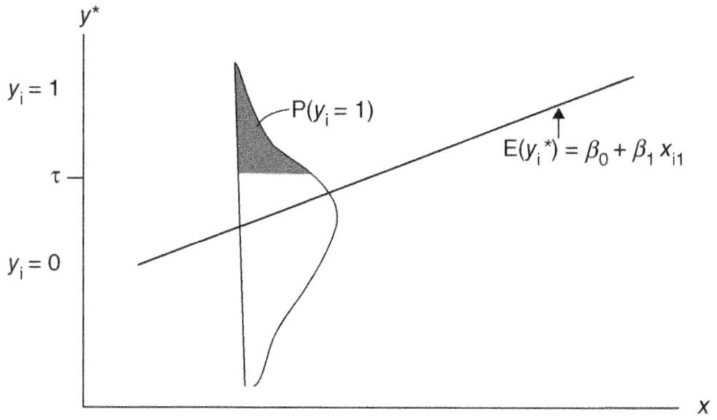

Figure 5.1 Threshold latent variable model, for which we observe $y_i = 1$ when underlying latent variable $y_i^* > \tau$.

constant, so we can take F to have a standard form with fixed variance, such as the standard normal cdf.

For the most common models, F corresponds to a pdf that is symmetric around 0, so $F(z) = 1 - F(-z)$ and

$$P(y_i = 1) = F\left(\sum_{j=1}^{p} \beta_j x_{ij}\right), \quad \text{and} \quad F^{-1}[P(y_i = 1)] = \sum_{j=1}^{p} \beta_j x_{ij}. \tag{5.1}$$

That is, models for binary data naturally take the link function to be the inverse of the standard cdf for a family of continuous distributions for a latent variable.

5.1.3 Probit, Logistic, and Linear Probability Models

When F is the standard normal cdf, the link function F^{-1} is called the *probit link* and model (5.1) is called the *probit model*. We discuss probit models in Section 5.6. A model that has a similar fit but a simpler form of link function uses the standard cdf of the *logistic distribution*,

$$F(z) = e^z/(1 + e^z).$$

Like the standard normal, the standard logistic distribution is defined over the entire real line and has a bell-shaped density function with mean 0. The model (5.1) is then the *logistic regression model*. Its link function F^{-1} is the logit.

Occasionally the identity link function is used, corresponding to F^{-1} for a uniform cdf. The model for the binomial parameter π_i for observation i,

$$\pi_i = \sum_{j=1}^{p} \beta_j x_{ij},$$

is called the *linear probability model*. This model has the awkward aspect that the linear predictor must fall between 0 and 1 for the model to generate legitimate probability values. Because of this and because in practice an S-shaped curve for which π_i very gradually approaches 0 and 1 is more plausible, linear probability models are not commonly used.

5.2 LOGISTIC REGRESSION: PROPERTIES AND INTERPRETATIONS

Next we present properties and interpretations of model parameters for logistic regression. The model has two formulations.

Logistic regression model formulas:

$$\pi_i = \frac{\exp\left(\sum_{j=1}^{p} \beta_j x_{ij}\right)}{1 + \exp\left(\sum_{j=1}^{p} \beta_j x_{ij}\right)} \quad \text{or} \quad \text{logit}(\pi_i) = \log\left(\frac{\pi_i}{1 - \pi_i}\right) = \sum_{j=1}^{p} \beta_j x_{ij}. \qquad (5.2)$$

5.2.1 Interpreting β: Effects on Probabilities and on Odds

For a single quantitative x with $\beta > 0$, the curve for π_i has the shape of the cdf of a logistic distribution. Since the logistic density is symmetric, as x_i changes, π_i approaches 1 at the same rate that it approaches 0. With multiple explanatory variables, since $1 - \pi_i = [1 + \exp(\sum_j \beta_j x_{ij})]^{-1}$, π_i is monotone in each explanatory variable according to the sign of its coefficient. The rate of climb or descent increases as $|\beta_j|$ increases. When $\beta_j = 0$, y is conditionally independent of x_j, given the other explanatory variables.

How do we interpret the magnitude of β_j? For a quantitative explanatory variable, a straight line drawn tangent to the curve at any particular value describes the instantaneous rate of change in π_i at that point. Specifically,

$$\frac{\partial \pi_i}{\partial x_{ij}} = \beta_j \frac{\exp\left(\sum_j \beta_j x_{ij}\right)}{\left[1 + \exp\left(\sum_j \beta_j x_{ij}\right)\right]^2} = \beta_j \pi_i (1 - \pi_i).$$

The slope is steepest (and equals $\beta_j/4$) at a value of x_{ij} for which $\pi_i = 1/2$, and the slope decreases toward 0 as π_i moves toward 0 or 1.

How do we interpret β_j for a qualitative explanatory variable? Consider first a single binary indicator x. The model, $\text{logit}(\pi_i) = \beta_0 + \beta_1 x_i$, then describes a 2×2 contingency table. For it,

$$\text{logit}[P(y = 1 \mid x = 1)] - \text{logit}[P(y = 1 \mid x = 0)] = [\beta_0 + \beta_1(1)] - [\beta_0 + \beta_1(0)] = \beta_1.$$

It follows that e^{β_1} is the *odds ratio* (Yule 1900, 1912),

$$e^{\beta_1} = \frac{P(y = 1 \mid x = 1)/[1 - P(y = 1 \mid x = 1)]}{P(y = 1 \mid x = 0)/[1 - P(y = 1 \mid x = 0)]}.$$

With multiple explanatory variables, exponentiating both sides of the equation for the logit shows that the odds $\pi_i/(1 - \pi_i)$ are an exponential function of x_j. The odds multiply by e^{β_j} per unit increase in x_j, adjusting for the other explanatory variables in the model. For example, e^{β_j} is a conditional odds ratio—the odds at $x_j = u + 1$ divided by the odds at $x_j = u$, adjusting for the other $\{x_k\}$.

It is simpler to understand the effects presented on a probability scale than as odds ratios. To summarize the effect of a quantitative explanatory variable, we could compare $P(y = 1)$ at extreme values of that variable, with other explanatory variables set at their means. This type of summary is sensible when the distribution of the data indicate that such extreme values can occur at mean values for the other explanatory variables. With a continuous variable, however, this summary can be sensitive to an outlier. So the comparison could instead use its quartiles, thus showing the change in $P(y = 1)$ over the middle half of the explanatory variable's range of observations. The data can more commonly support such a comparison.

5.2.2 Logistic Regression with Case-Control Studies

In case-control studies, y is known, and researchers look into the past to observe $\boldsymbol{x}$ as the random variable. For example, for cases of a particular type of cancer ($y = 1$) and disease-free controls ($y = 0$), a study might observe $x =$ whether the person has been a significant smoker. For 2×2 tables, we just observed that e^{β} is the odds ratio with y as the response. But, from Bayes' theorem,

$$e^{\beta} = \frac{P(y = 1 \mid x = 1)/P(y = 0 \mid x = 1)}{P(y = 1 \mid x = 0)/P(y = 0 \mid x = 0)}$$
$$= \frac{P(x = 1 \mid y = 1)/P(x = 0 \mid y = 1)}{P(x = 1 \mid y = 0)/P(x = 0 \mid y = 0)}.$$

So it is possible to estimate the odds ratio in retrospective studies that sample x, for given y. More generally, with logistic regression we can estimate effects in studies for which the research design reverses the roles of x and y as response and explanatory, and the effect parameters still have interpretations as log odds ratios.

Here is a formal justification: let z indicate whether a subject is sampled ($1 =$ yes, $0 =$ no). Even though the conditional distribution of y given $\boldsymbol{x}$ is not sampled, we need a model for $P(y = 1 \mid z = 1, \boldsymbol{x})$, assuming that $P(y = 1 \mid \boldsymbol{x})$ follows the logistic model. By Bayes' theorem,

$$P(y = 1 \mid z = 1, \boldsymbol{x}) = \frac{P(z = 1 \mid y = 1, \boldsymbol{x})P(y = 1 \mid \boldsymbol{x})}{\sum_{j=0}^{1}[P(z = 1 \mid y = j, \boldsymbol{x})P(y = j \mid \boldsymbol{x})]}. \quad (5.3)$$

Now, suppose that $P(z = 1 \mid y, \boldsymbol{x}) = P(z = 1 \mid y)$ for $y = 0$ and 1; that is, for each y, the sampling probabilities do not depend on $\boldsymbol{x}$. For instance, for cases and for controls, the probability of being sampled is the same for smokers and nonsmokers. Under this assumption, substituting $\rho_1 = P(z = 1 \mid y = 1)$ and $\rho_0 = P(z = 1 \mid y = 0)$ in Equation (5.3) and dividing the numerator and denominator by $P(y = 0 \mid \boldsymbol{x})$,

$$P(y = 1 \mid z = 1, \boldsymbol{x}) = \frac{\rho_1 \exp\left(\sum_j \beta_j x_j\right)}{\rho_0 + \rho_1 \exp\left(\sum_j \beta_j x_j\right)}.$$

Then, letting $\beta_0^* = \beta_0 + \log(\rho_1/\rho_0)$,

$$\text{logit}[P(y = 1 \mid z = 1, \boldsymbol{x})] = \beta_0^* + \beta_1 x_1 + \cdots.$$

The logistic regression model holds with the same effect parameters as in the model for $P(y = 1 \mid \boldsymbol{x})$. With a case-control study we can estimate those effects but we cannot estimate the intercept term, because the data do not supply information about the relative numbers of $y = 1$ and $y = 0$ observations.

5.2.3 Logistic Regression is Implied by Normal Explanatory Variables

Regardless of the sampling design, suppose the explanatory variables are continuous and have a normal distribution, for each response outcome. Specifically, given y, suppose $\boldsymbol{x}$ has an $N(\boldsymbol{\mu}_y, \boldsymbol{V})$ distribution, $y = 0, 1$. Then, by Bayes' theorem, $P(y = 1 \mid \boldsymbol{x})$ satisfies the logistic regression model with $\boldsymbol{\beta} = \boldsymbol{V}^{-1}(\boldsymbol{\mu}_1 - \boldsymbol{\mu}_0)$ (Warner 1963).

For example, in a health study of senior citizens, suppose y = whether a person has ever had a heart attack and x = cholesterol level. Suppose those who have had a heart attack have an approximately normal distribution on x and those who have not had one also have an approximately normal distribution on x, with similar variance. Then, the logistic regression function approximates well the curve for $P(y = 1 \mid x)$. The effect is greater when the groups' mean cholesterol levels are farther apart. If the distributions are normal but with different variances, the logistic model applies, but having a quadratic term (Exercise 5.1).

5.2.4 Summarizing Predictive Power: Classification Tables and ROC Curves

A *classification table* cross-classifies the binary response y with a prediction $\hat{y}$ of whether $y = 0$ or 1 (see Table 5.1). For a model fit to ungrouped data, the prediction for observation i is $\hat{y}_i = 1$ when $\hat{\pi}_i > \pi_0$ and $\hat{y}_i = 0$ when $\hat{\pi}_i \leq \pi_0$, for a selected cutoff π_0. Common cutoffs are (1) $\pi_0 = 0.50$, (2) the sample proportion of $y = 1$ outcomes, which each $\hat{\pi}_i$ equals for the model containing only an intercept term. Rather than using $\hat{\pi}_i$ from the model fitted to the dataset that includes y_i, it is better to make the prediction with the "leave-one-out" cross-validation approach, which bases $\hat{\pi}_i$ on the

Table 5.1 A Classification Table

	Prediction $\hat{y}$	
y	0	1
0		
1		

Cell counts in such tables yield estimates of sensitivity = $P(\hat{y} = 1 \mid y = 1)$ and specificity = $P(\hat{y} = 0 \mid y = 0)$.

model fitted to the other $n - 1$ observations. For a particular cutoff, summaries of the predictive power from the classification table are estimates of

$$\text{sensitivity} = P(\hat{y} = 1 \mid y = 1) \quad \text{and} \quad \text{specificity} = P(\hat{y} = 0 \mid y = 0).$$

A disadvantage of a classification table is that its cell entries depend strongly on the cutoff π_0 for predictions. A more informative approach considers the estimated sensitivity and specificity for all the possible π_0. The sensitivity is the *true positive rate* (tpr), and $P(\hat{y} = 1 \mid y = 0) = (1 - \text{specificity})$ is the *false positive rate* (fpr). A plot of the true positive rate as a function of the false positive rate as π_0 decreases from 1 to 0 is called a *receiver operating characteristic* (ROC) curve. When π_0 is near 1, almost all predictions are $\hat{y}_i = 0$; then, the point (fpr, tpr) $\approx$ (0, 0). When π_0 is near 0, almost all predictions are $\hat{y}_i = 1$; then, (fpr, tpr) $\approx$ (1, 1). For a given specificity, better predictive power corresponds to higher sensitivity. So, the better the predictive power, the higher the ROC curve and the greater the area under it. A ROC curve usually has a concave shape connecting the points (0, 0) and (1, 1), as illustrated by Figure 5.2.

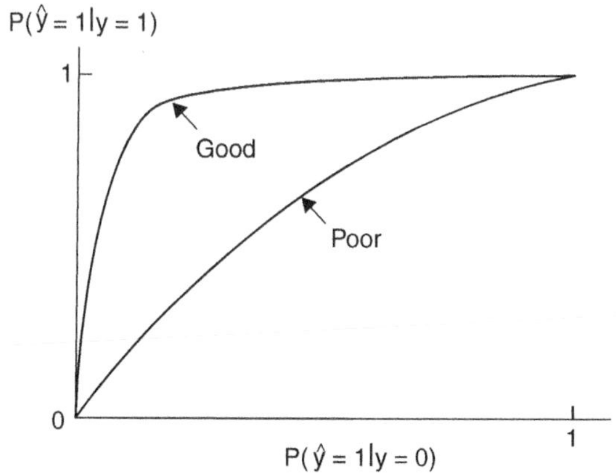

Figure 5.2 ROC curves for a binary GLM having good predictive power and for a binary GLM having poor predictive power.

The area under a ROC curve equals a measure of predictive power called the *concordance index* (Hanley and McNeil 1982). Consider all pairs of observations (i, j) for which $y_i = 1$ and $y_j = 0$. The concordance index c is the proportion of the pairwise predictions that are concordant with the outcomes, having $\hat{\pi}_i > \hat{\pi}_j$. A pair having $\hat{\pi}_i = \hat{\pi}_j$ contributes $\frac{1}{2}$ to the count of such pairs. The "no effect" value of $c = 0.50$ occurs when the ROC curve is a straight line connecting the points (0, 0) and (1, 1).

5.2.5 Summarizing Predictive Power: Correlation Measures

An alternative measure of predictive power is the correlation between the observed responses $\{y_i\}$ and the model's fitted values $\{\hat{\mu}_i\}$. This generalization of the multiple correlation for linear models is applicable for any GLM (Section 4.6.4). In logistic regression with ungrouped data, $\text{corr}(\boldsymbol{y}, \hat{\boldsymbol{\mu}})$ is the correlation between the N binary $\{y_i\}$ observations (1 or 0 for each) and the estimated probabilities. The highly discrete nature of y constrains the range of possible correlation values. A related measure estimates $\text{corr}(\boldsymbol{y}^*, \hat{\boldsymbol{\mu}})$ for the latent continuous variable for the underlying latent variable model. The square of this measure is an R^2 analog (McKelvey and Zavoina 1975) that divides the estimated variance of $\hat{y}^*$ by the estimated variance of y^*, where $\hat{y}_i^* = \sum_j \hat{\beta}_j x_{ij}$ is the same as the estimated linear predictor. The estimated variance of y^* equals the estimated variance of $\hat{y}^*$ plus the variance of ϵ in the latent variable model. For the probit latent model with standard normal error, $\text{var}(\epsilon) = 1$. For the corresponding logistic model, $\text{var}(\epsilon) = \pi^2/3 = 3.29$, the variance of the standard logistic distribution.

Such correlation measures are useful for comparing fits of different models for the same data. They can distinguish between models when the concordance index does not. For instance, with a single explanatory variable, c takes the same value for every link function that gives a monotone relationship of the same sign between x and $\hat{\pi}$.

5.3 INFERENCE ABOUT PARAMETERS OF LOGISTIC REGRESSION MODELS

The mechanics of ML estimation and model fitting for logistic regression are special cases of the GLM fitting results of Sections 4.1 and 4.5. From (4.10), the likelihood equations for a GLM are

$$\sum_{i=1}^{N} \frac{(y_i - \mu_i)x_{ij}}{\text{var}(y_i)} \frac{\partial \mu_i}{\partial \eta_i} = 0, \quad j = 1, 2, \ldots, p.$$

For a GLM for binary data, $n_i y_i \sim \text{bin}(n_i, \pi_i)$ with $\pi_i = \mu_i = F(\sum_j \beta_j x_{ij}) = F(\eta_i)$ for some standard cdf F. Thus, $\partial \mu_i / \partial \eta_i = f(\eta_i)$ where f is the pdf corresponding to F.

Since the binomial proportion y_i has $\text{var}(y_i) = \pi_i(1-\pi_i)/n_i$, the likelihood equations are

$$\sum_{i=1}^{N} \frac{n_i(y_i - \pi_i)x_{ij}}{\pi_i(1-\pi_i)} f(\eta_i) = 0, \quad j = 1, 2, \ldots, p.$$

That is, in terms of $\boldsymbol{\beta}$,

$$\sum_{i=1}^{N} \frac{n_i \left[y_i - F\left(\sum_j \beta_j x_{ij}\right)\right] x_{ij} f\left(\sum_j \beta_j x_{ij}\right)}{F\left(\sum_j \beta_j x_{ij}\right)\left[1 - F\left(\sum_j \beta_j x_{ij}\right)\right]} = 0, \quad j = 1, 2, \ldots, p. \tag{5.4}$$

5.3.1 Logistic Regression Likelihood Equations

For logistic regression models for binary data,

$$F(z) = \frac{e^z}{1+e^z}, \quad f(z) = \frac{e^z}{(1+e^z)^2} = F(z)[1 - F(z)].$$

The likelihood equations then simplify to

$$\sum_{i=1}^{N} n_i(y_i - \pi_i)x_{ij} = 0, \quad j = 1, \ldots, p. \tag{5.5}$$

Let $\boldsymbol{X}$ denote the $N \times p$ model matrix of values of $\{x_{ij}\}$. Let $\boldsymbol{s}$ denote the binomial vector of "success" totals with elements $s_i = n_i y_i$. The likelihood equations (5.5) have the form

$$\boldsymbol{X}^{\mathrm{T}}\boldsymbol{s} = \boldsymbol{X}^{\mathrm{T}}E(\boldsymbol{s}).$$

This equation illustrates the fundamental result for GLMs with canonical link function, shown in Equation 4.27, that the likelihood equations equate the sufficient statistics to their expected values.

5.3.2 Covariance Matrix of Logistic Parameter Estimators

The ML estimator $\hat{\boldsymbol{\beta}}$ has a large-sample normal distribution around $\boldsymbol{\beta}$ with covariance matrix equal to the inverse of the information matrix. From (4.13), the information matrix for a GLM has the form $\boldsymbol{\mathcal{J}} = \boldsymbol{X}^{\mathrm{T}}\boldsymbol{W}\boldsymbol{X}$, where $\boldsymbol{W}$ is the diagonal matrix with elements

$$w_i = (\partial\mu_i/\partial\eta_i)^2/\text{var}(y_i).$$

For binomial observations, $\mu_i = \pi_i$ and $\text{var}(y_i) = \pi_i(1 - \pi_i)/n_i$. For the logistic regression model, $\eta_i = \log[\pi_i/(1 - \pi_i)]$, so that $\partial\eta_i/\partial\pi_i = 1/[\pi_i(1 - \pi_i)]$. Thus, $w_i = n_i\pi_i(1 - \pi_i)$, and for large samples, the estimated covariance matrix of $\hat{\boldsymbol{\beta}}$ is

$$\widehat{\text{var}}(\hat{\boldsymbol{\beta}}) = \{\boldsymbol{X}^{\text{T}}\widehat{\boldsymbol{W}}\boldsymbol{X}\}^{-1} = \{\boldsymbol{X}^{\text{T}}\textbf{Diag}[n_i\hat{\pi}_i(1 - \hat{\pi}_i)]\boldsymbol{X}\}^{-1}, \tag{5.6}$$

where $\widehat{\boldsymbol{W}} = \textbf{Diag}[n_i\hat{\pi}_i(1 - \hat{\pi}_i)]$ denotes the $N \times N$ diagonal matrix having $\{n_i\hat{\pi}_i(1 - \hat{\pi}_i)\}$ on the main diagonal. "Large samples" here means a large number of Bernoulli trials, that is, large N for ungrouped data and large $n = \sum_i n_i$ for grouped data, in each case with p fixed. The square roots of the main diagonal elements of Equation (5.6) are estimated standard errors of $\hat{\boldsymbol{\beta}}$.

5.3.3 Statistical Inference: Wald Method is Suboptimal

For statistical inference for logistic regression models, we can use the Wald, likelihood-ratio, or score methods introduced in Section 4.3. For example, to test H_0: $\beta_j = 0$, the Wald chi-squared ($df = 1$) uses $(\hat{\beta}_j/SE_j)^2$, whereas the likelihood-ratio chi-squared uses the difference between the deviances for the simpler model with $\beta_j = 0$ and the full model.

These methods usually give similar results for large sample sizes. However, the Wald method has two disadvantages. First, its results depend on the scale for parameterization. To illustrate, for the null model, $\text{logit}(\pi) = \beta_0$, consider testing H_0: $\beta_0 = 0$ (i.e., $\pi = 0.50$) when ny has a $\text{bin}(n, \pi)$ distribution. From the delta method, the asymptotic variance of $\hat{\beta}_0 = \text{logit}(y)$ is $[n\pi(1 - \pi)]^{-1}$. The Wald chi-squared test statistic, which uses the ML estimate of the asymptotic variance, is $(\hat{\beta}_0/SE)^2 = [\text{logit}(y)]^2[ny(1 - y)]$. On the proportion scale, the Wald test statistic is $(y - 0.50)^2/[y(1 - y)/n]$. These are not the same. Evaluations reveal that the logit-scale statistic is too conservative[1] and the proportion-scale statistic is too liberal. A second disadvantage is that when a true effect in a binary regression model is very large, the Wald test is less powerful than the other methods and can show aberrant behavior. For this single-binomial example, suppose $n = 25$. Then, $y = 24/25$ is stronger evidence against H_0: $\pi = 0.50$ than $y = 23/25$, yet the logit Wald statistic equals 9.7 when $y = 24/25$ and 11.0 when $y = 23/25$. For comparison, the likelihood-ratio statistics are 26.3 and 20.7. As the true effect in a binary regression model increases, for a given sample size the information decreases so quickly that the standard error grows faster than the effect.[2] The Wald method fails completely when $\hat{\beta}_j = \pm\infty$, a case we discuss in Section 5.4.2.

5.3.4 Conditional Logistic Regression to Eliminate Nuisance Parameters

The total number of binary observations is $n = \sum_{i=1}^{N} n_i$ for grouped data and $n = N$ for ungrouped data. ML estimators of the p parameters of the logistic regression

[1]When H_0 is true, the probability a test of nominal size α rejects H_0 is *less* than α.
[2]See Davison (2003, p. 489), Hauck and Donner (1977), and Exercise 5.7.

model and standard methods of inference perform well when n is large compared with p. Sometimes n is small. Sometimes p grows as n grows, as in highly stratified data in which each stratum has its own model parameter. In either case, improved inference results from using *conditional maximum likelihood*. This method reduces the parameter space, eliminating nuisance parameters from the likelihood function by conditioning on their sufficient statistics. Inference based on the conditional likelihood can use large-sample asymptotics or small-sample distributions.

We illustrate with a simple case: logistic regression with a single binary explanatory variable x and small n. For subject i in an ungrouped data file,

$$\text{logit}[P(y_i = 1)] = \beta_0 + \beta_1 x_i, \quad i = 1, \dots, N, \tag{5.7}$$

where $x_i = 1$ or $x_i = 0$. Usually the log odds ratio β_1 is the parameter of interest, and β_0 is a nuisance parameter. For the exponential dispersion family (4.1) with $a(\phi) = 1$, the kernel of the log-likelihood function is $\sum_i y_i \theta_i$. For the logistic model, this is

$$\sum_{i=1}^{N} y_i \theta_i = \sum_{i=1}^{N} y_i(\beta_0 + \beta_1 x_i) = \beta_0 \sum_{i=1}^{N} y_i + \beta_1 \sum_{i=1}^{N} x_i y_i.$$

The sufficient statistics are $\sum_i y_i$ for β_0 and $\sum_i x_i y_i$ for β_1. The grouped form of the data is summarized with a 2×2 contingency table. Denote the two independent binomial "success" totals in the table by s_1 and s_2, having bin(n_1, π_1) and bin(n_2, π_2) distributions, as Table 5.2 shows. To conduct conditional inference about β_1 while eliminating β_0, we use the distribution of $\sum_i x_i y_i = s_1$, conditional on $\sum_i y_i = s_1 + s_2$.

Consider testing H_0: $\beta_1 = 0$, which corresponds to H_0: $\pi_1 = \pi_2$. Under H_0, let $\pi = e^{\beta_0}/(1 + e^{\beta_0})$ denote the common value. We eliminate β_0 by finding $P(s_1 = t \mid s_1 + s_2 = v)$. By the independence of the binomial variates and the fact that their sum is also binomial, under H_0

$$P(s_1 = t, s_2 = u) = \binom{n_1}{t} \pi^t (1-\pi)^{n_1 - t} \binom{n_2}{u} \pi^u (1-\pi)^{n_2 - u}, \; t = 0, \dots, n_1, \; u = 0, \dots, n_2$$

$$P(s_1 + s_2 = v) = \binom{n_1 + n_2}{v} \pi^v (1-\pi)^{n_1 + n_2 - v}, \quad v = 0, 1, \dots, n_1 + n_2.$$

Table 5.2 A 2×2 Table for Binary Response and Explanatory Variables

	y		
x	1	0	Total
1	s_1	$n_1 - s_1$	n_1
0	s_2	$n_2 - s_2$	n_2

So the conditional probability is

$$P(s_1 = t \mid s_1 + s_2 = v) = \frac{\binom{n_1}{t}\pi^t(1-\pi)^{n_1-t}\binom{n_2}{v-t}\pi^{v-t}(1-\pi)^{n_2-(v-t)}}{\binom{n_1+n_2}{v}\pi^v(1-\pi)^{n_1+n_2-v}}$$

$$= \frac{\binom{n_1}{t}\binom{n_2}{v-t}}{\binom{n_1+n_2}{v}}, \quad \max(0, v-n_2) \leq t \leq \min(n_1, v).$$

This is the *hypergeometric distribution*. To test H_0: $\beta_1 = 0$ against H_1: $\beta_1 > 0$, the P-value is $P(s_1 \geq t \mid s_1 + s_2)$, for observed value t for s_1. This probability does not depend on β_0. We can find it exactly rather than rely on a large-sample approximation. This test was proposed by R. A. Fisher (1935) and is called *Fisher's exact test* (see Exercise 5.31).

The conditional approach has the limitation of requiring sufficient statistics for the nuisance parameters. Reduced sufficient statistics exist only with GLMs that use the canonical link. Thus, the conditional approach works for the logistic model but not for binary GLMs that use other link functions. Another limitation is that when some explanatory variables are continuous, the $\{y_i\}$ values may be completely determined by the given sufficient statistics, making the conditional distribution degenerate.

5.4 LOGISTIC REGRESSION MODEL FITTING

We can use standard iterative methods to solve the logistic regression likelihood equations (5.5). In certain cases, however, some or all ML estimates may be infinite or may not even exist.

5.4.1 Iterative Fitting of Logistic Regression Models

The Newton–Raphson iterative method (Section 4.5.1) is equivalent to Fisher scoring, because the logit link is the canonical link. Using expressions (4.8) and the inverse of Equation (5.6), in terms of the binomial "success" counts $\{s_i = n_i y_i\}$, let

$$u_j^{(t)} = \left.\frac{\partial L(\boldsymbol{\beta})}{\partial \beta_j}\right|_{\boldsymbol{\beta}^{(t)}} = \sum_i (s_i - n_i\pi_i^{(t)})x_{ij}$$

$$h_{ab}^{(t)} = \left.\frac{\partial^2 L(\boldsymbol{\beta})}{\partial \beta_a \, \partial \beta_b}\right|_{\boldsymbol{\beta}^{(t)}} = -\sum_i x_{ia}x_{ib}n_i\pi_i^{(t)}(1-\pi_i^{(t)}).$$

Here $\boldsymbol{\pi}^{(t)}$, approximation t for $\hat{\boldsymbol{\pi}}$, is obtained from $\boldsymbol{\beta}^{(t)}$ through

$$\pi_i^{(t)} = \frac{\exp\left(\sum_{j=1}^p \beta_j^{(t)}x_{ij}\right)}{1+\exp\left(\sum_{j=1}^p \beta_j^{(t)}x_{ij}\right)}. \tag{5.8}$$

We use $\boldsymbol{u}^{(t)}$ and $\boldsymbol{H}^{(t)}$ with formula (4.23) to obtain the next value, $\boldsymbol{\beta}^{(t+1)}$, which in this context is

$$\boldsymbol{\beta}^{(t+1)} = \boldsymbol{\beta}^{(t)} + \left\{\boldsymbol{X}^{\mathrm{T}}\mathbf{Diag}\left[n_i\pi_i^{(t)}(1-\pi_i^{(t)})\right]\boldsymbol{X}\right\}^{-1}\boldsymbol{X}^{\mathrm{T}}(\boldsymbol{s}-\boldsymbol{\mu}^{(t)}), \qquad (5.9)$$

where $\mu_i^{(t)} = n_i\pi_i^{(t)}$. This is used to obtain $\boldsymbol{\pi}^{(t+1)}$, and so forth.

With an initial guess $\boldsymbol{\beta}^{(0)}$, Equation (5.8) yields $\boldsymbol{\pi}^{(0)}$, and for $t > 0$ the iterations proceed as just described using Equations (5.9) and (5.8). In the limit, $\boldsymbol{\pi}^{(t)}$ and $\boldsymbol{\beta}^{(t)}$ converge to the ML estimates $\hat{\boldsymbol{\pi}}$ and $\hat{\boldsymbol{\beta}}$, except for certain data configurations for which at least one estimate is infinite or does not exist (Section 5.4.2). The $\boldsymbol{H}^{(t)}$ matrices converge to $\widehat{\boldsymbol{H}} = -\boldsymbol{X}^{\mathrm{T}}\mathbf{Diag}[n_i\hat{\pi}_i(1-\hat{\pi}_i)]\boldsymbol{X}$. By Equation (5.6) the estimated asymptotic covariance matrix of $\hat{\boldsymbol{\beta}}$ is a by-product of the model fitting, namely $-\widehat{\boldsymbol{H}}^{-1}$.

From Section 4.5.4, $\boldsymbol{\beta}^{(t+1)}$ has the iteratively reweighted least squares form $(\boldsymbol{X}^{\mathrm{T}}\boldsymbol{V}_t^{-1}\boldsymbol{X})^{-1}\boldsymbol{X}^{\mathrm{T}}\boldsymbol{V}_t^{-1}\boldsymbol{z}^{(t)}$, where $\boldsymbol{z}^{(t)}$ has elements

$$z_i^{(t)} = \log\frac{\pi_i^{(t)}}{1-\pi_i^{(t)}} + \frac{s_i - n_i\pi_i^{(t)}}{n_i\pi_i^{(t)}\left(1-\pi_i^{(t)}\right)},$$

and where $\boldsymbol{V}_t = (\boldsymbol{W}^{(t)})^{-1}$ is a diagonal matrix with elements $\{1/[n_i\pi_i^{(t)}(1-\pi_i^{(t)})]\}$. In this expression, $\boldsymbol{z}^{(t)}$ is the linearized form of the logit link function for the sample data, evaluated at $\boldsymbol{\pi}^{(t)}$ (see (4.25)). The limit $\widehat{\boldsymbol{V}}$ of $\boldsymbol{V}_t$ has diagonal elements that estimate the variances of the approximate normal distributions[3] of the sample logits for large $\{n_i\}$, by the delta method.

5.4.2 Infinite Parameter Estimates in Logistic Regression

The Hessian matrix for logistic regression models is negative-definite, and the log-likelihood function is concave. ML estimates exist and are finite except when a hyperplane separates the set of explanatory variable values having $y = 0$ from the set having $y = 1$ (Albert and Anderson 1984).

For example, with a single explanatory variable and six observations, suppose $y = 1$ at $x = 1, 2, 3$ and $y = 0$ at $x = 4, 5, 6$ (see Figure 5.3). For the model $\text{logit}(\pi_i) = \beta_0 + \beta_1 x_i$ with observations in increasing order on x, the likelihood equations (5.5) are $\sum_i \hat{\pi}_i = \sum_i y_i$ and $\sum_i x_i\hat{\pi}_i = \sum_i x_i y_i$, or

$$\sum_{i=1}^{6} \hat{\pi}_i = 3 \quad \text{and} \quad \sum_{i=1}^{6} i\hat{\pi}_i = (1+2+3)1 + (4+5+6)0 = 6.$$

A solution is $\hat{\pi}_i = 1$ for $i = 1, 2, 3$ and $\hat{\pi}_i = 0$ for $i = 4, 5, 6$. Any other set of $\{\hat{\pi}_i\}$ having $\sum_i \hat{\pi}_i = 3$ would have $\sum_i i\hat{\pi}_i > 6$, so this is the unique solution. By letting $\hat{\beta}_1 \to -\infty$ and, for fixed $\hat{\beta}_1$, letting $\hat{\beta}_0 = -3.5\hat{\beta}_1$ so that $\hat{\pi} = 0.50$ at $x = 3.5$, we can

[3]The actual variance does not exist, because with positive probability the sample proportion $y_i = 1$ or 0 and the sample logit $= \pm\infty$.

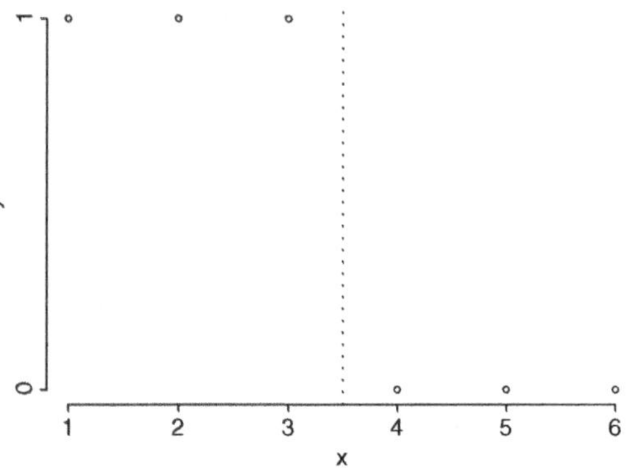

Figure 5.3 Complete separation of explanatory variable values, such as $y = 1$ when $x < 3.5$ and $y = 0$ when $x > 3.5$, causes an infinite ML effect estimate.

generate a sequence with ever-increasing value of the likelihood function that comes successively closer to satisfying these equations and giving a perfect fit.

In practice, software may fail to recognize when an ML estimate is actually infinite. After a certain number of cycles of iterative fitting, the log-likelihood looks flat at the working estimate, because the log-likelihood approaches a limiting value as the parameter value grows unboundedly. So, convergence criteria are satisfied, and software reports estimated. Because the log-likelihood is so flat and because the variance of $\hat{\beta}_j$ comes from its curvature as described by the negative inverse of the matrix of second partial derivatives, software typically reports huge standard errors.

```
---------------------------------------------------------------------
> x <- c(1,2,3,4,5,6);  y <- c(1,1,1,0,0,0)  # complete separation
> fit <- glm(y ~ x, family = binomial(link = logit))
> summary(fit)
Coefficients:
             Estimate  Std. Error  z value Pr(>|z|)
(Intercept)    165.32   407521.43        0        1 # x estimate is
x              -47.23   115264.41        0        1 # actually -infinity

Number of Fisher Scoring iterations: 25   # unusually large
> logLik(fit)
'log Lik.' -1.107576e-10 (df=2) # maximized log-likelihood = 0
---------------------------------------------------------------------
```

The space of explanatory variable values is said to have *complete separation* when a hyperplane can pass through that space such that on one side of that hyperplane $y_i = 0$ for all observations, whereas on the other side $y_i = 1$ always, as in Figure 5.3. There is then *perfect discrimination*, as we can predict the sample outcomes perfectly by knowing the explanatory variable values. In practice, we have an indication of complete separation when the fitted prediction equation perfectly predicts the response outcome for the entire dataset; that is, $\hat{\pi}_i = 1.0$ (to many decimal places) whenever

$y_i = 1$ and $\hat{\pi}_i = 0.0$ whenever $y_i = 0$. A related indication is that the reported maximized log-likelihood value is 0 to many decimal places. Another warning signal is standard errors that seem unnaturally large.

A weaker condition that causes at least one estimate to be infinite, called *quasi-complete separation*, occurs when a hyperplane separates explanatory variable values with $y_i = 1$ and with $y_i = 0$, but cases exist with both outcomes on that hyperplane. For example, this toy example of six observations has quasi-complete separation if we add two observations at $x = 3.5$, one with $y = 1$ and one with $y = 0$. Quasi-complete separation is more likely to happen with qualitative predictors than with quantitative predictors. If any category of a qualitative predictor has either no cases with $y = 0$ or no cases with $y = 1$, quasi-complete separation occurs when that variable is entered as a factor in the model (i.e., using an indicator variable for that category). With quasi-complete separation, there is not perfect discrimination for all observations. The maximized log-likelihood is then strictly less than 0. However, a warning signal is again reported standard errors that seem unnaturally large.

What inference can you conduct when the data have complete or quasi-complete separation? With an infinite estimate, you can still compute likelihood-ratio tests. The log-likelihood has a maximized value at the infinite estimate for a parameter, so you can compare it with the value when the parameter is equated to some fixed value such as zero. Likewise, you can invert the test to construct a confidence interval. If $\hat{\beta} = \infty$, for example, a 95% profile likelihood confidence interval has the form (L, ∞), where L is such that the likelihood-ratio test of H_0: $\beta = L$ has P-value $= 0.05$. With quasi-complete separation, some parameter estimates and *SE* values may be unaffected, and even Wald inference methods are available with them.

Alternatively, you can make some adjustment so that all estimates are finite. Some approaches smooth the data, thus producing finite estimates. The Bayesian approach is one way to do that (Section 10.3). A related way maximizes a *penalized likelihood* function. This adds a term to the ordinary log-likelihood function such that maximizing the amended function smooths the estimates by shrinking them toward 0 (Section 11.1.7).

5.5 DEVIANCE AND GOODNESS OF FIT FOR BINARY GLMS

For grouped or ungrouped binary data, one way to detect lack of fit uses a likelihood-ratio test to compare the model with more complex ones. If more complex models do not fit better, this provides some assurance that the model chosen is reasonable. Other approaches to detecting lack of fit search for *any* way that the model fails, using global statistics such as the deviance or Pearson statistics.

5.5.1 Deviance and Pearson Goodness-of-Fit Statistics

From Section 4.4.3, for binomial GLMs the deviance is the likelihood-ratio statistic comparing the model to the unrestricted (saturated model) alternative. The saturated model has the perfect fit $\tilde{\pi}_i = y_i$. The likelihood-ratio statistic comparing this to the ML model fit $\hat{\pi}_i$ for all i is

$$-2\log\left\{\left[\prod_{i=1}^{N}\hat{\pi}_i^{n_iy_i}(1-\hat{\pi}_i)^{n_i-n_iy_i}\right]\bigg/\left[\prod_{i=1}^{N}\tilde{\pi}_i^{n_iy_i}(1-\tilde{\pi}_i)^{n_i-n_iy_i}\right]\right\}$$
$$= 2\sum_i n_iy_i\log\frac{n_iy_i}{n_i\hat{\pi}_i} + 2\sum_i(n_i-n_iy_i)\log\frac{n_i-n_iy_i}{n_i-n_i\hat{\pi}_i}.$$

At setting i of the explanatory variables, n_iy_i is the number of successes and $(n_i - n_iy_i)$ is the number of failures, $i = 1, \ldots, N$. Thus, the deviance is a sum over the $2N$ success and failure totals at the N settings, having the form

$$D(y;\hat{\mu}) = 2\sum \text{observed} \times \log(\text{observed}/\text{fitted}).$$

This has the same form as the deviance (4.16) for Poisson loglinear models with intercept term. In either case, we denote it by G^2.

For naturally grouped data (e.g., solely categorical explanatory variables), the data file can be expressed in grouped or in ungrouped form. The deviance differs[4] in the two cases. For grouped data, the saturated model has a parameter at each setting for the explanatory variables. For ungrouped data, by contrast, it has a parameter for each subject.

For grouped data, a Pearson statistic also summarizes goodness of fit. It is the sum over the $2N$ cells of successes and failures,

$$X^2 = \sum\frac{(\text{observed} - \text{fitted})^2}{\text{fitted}}$$
$$= \sum_{i=1}^{N}\frac{(n_iy_i - n_i\hat{\pi}_i)^2}{n_i\hat{\pi}_i} + \sum_{i=1}^{N}\frac{[(n_i-n_iy_i)-(n_i-n_i\hat{\pi}_i)]^2}{n_i(1-\hat{\pi}_i)}$$
$$= \sum_{i=1}^{N}\frac{(n_iy_i - n_i\hat{\pi}_i)^2}{n_i\hat{\pi}_i(1-\hat{\pi}_i)} = \sum_{i=1}^{N}\frac{(y_i-\hat{\pi}_i)^2}{\hat{\pi}_i(1-\hat{\pi}_i)/n_i}. \quad (5.10)$$

In the form of Equation (5.10), this statistic is a special case of the score statistic for GLMs introduced in (4.17), having variance function in the denominator.

5.5.2 Chi-Squared Tests of Fit and Model Comparisons

When the data are grouped, the deviance G^2 and Pearson X^2 are goodness-of-fit test statistics for testing H_0 that the model truly holds. Under H_0, they have limiting chi-squared distributions as the overall sample size n increases, by $\{n_i\}$ increasing

[4]Exercise 5.17 shows a numerical example.

(i.e., small-dispersion asymptotics). Grouped data have a fixed number of settings N of the explanatory variables and hence a fixed number of parameters for the saturated model, so the *df* for the chi-squared distribution is the difference between the numbers of parameters in the two models, $df = N - p$. The X^2 statistic results[5] from summing the terms up to second-order in a Taylor series expansion of G^2, and $(X^2 - G^2)$ converges in probability to 0 under H_0. As n increases, the X^2 statistic converges to chi-squared more quickly than G^2 and has a more trustworthy P-value when some expected success or failure totals are less than about five.

The chi-squared limiting distribution does not occur for ungrouped data. In fact, G^2 and X^2 can be uninformative about lack of fit (Exercises 5.14 and 5.16). The chi-squared approximation is also poor with grouped data having a large N with relatively few observations at each setting, such as when there are many explanatory variables or one of them is nearly continuous in measurement (e.g., a person's age). For ungrouped data, G^2 and X^2 can be applied in an approximate manner to grouped observed and fitted values for a partition of the space of $\boldsymbol{x}$ values (Tsiatis 1980) or for a partition of the estimated probabilities of success (Hosmer and Lemeshow 1980). However, a large value of any global fit statistic merely indicates *some* lack of fit but provides no insight about its nature. The approach of comparing the working model with a more complex one is more useful from a scientific perspective, since it investigates lack of fit of a particular type.

Although the deviance is not useful for testing model fit when the data are ungrouped or nearly so, it remains useful for comparing models. For either grouped or ungrouped data, we can compare two nested models using the difference of deviances (Section 4.4.3). Suppose model M_0 has p_0 parameters and the more complex model M_1 has $p_1 > p_0$ parameters. Then the difference of deviances is the likelihood-ratio test statistic for comparing the models. If model M_0 holds, this difference has an approximate chi-squared distribution with $df = p_1 - p_0$. One can also compare the models using the Pearson comparison statistic (4.18).

5.5.3 Residuals: Pearson, Deviance, and Standardized

After a preliminary choice of model, such as with a global goodness-of-fit test or by comparing pairs of models, we obtain further insight by switching to a microscopic mode of analysis. With grouped data, it is useful to form residuals to compare observed and fitted proportions.

For observation i with sample proportion y_i and model fitted proportion $\hat{\pi}_i$, the Pearson residual (4.20) is

$$e_i = \frac{y_i - \hat{\pi}_i}{\sqrt{\widehat{\text{var}}(y_i)}} = \frac{y_i - \hat{\pi}_i}{\sqrt{\hat{\pi}_i(1 - \hat{\pi}_i)/n_i}}.$$

Equivalently, this divides the raw residual $(n_i y_i - n_i \hat{\pi}_i)$ comparing the observed and fitted number of successes by the estimated binomial standard deviation of $n_i y_i$. From

[5]For details, see Agresti (2013, p. 597).

Equation (5.10) these residuals satisfy

$$X^2 = \sum_{i=1}^{N} e_i^2,$$

for the Pearson statistic for testing the model fit. An alternative *deviance residual*, introduced for GLMs in (4.21), uses components of the deviance.

As explained in Section 4.4.6, the Pearson residuals have standard deviations less than 1. The standardized residual divides $(y_i - \hat{\pi}_i)$ by its estimated standard error. This uses the leverage $\hat{h}_{ii}$ from the diagonal of the GLM estimated hat matrix

$$\hat{\boldsymbol{H}}_W = \hat{\boldsymbol{W}}^{1/2} \boldsymbol{X}(\boldsymbol{X}^{\mathrm{T}}\hat{\boldsymbol{W}}\boldsymbol{X})^{-1}\boldsymbol{X}^{\mathrm{T}}\hat{\boldsymbol{W}}^{1/2},$$

in which the weight matrix $\hat{\boldsymbol{W}}$ is diagonal with element $\hat{w}_{ii} = n_i\hat{\pi}_i(1 - \hat{\pi}_i)$. For observation i, the standardized residual is

$$r_i = \frac{e_i}{\sqrt{1 - \hat{h}_{ii}}} = \frac{y_i - \hat{\pi}_i}{\sqrt{[\hat{\pi}_i(1 - \hat{\pi}_i)(1 - \hat{h}_{ii})]/n_i}}.$$

Compared with the Pearson and deviance residuals, it has the advantages of having an approximate $N(0, 1)$ distribution when the model holds (with large n_i) and appropriately recognizing redundancies in the data (Exercise 5.12). Absolute values larger than about 2 or 3 provide evidence of lack of fit.

Plots of residuals against explanatory variables or linear predictor values help to highlight certain types of lack of fit. When fitted success or failure totals are very small; however, just as X^2 and G^2 lose relevance, so do residuals. As an extreme case, for ungrouped data, $n_i = 1$ at each setting. Then y_i can equal only 0 or 1, and a residual can take only two values. One must then be cautious about regarding either outcome as extreme, and a single residual is essentially uninformative. When $\hat{\pi}_i$ is near 1, for example, residuals are necessarily either small and positive or large and negative. Plots of residuals also then have limited use. For example, suppose an explanatory variable x has a strong positive effect. Then, necessarily for small values of x, an observation with $y_i = 1$ will have a relatively large positive residual, whereas for large x an observation with $y_i = 0$ will have a relatively large negative residual. When raw residuals are plotted against fitted values, the plot consists merely of two nearly parallel lines of points. (Why?) When explanatory variables are categorical, so data can have grouped or ungrouped form, it is better to compute residuals and the deviance for the grouped data.

5.5.4 Influence Diagnostics for Logistic Regression

Other regression diagnostic tools also help in assessing fit. These include analyses that describe an observation's influence on parameter estimates and fit statistics.

However, a single observation can have a much greater influence in ordinary least squares regression than in logistic regression, because ordinary regression has no bound on the distance of y_i from its expected value. Also, the estimated hat matrix $\hat{\boldsymbol{H}}_W$ for a binary GLM depends on the fit as well as the model matrix $\boldsymbol{X}$. Points that have extreme predictor values need not have high leverage. In fact, the leverage can be relatively small if $\hat{\pi}_i$ is close to 0 or 1.

Several measures describe the effect of removing an observation from the dataset (Pregibon 1981; Williams 1987). These include the change in X^2 or G^2 goodness-of-fit statistics and analogs of influence measures for ordinary linear models, such as Cook's distance ($r_i^2[\hat{h}_{ii}/p(1-\hat{h}_{ii})]$) using the leverage and standardized residual.

5.6 PROBIT AND COMPLEMENTARY LOG–LOG MODELS

In this section we present two alternatives to the logistic regression model for binary responses. Instead of using the logistic distribution for the cdf inverted to get the link function, one uses the normal distribution and the other uses a skewed distribution.

5.6.1 Probit Models: Interpreting Effects

The binary-response model that takes the link function to be the inverse of the standard normal cdf Φ is called the *probit model*. For the binomial parameter π_i for observation i, the model is

$$\Phi^{-1}(\pi_i) = \sum_{j=1}^{p} \beta_j x_{ij}, \quad \text{or} \quad \pi_i = \Phi\left(\sum_{j=1}^{p} \beta_j x_{ij}\right).$$

For the probit model, the instantaneous rate of change in π_i as predictor j changes, adjusting for the other predictors, is $\partial\pi_i/\partial x_{ij} = \beta_j \phi(\sum_j \beta_j x_{ij})$, where $\phi(\cdot)$ is the standard normal density function. The rate is highest when $\sum_j \beta_j x_{ij} = 0$, at which $\pi_i = \frac{1}{2}$ and the rate equals $0.40\beta_j$. As a function of predictor j, the probit response curve for π_i (or for $1-\pi_i$, when $\beta_j < 0$) has the appearance of a normal cdf with standard deviation $1/|\beta_j|$. By comparison, in logistic regression the rate of change at $\pi_i = \frac{1}{2}$ is $0.25\beta_j$, and the logistic curve for π_i as a function of predictor j has standard deviation $\pi/|\beta_j|\sqrt{3}$ (for $\pi = 3.14\ldots$). The rates of change at $\pi_i = \frac{1}{2}$ are the same for the cdf's corresponding to the probit and logistic curves when the logistic β_j is $0.40/0.25 = 1.60$ times the probit β_j. The standard deviations for the response curves are the same when the logistic β_j is $\pi/\sqrt{3} = 1.81$ times the probit β_j. When both models fit well, ML parameter estimates in logistic regression are about 1.6–1.8 times those in probit models. Although probit model parameters are on a different scale than logistic model parameters, the probability summaries of effects are similar.

Another way to interpret parameters in probit models uses effects in the latent variable threshold model of Section 5.1.2. Since $y_i^* = \sum_j \beta_j x_{ij} + \epsilon_i$ where $\epsilon_i \sim N(0, 1)$

has cdf Φ, a 1-unit increase in x_{ij} corresponds to a change of β_j in $E(y_i^*)$, adjusted for the other explanatory variables. We interpret the magnitude of β_j in terms of the conditional standard deviation of 1 for y_i^*, so β_j represents a fraction or multiple of a standard deviation increase. Summary measures of model predictive power include the area under the ROC curve and corr($\boldsymbol{y}, \hat{\boldsymbol{\mu}}$), as described in Sections 5.2.4 and 5.2.5.

5.6.2 Probit Model Fitting

The likelihood equations for a probit model substitute Φ and ϕ in the general equations (5.4) for GLMs for binary data. The estimated large-sample covariance matrix of $\hat{\boldsymbol{\beta}}$ has the GLM form (4.14),

$$\widehat{\text{var}}(\hat{\boldsymbol{\beta}}) = (\boldsymbol{X}^{\text{T}}\widehat{\boldsymbol{W}}\boldsymbol{X})^{-1},$$

where $\widehat{\boldsymbol{W}}$ is the diagonal matrix with estimates of $w_i = (\partial\mu_i/\partial\eta_i)^2/\text{var}(y_i)$. Since $\mu_i = \pi_i = \Phi(\eta_i) = \Phi(\sum_j \beta_j x_{ij})$,

$$\hat{w}_i = n_i \left[\phi\left(\sum_{j=1}^{p} \hat{\beta}_j x_{ij}\right)\right]^2 \Bigg/ \left\{\Phi\left(\sum_{j=1}^{p} \hat{\beta}_j x_{ij}\right)\left[1 - \Phi\left(\sum_{j=1}^{p} \hat{\beta}_j x_{ij}\right)\right]\right\}.$$

We can solve the likelihood equations using the Fisher scoring algorithm for GLMs or the Newton–Raphson algorithm. They both yield the ML estimates but the Newton–Raphson algorithm gives slightly different standard errors because it inverts the observed information matrix to estimate the covariance matrix, whereas Fisher scoring uses expected information. These differ for link functions other than the canonical link.

5.6.3 Log–Log and Complementary Log–Log Link Models

The logit and probit links are symmetric about 0.50, in the sense that

$$\text{link}(\pi_i) = -\text{link}(1 - \pi_i).$$

To illustrate,

$$\text{logit}(\pi_i) = \log[\pi_i/(1 - \pi_i)] = -\log[(1 - \pi_i)/\pi_i] = -\text{logit}(1 - \pi_i).$$

This means that the response curve for π_i has a symmetric appearance about the point where $\pi_i = 0.50$. Logistic models and probit models are inappropriate when this is badly violated.

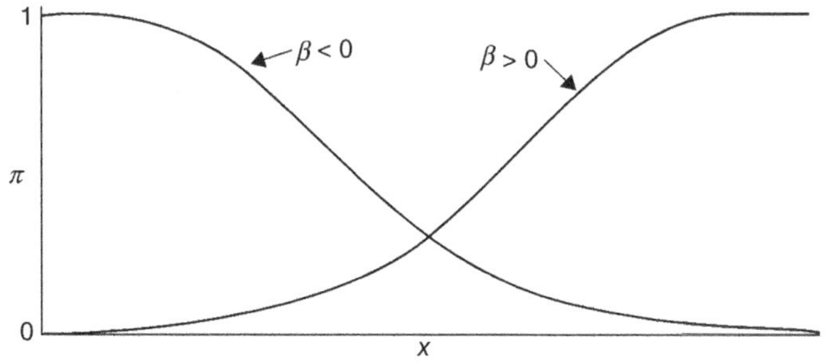

Figure 5.4 GLM for binary data using complementary log–log link function.

A different shape of response curve is given by the model

$$\pi_i = 1 - \exp\left[-\exp\left(\sum_{j=1}^{p} \beta_j x_{ij}\right)\right]. \tag{5.11}$$

With a single explanatory variable, this has the shape shown in Figure 5.4. The curve is asymmetric, π_i approaching 0 slowly but approaching 1 rather sharply. For this model,

$$\log[-\log(1 - \pi_i)] = \sum_{j=1}^{p} \beta_j x_{ij}.$$

The link function for this GLM is called the *complementary log–log* link, since the log–log link applies to the complement of π_i.

A related model to Equation (5.11) is

$$\pi_i = \exp\left[-\exp\left(-\sum_{j=1}^{p} \beta_j x_{ij}\right)\right].$$

In GLM form it uses the *log–log* link function.

$$-\log[-\log(\pi_i)] = \sum_{j=1}^{p} \beta_j x_{ij}.$$

For it, π_i approaches 0 sharply but approaches 1 slowly. When the log–log model holds for the probability of a success, the complementary log–log model holds for the probability of a failure, but with a reversal in sign of $\{\hat{\beta}_j\}$.

The log–log link is a special case of an inverse cdf link using the cdf of the *Type I extreme-value* distribution (also called the *Gumbel* distribution). The cdf equals

$$F(x) = \exp\{-\exp[-(x - a)/b]\}$$

for parameters $b > 0$ and $-\infty < a < \infty$. The distribution has mode a, mean $a + 0.577b$, and standard deviation $1.283b$, and is highly skewed to the right. The term *extreme value* refers to its being the limit distribution of the maximum of a sequence of independent and identically distributed continuous random variables.

Models with log–log link can be fitted by using the Fisher scoring algorithm for GLMs. How do we interpret effects in such models? Consider the complementary log–log link model (5.11) with a single explanatory variable x. As x increases, the curve is monotone increasing when $\beta > 0$. The complement probability at $x + 1$ equals the complement probability at x raised to the $\exp(\beta)$ power. We illustrate in the following example.

How can we evaluate the suitability of various possible link functions for a dataset? Measures such as the deviance and AIC provide some information. It is challenging to provide graphical portrayals of relations, especially for ungrouped data, since only $y = 1$ and $y = 0$ values appear on the graph. Plotting response proportions for grouped data can be helpful, as illustrated in the following example. Smoothing methods presented in Section 11.3 are also helpful for portraying the effects.

5.7 EXAMPLES: BINARY DATA MODELING

In this section we analyze two datasets. The first illustrates a logistic regression analysis for which one parameter has an infinite ML estimate. The second is a classic dose–response example from Bliss (1935), the first article to use ML fitting of a probit model.

5.7.1 Example: Risk Factors for Endometrial Cancer Grade

Heinze and Schemper (2002) described a study about endometrial cancer that analyzed how y = histology of 79 cases (0 = low grade for 30 patients, 1 = high grade for 49 patients) relates to three risk factors: x_1 = neovasculation (1 = present for 13 patients, 0 = absent for 66 patients), x_2 = pulsatility index of arteria uterina (ranging from 0 to 49), and x_3 = endometrium height (ranging from 0.27 to 3.61). Table 5.3 shows some of the data.

For these data, consider the main effects model

$$\text{logit}[P(y_i = 1)] = \beta_0 + \beta_1 x_{i1} + \beta_2 x_{i2} + \beta_3 x_{i3}.$$

Table 5.3 Part of Endometrial Cancer Dataset[a]

HG	NV	PI	EH	HG	NV	PI	EH	HG	NV	PI	EH
0	0	13	1.64	0	0	16	2.26	0	0	8	3.14
...											
1	1	21	0.98	1	0	5	0.35	1	1	19	1.02

Source: Data courtesy of Ella Asseryanis, Georg Heinze, and Michael Schemper. Complete data ($n = 79$) are in the file `Endometrial.dat` at `www.stat.ufl.edu/~aa/glm/data`.

[a]HG = histology grade, NV = neovasculation, PI = pulsatility index, EH = endometrium height.

When $x_{i1} = 0$ both response outcomes occur, but for all 13 patients having $x_{i1} = 1$ the outcome is $y_i = 1$, so there is quasi-complete separation. The ML estimate $\hat{\beta}_1 = \infty$.

```
-----------------------------------------------------------------------
> Endometrial
   NV PI   EH HG
1   0 13 1.64  0
2   0 16 2.26  0
...
79  1 19 1.02  1
> attach(Endometrial)
> table(NV,HG) # quasi-complete separation: When NV=1, no HG=0 cases
        HG
   NV    0   1
     0  49  17
     1   0  13
> fit <- glm(HG ~ NV + PI + EH, family=binomial) # logit default link
> summary(fit)
Coefficients:
            Estimate  Std. Error  z value  Pr(>|z|)
(Intercept)  4.305        1.637    2.629    0.0086
NV          18.186     1715.751    0.011    0.9915 # 18.186 and 1715.751
PI          -0.042        0.044   -0.952    0.3413 # should be infinity
EH          -2.903        0.846   -3.433    0.0006
---
    Null deviance: 104.903  on 78  degrees of freedom
Residual deviance:  55.393  on 75  degrees of freedom
> logLik(fit) # not exactly 0 because separation is quasi, not complete
'log Lik.' -27.69663 (df=4)
-----------------------------------------------------------------------
```

Despite $\hat{\beta}_1 = \infty$, inference is possible[6] about β_1. The likelihood-ratio statistic for H_0: $\beta_1 = 0$ equals 9.36 with $df = 1$ and has P-value $= 0.002$. The 95% profile likelihood confidence interval for β_1 is $(1.28, \infty)$.

```
-----------------------------------------------------------------------
> deviance(glm(HG ~ PI + EH, family=binomial)) - deviance(fit)
[1] 9.357643 # likelihood-ratio (LR) stat. with df=1 for H0: beta1 = 0

> library(ProfileLikelihood)
> xx <- profilelike.glm(HG~1+PI+EH,data=Endometrial,family=binomial,
+ profile.theta="NV",method="ML",lo.theta=-5,hi.theta=10,length=500,
+ round=3)
> profilelike.plot(theta=xx$theta, profile.lik.norm=xx$profile.lik.norm,
+ round=2)
> profilelike.summary(k=6.82,theta=xx$theta,
+ profile.lik.norm=xx$profile.lik.norm)
$LI.norm # LR = 6.82 gives 2log(6.82)=3.84 = 95 chi-sq percentile
[1]  1.283 10.000 # 10 was initial upper bound, correct upper limit
# is infinity but numerical instability occurs for beta1 values above 10
-----------------------------------------------------------------------
```

[6]We present other inferences for β_1 using Bayesian methods in Section 10.3.2 and using penalized likelihood methods in Section 11.1.8.

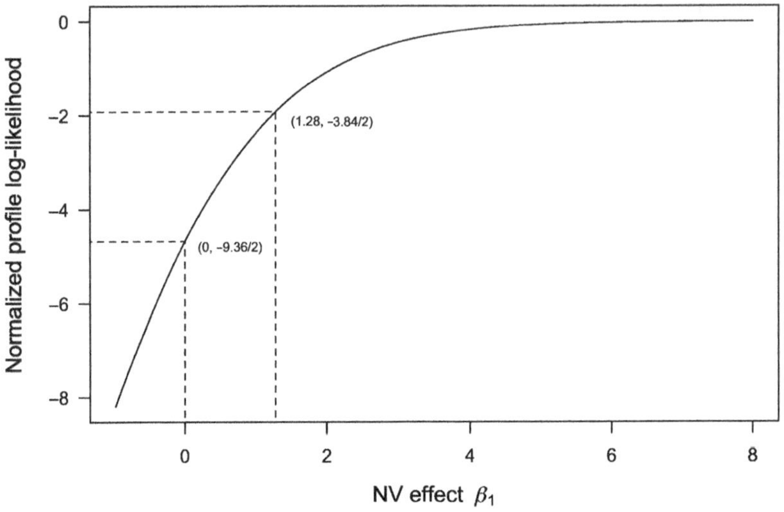

Figure 5.5 Normalized profile log-likelihood function $L(\beta_1) - L(\hat{\beta}_1)$ for NV effect in main-effects logistic model. Double the log-likelihood increases by 9.36 between $\beta_1 = 0$ and $\hat{\beta}_1 = \infty$ and by 3.84 between $\beta_1 = 1.28$ and $\hat{\beta}_1 = \infty$ (the 95% profile likelihood confidence interval). Figure constructed by Alessandra Brazzale with cond R package for higher-order likelihood-based conditional inference for logistic models.

We can conclude that $\beta_1 > 0$ (despite what the Wald P-value shows on the R output!) and that the effect is substantial. Figure 5.5 shows the normalized profile log-likelihood function for β_1.

The other ML estimates are not affected by the quasi-complete separation. Most of the predictive power is provided by the EH predictor: $\text{corr}(\boldsymbol{y}, \hat{\boldsymbol{\mu}}) = 0.745$ for the full model and 0.692 for the model with EH as the sole predictor; the areas under the ROC curves are 0.907 and 0.895. More complex models (not shown here) do not provide an improved fit.

5.7.2 Example: Dose–Response Study

From a dose–response study, Table 5.4 reports, in grouped-data form, the number of adult flour beetles that died after 5 hours of exposure to gaseous carbon disulfide at various dosages. Figure 5.6 plots the proportion killed against $x = \log_{10}(\text{dose})$. The proportion jumps up at about $x = 1.81$, and it is close to 1 above there.

To let the response curve take the shape of a normal cdf, Bliss (1935) used the probit model. The ML fit is

$$\Phi^{-1}(\hat{\pi}_i) = -34.96 + 19.74x_i, \quad i = 1, \ldots, 8.$$

Now $\hat{\pi} = 0.50$ when $\hat{\beta}_0 + \hat{\beta}_1 x = 0$, which for this fit is at $x = 34.96/19.74 = 1.77$. The fit corresponds to a normal cdf with $\mu = 1.77$ and $\sigma = 1/19.74 = 0.05$. Figure 5.6

Table 5.4 Beetles Killed after Exposure to Carbon Disulfide

			Fitted Number Dead		
Log Dosage	Number of Beetles	Number Dead	Comp. Log–Log	Probit	Logit
1.691	59	6	5.6	3.4	3.5
1.724	60	13	11.3	10.7	9.8
1.755	62	18	21.0	23.5	22.5
1.784	56	28	30.4	33.8	33.9
1.811	63	52	47.8	49.6	50.1
1.837	59	53	54.1	53.3	53.3
1.861	62	61	61.1	59.7	59.2
1.884	60	60	59.9	59.2	58.7

Source: Data file `Beetles2.dat` at text website, reprinted from Bliss (1935) with permission of John Wiley & Sons, Inc.

shows the fit. As x increases from 1.691 to 1.884, $\hat{\pi}$ increases from 0.058 to 0.987. For a 0.10-unit increase in x, such as from 1.70 to 1.80, the estimated conditional distribution of the latent variable y^* shifts up by $0.10(19.74) \approx 2$ standard deviations. The following R code enters the data in grouped-data form:

```
---------------------------------------------------------------------------
> logdose <- c(1.691, 1.724, 1.755, 1.784, 1.811, 1.837, 1.861, 1.884)
> dead <- c(6, 13, 18, 28, 52, 53, 61, 60)  # numbers dead
> n <- c(59, 60, 62, 56, 63, 59, 62, 60)    # binomial sample sizes
> alive <- n - dead                         # numbers not dead
> data <- matrix(append(dead,alive),ncol=2) # matrix of binomial counts
> fit.probit <- glm(data ~ logdose, family=binomial(link=probit))
> summary(fit.probit)
            Estimate  Std. Error  z value  Pr(>|z|)
(Intercept)  -34.956       2.649   -13.20   <2e-16
```

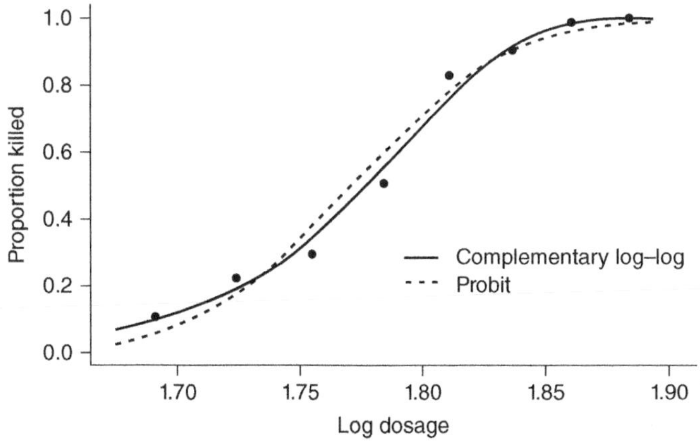

Figure 5.6 Proportion of dead beetles versus log dosage of gaseous carbon disulfide, with fits of probit and complementary log–log models.

```
logdose       19.741       1.488    13.27    <2e-16
---
    Null deviance: 284.202 on 7 degrees of freedom
Residual deviance:   9.987 on 6 degrees of freedom # (df = N-p = 8-2)
AIC: 40.185
> sum(resid(fit.probit, type="pearson")^2) # Pearson chi-squared, df=6
[1] 9.368992
> 1 - pchisq(9.368992, 6) # P-value for Pearson goodness-of-fit test
[1] 0.1538649
-----------------------------------------------------------------------
```

The deviance $G^2 = 9.99$ and Pearson $X^2 = 9.37$ (the sum of the squared Pearson residuals) have $df = 8 - 2 = 6$ and show slight evidence of lack of fit (P-value = 0.15 for X^2). The ML estimates are the same for grouped and ungrouped data, but the goodness-of-fit statistics apply only for the grouped data. The following R code shows the fit for the ungrouped data (file `Beetles.dat` at the text website).

```
-----------------------------------------------------------------------
Beetles <- read.table("Beetles.dat",header=TRUE)
> Beetles # ungrouped data at www.stat.ufl.edu/~aa/glm/data/Beetles.dat
         x  y # y=1 for dead, y=0 for alive
1     1.691  1
2     1.691  1
...
481   1.884  1
> attach(Beetles)
> fit.probit2 <- glm(y ~ x, family=binomial(link=probit))
> summary(fit.probit2)
              Estimate  Std. Error  z value  Pr(>|z|)
(Intercept)   -34.956       2.649   -13.20    <2e-16
x              19.741       1.488    13.27    <2e-16
---
    Null deviance: 645.44 on 480 degrees of freedom  # very different
Residual deviance: 371.23 on 479 degrees of freedom  # from grouped data
-----------------------------------------------------------------------
```

To summarize predictive power, for the ungrouped data corr$(y, \hat{\mu}) = 0.696$. The following code shows this and shows how to use an R package to construct the ROC curve for the model fit. For that curve, shown in Figure 5.7, the estimated concordance index $c = 0.901$.

```
-----------------------------------------------------------------
> cor(y, fitted(fit.probit2))
[1] 0.696391
> library(ROCR) # to construct ROC curve
> pred <- prediction(fitted(fit.probit2), y)
> perf <- performance(pred, "tpr", "fpr")
> plot(perf)
> performance(pred, "auc")
[1] 0.9010852     # concordance index = area under ROC curve (auc)
-----------------------------------------------------------------
```

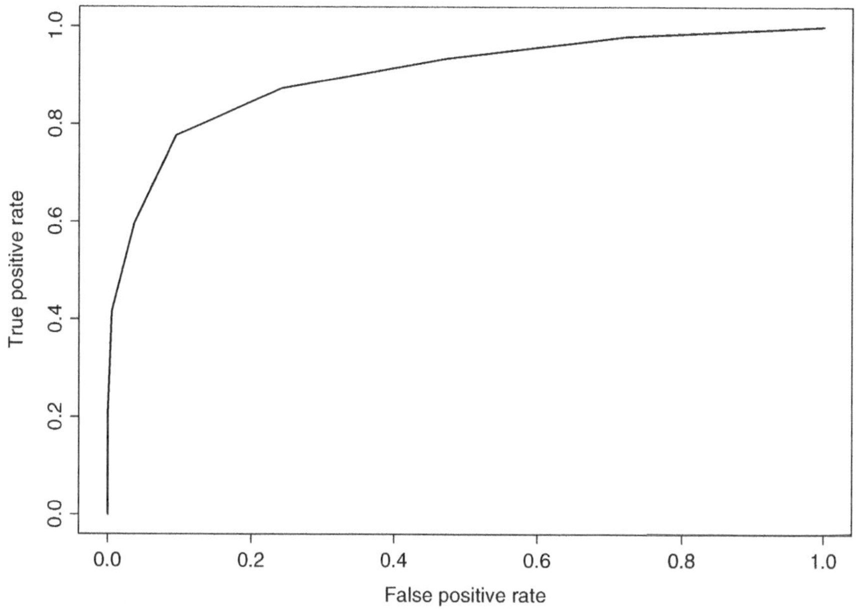

Figure 5.7 ROC curve for probit model fitted to beetle mortality data.

For comparison, we fit the corresponding logistic model. The ratio of $\hat{\beta}_1$ estimates for logit/probit is $34.29/19.74 = 1.74$. At dosage x_i with n_i beetles, $n_i\hat{\pi}_i$ is the fitted death count. Table 5.4 reports the fitted values for the grouped data. The logistic and probit models fit similarly.

```
-----------------------------------------------------------------------
> fit.logit <- glm(data ~ logdose, family = binomial(link=logit))
> summary(fit.logit) # grouped data
             Estimate  Std. Error  z value  Pr(>|z|)
(Intercept)   -60.740       5.182   -11.72    <2e-16
logdose        34.286       2.913    11.77    <2e-16
---
    Null deviance: 284.202  on 7  degrees of freedom
Residual deviance:  11.116  on 6  degrees of freedom
AIC: 41.314
-----------------------------------------------------------------------
```

The model with complementary log–log link has $\log[-\log(1-\hat{\pi}_i)] = -39.52 + 22.015x_i$. At dosage $x = 1.70$, the fitted probability of survival is $\exp\{-\exp[-39.52 + 22.015(1.70)]\} = 0.885$, whereas at $x = 1.80$ it is 0.330 and at $x = 1.90$ it is 4×10^{-5}. The probability of survival at dosage $x + 0.10$ equals the probability of survival at dosage x raised to the $e^{0.10(22.015)} = 9.04$ power. For instance, $0.330 = (0.885)^{9.04}$. Table 5.4 shows the fitted values for the grouped data, and Figure 5.6 shows the fit, which seems adequate (deviance $G^2 = 3.51, df = 6$). The code also shows the use of the `confint` function for obtaining profile-likelihood confidence intervals for the model parameters.

```
-----------------------------------------------------------------------
> fit.cloglog <- glm(data ~ logdose, family = binomial(link=cloglog))
> summary(fit.cloglog) # grouped data
            Estimate  Std. Error  z value   Pr(>|z|)
(Intercept)  -39.522       3.236   -12.21   <2e-16
logdose       22.015       1.797    12.25   <2e-16
---
    Null deviance: 284.2024  on 7  degrees of freedom
Residual deviance:   3.5143  on 6  degrees of freedom
AIC: 33.712
> sum(resid(fit.cloglog, type="pearson")^2) # Pearson chi-squared stat.
[1] 3.35924
> confint(fit.cloglog) # profile likelihood confidence intervals
               2.5 %    97.5 %
(Intercept)  -46.140   -33.499
logdose       18.669    25.689
-----------------------------------------------------------------------
```

By contrast, the log–log link yields a very poor fit. To use $-\log[-\log(\pi_i)]$ instead of $\log[-\log(\pi_i)]$ as the link function, corresponding to the inverse of the extreme-value cdf, we take the negative of the estimates reported here in the output for the model object called *fit.loglog*.

```
-----------------------------------------------------------------------
> data2 <- matrix(append(alive,dead),ncol=2) # reverse for log-log link
> fit.loglog <- glm(data2 ~ logdose, family=binomial(link=cloglog))
> summary(fit.loglog) # much poorer fit than complementary log-log link
            Estimate  Std. Error  z value  Pr(>|z|)
(Intercept)   37.661       2.949    12.77    <2e-16
logdose      -21.583       1.680   -12.85    <2e-16
---
    Null deviance: 284.202  on 7  degrees of freedom
Residual deviance:  27.573  on 6  degrees of freedom # grouped data
AIC: 57.771
-----------------------------------------------------------------------
```

The models with different link functions are not nested, so we cannot compare them with likelihood-ratio tests. The AIC values for the grouped data are 41.3 for the logit link, 40.2 for the probit model, 33.7 for the complementary log–log link, and 57.8 for the log–log link, showing a clear preference for the complementary log–log link. By contrast, the ROC curve is identical for the four link functions. The corr$(\boldsymbol{y}, \hat{\boldsymbol{\mu}})$ values for the ungrouped data are 0.684 for the log–log link, 0.696 for the probit link, 0.697 for the logit link, and 0.701 for the complementary log–log link.

Next, we perform a residual analysis for the complementary log–log link model applied to the grouped data. According to the standardized residuals, no observation exhibits lack of fit. Finally, Figure 5.8 plots the sample proportions dead, the fitted

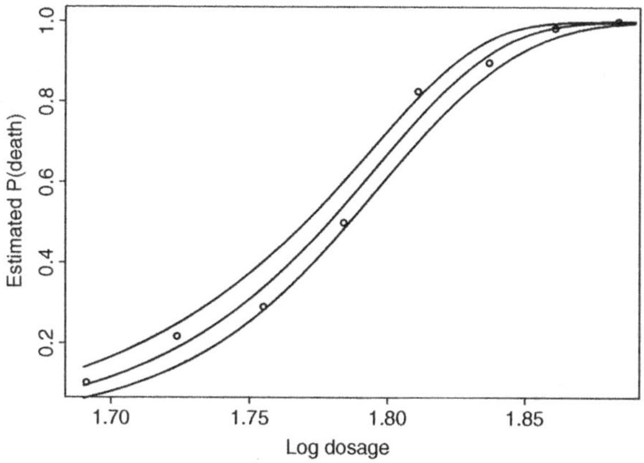

Figure 5.8 Plot of sample proportions, fitted complementary log–log model, and model-based confidence intervals for probability of death as function of log dosage.

values for the model, and 95% pointwise confidence bands for the true probabilities (assuming the model).

```
---------------------------------------------------------------------------
> pearson.res <- resid(fit.cloglog, type="pearson") # Pearson residuals
> std.res <- rstandard(fit.cloglog, type="pearson") # standardized res.
> cbind(logdose, dead/n, fitted(fit.cloglog), pearson.res, std.res)
  logdose                    pearson.res     std.res  # grouped data
1   1.691  0.102  0.096          0.153       0.177
2   1.724  0.217  0.188          0.568       0.669
3   1.755  0.290  0.338         -0.790      -0.922
4   1.784  0.500  0.542         -0.627      -0.704
5   1.811  0.825  0.757          1.268       1.486
6   1.837  0.898  0.918         -0.565      -0.702
7   1.861  0.984  0.986         -0.125      -0.149
8   1.884  1.000  0.999          0.228       0.237
> plot(logdose, dead/n)
> lines(logdose, fitted(fit.cloglog))
> fv <- predict(fit.cloglog, se.fit = TRUE)
>  U <- fv$fit + 1.96*fv$se.fit;  L <- fv$fit - 1.96*fv$se.fit
> lines(logdose, 1 - exp(-exp(U))); lines(logdose, 1 - exp(-exp(L)))
---------------------------------------------------------------------------
```

CHAPTER NOTES

Section 5.1: Link Functions for Binary Data

5.1 **Other link functions**: Other link functions for binary data include the inverse cdf of a t distribution (the probit being the limit as $df \to \infty$); a log-gamma link (Genter and Farewell 1985), for which probit, complementary log–log and log–log are special cases; a family of link functions that includes the logit (Pregibon 1980); and extensions with shape parameters that modify the logistic curve in extreme probability regions (Aranda-Ordaz 1981; Stukel 1988).

Section 5.3: Inference about Parameters of Logistic Regression Models

5.2 **Conditional logistic**: For more details about case-control studies and conditional logistic regression, see Breslow and Day (1980, Chapter 7). For more on "exact" inference using conditional distributions with logistic models, see Mehta and Patel (1995). Fisher's exact test extends to $r \times c$ tables and to stratified tables (Agresti 1992).

5.3 **Propensity scores**: Rosenbaum and Rubin (1983) proposed methods of comparing $E(y)$ for two groups in observational studies while adjusting for possibly confounding variables $\boldsymbol{x}$. They defined the *propensity* as the probability of being in one group, as a function of $\boldsymbol{x}$. They used logistic regression to estimate how propensity depends on $\boldsymbol{x}$. Their method takes into account differing distributions of the groups on $\boldsymbol{x}$ by using the estimated propensity to match samples from the groups or to subclassify subjects into intervals of propensity scores or to adjust directly by entering the propensity in the model.

Section 5.6: Probit and Complementary Log–Log Models

5.4 **Binary GLM history**: The probit model was presented by Bliss (1935) and popularized in three editions of Finney (1971). Logistic regression was proposed by Berkson (1944) as a model that has similar fit as a probit model but has closed form for the link function. Yates (1955) proposed the complementary log–log link. The logistic model became more popular following publication of an influential article (1958) and text (1970) by D. R. Cox, because of its direct interpretation in terms of odds ratios, validity in case-control studies, and availability of the conditional approach to eliminate nuisance parameters.

EXERCISES

5.1 For the population having value y on a binary response, suppose x has an $N(\mu_y, \sigma^2)$ distribution, $y = 0, 1$.

a. Using Bayes' theorem, show that $P(y = 1 \mid x)$ satisfies the logistic regression model with $\beta_1 = (\mu_1 - \mu_0)/\sigma^2$.

b. Suppose that $(x \mid y) \sim N(\mu_y, \sigma_y^2)$ with $\sigma_0 \neq \sigma_1$. Show that the logistic model holds with a quadratic term (Anderson 1975).

c. Suppose that $(x \mid y)$ has natural exponential family density

$$f(x; \theta_y) = h(x) \exp[x\theta_y - b(\theta_y)].$$

Show that $P(y = 1 \mid x)$ satisfies the logistic model with $\beta_1 = (\theta_1 - \theta_0)$.

5.2 Refer to Note 1.5. For a logistic model, show that the average estimated rate of change in the response probability as a function of explanatory variable j, adjusting for the others, satisfies $\frac{1}{n}\sum_i(\partial\hat{\pi}_i/\partial x_{ij}) = \hat{\beta}_j\frac{1}{n}\sum_i[\hat{\pi}_i(1 - \hat{\pi}_i)]$.

5.3 Construct the ROC curves for (**a**) the toy example in Section 5.4.2 with complete separation and (**b**) the dataset ($n = 8$) that adds two observations at $x = 3.5$, one with $y = 1$ and one with $y = 0$. In each case, report the area under the curve and summarize predictive power. For contrast, construct a toy dataset with $n = 8$ for which the area under the ROC curve equals 0.50.

5.4 From the likelihood equation (5.5) for a logistic regression intercept parameter, show that the overall sample proportion of successes equals the sample mean of the fitted success probabilities. Is this true for other binary GLMs?

5.5 Suppose that $n_i y_i$ has a bin(n_i, π_i) distribution. Consider a binary GLM $\pi_i = F(\sum_j \beta_j x_{ij})$ with F the standard cdf of some family of continuous distributions. Find w_i in $w_i = (\partial\mu_i/\partial\eta_i)^2/\text{var}(y_i)$ and hence var$(\hat{\boldsymbol{\beta}})$.

5.6 Explain how expression (5.6) for $\widehat{\text{var}}(\hat{\boldsymbol{\beta}})$ in logistic regression suggests that the standard errors of $\{\hat{\beta}_j\}$ tend to be smaller as you obtain more data. Answer this for (**a**) grouped data with $\{n_i\}$ increasing, (**b**) ungrouped data with N increasing.

5.7 Assuming the model logit$[P(y_i = 1)] = \beta x_i$, you take all n observations at x_0. Find $\hat{\beta}$ and the large-sample var$(\hat{\beta})$. For the Wald test, explain why the chi-squared noncentrality is $\beta^2/\text{var}(\hat{\beta})$, and evaluate it as $\beta \to \infty$. Explain how this illustrates that the Wald test in logistic regression has poor behavior when the effect is strong.

5.8 For a $2 \times 2 \times \ell$ contingency table that cross classifies y with a binary treatment variable x and an adjustment factor z, specify a logistic model with a lack of interaction between x and z. Construct the likelihood function, and explain the conditioning required to generate an exact conditional test for the effect of x. Explain how you would form a P-value for a one-sided alternative of a positive effect of x.

5.9 To use conditional logistic regression to test H_0: $\beta_1 = 0$ against H_1: $\beta_1 < 0$ for the toy example in Section 5.4.2, find the conditional distribution of $\sum_i x_i y_i$, given $\sum_i y_i$. Find the exact small-sample P-value.

5.10 The calibration problem is that of estimating x_0 at which $P(y = 1) = \pi_0$ for some fixed π_0 such as 0.50. For the logistic model with a single explanatory variable, explain why a confidence interval for x_0 is the set of x values for which

$$|\hat{\beta}_0 + \hat{\beta}_1 x - \text{logit}(\pi_0)|/[\text{var}(\hat{\beta}_0) + x^2\text{var}(\hat{\beta}_1) + 2x\,\text{cov}(\hat{\beta}_0, \hat{\beta}_1)]^{1/2} < z_{\alpha/2}.$$

How could you invert a likelihood-ratio test to form an interval?

5.11 Construct the log-likelihood function for the model logit$(\pi_i) = \beta_0 + \beta_1 x_i$ with independent binomial proportions of y_1 successes in n_1 trials at $x_1 = 0$ and y_2

successes in n_2 trials at $x_2 = 1$. Derive the likelihood equations, and show that $\hat{\beta}_1$ is the sample log odds ratio.

5.12 Refer to the previous exercise. Denote the cell counts in the 2×2 table by $\{n_{ij}\}$. For the case $\beta_1 = 0$ (the *independence model*), the fitted values in the cells of that table are $\{\hat{\mu}_{ij} = n_{i+}n_{+j}/n\}$. These have a common value for the four $|n_{ij} - \hat{\mu}_{ij}|$.

a. Construct the Pearson residuals. Explain why all four may differ in absolute value.

b. The standardized residuals in this case are

$$r_{ij} = (n_{ij} - \hat{\mu}_{ij})/\sqrt{\hat{\mu}_{ij}[1 - (n_{i+}/n)][1 - (n_{+j}/n)]}.$$

Show that all four are identical in absolute value, thus appropriately recognizing that residual $df = 1$ for the independence model.

5.13 Suppose the logistic model holds in which x is uniformly distributed between 0 and 100, and $\text{logit}(\pi_i) = -2.0 + 0.04x_i$. Randomly generate 100 independent observations from this model. Plot the residuals against x and against the fitted values. Why do residual plots for binary data have this appearance?

5.14 Let $n_i y_i$ be a bin(n_i, π_i) variate for group i, $i = 1, \ldots, N$, with $\{y_i\}$ independent. Consider the null model, for which $\pi_1 = \cdots = \pi_N$. Show that $\hat{\pi} = (\sum_i n_i y_i)/(\sum_i n_i)$. When all $n_i = 1$, for testing goodness of fit of the null model in the $N \times 2$ table, show that $X^2 = N$.

5.15 Let y_i be a bin$(1, \pi_i)$ variate, $i = 1, \ldots, N$. For the model $\text{logit}(\pi_i) = \beta_0 + \beta_1 x_i$, show that the deviance depends on $\hat{\pi}_i$ but not y_i. Hence, it is not useful for checking model fit. (This exercise and the previous one show that goodness-of-fit statistics are uninformative for ungrouped data.)

5.16 A study has n_i independent binary observations $\{y_{i1}, \ldots, y_{in_i}\}$ at x_i, $i = 1, \ldots, N$, with $n = \sum_i n_i$. Consider the model $\text{logit}(\pi_i) = \beta_0 + \beta_1 x_i$, where $\pi_i = P(y_{ij} = 1)$.

a. Show that the kernel of the likelihood function is the same if treating the data as n Bernoulli observations or N binomial observations.

b. For the saturated model, explain why the likelihood function is different for these two data forms. Hence, the deviance reported by software depends on the form of data entry.

c. Explain why the difference between deviances for two unsaturated models does not depend on the form of data entry.

5.17 Use the following toy data to illustrate comments in Section 5.5 about grouped versus ungrouped binary data in the effect on the deviance:

```
-------------------------------------------------------
 x      Number of trials    Number of successes
 0             4                   1
 1             4                   2
 2             4                   4
-------------------------------------------------------
```

Denote by M_0 the null model $\text{logit}(\pi_i) = \beta_0$ and by M_1 the model $\text{logit}(\pi_i) = \beta_0 + \beta_1 x_i$.

a. Create a data file in two ways, entering the data as (i) ungrouped data: $n_i = 1$, $i = 1, \ldots, 12$, (ii) grouped data: $n_i = 4$, $i = 1, 2, 3$. Fit M_0 and M_1 for each data file. Show that the deviances for M_0 and M_1 differ for the two forms of data entry. Why is this?

b. Show that the difference between the deviances for M_0 and M_1 is the same for each form of data entry. Why is this? (Thus, the data file format does not matter for inference, but it does matter for goodness-of-fit testing.)

5.18 Refer to the deviance comparison statistic $G^2(M_0 \mid M_1)$ introduced in Section 4.4.3. For a sequence of s nested binary response models $M_1, \ldots, M_s$, model M_s is the most complex. Let v denote the difference in residual *df* between M_1 and M_s.

a. Explain why for $j < k$, $G^2(M_j \mid M_k) \leq G^2(M_j \mid M_s)$.

b. Assume model M_j, so that M_k also holds when $k > j$. For all $k > j$, as $n \to \infty$, explain why $P[G^2(M_j \mid M_k) > \chi^2_v(\alpha)] \leq \alpha$.

c. Gabriel (1966) suggested a simultaneous testing procedure in which, for each pair of models, the critical value for differences between G^2 values is $\chi^2_v(\alpha)$. The final model accepted must be more complex than any model rejected in a pairwise comparison. Since part (**b**) is true for all $j < k$, argue that Gabriel's procedure has type I error probability no greater than α.

5.19 In a football league, for matches involving teams a and b, let π_{ab} be the probability that a defeats b. Suppose $\pi_{ab} + \pi_{ba} = 1$ (i.e., ties cannot occur). Bradley and Terry (1952) proposed the model

$$\log(\pi_{ab}/\pi_{ba}) = \beta_a - \beta_b.$$

For $a < b$, let N_{ab} denote the number of matches between teams a and b, with team a winning n_{ab} times and team b winning n_{ba} times.

a. Find the log-likelihood, treating n_{ab} as a binomial variate for N_{ab} trials. Show that sufficient statistics are $\{n_{a+}\}$, so that "victory totals" determine the estimated ranking of teams.

b. Generalize the model to allow a "home-team advantage," with a team's chance of winning possibly increasing when it plays at its home city. Interpret parameters.

5.20 Let y_i, $i = 1, \ldots, N$, denote N independent binary random variables.

a. Derive the log-likelihood for the probit model $\Phi^{-1}[\pi(\mathbf{x}_i)] = \sum_j \beta_j x_{ij}$.

b. Show that the likelihood equations for the logistic and probit regression models are

$$\sum_{i=1}^{N}(y_i - \hat{\pi}_i)z_i x_{ij} = 0, \qquad j = 1, \ldots, p,$$

where $z_i = 1$ for the logistic case and $z_i = \phi(\sum_j \hat{\beta}_j x_{ij})/\hat{\pi}_i(1 - \hat{\pi}_i)$ for the probit case.

5.21 An alternative latent variable model results from early applications of binary response models to toxicology studies (such as Table 5.4) of the effect of dosagc of a toxin on whether a subject dies, with an unobserved *tolerance distribution*. For a randomly selected subject, let x_i denote the dosage level and let $y_i = 1$ if the subject dies. Suppose that the subject has a latent tolerance threshold T_i for the dosage, with $(y_i = 1)$ equivalent to $(T_i \leq x_i)$. Let $F(t) = P(T \leq t)$.

a. For fixed dosage x_i, explain why $P(y_i = 1 \mid x_i) = F(x_i)$.

b. Suppose F belongs to the normal parametric family, for some μ and σ. Explain why the model has the form

$$\Phi^{-1}(\pi_i) = \beta_0 + \beta_1 x_i$$

and relate β_0 and β_1 to μ and σ.

5.22 Consider the choice between two options, such as two product brands. Let U_y denote the *utility* of outcome y, for $y = 0$ and $y = 1$. Suppose $U_y = \beta_{y0} + \beta_{y1}x + \epsilon_y$, using a scale such that ϵ_y has some standardized distribution. A subject selects $y = 1$ if $U_1 > U_0$ for that subject.

a. If ϵ_0 and ϵ_1 are independent $N(0, 1)$ random variables, show that $P(y = 1)$ satisfies the probit model.

b. If ϵ_y are independent extreme-value random variables, with cdf $F(\epsilon) = \exp[-\exp(-\epsilon)]$, show that $P(y = 1)$ satisfies the logistic regression model (McFadden 1974).

5.23 When $\Phi^{-1}(\pi_i) = \beta_0 + \beta_1 x_i$, explain why the response curve for π_i [or for $1 - \pi_i$, when $\beta_1 < 0$] has the appearance of a normal cdf with mean $\mu = -\beta_0/\beta_1$ and standard deviation $\sigma = 1/|\beta_1|$. By comparison, explain why the logistic regression curve for π_i has mean $\mu = -\beta_0/\beta_1$ and standard deviation $\pi/|\beta_1|\sqrt{3}$. What does this suggest about relative magnitudes of estimates in logistic and probit models?

5.24 Consider binary GLM $F^{-1}(\pi_i) = \beta_0 + \beta_1 x_i$, where F is a cdf corresponding to a pdf f that is symmetric around 0. Show that x_i at which $\pi_i = 0.50$ is $x_i = -\beta_0/\beta_1$. Show that the rate of change in π_i when $\pi_i = 0.50$ is $\beta_1 f(0)$, and find this for the logit and probit links. What does this suggest about relative magnitudes of estimates in logistic and probit models?

5.25 For the model $\log[-\log(1-\pi_i)] = \beta_0 + \beta_1 x_i$, find x_i at which $\pi_i = \frac{1}{2}$. Show that the greatest rate of change of π occurs at $x = -\beta_0/\beta_1$, and find π at that point. Give the corresponding result for the model with log–log link, and compare with the logistic and probit models.

5.26 In a study of the presence of tumors in animals, suppose $\{y_i\}$ are independent counts that satisfy a Poisson loglinear model, $\log(\mu_i) = \sum_j \beta_j x_{ij}$. However, the observed response merely indicates whether each y_i is positive, $z_i = I(y_i > 0)$, for the indicator function I. Show that $\{z_i\}$ satisfy a binary GLM with complementary log–log link (Dunson and Herring 2005).

5.27 Suppose $y = 0$ at $x = 10, 20, 30, 40$ and $y = 1$ at $x = 60, 70, 80, 90$. Using software, what do you get for estimates and standard errors when you fit the logistic regression model (**a**) to these data? (**b**) to these eight observations and two observations at $x = 50$, one with $y = 1$ and one with $y = 0$? (**c**) to these eight observations and observations at $x = 49.9$ with $y = 1$ and at $x = 50.1$ with $y = 0$? In cases (**a**) and (**b**), explain why actually the ML estimate $\hat{\beta} = \infty$. Why does software report such a large *SE* for $\hat{\beta}$? In case (**a**), what is the reported maximized log-likelihood value. Why?

5.28 For the logistic model (5.7) for a 2×2 table, give an example of cell counts corresponding to (**a**) complete separation and $\hat{\beta}_1 = \infty$, (**b**) quasi-complete separation and $\hat{\beta}_1 = \infty$, (**c**) non-existence of $\hat{\beta}_1$.

5.29 You plan to study the relation between x = age and y= whether belong to a social network such as Facebook (1 = yes). A priori, you predict that $P(y = 1)$ is currently between about 0.80 and 0.90 at $x = 18$ and between about 0.20 and 0.30 at $x = 65$. If the logistic regression model describes this relation well, what is a plausible range of values for the effect β_1 of x in the model?

5.30 In one of the first studies of the link between lung cancer and smoking[7], Richard Doll and Austin Bradford Hill collected data from 20 hospitals in London, England. Each patient admitted with lung cancer in the preceding year was queried about their smoking behavior. For each of the 709 patients admitted, they recorded the smoking behavior of a noncancer patient at the same hospital of the same gender and within the same 5-year grouping on age. A smoker was defined as a person who had smoked at least one cigarette a day for at least a year. Of the 709 cases having lung cancer, 688 reported being smokers. Of the 709 controls, 650 reported being smokers. Specify a relevant logistic regression model, explain what can be estimated and what cannot (and why), and conduct a statistical analysis.

[7]See *British Med. J.*, Sept. 30, 1950, pp. 739–748.

5.31 To illustrate Fisher's exact test, Fisher (1935) described the following experiment: a colleague of his claimed that, when drinking tea, she could distinguish whether milk or tea was added to the cup first (she preferred milk first). To test her claim, Fisher asked her to taste eight cups of tea, four of which had milk added first and four of which had tea added first. She knew there were four cups of each type and had to predict which four had the milk added first. The order of presenting the cups to her was randomized. For the 2×2 table relating what was actually poured first to the guess of what was poured first, explain how to use Fisher's exact test to evaluate whether her ability to distinguish the order of pouring was better than with random guessing. Find the P-value if she guesses correctly for three of the four cups that had milk poured first.

5.32 For the horseshoe crab dataset (`Crabs.dat` at the text website) introduced in Section 4.4.3, let $y = 1$ if a female crab has at least one satellite, and let $y = 0$ if a female crab does not have any satellites. Fit a main-effects logistic model using color and weight as explanatory variables. Interpret and show how to conduct inference about the color and weight effects. Next, allow interaction between color and weight in their effects on y, and test whether this model provides a significantly better fit.

5.33 The dataset `Crabs2.dat` at the text website collects several variables that may be associated with $y =$ whether a female horseshoe crab is monandrous (eggs fertilized by a single male crab) or polyandrous (eggs fertilized by multiple males). A probit model that uses as explanatory variables $Fcolor =$ the female crab's color (1 = dark, 3 = medium, 5 = light) and $Fsurf =$ her surface condition (values 1, 2, 3, 4, 5 with lower values representing worse) has the output shown. Interpret the parameter estimates and the inferential results. Approximately what values would you expect for the ML estimate of the Fsurf effect and its SE if you fitted the corresponding logistic model?

```
-----------------------------------------------------------------
                   Estimate  Std. Error  z value  Pr(>|z|)
(Intercept)         -0.3378      0.1217   -2.775  0.005522
factor(Fcolor)3      0.4797      0.1065    4.504  6.66e-06
factor(Fcolor)5      0.1651      0.1158    1.426  0.153902
Fsurf               -0.1360      0.0376   -3.619  0.000296
---
    Null deviance: 1633.8  on 1344  degrees of freedom
Residual deviance: 1587.8  on 1341  degrees of freedom
-----------------------------------------------------------------
```

5.34 Refer to the previous exercise. Download the file from the text website. Using *year* of observation, *Fcolor*, *Fsurf*, *FCW* = female's carapace width, *AMCW* = attached male's carapace width, *AMcolor* = attached male's color, and *AMsurf* = attached male's surface condition, conduct a logistic model-building process, including descriptive and inferential analyses. Prepare a report summarizing this process (with edited software output as an appendix), also interpreting results for your chosen model.

5.35 *The New York Times* reported results of a study on the effects of AZT in slowing the development of AIDS symptoms (February 15, 1991). Veterans whose immune symptoms were beginning to falter after infection with HIV were randomly assigned to receive AZT immediately or wait until their T cells showed severe immune weakness. During the 3-year study, of those who received AZT, 11 of 63 black subjects and 14 of 107 white subjects developed AIDS symptoms. Of those who did not receive AZT, 12 of 55 black subjects and 32 of 113 white subjects developed AIDS symptoms. Use model building, including checking fit and interpreting effects and inference, to analyze these data.

5.36 Download the data for the example in Section 5.7.1. Fit the main effects model. What does your software report for $\hat{\beta}_1$ and its *SE*? How could you surmise from the output that actually $\hat{\beta}_1 = \infty$?

5.37 Refer to the previous exercise. For these data, what, if anything, can you learn about potential interactions for pairs of the explanatory variables? Conduct the likelihood-ratio test of the hypothesis that all three interaction terms are 0.

5.38 Table 5.5 shows data, the file `SoreThroat.dat` at the text website, from a study about y = whether a patient having surgery experienced a sore throat on waking (1 = yes, 0 = no) as a function of d = duration of the surgery (in minutes) and t = type of device used to secure the airway (1 = tracheal tube, 0 = laryngeal mask airway). Use a model-building strategy to select a GLM for binary data. Interpret parameter estimates and conduct inference about the effects.

Table 5.5 Data for Exercise 5.38 on Surgery and Sore Throats

Patient	d	t	y	Patient	d	t	y	Patient	d	t	y
1	45	0	0	13	50	1	0	25	20	1	0
2	15	0	0	14	75	1	1	26	45	0	1
3	40	0	1	15	30	0	0	27	15	1	0
4	83	1	1	16	25	0	1	28	25	0	1
5	90	1	1	17	20	1	0	29	15	1	0
6	25	1	1	18	60	1	1	30	30	0	1
7	35	0	1	19	70	1	1	31	40	0	1
8	65	0	1	20	30	0	1	32	15	1	0
9	95	0	1	21	60	0	1	33	135	1	1
10	35	0	1	22	61	0	0	34	20	1	0
11	75	0	1	23	65	0	1	35	40	1	0
12	45	1	1	24	15	1	0				

Source: Data from Collett (2005) with permission of John Wiley & Sons, Inc.

CHAPTER 6

Multinomial Response Models

In Chapter 5 we presented generalized linear models (GLMs) for binary response variables that assume a *binomial* random component. GLMs for multicategory response variables assume a *multinomial* random component. In this chapter we present generalizations of logistic regression for multinomial response variables. Separate models are available for nominal response variables and for ordinal response variables.

In Section 6.1 we present a model for nominal response variables. It uses a separate binary logistic equation for each pair of response categories. An important type of application analyzes effects of explanatory variables on a person's choice from a discrete set of options, such as a choice of product brand to buy. In Section 6.2 we present a model for ordinal response variables. It applies the logit or some other link simultaneously to all the cumulative response probabilities, such as to model whether the importance of religion to a person is below or above some point on a scale (unimportant, slightly important, moderately important, very important). A parsimonious version of the model uses the same effect parameters for each logit. Section 6.3 presents examples and discusses model selection for multicategory responses.

We denote the number of response categories by c. For subject i, let π_{ij} denote the probability of response in category j, with $\sum_{j=1}^{c} \pi_{ij} = 1$. The category choice is the result of a single multinomial trial. Let $\boldsymbol{y}_i = (y_{i1}, \dots, y_{ic})$ represent the multinomial trial for subject i, $i = 1, \dots, N$, where $y_{ij} = 1$ when the response is in category j and $y_{ij} = 0$ otherwise. Then $\sum_j y_{ij} = 1$, and the multinomial probability distribution for that subject is

$$p(y_{i1}, \dots, y_{ic}) = \pi_{i1}^{y_{i1}} \cdots \pi_{ic}^{y_{ic}}.$$

In this chapter we express models in terms of such ungrouped data. As with binary data, however, with discrete explanatory variables it is better to group the N observations according to their multicategory trial indices $\{n_i\}$ before forming the deviance and other goodness-of-fit statistics and residuals.

Foundations of Linear and Generalized Linear Models, First Edition. Alan Agresti.

6.1 NOMINAL RESPONSES: BASELINE-CATEGORY LOGIT MODELS

For nominal-scale response variables having c categories, multicategory logistic models simultaneously describe the log odds for all $c(c-1)/2$ pairs of categories. Given a certain choice of $c-1$ of these, the rest are redundant.

6.1.1 Baseline-Category Logits

We construct a multinomial logistic model by pairing each response category with a baseline category, such as category c, using

$$\log\frac{\pi_{i1}}{\pi_{ic}},\ \log\frac{\pi_{i2}}{\pi_{ic}},\ldots,\log\frac{\pi_{i,c-1}}{\pi_{ic}}.$$

The jth *baseline-category logit*, $\log(\pi_{ij}/\pi_{ic})$, is the logit of a conditional probability,

$$\begin{aligned}&\text{logit}[P(y_{ij}=1\mid y_{ij}=1\text{ or }y_{ic}=1)]\\&\quad=\log\left[\frac{P(y_{ij}=1\mid y_{ij}=1\text{ or }y_{ic}=1)}{1-P(y_{ij}=1\mid y_{ij}=1\text{ or }y_{ic}=1)}\right]=\log\frac{\pi_{ij}}{\pi_{ic}}.\end{aligned}$$

Let $\boldsymbol{x}_i=(x_{i1},\ldots,x_{ip})$ denote explanatory variable values for subject i, and let $\boldsymbol{\beta}_j=(\beta_{j1},\ldots,\beta_{jp})^{\mathrm{T}}$ denote parameters for the jth logit.

Baseline-category logit model:

$$\log\frac{\pi_{ij}}{\pi_{ic}}=\boldsymbol{x}_i\boldsymbol{\beta}_j=\sum_{k=1}^{p}\beta_{jk}x_{ik},\quad j=1,\ldots,c-1. \tag{6.1}$$

This model, also often called the *multinomial logit model*, simultaneously describes the effects of $\boldsymbol{x}$ on the $c-1$ logits. The effects vary according to the response paired with the baseline.

These $c-1$ equations determine equations for logits with other pairs of response categories, since

$$\log\frac{\pi_{ia}}{\pi_{ib}}=\log\frac{\pi_{ia}}{\pi_{ic}}-\log\frac{\pi_{ib}}{\pi_{ic}}=\boldsymbol{x}_i(\boldsymbol{\beta}_a-\boldsymbol{\beta}_b).$$

As in other models, typically $x_{i1}=1$ for the coefficient of an intercept term, which also differs for each logit. The model treats the response variable as nominal scale, in the following sense: if the model holds and the outcome categories are permuted in any way, the model still holds with the corresponding permutation of the effects.

We can express baseline-category logit models directly in terms of response probabilities $\{\pi_{ij}\}$ by

$$\pi_{ij} = \frac{\exp(\boldsymbol{x}_i\boldsymbol{\beta}_j)}{1 + \sum_{h=1}^{c-1}\exp(\boldsymbol{x}_i\boldsymbol{\beta}_h)} \tag{6.2}$$

with $\boldsymbol{\beta}_c = \mathbf{0}$. (The parameters also equal zero for a baseline category for identifiability reasons; see Exercise 6.2.) The numerators in Equation (6.2) for various j sum to the denominator, so $\sum_{j=1}^{c}\pi_{ij} = 1$ for each i. For $c = 2$, this formula simplifies to the binary logistic regression probability formula (5.2).

Interpretation of effects overall rather than conditional on response in category j or c is not simple, because Equation (6.2) shows that all $\{\boldsymbol{\beta}_h\}$ contribute to π_{ij}. The relation $\partial\pi_i/\partial x_{ik} = \beta_k\pi_i(1-\pi_i)$ for binary logistic regression generalizes to

$$\frac{\partial\pi_{ij}}{\partial x_{ik}} = \pi_{ij}\left(\beta_{jk} - \sum_{j'}\pi_{ij'}\beta_{j'k}\right). \tag{6.3}$$

In particular, this rate of change need not have the same sign as β_{jk}, and the curve for π_{ij} as a function of x_{ik} may change direction as the value of x_{ik} increases (see Exercise 6.4).

6.1.2 Baseline-Category Logit Model is a Multivariate GLM

The GLM $g(\mu_i) = \boldsymbol{x}_i\boldsymbol{\beta}$ for a univariate response variable extends to a *multivariate generalized linear model*. The model applies to random components that have distribution in a multivariate generalization of the exponential dispersion family,

$$f(\boldsymbol{y}_i;\boldsymbol{\theta}_i,\phi) = \exp\left\{\left[\boldsymbol{y}_i^{\mathrm{T}}\boldsymbol{\theta}_i - b(\boldsymbol{\theta}_i)\right]/a(\phi) + c(\boldsymbol{y}_i,\phi)\right\},$$

where $\boldsymbol{\theta}_i$ is the natural parameter. For response vector $\boldsymbol{y}_i$ for subject i, with $\boldsymbol{\mu}_i = E(\boldsymbol{y}_i)$, let $\boldsymbol{g}$ be a vector of link functions. The multivariate GLM has the form

$$\boldsymbol{g}(\boldsymbol{\mu}_i) = \boldsymbol{X}_i\boldsymbol{\beta}, \quad i = 1,\ldots,N, \tag{6.4}$$

where row j of the model matrix $\boldsymbol{X}_i$ for observation i contains values of explanatory variables for response component y_{ij}.

The multinomial distribution is a member of the multivariate exponential dispersion family. The baseline-category logit model is a multivariate GLM. For this representation, we let $\boldsymbol{y}_i = (y_{i1},\ldots,y_{i,c-1})^{\mathrm{T}}$, since $y_{ic} = 1 - (y_{i1} + \cdots + y_{i,c-1})$ is redundant, $\boldsymbol{\mu}_i = (\pi_{i1},\ldots,\pi_{i,c-1})^{\mathrm{T}}$, and

$$g_j(\boldsymbol{\mu}_i) = \log\{\mu_{ij}/[1 - (\mu_{i1} + \cdots + \mu_{i,c-1})]\}.$$

With $(c-1) \times (c-1)p$ model matrix $\boldsymbol{X}_i$ for observation i,

$$\boldsymbol{X}_i\boldsymbol{\beta} = \begin{pmatrix} \boldsymbol{x}_i & \boldsymbol{0} & \cdots & \boldsymbol{0} \\ \boldsymbol{0} & \boldsymbol{x}_i & \cdots & \boldsymbol{0} \\ \vdots & \vdots & \ddots & \vdots \\ \boldsymbol{0} & \boldsymbol{0} & \cdots & \boldsymbol{x}_i \end{pmatrix} \begin{pmatrix} \boldsymbol{\beta}_1 \\ \boldsymbol{\beta}_2 \\ \vdots \\ \boldsymbol{\beta}_{c-1} \end{pmatrix},$$

where $\boldsymbol{0}$ is a $1 \times p$ vector of 0 elements.

6.1.3 Fitting Baseline-Category Logit Models

Maximum likelihood (ML) fitting of baseline-category logit models maximizes the multinomial likelihood subject to $\{\pi_{ij}\}$ simultaneously satisfying the $c-1$ equations that specify the model. The contribution to the log-likelihood from subject i is

$$\log\left(\prod_{j=1}^{c} \pi_{ij}^{y_{ij}}\right) = \sum_{j=1}^{c-1} y_{ij} \log \pi_{ij} + \left(1 - \sum_{j=1}^{c-1} y_{ij}\right) \log \pi_{ic}$$

$$= \sum_{j=1}^{c-1} y_{ij} \log \frac{\pi_{ij}}{\pi_{ic}} + \log \pi_{ic}.$$

Thus, the baseline-category logits are the natural parameters for the multinomial distribution. They are the canonical link functions for multinomial GLMs.

Next we construct the likelihood equations for N independent observations. In the last expression above, we substitute $\boldsymbol{x}_i\boldsymbol{\beta}_j$ for $\log(\pi_{ij}/\pi_{ic})$ and

$$\pi_{ic} = 1 \Bigg/ \left[1 + \sum_{j=1}^{c-1} \exp(\boldsymbol{x}_i\boldsymbol{\beta}_j)\right].$$

Then the log-likelihood function is

$$L(\boldsymbol{\beta}; \boldsymbol{y}) = \log\left[\prod_{i=1}^{N}\left(\prod_{j=1}^{c} \pi_{ij}^{y_{ij}}\right)\right]$$

$$= \sum_{i=1}^{N}\left\{\sum_{j=1}^{c-1} y_{ij}(\boldsymbol{x}_i\boldsymbol{\beta}_j) - \log\left[1 + \sum_{j=1}^{c-1} \exp(\boldsymbol{x}_i\boldsymbol{\beta}_j)\right]\right\}$$

$$= \sum_{j=1}^{c-1}\left[\sum_{k=1}^{p} \beta_{jk}\left(\sum_{i=1}^{N} x_{ik}y_{ij}\right)\right] - \sum_{i=1}^{N} \log\left[1 + \sum_{j=1}^{c-1} \exp(\boldsymbol{x}_i\boldsymbol{\beta}_j)\right].$$

The sufficient statistic for β_{jk} is $\sum_i x_{ik} y_{ij}$. When all $x_{i1} = 1$ for an intercept term, the sufficient statistic for β_{j1} is $\sum_i x_{i1} y_{ij} = \sum_i y_{ij}$, which is the total number of observations in category j. Since

$$\frac{\partial L(\boldsymbol{\beta}, \mathbf{y})}{\partial \beta_{jk}} = \sum_{i=1}^{N} x_{ik} y_{ij} - \sum_{i=1}^{N} \left[\frac{x_{ik} \exp(\boldsymbol{x}_i \boldsymbol{\beta}_j)}{1 + \sum_{\ell=1}^{c-1} \exp(\boldsymbol{x}_i \boldsymbol{\beta}_\ell)} \right] = \sum_{i=1}^{N} x_{ik}(y_{ij} - \pi_{ij}),$$

the likelihood equations are

$$\sum_{i=1}^{N} x_{ik} y_{ij} = \sum_{i=1}^{N} x_{ik} \pi_{ij},$$

with π_{ij} as expressed in Equation (6.2). As with canonical link functions for univariate GLMs, the likelihood equations equate the sufficient statistics to their expected values.

Differentiating again, you can check that

$$\frac{\partial^2 L(\boldsymbol{\beta}, \mathbf{y})}{\partial \beta_{jk} \partial \beta_{jk'}} = - \sum_{i=1}^{N} x_{ik} x_{ik'} \pi_{ij}(1 - \pi_{ij}),$$

and for $j \neq j'$,

$$\frac{\partial^2 L(\boldsymbol{\beta}, \mathbf{y})}{\partial \beta_{jk} \partial \beta_{j'k'}} = \sum_{i=1}^{N} x_{ik} x_{ik'} \pi_{ij} \pi_{ij'}.$$

The information matrix consists of $(c-1)^2$ blocks of size $p \times p$,

$$-\frac{\partial^2 L(\boldsymbol{\beta}, \mathbf{y})}{\partial \boldsymbol{\beta}_j \partial \boldsymbol{\beta}_{j'}^{\mathrm{T}}} = \sum_{i=1}^{N} \pi_{ij}[I(j = j') - \pi_{ij'}] \boldsymbol{x}_i^{\mathrm{T}} \boldsymbol{x}_i,$$

where $I(\cdot)$ is the indicator function. The Hessian is negative-definite, so the log-likelihood function is concave and has a unique maximum. The observed and expected information are identical, so the Newton–Raphson method is equivalent to Fisher scoring for finding the ML parameter estimates, a consequence of the link function being the canonical one. Convergence is usually fast unless at least one estimate is infinite or does not exist (see Note 6.2).

6.1.4 Deviance and Inference for Multinomial Models

For baseline-category logit models, the ML estimator $\hat{\boldsymbol{\beta}}$ has a large-sample normal distribution. As usual, standard errors are square roots of diagonal elements of the inverse information matrix. The $\{\hat{\boldsymbol{\beta}}_j\}$ are correlated. The estimate $(\hat{\boldsymbol{\beta}}_a - \hat{\boldsymbol{\beta}}_b)$ of the effects in the linear predictor for $\log(\pi_{ia}/\pi_{ib})$ does not depend on which category is the baseline.

Statistical inference can use likelihood-ratio, Wald, and score inference methods for GLMs. For example, the likelihood-ratio test for the effect of explanatory variable k tests H_0: $\beta_{1k} = \beta_{2k} = \cdots = \beta_{c-1,k} = 0$ by treating double the change in the maximized log-likelihood from adding that variable to the model as having a null chi-squared distribution with $df = c - 1$. The likelihood-ratio test statistic equals the difference in the deviance values for comparing the models.

The derivation of the deviance shown in Section 5.5.1 for binomial GLMs generalizes directly to multinomial GLMs. For grouped data with n_i trials for the observations at setting i of the explanatory variables, let y_{ij} now denote the *proportion* of observations in category j. The deviance is the likelihood-ratio statistic comparing double the log of the multinomial likelihood $\prod_i \left(\prod_j \pi_{ij}^{n_i y_{ij}}\right)$ evaluated for the model fitted probabilities $\{\hat{\pi}_{ij}\}$ and the unrestricted (saturated model) alternative $\{\tilde{\pi}_{ij} = y_{ij}\}$. The deviance and the Pearson statistic equal

$$G^2 = 2\sum_{i=1}^{N}\sum_{j=1}^{c} n_i y_{ij} \log \frac{n_i y_{ij}}{n_i \hat{\pi}_{ij}}, \quad X^2 = \sum_{i=1}^{N}\sum_{j=1}^{c} \frac{(n_i y_{ij} - n_i \hat{\pi}_{ij})^2}{n_i \hat{\pi}_{ij}}. \tag{6.5}$$

These have the form seen in Section 4.4.4 for Poisson GLMs and in Section 5.5.1 for binomial GLMs of

$$G^2 = 2\sum \text{observed} \log\left(\frac{\text{observed}}{\text{fitted}}\right), \quad X^2 = \sum \frac{(\text{observed} - \text{fitted})^2}{\text{fitted}},$$

with sums taken over all observed counts $\{n_i y_{ij}\}$ and fitted counts $\{n_i \hat{\pi}_{ij}\}$.

As in the binary case, with categorical explanatory variables and the grouped form of the data, G^2 and X^2 are goodness-of-fit statistics that provide a global model check. They have approximate chi-squared null distributions when the expected cell counts mostly exceed about 5. The df equal the number of multinomial probabilities modeled, which is $N(c - 1)$, minus the number of model parameters. The residuals of Section 5.5.3 are useful for follow-up information about poorly fitting models. For ungrouped data (i.e., all $\{n_i = 1\}$), such as when at least one explanatory variable is continuous, formula (6.5) for G^2 remains valid and is used to compare nested unsaturated models.

6.1.5 Discrete-Choice Models

Some applications of multinomial logit models relate to determining effects of explanatory variables on a subject's choice from a discrete set of options—for instance, transportation system to take to work (driving alone, carpooling, bus, subway, walk, bicycle), housing (house, condominium, rental, other), primary shopping location (downtown, mall, catalogs, internet), or product brand. Models for response variables consisting of a discrete set of choices are called *discrete-choice models*.

In most discrete-choice applications, some explanatory variables take different values for different response choices. As predictors of choice of transportation system, the cost and time to reach the destination take different values for each option. As a predictor of choice of product brand, the price varies according to the option. Explanatory variables of this type are called *characteristics of the choices*. They differ from the usual ones, for which values remain constant across the choice set. Such *characteristics of the chooser* include demographic and socioeconomic variables such as gender, race, annual income, and educational attainment.

We introduce the discrete-choice model for the case that the p explanatory variables are all characteristics of the choices. For subject i and response choice j, let $\boldsymbol{x}_{ij} = (x_{ij1}, \ldots, x_{ijp})$ denote the values of those variables. The discrete choice model for the probability of selecting option j is

$$\pi_{ij} = \frac{\exp(\boldsymbol{x}_{ij}\boldsymbol{\beta})}{\sum_{h=1}^{c} \exp(\boldsymbol{x}_{ih}\boldsymbol{\beta})}. \tag{6.6}$$

For each pair of choices a and b, this model has the logit form for conditional probabilities,

$$\log(\pi_{ia}/\pi_{ib}) = (\boldsymbol{x}_{ia} - \boldsymbol{x}_{ib})\boldsymbol{\beta}. \tag{6.7}$$

Conditional on the choice being a or b, a variable's influence depends on the distance between the subject's values of that variable for those choices. If the values are the same, the model asserts that the variable has no influence on the choice between a and b. The effects $\boldsymbol{\beta}$ are identical for each pair of choices.

From Equation (6.7), the odds of choosing a over b do not depend on the other alternatives in the choice set or on their values of the explanatory variables. This property is referred to as *independence from irrelevant alternatives*. For this to be at all realistic, the model should be used only when the alternatives are distinct and regarded separately by the person making the choice.

A more general version of the model permits the choice set to vary among subjects. For instance, in a study of the choice of transportation system to take to work, some people may not have the subway as an option. In the denominator of Equation (6.6), the sum is then taken over the choice set for subject i.

6.1.6 Baseline-Category Logit Model as a Discrete-Choice Model

Discrete-choice models can also include characteristics of the chooser. A baseline-category logit model (6.2) with such explanatory variables can be expressed in the discrete-choice form (6.6) when we replace each explanatory variable by c artificial variables. The jth is the product of the explanatory variable with an indicator variable that equals 1 when the response choice is j. For instance, for a single explanatory variable with value x_i for subject i and linear predictor $\beta_{0j} + \beta_{1j}x_i$ for the jth logit, we form the $1 \times 2c$ vectors

$$z_{i1} = (1, 0, \dots, 0, x_i, 0, \dots, 0), \dots, z_{ic} = (0, 0, \dots, 1, 0, 0, \dots, x_i).$$

Let $\boldsymbol{\beta} = (\beta_{01}, \dots, \beta_{0c}, \beta_{11}, \dots, \beta_{1c})^{\mathrm{T}}$. Then $z_{ij}\boldsymbol{\beta} = \beta_{0j} + \beta_{1j}x_i$, and Equation (6.2) is (with $\beta_{0c} = \beta_{1c} = 0$ for identifiability)

$$\pi_{ij} = \frac{\exp(\beta_{0j} + \beta_{1j}x_i)}{\exp(\beta_{01} + \beta_{11}x_i) + \cdots + \exp(\beta_{0c} + \beta_{1c}x_i)} = \frac{\exp(z_{ij}\boldsymbol{\beta})}{\exp(z_{i1}\boldsymbol{\beta}) + \cdots + \exp(z_{ic}\boldsymbol{\beta})}.$$

This has the discrete-choice model form (6.6).

This model extends directly to having multiple explanatory variables of each type. With this approach, the discrete-choice model is very general. The ordinary baseline-category logit model is a special case.

6.2 ORDINAL RESPONSES: CUMULATIVE LOGIT AND PROBIT MODELS

For ordinal response variables, models have terms that reflect ordinal characteristics such as a monotone trend, whereby responses tend to fall in higher (or lower) categories as the value of an explanatory variable increases. Such models are more parsimonious than models for nominal responses, because potentially they have many fewer parameters. In this section we introduce logistic and probit models for ordinal responses.

6.2.1 Cumulative Logit Models: Proportional Odds

Let y_i denote the response outcome category for subject i. That is, $y_i = j$ means that $y_{ij} = 1$ and $y_{ik} = 0$ for $k \neq j$, for the c multinomial indicators. To use the category ordering, we express models in terms of the cumulative probabilities,

$$P(y_i \leq j) = \pi_{i1} + \cdots + \pi_{ij}, \quad j = 1, \dots, c.$$

The *cumulative logits* are logits of these cumulative probabilities,

$$\text{logit}[P(y_i \leq j)] = \log \frac{P(y_i \leq j)}{1 - P(y_i \leq j)} = \log \frac{\pi_{i1} + \cdots + \pi_{ij}}{\pi_{i,j+1} + \cdots + \pi_{ic}}, \quad j = 1, \dots, c-1.$$

Each cumulative logit uses all c response categories.

A model for logit$[P(y_i \leq j)]$ alone is an ordinary logistic model for a binary response in which categories 1 to j represent "success" and categories $j+1$ to c represent "failure." Here is a parsimonious model that simultaneously uses all $(c-1)$ cumulative logits:

Cumulative logit model:

$$\text{logit}[P(y_i \leq j)] = \alpha_j + \boldsymbol{x}_i\boldsymbol{\beta}, \quad j = 1, \ldots, c-1. \tag{6.8}$$

Each cumulative logit has its own intercept. The $\{\alpha_j\}$ are increasing in j, because $P(y_i \leq j)$ increases in j at any fixed $\boldsymbol{x}_i$, and the logit is an increasing function of $P(y_i \leq j)$. We use separate notation α_j and show the intercept terms by themselves in the linear predictor, because they depend on j but the other effects do not. This model states that the effects $\boldsymbol{\beta}$ of the explanatory variables are the same for each cumulative logit. For a single continuous explanatory variable x, Figure 6.1 depicts the model when $c = 4$. The curves for $j = 1$, 2, and 3 have exactly the same shape and do not cross.

This model treats the response variable as ordinal scale, in the following sense: if the model holds and the order of the outcome categories is reversed, the model still holds with a change in the sign of $\boldsymbol{\beta}$; however, the model need not hold if the outcome categories are permuted in any other way.

The cumulative logit model (6.8) satisfies

$$\text{logit}[P(y_i \leq j \mid \boldsymbol{x}_i = \boldsymbol{u})] - \text{logit}[P(y_i \leq j \mid \boldsymbol{x}_i = \boldsymbol{v})]$$
$$= \log \frac{P(y_i \leq j \mid \boldsymbol{x}_i = \boldsymbol{u})/P(y_i > j \mid \boldsymbol{x}_i = \boldsymbol{u})}{P(y_i \leq j \mid \boldsymbol{x}_i = \boldsymbol{v})/P(y_i > j \mid \boldsymbol{x}_i = \boldsymbol{v})} = (\boldsymbol{u} - \boldsymbol{v})\boldsymbol{\beta}.$$

The odds that the response $\leq j$ at $\boldsymbol{x}_i = \boldsymbol{u}$ are $\exp[(\boldsymbol{u} - \boldsymbol{v})\boldsymbol{\beta}]$ times the odds at $\boldsymbol{x}_i = \boldsymbol{v}$. An odds ratio of cumulative probabilities is called a *cumulative odds ratio*. The log cumulative odds ratio is proportional to the distance between $\boldsymbol{u}$ and $\boldsymbol{v}$. For each j, the odds that $y_i \leq j$ multiply by $\exp(\beta_k)$ per 1-unit increase in x_{ik}, adjusting for the other explanatory variables. The same proportionality constant applies to all $c - 1$

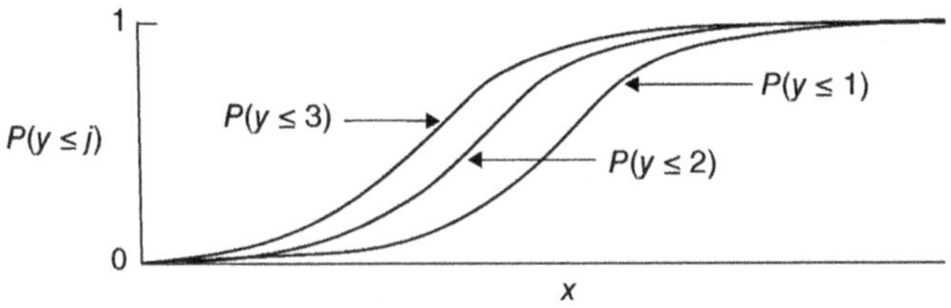

Figure 6.1 Cumulative logit model with the same effect of x on each of three cumulative probabilities, for an ordinal response variable with $c = 4$ categories.

cumulative logits; that is, the effect is β_k, not β_{jk}. This property of a common effect for all the cumulative probabilities is referred to as *proportional odds*.

6.2.2 Latent Variable Motivation for Proportional Odds Structure

A linear model for a latent continuous variable assumed to underlie y motivates the common effect $\boldsymbol{\beta}$ for different j in the proportional odds form of the cumulative logit model. Let y_i^* denote this underlying latent variable for subject i. Suppose it has cdf $G(y_i^* - \mu_i)$, where values of y^* vary around a mean that depends on $\boldsymbol{x}$ through $\mu_i = \boldsymbol{x}_i\boldsymbol{\beta}$. Suppose that the continuous scale has *cutpoints* $-\infty = \alpha_0 < \alpha_1 < \cdots < \alpha_c = \infty$ such that we observe

$$y_i = j \quad \text{if } \alpha_{j-1} < y_i^* \leq \alpha_j.$$

That is, y_i falls in category j when the latent variable falls in the jth interval of values, as Figure 6.2 depicts. Then

$$P(y_i \leq j) = P\left(y_i^* \leq \alpha_j\right) = G(\alpha_j - \mu_i) = G(\alpha_j - \boldsymbol{x}_i\boldsymbol{\beta}).$$

The model for y implies that the link function G^{-1}, the inverse of the cdf for y^*, applies to $P(y_i \leq j)$. If $y_i^* = \boldsymbol{x}_i\boldsymbol{\beta} + \epsilon_i$, where the cdf G of ϵ_i is the standard logistic (Section 5.1.3), then G^{-1} is the logit link, and the cumulative logit model with proportional odds structure results. Normality for ϵ_i implies a probit link (Section 6.2.3) for the cumulative probabilities.

Using a cdf of the form $G(y_i^* - \mu_i)$ for the latent variable results in the linear predictor $\alpha_j - \boldsymbol{x}_i\boldsymbol{\beta}$ rather than $\alpha_j + \boldsymbol{x}_i\boldsymbol{\beta}$. With this alternate parameterization, when $\beta_k > 0$, as x_{ik} increases, each cumulative logit decreases, so each cumulative probability decreases and relatively less probability mass falls at the low end of the y scale. Thus, y_i tends to be larger at higher values of x_{ik}. Then the sign of β_k has the usual

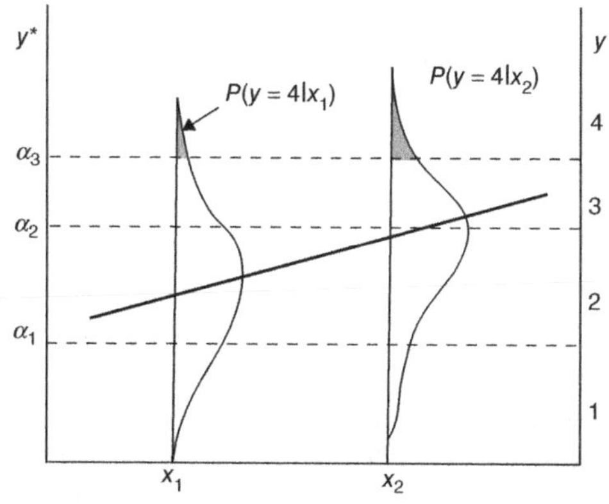

Figure 6.2 Ordinal measurement and underlying linear model for a latent variable.

meaning of a positive or negative effect. When you use software to fit the model, you should check whether it parameterizes the linear predictor as $\alpha_j + \boldsymbol{x}_i\boldsymbol{\beta}$ or as $\alpha_j - \boldsymbol{x}_i\boldsymbol{\beta}$, as signs of estimated effects differ accordingly.

In the latent variable derivation, the same parameters $\boldsymbol{\beta}$ occur for the effects regardless of how the cutpoints $\{\alpha_j\}$ chop up the scale for y^* and regardless of the number of categories. The effect parameters are invariant to the choice of categories for y. This feature makes it possible to compare $\hat{\boldsymbol{\beta}}$ from studies using different response scales.

6.2.3 Cumulative Probit and Other Cumulative Link Models

As in binary GLMs, other link functions are possible for the cumulative probabilities. Let G^{-1} denote a link function that is the inverse of the continuous cdf G. The *cumulative link model*

$$G^{-1}[P(y_i \le j)] = \alpha_j + \boldsymbol{x}_i\boldsymbol{\beta} \tag{6.9}$$

links the cumulative probabilities to the linear predictor. As in the cumulative logit model with proportional odds form (6.8), effects are the same for each cumulative probability. This assumption holds when a latent variable y^* satisfies a linear model with standard cdf G for the error term. Thus, we can regard cumulative link models as linear models that use a linear predictor $\boldsymbol{x}_i\boldsymbol{\beta}$ to describe effects of explanatory variables on a crude ordinal measurement y_i of y_i^*.

The *cumulative probit model* is the cumulative link model that uses the standard normal cdf Φ for G. This generalizes the binary probit model (Section 5.6) to ordinal responses. Cumulative probit models provide fits similar to cumulative logit models. They have smaller estimates and standard errors because the standard normal distribution has standard deviation 1.0 compared with 1.81 for the standard logistic. When we expect an underlying latent variable to be highly skewed, such as an extreme-value distribution, we can generalize the binary model with log–log or complementary log–log link (Section 5.6.3) to ordinal responses.

Effects in cumulative link models can be interpreted in terms of the underlying latent variable model. The cumulative probit model is appropriate when the latent variable model holds with a normal conditional distribution for y^*. Consider the parameterization $\Phi^{-1}[P(y_i \le j)] = \alpha_j - \boldsymbol{x}_i\boldsymbol{\beta}$. From Section 6.2.2, $y_i^* = \boldsymbol{x}_i\boldsymbol{\beta} + \epsilon_i$ where $\epsilon_i \sim N(0, 1)$ has cdf Φ. A 1-unit increase in x_{ik} corresponds to a β_k increase (and thus an increase of β_k standard deviations) in $E(y_i^*)$, adjusting for the other explanatory variables.

To describe predictive power, an analog of the multiple correlation for linear models is the correlation between the latent variable y^* and the fitted linear predictor values. This directly generalizes the measure for binary data presented in Section 5.2.5. Its square is an R^2 analog (McKelvey and Zavoina 1975). This R^2 measure equals the estimated variance of $\hat{y}^*$ divided by the estimated variance of y^*. Here, $\hat{y}_i^* = \sum_j \hat{\beta}_j x_{ij}$ is the same as the estimated linear predictor without the intercept term, and the estimated variance of y^* equals this plus the variance of ϵ in the latent variable model (1 for the probit link and $\pi^2/3 = 3.29$ for the logit link). It is also helpful to summarize the effect of an explanatory variable in terms of the change in the probability of the highest (or the lowest) category of the ordinal scale over the range or interquartile range of that variable, at mean values of other explanatory variables.

6.2.4 Fitting and Checking Cumulative Link Models

For multicategory indicator $(y_{i1}, \ldots, y_{ic})$ of the response for subject i, the multinomial likelihood function for the cumulative link model $G^{-1}[P(y_i \leq j)] = \alpha_j + \boldsymbol{x}_i\boldsymbol{\beta}$ is

$$\prod_{i=1}^{N}\left(\prod_{j=1}^{c}\pi_{ij}^{y_{ij}}\right) = \prod_{i=1}^{N}\left\{\prod_{j=1}^{c}[P(y_i \leq j) - P(y_i \leq j-1)]^{y_{ij}}\right\}$$

viewed as a function of $(\{\alpha_j\}, \boldsymbol{\beta})$, where $P(y_i \leq 0) = 0$. The log-likelihood function is

$$L(\boldsymbol{\alpha}, \boldsymbol{\beta}) = \sum_{i=1}^{N}\sum_{j=1}^{c} y_{ij} \log[G(\alpha_j + \boldsymbol{x}_i\boldsymbol{\beta}) - G(\alpha_{j-1} + \boldsymbol{x}_i\boldsymbol{\beta})].$$

Let g denote the derivative of G, that is, the pdf corresponding to the cdf G, and let δ_{jk} denote the Kronecker delta, $\delta_{jk} = 1$ if $j = k$ and $\delta_{jk} = 0$ otherwise. Then the likelihood equations are

$$\frac{\partial L}{\partial \beta_k} = \sum_{i=1}^{N}\sum_{j=1}^{c} y_{ij}x_{ik}\frac{g(\alpha_j + \boldsymbol{x}_i\boldsymbol{\beta}) - g(\alpha_{j-1} + \boldsymbol{x}_i\boldsymbol{\beta})}{G(\alpha_j + \boldsymbol{x}_i\boldsymbol{\beta}) - G(\alpha_{j-1} + \boldsymbol{x}_i\boldsymbol{\beta})} = 0,$$

and

$$\frac{\partial L}{\partial \alpha_k} = \sum_{i=1}^{N}\sum_{j=1}^{c} y_{ij}\frac{\delta_{jk}g(\alpha_j + \boldsymbol{x}_i\boldsymbol{\beta}) - \delta_{j-1,k}g(\alpha_{j-1} + \boldsymbol{x}_i\boldsymbol{\beta})}{G(\alpha_j + \boldsymbol{x}_i\boldsymbol{\beta}) - G(\alpha_{j-1} + \boldsymbol{x}_i\boldsymbol{\beta})} = 0.$$

The Hessian matrix is rather messy[1] and not shown here. The likelihood equations can be solved using Fisher scoring or the Newton–Raphson method. The *SE* values differ somewhat for the two methods, because the expected and observed information matrices are not the same for this noncanonical link model.

Since the latent variable model on which the cumulative link model is based describes location effects while assuming constant variability, settings of the explanatory variables are *stochastically ordered* on the response: for the observed

[1]This is because $\boldsymbol{\alpha}$ and $\boldsymbol{\beta}$ are not orthogonal. See Agresti (2010, Section 5.1.2).

response with any pair $\boldsymbol{u}$ and $\boldsymbol{v}$ of potential explanatory variable values, either $P(y_i \leq j \mid \boldsymbol{x}_i = \boldsymbol{u}) \leq P(y_i \leq j \mid \boldsymbol{x}_i = \boldsymbol{v})$ for all j or $P(y_i \leq j \mid \boldsymbol{x}_i = \boldsymbol{u}) \geq P(y_i \leq j \mid \boldsymbol{x}_i = \boldsymbol{v})$ for all j. When this is violated and such models fit poorly, often it is because the response variability also varies with $\boldsymbol{x}$. For example, with $c = 4$ and response probabilities (0.3, 0.2, 0.2, 0.3) at $\boldsymbol{u}$ and (0.1, 0.4, 0.4, 0.1) at $\boldsymbol{v}$, $P(y_i \leq 1 \mid \boldsymbol{x}_i = \boldsymbol{u}) > P(y_i \leq 1 \mid \boldsymbol{x}_i = \boldsymbol{v})$ but $P(y_i \leq 3 \mid \boldsymbol{x}_i = \boldsymbol{u}) < P(y_i \leq 3 \mid \boldsymbol{x}_i = \boldsymbol{v})$. At $\boldsymbol{u}$ the responses concentrate more in the extreme categories than at $\boldsymbol{v}$.

An advantage of the simple structure of the same effects $\boldsymbol{\beta}$ for different cumulative probabilities is that effects are simple to summarize and are parsimonious, requiring only a single parameter for each explanatory variable. The models generalize to include separate effects, replacing $\boldsymbol{\beta}$ in Equation (6.9) by $\boldsymbol{\beta}_j$. This implies nonparallelism of curves for different cumulative probabilities. Curves may then cross for some $\boldsymbol{x}$ values, violating the proper order among the cumulative probabilities (Exercise 6.13).

When we can fit the more general model with effects $\{\boldsymbol{\beta}_j\}$, a likelihood-ratio test checks whether that model fits significantly better. Often though, convergence fails in fitting the model, because the cumulative probabilities are out of order. We can then use a score test of whether the $\{\boldsymbol{\beta}_j\}$ takes a common value, because the score test uses the likelihood function only at the null (i.e., where common $\boldsymbol{\beta}$ holds). Some software for the model provides this score test. It is often labelled as a "test of the proportional odds assumption," because that is the name for the simple structure when we use the logit link. When there is strong evidence against the common $\boldsymbol{\beta}$, the simpler model is still often useful for describing overall effects, with the fit of the more general model pointing out ways to fine-tune the description of effects. With categorical predictors and grouped data with most expected cell counts exceeding about 5, the deviance and Pearson statistics in Equation (6.5) provide global chi-squared goodness-of-fit tests.

The log-likelihood function is concave for many cumulative link models, including the logit and probit. Iterative algorithms such as Fisher scoring usually converge rapidly to the ML estimates.

6.2.5 Why not Use OLS Regression to Model Ordinal Responses?

Many methodologists analyze ordinal response data by ignoring the categorical nature of y and assigning numerical scores to the ordered categories and using ordinary least squares (OLS) methods such as linear regression and ANOVA. That approach can identify variables that clearly affect y and provide simple description, but it has limitations. First, usually there is no clear-cut choice for the scores. How would you assign scores to categories such as (never, rarely, sometimes, always)? Second, a particular ordinal outcome is consistent with a range of values for some underlying latent variable. Ordinary linear modeling does not allow for the measurement error that results from replacing such a range by a single numerical value. Third, the approach does not yield estimated probabilities for the ordinal categories at fixed settings of the explanatory variables. Fourth, the approach ignores that the variability of y is naturally nonconstant for categorical data: an ordinal response has little variability at explanatory variable values for which observations fall mainly in the highest category (or mainly in the lowest category), but considerable variability at values for which observations are spread among the categories.

Related to the second and fourth limitations, the ordinary linear modeling approach does not account for "ceiling effects" and "floor effects" that occur because of the upper and lower limits for y. Such effects can cause ordinary linear modeling to give misleading[2] results. To illustrate, we apply a normal linear model to simulated data with an ordinal response variable y based on an underlying normal latent variable y^*. We generated 100 observations as follows: x_i values were independent uniform variates between 0 and 100, z_i values were independent with $P(z_i = 0) = P(z_i = 1) = 0.50$, and the latent y_i^* was a normal variate with mean

$$E\left(y_i^*\right) = 20.0 + 0.6x_i - 40.0z_i$$

and standard deviation 10. The first scatterplot in Figure 6.3 shows the 100 observations on y_i^* and x_i, each data point labelled by the category for z_i. The plot also shows the OLS fit for this model.

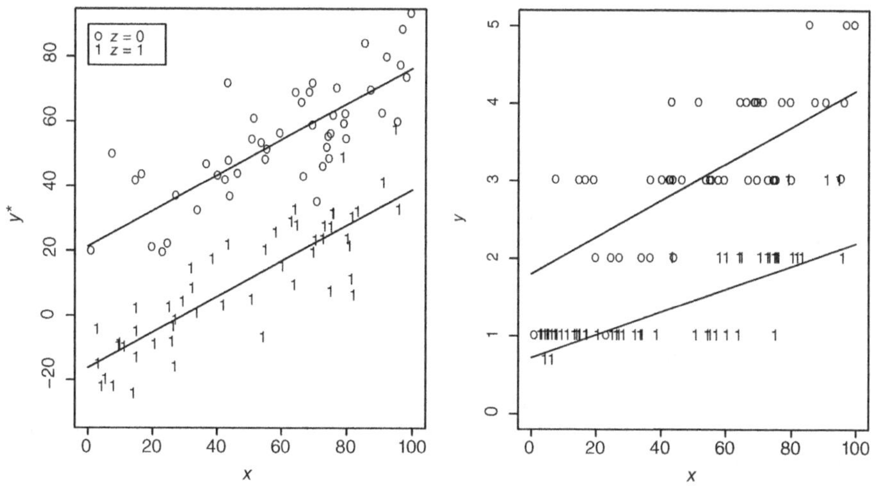

Figure 6.3 Ordinal data (in second panel) for which ordinary linear modeling suggests interaction, because of a floor effect, but ordinal modeling does not. The data were generated (in first panel) from a normal linear model with continuous (x) and binary (z) explanatory variables. When the continuous y^* is categorized and y is measured as (1, 2, 3, 4, 5), the observations labeled "1" for the category of z have a linear x effect with only half the slope of the observations labeled "0."

[2]These effects also result in substantial correlation between the values of residuals and the values of quantitative explanatory variables.

We then categorized the 100 generated values on y^* into five categories to create observations for an ordinal variable y, as follows:

$$y_i = 1 \text{ if } y_i^* \leq 20, \quad y_i = 2 \text{ if } 20 < y_i^* \leq 40, \quad y_i = 3 \text{ if } 40 < y_i^* \leq 60,$$
$$y_i = 4 \text{ if } 60 < y_i^* \leq 80, \quad y_i = 5 \text{ if } y_i^* > 80.$$

The second scatterplot in Figure 6.3 shows the data as they would actually be observed. Using OLS with scores (1, 2, 3, 4, 5) for y suggests a better fit when the model has an interaction term, allowing different slopes relating $E(y_i)$ to x_i when $z_i = 0$ and when $z_i = 1$. The second scatterplot shows the OLS fit of the linear model $E(y_i) = \beta_0 + \beta_1 x_i + \beta_2 z_i + \beta_3(x_i \times z_i)$. The slope of the line is about twice as high when $z_i = 0$ as when $z_i = 1$. This interaction effect is caused by the observations when $z_i = 1$ tending to fall in category $y_i = 1$ whenever x_i takes a relatively low value. As x_i gets lower, the underlying value y_i^* can continue to tend to get lower, but y_i cannot fall below 1. So at low x_i values there is a floor effect, particularly for observations with $z_i = 1$.

Standard ordinal models such as the cumulative logit model with proportional odds structure fit these data well without the need for an interaction term. In fact, by the latent variable model of Section 6.2.2, the true structure is that of a cumulative probit model with common $\boldsymbol{\beta}$. Such models allow for underlying values of y^* when $z = 1$ to be below those when $z = 0$ even if x is so low that y is very likely to be in the first category at both levels of z.

6.3 EXAMPLES: NOMINAL AND ORDINAL RESPONSES

In this chapter we have presented the most popular models for multinomial response variables. End-of-chapter exercises briefly introduce other models for nominal and ordinal responses that are beyond our scope.

6.3.1 Issues in Selecting Multinomial Models

With a nominal-scale response variable, the baseline category logit form of model is usually the default choice. It is not sensible to force similar effects for different logits, so the model will necessarily contain a large number of parameters unless c and p are small.

With ordinal response variables, the choice is not so clear, because we can treat y as nominal when cumulative link models that assume a common value $\boldsymbol{\beta}$ for $\{\boldsymbol{\beta}_j\}$ fit poorly. Especially with large samples, it is not surprising to obtain a small P-value in comparing a model with common $\boldsymbol{\beta}$ to one with separate $\{\boldsymbol{\beta}_j\}$. However, it is poor practice to be guided merely by statistical significance in selecting a model. Even if a more complex model fits significantly better, for reasons of parsimony the simpler model might be preferable when $\{\hat{\boldsymbol{\beta}}_j\}$ are not substantially different in practical terms. It is helpful to check whether the violation of the common $\boldsymbol{\beta}$ property is substantively important, by comparing $\hat{\boldsymbol{\beta}}$ to $\{\hat{\boldsymbol{\beta}}_j\}$ obtained from fitting the more general model (when possible) or from separate fits to the binary collapsings of the response.

Although effect estimators using the simpler model are biased, because of the bias–variance tradeoff (Section 4.6.2) they may have smaller overall mean squared error for estimating the category probabilities. This is especially true when c is large, as the difference is then quite large between the numbers of parameters in the two models.

If a model with common $\boldsymbol{\beta}$ fits poorly in practical terms, alternative strategies exist. For example, you can report $\{\hat{\boldsymbol{\beta}}_j\}$ from the more general model, to investigate the nature of the lack of fit and to describe effects separately for each cumulative probability. Or, you could use an alternative ordinal logit model for which the more complex nonproportional odds form is also valid[3]. Or you can fit baseline-category logit models and use the ordinality in an informal way in interpreting the associations.

6.3.2 Example: Baseline-Category Logit Model for Alligator Food Choice

Table 6.1 comes from a study of factors influencing the primary food choice of alligators. The study captured 219 alligators in four Florida lakes. The nominal-scale response variable is the primary food type, in volume, found in an alligator's stomach. Table 6.1 classifies the primary food choice according to the lake of capture and the size of the alligator. Here, size is binary, distinguishing between nonadult (length $\leq$ 2.3 meters) and adult alligators.

We use baseline-category logit models to investigate the effects of size and lake on the primary food choice. Fish was the most frequent primary food choice, and we use it as the baseline category. We estimate the effects on the odds that alligators select other primary food types instead of fish. Let $s = 1$ for alligator size ≤ 2.3 meters and 0 otherwise, and let z^H, z^O, z^{T}, and z^G be indicator variables for the lakes ($z = 1$ for alligators in that lake and 0 otherwise). The model with main effects is

$$\log(\pi_{ij}/\pi_{i1}) = \beta_{j0} + \beta_{j1}s_i + \beta_{j2}z_i^O + \beta_{j3}z_i^{\mathrm{T}} + \beta_{j4}z_i^G, \quad \text{for } j = 2, 3, 4, 5.$$

In R the `vglm` function in the `VGAM` package can fit this model[4].

Table 6.1 Primary Food Choice of Alligators, by Lake and Size of the Alligator

Lake	Size (meters)	Primary Food Choice				
		Fish	Invertebrate	Reptile	Bird	Other
Hancock	≤ 2.3	23	4	2	2	8
	> 2.3	7	0	1	3	5
Ocklawaha	≤ 2.3	5	11	1	0	3
	> 2.3	13	8	6	1	0
Trafford	≤ 2.3	5	11	2	1	5
	> 2.3	8	7	6	3	5
George	≤ 2.3	16	19	1	2	3
	> 2.3	17	1	0	1	3

Source: Data courtesy of Mike Delany and Clint Moore. For details, see Delany et al. (1999).

[3]See Exercise 6.8 and Agresti (2010, Chapter 4).

[4]The `multinom` function in the `nnet` R package also fits it. The VGAM package fits many models for discrete data. See `www.stat.auckland.ac.nz/~yee/VGAM/doc/VGAM.pdf`.

```
-----------------------------------------------------------------------------
> Alligators # file Alligators.dat at www.stat.ufl.edu/~aa/glm/data
  lake size y1 y2 y3 y4 y5
1    1    1 23  4  2  2  8
2    1    0  7  0  1  3  5
...
8    4    0 17  1  0  1  3
> attach(Alligators)
> library(VGAM)
> fit <- vglm(formula = cbind(y2,y3,y4,y5,y1) ~ size + factor(lake),
+   family=multinomial, data=Alligators) # fish=1 is baseline category
> summary(fit)
                   Estimate  Std. Error   z value
(Intercept):1       -3.2074      0.6387   -5.0215
(Intercept):2       -2.0718      0.7067   -2.9315
(Intercept):3       -1.3980      0.6085   -2.2973
(Intercept):4       -1.0781      0.4709   -2.2893
size:1               1.4582      0.3959    3.6828
size:2              -0.3513      0.5800   -0.6056
size:3              -0.6307      0.6425   -0.9816
size:4               0.3315      0.4482    0.7397
factor(lake)2:1      2.5956      0.6597    3.9344
factor(lake)2:2      1.2161      0.7860    1.5472
factor(lake)2:3     -1.3483      1.1635   -1.1588
factor(lake)2:4     -0.8205      0.7296   -1.1247
factor(lake)3:1      2.7803      0.6712    4.1422
factor(lake)3:2      1.6925      0.7804    2.1686
factor(lake)3:3      0.3926      0.7818    0.5023
factor(lake)3:4      0.6902      0.5597    1.2332
factor(lake)4:1      1.6584      0.6129    2.7059
factor(lake)4:2     -1.2428      1.1854   -1.0484
factor(lake)4:3     -0.6951      0.7813   -0.8897
factor(lake)4:4     -0.8262      0.5575   -1.4819
---
Residual deviance: 17.0798 on 12 degrees of freedom
Log-likelihood: -47.5138 on 12 degrees of freedom
> 1 - pchisq(17.0798, df=12)
[1] 0.146619  # P-value for deviance goodness-of-fit test
-----------------------------------------------------------------------------
```

The data are a bit sparse, but the deviance of 17.08 ($df = 12$) does not give much evidence against the main-effects model. The df value reflects that we have modeled 32 multinomial probabilities (4 at each combination of size and lake) using 20 parameters (5 for each logit). The more complex model allowing interaction between size and lake has 12 more parameters and is the saturated model. Removing size or lake from the main-effects model results in a significantly poorer fit: the deviance increases by 21.09 ($df = 4$) in removing size and 49.13 ($df = 12$) in removing lake.

The equations shown for fish as the baseline determine those for other primary food choice pairs. Viewing all these, we see that size has its greatest impact on whether invertebrates rather than fish are the primary food choice. The prediction equation for the log odds of selecting invertebrates instead of fish is

$$\log(\hat{\pi}_{i2}/\hat{\pi}_{i1}) = -3.207 + 1.458s_i + 2.596z_i^O + 2.780z_i^T + 1.658z_i^G.$$

For a given lake, for small alligators the estimated odds that primary food choice was invertebrates instead of fish are exp(1.458) = 4.30 times the estimated odds for large alligators. The estimated effect is imprecise, as the Wald 95% confidence interval is $\exp[1.458 \pm 1.96(0.396)] = (1.98, 9.34)$. The lake effects indicate that the estimated odds that the primary food choice was invertebrates instead of fish are relatively higher at lakes Ocklawaha, Trafford and George than they are at Lake Hancock.

The model parameter estimates yield fitted probabilities. For example, the estimated probability that a large alligator in Lake George has invertebrates as the primary food choice is

$$\hat{\pi}_{i2} = \frac{e^{-3.207+1.658}}{1 + e^{-3.207+1.658} + e^{-2.072-1.243} + e^{-1.398-0.695} + e^{-1.078-0.826}} = 0.14.$$

The estimated probabilities of (invertebrates, reptiles, birds, other, fish) for large alligators in that lake are (0.14, 0.02, 0.08, 0.10, 0.66).

```
-------------------------------------------------------------------
> fitted(fit)
      y2       y3       y4       y5       y1
1  0.0931   0.0475   0.0704   0.2537   0.5353
2  0.0231   0.0718   0.1409   0.1940   0.5702
...
8  0.1397   0.0239   0.0811   0.0979   0.6574
-------------------------------------------------------------------
```

6.3.3 Example: Cumulative Link Models for Mental Impairment

The data in Table 6.2 are based on a study of mental health for a random sample of adult residents of Alachua County, Florida[5]. Mental impairment is ordinal, with categories (1 = well, 2 = mild symptom formation, 3 = moderate symptom formation, 4 = impaired). The study related y = mental impairment to several explanatory variables, two of which are used here. The life events index x_1 is a composite measure of the number and severity of important life events that occurred to the subject within the past 3 years, such as the birth of a child, a new job, a divorce, or a death in the family. In this sample, x_1 has a mean of 4.3 and standard deviation of 2.7. Socioeconomic status (x_2 = SES) is measured here as binary (1 = high, 0 = low).

The cumulative logit model of the proportional odds form with main effects has ML fit

$$\text{logit}[\hat{P}(y_i \leq j)] = \hat{\alpha}_j - 0.319x_{i1} + 1.111x_{i2}.$$

[5]Thanks to Charles Holzer for the background for this study; the 40 observations analyzed here are merely reflective of patterns found with his much larger sample.

Table 6.2 Mental Impairment, Life Events Index, and Socioeconomic Status (SES), for 40 Adults in Alachua County, Florida

Subject	Mental Impairment	Life Events	SES	Subject	Mental Impairment	Life Events	SES
1	Well	1	1	21	Mild	9	1
2	Well	9	1	22	Mild	3	0
3	Well	4	1	23	Mild	3	1
4	Well	3	1	24	Mild	1	1
5	Well	2	0	25	Moderate	0	0
6	Well	0	1	26	Moderate	4	1
7	Well	1	0	27	Moderate	3	0
8	Well	3	1	28	Moderate	9	0
9	Well	3	1	29	Moderate	6	1
10	Well	7	1	30	Moderate	4	0
11	Well	1	0	31	Moderate	3	0
12	Well	2	0	32	Impaired	8	1
13	Mild	5	1	33	Impaired	2	1
14	Mild	6	0	34	Impaired	7	1
15	Mild	3	1	35	Impaired	5	0
16	Mild	1	0	36	Impaired	4	0
17	Mild	8	1	37	Impaired	4	0
18	Mild	2	1	38	Impaired	8	1
19	Mild	5	0	39	Impaired	8	0
20	Mild	5	1	40	Impaired	9	0

The estimated cumulative probability, starting at the "well" end of the mental impairment scale, decreases as life events increases and is higher at the higher level of SES, adjusted for the other variable. Given the life events score, at the high SES level the estimated odds of mental impairment below any fixed level are $e^{1.111} = 3.0$ times the estimated odds at the low SES level. The 95% Wald confidence interval[6] for this effect is $\exp[1.111 \pm 1.96(0.614)] = (0.91, 10.12)$. A null SES effect is plausible, but the SES effect could also be very strong. The Wald test shows strong evidence of a life events effect.

```
------------------------------------------------------------------------
> Mental # file Mental.dat at www.stat.ufl.edu/~aa/glm/data
   impair life ses  # impair has well=1, ... , impaired=4
1        1    1   1
2        1    9   1
...
40       4    9   0
> attach(Mental)
> library(VGAM) # Alternative is polr function in MASS package
> fit <- vglm(impair ~ life + ses, family=cumulative(parallel=TRUE),
```

[6]The `ProfileLikelihood` package in R has a function for profile likelihood intervals for this model.

```
+      data=Mental)  # parallel=TRUE imposes proportional odds structure
> summary(fit)
                 Estimate  Std. Error  z value
(Intercept):1    -0.2818       0.6230  -0.4522 # c-1 = 3 intercepts for
(Intercept):2     1.2129       0.6512   1.8626 # c=4 response categories
(Intercept):3     2.2095       0.7172   3.0807
life             -0.3189       0.1194  -2.6697
ses               1.1111       0.6143   1.8088
Residual deviance: 99.0979 on 115 degrees of freedom
Log-likelihood: -49.54895 on 115 degrees of freedom
-----------------------------------------------------------------------
```

To help us interpret the effects, we can estimate response category probabilities. First, consider the SES effect. At the mean life events of 4.3, $\hat{P}(y = 1) = 0.37$ at high SES (i.e., $x_2 = 1$) and $\hat{P}(y = 1) = 0.16$ at low SES ($x_2 = 0$). Next, consider the life events effect. For high SES, $\hat{P}(y = 1)$ changes from 0.70 to 0.12 between the sample minimum of 0 and maximum of 9 life events; for low SES, it changes from 0.43 to 0.04. Comparing 0.70 to 0.43 at the minimum life events and 0.12 to 0.04 at the maximum provides a further description of the SES effect. The sample effect is substantial for each predictor. The following output shows estimated response category probabilities for a few subjects in the sample. Subjects (such as subject 40) having low SES and relatively high life events have a relatively high estimated probability of being mentally impaired.

```
-----------------------------------------------------------------------
> fitted(fit)
         1       2       3       4
1   0.6249  0.2564  0.0713  0.0473 # (for life=1, ses=1)
2   0.1150  0.2518  0.2440  0.3892 # (for life=9, ses=1)
...
40  0.0410  0.1191  0.1805  0.6593 # (for life=9, ses=0)
-----------------------------------------------------------------------
```

To check the proportional odds structure, we can fit a more-complex model that permits effects to vary for the three cumulative logits. The fit is not significantly better, the likelihood-ratio test having P-value = 0.67. Estimated effects are similar for each cumulative logit, taking into account sampling error, with positive effects for SES and negative effects for life events.

```
-----------------------------------------------------------------------
> fit.nonpo <- vglm(impair ~ life + ses, family=cumulative, data=Mental)
> summary(fit.nonpo) # not using parallel=true option for propor. odds
                Estimate  Std. Error  z value
(Intercept):1   -0.1929       0.7387  -0.2611 # first cumulative logit
(Intercept):2    0.8281       0.7037   1.1768 # second cumulative logit
(Intercept):3    2.8037       0.9615   2.9160 # third cumulative logit
life:1          -0.3182       0.1597  -1.9928
life:2          -0.2740       0.1372  -1.9972
life:3          -0.3962       0.1592  -2.4883
```

```
ses:1             0.9732       0.7720    1.2605
ses:2             1.4960       0.7460    2.0055
ses:3             0.7522       0.8358    0.8999
Residual deviance: 96.7486 on 111 degrees of freedom
Log-likelihood: -48.3743 on 111 degrees of freedom
> 1 - pchisq(2*(logLik(fit.nonpo)-logLik(fit)),
+             df=df.residual(fit)-df.residual(fit.nonpo))
[1] 0.6718083 # P-value comparing to model assuming proportional odds
-------------------------------------------------------------------------
```

When we add an interaction term with the proportional odds structure, the fit suggests that the life events effect may be weaker at higher SES. However, the fit is not significantly better (P-value = 0.44).

```
-------------------------------------------------------------------------
> fit.interaction <- vglm(impair ~ life + ses + life:ses,
+                    family=cumulative(parallel=TRUE), data=Mental)
> summary(fit.interaction)
               Estimate  Std. Error  z value
(Intercept):1    0.0981      0.8110   0.1209
(Intercept):2    1.5925      0.8372   1.9022
(Intercept):3    2.6066      0.9097   2.8655
life            -0.4204      0.1903  -2.2093
ses              0.3709      1.1302   0.3282
life:ses         0.1813      0.2361   0.7679
Residual deviance: 98.5044 on 114 degrees of freedom
Log-likelihood: -49.2522 on 114 degrees of freedom
> 1 - pchisq(2*(logLik(fit.interaction)-logLik(fit)),
+      df=df.residual(fit)-df.residual(fit.interaction))
[1] 0.44108 # P-value for LR test comparing to model without interaction
-------------------------------------------------------------------------
```

We obtain similar substantive results with a cumulative probit model. In the underlying latent variable model, we estimate the difference between mean mental impairment at low and high levels of SES to be 0.68 standard deviations, adjusting for life events. The estimated multiple correlation for the underlying latent variable model equals 0.513. With the addition of an interaction term (not shown), this increases only to 0.531.

```
-------------------------------------------------------------------------
> fit.probit <- vglm(impair ~ life + ses,
+             family=cumulative(link=probit,parallel=TRUE),data=Mental)
> summary(fit.probit) # cumulative probit model
              Estimate  Std. Error  z value
(Intercept):1  -0.1612      0.3755  -0.4293
(Intercept):2   0.7456      0.3864   1.9299
(Intercept):3   1.3392      0.4123   3.2484
life           -0.1953      0.0692  -2.8236
ses             0.6834      0.3627   1.8843
Residual deviance: 98.8397 on 115 degrees of freedom
```

```
Log-likelihood:    -49.4198 on 115 degrees of freedom

> lp <- -0.1953*life + 0.6834*ses # linear predictor for latent model
> sqrt(var(lp)/(var(lp) + 1)) # corr(y*, fitted) for latent variable
[1] 0.513
--------------------------------------------------------------------------
```

CHAPTER NOTES

Section 6.1: Nominal Responses: Baseline-Category Logit Models

6.1 **BCL and multivariate GLM**: After Mantel (1966), early development and application of baseline-category logit models were primarily in the econometrics literature (e.g., Theil 1969). For details about multivariate GLMs, see Fahrmeir and Tutz (2001).

6.2 **Infinite estimates**: When a choice of baseline category causes complete or quasi-complete separation (Section 5.4.2) to occur for each logit when paired with that category, some ML estimates and *SE* values are infinite or do not exist. Approaches to produce finite estimates include the Bayesian (Note 10.7) and a generalization of a penalized likelihood approach (Kosmidis and Firth 2011) presented in Section 11.1.7 for binary data.

6.3 **Discrete choice**: Daniel McFadden (1974) proposed the discrete-choice model, incorporating explanatory variables that are characteristics of the choices. In 2000, McFadden won the Nobel Prize in Economic Sciences, partly for this work. Greene (2011, Chapter 17–18) and Train (2009) surveyed many generalizations of the model since then, such as to handle nested choice structure.

Section 6.2: Ordinal Responses: Cumulative Logit and Probit Models

6.4 **Proportional odds**: Although not the first to use the proportional odds form of cumulative logit model, the landmark article on modeling ordinal data by McCullagh (1980) popularized it. Peterson and Harrell (1990) proposed a *partial proportional odds model* in which a subset of the explanatory variables have that structure[7]. McKelvey and Zavoina (1975) presented the latent variable motivation for the cumulative probit model. Agresti (2010) reviewed ways of modeling ordinal responses.

6.5 **Infinite estimates**: When at least some ML estimates are infinite in an ordinal model, approaches to produce finite estimates include the Bayesian (Note 10.7) and a reduced-bias solution that corresponds to a parameter-dependent adjustment of the multinomial counts (Kosmidis 2014).

EXERCISES

6.1 Show that the multinomial variate $\boldsymbol{y} = (y_1, \dots, y_{c-1})^{\mathrm{T}}$ (with $y_j = 1$ if outcome j occurred and 0 otherwise) for a single trial with parameters $(\pi_1, \dots, \pi_{c-1})$

[7]This model can be fitted with the `vglm` function in the `VGAM` R package.

has distribution in the $(c-1)$-parameter exponential dispersion family, with baseline-category logits as natural parameters.

6.2 For the baseline-category logit model without constraints on parameters,

$$\pi_{ij} = \frac{\exp(\boldsymbol{x}_i\boldsymbol{\beta}_j)}{\sum_{h=1}^{c}\exp(\boldsymbol{x}_i\boldsymbol{\beta}_h)},$$

show that dividing numerator and denominator by $\exp(\boldsymbol{x}_i\boldsymbol{\beta}_c)$ yields new parameters $\boldsymbol{\beta}_j^* = \boldsymbol{\beta}_j - \boldsymbol{\beta}_c$ that satisfy $\boldsymbol{\beta}_c^* = \mathbf{0}$. Thus, without loss of generality, we can take $\boldsymbol{\beta}_c = \mathbf{0}$.

6.3 Derive Equation (6.3) for the rate of change. Show how the equation for binary models is a special case.

6.4 With three outcome categories and a single explanatory variable, suppose

$$\pi_{ij} = \exp(\beta_{j0} + \beta_j x_i)/[1 + \exp(\beta_{10} + \beta_1 x_i) + \exp(\beta_{20} + \beta_2 x_i)],$$

$j = 1, 2$. Show that π_{i3} is (**a**) decreasing in x_i if $\beta_1 > 0$ and $\beta_2 > 0$, (**b**) increasing in x_i if $\beta_1 < 0$ and $\beta_2 < 0$, and (**c**) nonmonotone when β_1 and β_2 have opposite signs.

6.5 Derive the deviance expression in Equation (6.5) by deriving the corresponding likelihood-ratio test.

6.6 For a multinomial response, let u_{ij} denote the utility of response outcome j for subject i. Suppose that

$$u_{ij} = \boldsymbol{x}_i\boldsymbol{\beta}_j + \epsilon_{ij},$$

and the response outcome for subject i is the value of j having maximum utility. When $\{\epsilon_{ij}\}$ are assumed to be iid standard normal, this model is the simplest form of the *multinomial probit model* (Aitchison and Bennett 1970).

a. Explain why $(\beta_{ak} - \beta_{bk})$ describes the effect of a 1-unit increase in explanatory variable k on the difference in mean utilities, as measured in terms of the number of standard deviations of the utility distribution.

b. For observation i, explain why the probability of outcome in category j is

$$\pi_{ij} = \int \phi(u_{ij} - \boldsymbol{x}_i\boldsymbol{\beta}_j)\prod_{k\neq j}\Phi(u_{ij} - \boldsymbol{x}_i\boldsymbol{\beta}_k)du_{ij},$$

for the standard normal pdf ϕ and cdf Φ. Explain how to form the likelihood function.

6.7 Derive the likelihood equations and the information matrix for the discrete-choice model (6.6).

6.8 Consider the baseline-category logit model (6.1).

a. Suppose we impose the structure $\boldsymbol{\beta}_j = j\boldsymbol{\beta}$, for $j = 1, \ldots, c-1$. Does this model treat the response as ordinal or nominal? Explain.

b. Show that the model in (**a**) has proportional odds structure when the $c-1$ logits are formed using pairs of adjacent categories.

6.9 Section 5.3.4 introduced Fisher's exact test for 2×2 contingency tables. For testing independence in a $r \times c$ table in which the data are c independent multinomials, derive a conditional distribution that does not depend on unknown parameters. Explain a way to use it to conduct a small-sample exact test.

6.10 Does it make sense to use the cumulative logit model of proportional odds form with a nominal-scale response variable? Why or why not? Is the model a special case of a baseline-category logit model? Explain.

6.11 Show how to express the cumulative logit model of proportional odds form as a multivariate GLM (6.4).

6.12 For a binary explanatory variable, explain why the cumulative logit model with proportional odds structure is unlikely to fit well if, for an underlying latent response, the two groups have similar location but very different dispersion.

6.13 Consider the cumulative logit model, $\text{logit}[P(y_i \leq j)] = \alpha_j + \beta_j x_i$.

a. With continuous x_i taking values over the real line, show that the model is improper, in that cumulative probabilities are misordered for a range of x_i values.

b. When x_i is a binary indicator, explain why the model is proper but requires constraints on $(\alpha_j + \beta_j)$ (as well as the usual ordering constraint on $\{\alpha_j\}$) and is then equivalent to the saturated model.

6.14 For the cumulative link model, $G^{-1}[P(y_i \leq j)] = \alpha_j + \boldsymbol{x}_i\boldsymbol{\beta}$, show that for $1 \leq j < k \leq c-1$, $P(y_i \leq k)$ equals $P(y_i \leq j)$ at $\boldsymbol{x}^*$, where $\boldsymbol{x}^*$ is obtained by increasing component h of $\boldsymbol{x}_i$ by $(\alpha_k - \alpha_j)/\beta_h$ for each h. Interpret.

6.15 For an ordinal multinomial response with c categories, let

$$\omega_{ij} = P(y_i = j \mid y_i \geq j) = \frac{\pi_{ij}}{\pi_{ij} + \cdots + \pi_{ic}}, \quad j = 1, \ldots, c-1.$$

The *continuation-ratio logit model* is

$$\text{logit}(\omega_{ij}) = \alpha_j + \boldsymbol{x}_i\boldsymbol{\beta}_j. \quad j = 1, \ldots, c-1.$$

a. Interpret (i) $\boldsymbol{\beta}_j$, (ii) $\boldsymbol{\beta}$ for the simpler model with proportional odds structure. Describe a survival application for which such sequential formation of logits might be natural.

b. Express the multinomial probability for $(y_{i1}, \ldots, y_{ic})$ in the form $p(y_{i1})p(y_{i2} \mid y_{i1}) \cdots p(y_{ic} \mid y_{i1}, \ldots, y_{i,c-1})$. Using this, explain why the $\{\hat{\boldsymbol{\beta}}_j\}$ are independent and how it is possible to fit the model using binary logistic GLMs.

6.16 Consider the null multinomial model, having the same probabilities $\{\pi_j\}$ for every observation. Let $\gamma = \sum_j b_j \pi_j$, and suppose that $\pi_j = f_j(\theta) > 0$, $j = 1, \ldots, c$. For sample proportions $\{p_j = n_j/N\}$, let $S = \sum_j b_j p_j$. Let $T = \sum_j b_j \hat{\pi}_j$, where $\hat{\pi}_j = f_j(\hat{\theta})$, for the ML estimator $\hat{\theta}$ of θ.

a. Show that $\text{var}(S) = [\sum_j b_j^2 \pi_j - (\sum_j b_j \pi_j)^2]/N$.

b. Using the delta method, show $\text{var}(T) \approx [\text{var}(\hat{\theta})][\sum_j b_j f_j'(\theta)]^2$.

c. By computing the information for $L(\theta) = \sum_j n_j \log[f_j(\theta)]$, show that $\text{var}(\hat{\theta})$ is approximately $[N \sum_j (f_j'(\theta))^2 / f_j(\theta)]^{-1}$.

d. Asymptotically, show that a consequence of model parsimony is that $\text{var}[\sqrt{N}(T - \gamma)] \le \text{var}[\sqrt{N}(S - \gamma)]$.

6.17 A response scale has the categories (strongly agree, mildly agree, mildly disagree, strongly disagree, do not know). A two-part model uses a logistic regression model for the probability of a don't know response and a separate ordinal model for the ordered categories conditional on response in one of those categories. Explain how to construct a likelihood function to fit the two parts simultaneously.

6.18 The file `Alligators2.dat` at the text website is an expanded version of Table 6.1 that also includes the alligator's gender. Using all the explanatory variables, use model-building methods to select a model for predicting primary food choice. Conduct inference and interpret effects in that model.

6.19 For 63 alligators caught in Lake George, Florida, the file `Alligators3.dat` at the text website classifies primary food choice as (fish, invertebrate, other) and shows alligator length in meters. Analyze these data.

6.20 The following R output shows output from fitting a cumulative logit model to data from the US 2008 General Social Survey. For subject i let y_i = belief in existence of heaven (1 = yes, 2 = unsure, 3 = no), x_{i1} = gender (1 = female, 0 = male) and x_{i2} = race (1 = black, 0 = white). State the model fitted here, and interpret the race and gender effects. Test goodness-of-fit and construct confidence intervals for the effects.

```
-----------------------------------------------------------------------
> cbind(race, gender, y1, y2, y3)
     race gender  y1  y2 y3
[1,]    1      1  88  16  2
[2,]    1      0  54   7  5
[3,]    0      1 397 141 24
[4,]    0      0 235 189 39
> summary(vglm(cbind(y1,y2,y3)~gender+race,family=cumulative(parallel=T)))
                Estimate  Std. Error   z value
(Intercept):1     0.0763      0.0896    0.8515
(Intercept):2     2.3224      0.1352   17.1749
gender            0.7696      0.1225    6.2808
race              1.0165      0.2106    4.8266
Residual deviance: 9.2542 on 4 degrees of freedom
Log-likelihood: -23.3814 on 4 degrees of freedom
-----------------------------------------------------------------------
```

6.21 Refer to the previous exercise. Consider the model

$$\log(\pi_{ij}/\pi_{i3}) = \alpha_j + \beta_j^G x_{i1} + \beta_j^R x_{i2}, \quad j = 1, 2.$$

a. Fit the model and report prediction equations for $\log(\pi_{i1}/\pi_{i3})$, $\log(\pi_{i2}/\pi_{i3})$, and $\log(\pi_{i1}/\pi_{i2})$.

b. Using the "yes"and "no" response categories, interpret the conditional gender effect using a 95% confidence interval for an odds ratio.

c. Conduct a likelihood-ratio test of the hypothesis that opinion is independent of gender, given race. Interpret.

6.22 Refer to Exercise 5.33. The color of the female crab is a surrogate for age, with older crabs being darker. Analyze whether any characteristics or combinations of characteristics of the attached male crab can help to predict a female crab's color. Prepare a short report that summarizes your analyses and findings.

6.23 A 1976 article by M. Madsen (*Scand. J. Stat.* **3**: 97–106) showed a $4 \times 2 \times 3 \times 3$ contingency table (the file `Satisfaction.dat` at the text website) that cross classifies a sample of residents of Copenhagen on the type of housing, degree of contact with other residents, feeling of influence on apartment management, and satisfaction with housing conditions. Treating satisfaction as the response variable, analyze these data.

6.24 At the website `sda.berkeley.edu/GSS` for the General Social Survey, download a contingency table relating the variable GRNTAXES (about paying higher taxes to help the environment) to two other variables, using the survey results from 2010 by specifying *year(2010)* in the "Selection Filter." Model the data, and summarize your analysis and interpretations.

CHAPTER 7

Models for Count Data

Many response variables have counts as their possible outcomes. Examples are the number of alcoholic drinks you had in the previous week, and the number of devices you own that can access the internet (laptops, smart cell phones, tablets, etc.). Counts also occur as entries in cells of contingency tables that cross-classify categorical variables, such as the number of people in a survey who are female, college educated, and agree that humans are responsible for climate change. In this chapter we introduce generalized linear models (GLMs) for count response variables.

Section 7.1 presents models that assume a Poisson distribution for a count response variable. The *loglinear model*, using a log link to connect the mean with the linear predictor, is most common. The model can be adapted to model a *rate* when the count is based on an index such as space or time. Section 7.2 shows how to use Poisson and related multinomial models for contingency tables to analyze conditional independence and association structure for a multivariate categorical response variable. For the Poisson distribution, the variance must equal the mean, and data often exhibit greater variability than this. Section 7.3 introduces GLMs that assume a negative binomial distribution, which handles such *overdispersion* in a natural way. Many datasets show greater frequencies of zero counts than standard models allow, often because some subjects can have a zero outcome by chance but some subjects necessarily have a zero outcome. Section 7.4 introduces models that handle such *zero-inflated data*, which we might expect for a variable such as the frequency of alcohol drinking. Three examples illustrate models—one for rate data (Section 7.1.7), one for associations in contingency tables (Section 7.2.6), and one for zero-inflated count data (Section 7.5).

Foundations of Linear and Generalized Linear Models, First Edition. Alan Agresti.

7.1 POISSON GLMS FOR COUNTS AND RATES

The simplest distribution for count data, placing its mass on the set of nonnegative integer values, is the *Poisson*. Its probabilities depend on a single parameter, the mean $\mu > 0$.

7.1.1 The Poisson Distribution

In equation (4.5) we observed that the Poisson probability mass function, $p(y; \mu) = e^{-\mu}\mu^y/y!$ for $y = 0, 1, 2, \ldots$, is in the exponential dispersion family with $E(y) = \text{var}(y) = \mu$. The Poisson distribution is unimodal with mode equal to the integer part of μ. Its skewness is described by $E(y - \mu)^3/\sigma^3 = 1/\sqrt{\mu}$. As μ increases, the Poisson distribution is less skewed and approaches normality, the approximation being fairly good when $\mu > 10$.

The Poisson distribution is often used for counts of events[1] that occur randomly over time or space at a particular rate, when outcomes in disjoint time periods or regions are independent. For example, a manufacturer of cell phones might find that the Poisson describes reasonably well the number of warranty claims received each week. The Poisson also applies as an approximation for the binomial when the number of trials n is large and π is very small, with $\mu = n\pi$. For the binomial, if $n \to \infty$ and $\pi \to 0$ such that $n\pi = \mu$ is fixed, then the binomial distribution converges to the Poisson. If a manufacturer has sold 5000 cell phones of a particular type, and each independently has probability 0.001 of having a warranty claim in a given week, then the number of such claims per week has approximately a Poisson distribution with mean $5000(0.001) = 5$.

7.1.2 Variance Stabilization and Least Squares with Count Data

Let $y_1, \ldots, y_n$ denote independent observations from Poisson distributions, with $\mu_i = E(y_i)$. In modeling count data, we could transform the counts so that, at least approximately, the variance is constant and ordinary least squares methods are valid. By the delta method, the linearization $g(y) - g(\mu) \approx (y - \mu)g'(\mu)$ implies that $\text{var}[g(y)] \approx [g'(\mu)]^2\text{var}(y)$. If y has a Poisson distribution, then $\sqrt{y}$ has

$$\text{var}(\sqrt{y}) \approx \left(\frac{1}{2\sqrt{\mu}}\right)^2 \mu = \frac{1}{4}.$$

The approximation holds better for larger μ, for which $\sqrt{y}$ is more closely linear in a neighborhood of μ.

Since $\sqrt{y}$ has approximately constant variance, we could model $\sqrt{y_i}$, $i = 1, \ldots, n$, using linear models fitted by ordinary least squares. However, the model is then linear

[1]See Karlin and Taylor (1975, pp. 23–25) for precise conditions and a derivation of the Poisson formula.

in $E(\sqrt{y_i})$, not $E(y_i)$. Also, a linear relation with the linear predictor may hold more poorly for $E(\sqrt{y_i})$ than for $E(y_i)$, $E[\log(y_i)]$, or some other transformation. It is more appealing to use GLM methods, which apply a link function to the mean response rather than the mean to a function of the response.

7.1.3 Poisson GLMs and Loglinear Models

We now present the GLM approach for Poisson response data. Since $\text{var}(y_i) = \mu_i$, the GLM likelihood equations (4.10) for n independent observations simplify for a Poisson response with linear predictor $\eta_i = g(\mu_i) = \sum_j \beta_j x_{ij}$ having link function g to

$$\sum_{i=1}^{n} \frac{(y_i - \mu_i)x_{ij}}{\text{var}(y_i)} \left(\frac{\partial \mu_i}{\partial \eta_i}\right) = \sum_{i=1}^{n} \frac{(y_i - \mu_i)x_{ij}}{\mu_i} \left(\frac{\partial \mu_i}{\partial \eta_i}\right) = 0.$$

Although a GLM can model a positive mean using the identity link, it is more common to model the log of the mean. Like the linear predictor, the log mean can take any real value. From Section 4.1.2, the log mean is the natural parameter for the Poisson distribution, and the log link is the canonical link for a Poisson GLM. The *Poisson loglinear model* is

$$\log \mu_i = \sum_{j=1}^{p} \beta_j x_{ij}, \quad \text{or} \quad \log \boldsymbol{\mu} = \boldsymbol{X\beta}$$

in terms of a model matrix and model parameters. For $\eta_i = \log \mu_i$, $\partial \mu_i / \partial \eta_i = \mu_i$, so the likelihood equations are

$$\sum_i (y_i - \mu_i)x_{ij} = 0, \qquad (7.1)$$

as we found in Section 4.2.2.

For a Poisson loglinear model, the mean satisfies the exponential relation

$$\mu_i = \exp\left(\sum_{j=1}^{p} \beta_j x_{ij}\right) = (e^{\beta_1})^{x_{i1}} \cdots (e^{\beta_p})^{x_{ip}}.$$

A 1-unit increase in x_{ij} has a multiplicative impact of e^{β_j}: the mean at $x_{ij} + 1$ equals the mean at x_{ij} multiplied by e^{β_j}, adjusting for the other explanatory variables.

7.1.4 Model Fitting and Goodness of Fit

Except for simple models such as for the one-way layout or balanced two-way layout, the likelihood equations have no closed-form solution. However, the log-likelihood function is concave, and the Newton–Raphson method (which is equivalent to Fisher scoring for the canonical log link) yields fitted values and estimates of corresponding model parameters. From Section 4.2.4, the estimated covariance matrix (4.14) of $\hat{\boldsymbol{\beta}}$ is

$$\widehat{\text{var}}(\hat{\boldsymbol{\beta}}) = (\boldsymbol{X}^{\text{T}}\widehat{\boldsymbol{W}}\boldsymbol{X})^{-1},$$

where with the log link $\boldsymbol{W}$ is the diagonal matrix with elements $w_i = (\partial\mu_i/\partial\eta_i)^2/\text{var}(y_i) = \mu_i$.

From Section 4.4.2, the deviance of a Poisson GLM is

$$D(\boldsymbol{y}, \hat{\boldsymbol{\mu}}) = 2\sum_{i=1}^{n}\left[y_i \log\left(\frac{y_i}{\hat{\mu}_i}\right) - y_i + \hat{\mu}_i\right]. \qquad (7.2)$$

When a model with log link has an intercept, its likelihood equation implies that $\sum_i \hat{\mu}_i = \sum_i y_i$, and so $D(\boldsymbol{y}, \hat{\boldsymbol{\mu}}) = 2\sum_i [y_i \log(y_i/\hat{\mu}_i)]$. This is often denoted by G^2. The corresponding Pearson statistic (Section 4.4.4) is

$$X^2 = \sum_{i=1}^{n} \frac{(y_i - \hat{\mu}_i)^2}{\hat{\mu}_i}.$$

In some cases we can use these statistics to test goodness of fit. Asymptotic chi-squared distributions result when the number n of Poisson observations is fixed and their means increase unboundedly. The standard case where this holds reasonably well is contingency tables with a fixed number of cells and large overall sample size, as we explain in Section 7.2.2. But such a test, having a global alternative, does not reveal how a model fails. It is more informative to check a model by comparing it with more-complex models (e.g., with interaction terms) and by investigating particular aspects of lack of fit. For example, we can check that the variance truly has the same order of magnitude as the mean by comparing the model with a more-complex model that does not assume this, such as the negative binomial model presented in Section 7.3.

We can also search for unusual observations or patterns in the residuals. In Section 4.4.6 we presented the Pearson and standardized residuals for Poisson GLMs. Like y_i, these have skewed distributions, less so as μ_i increases. Finally, an informal way to assess the Poisson assumption is to compare the overall sample proportion of (0, 1, 2, ...) observations to the average of the fitted response distributions for the n observations. Often this shows that a Poisson model does not permit sufficient variability, underpredicting 0 outcomes and relatively high outcomes.

7.1.5 Example: One-Way Layout Comparing Poisson Means

For the one-way layout for a count response, let y_{ij} be observation j of a count variable for group i, $i = 1, \ldots, c$, $j = 1, \ldots, n_i$, with $n = \sum_i n_i$. Suppose that $\{y_{ij}\}$ are independent Poisson with $E(y_{ij}) = \mu_{ij}$. The model $\log(\mu_{ij}) = \beta_0 + \beta_i$ has a common mean within groups. For group means $\{\mu_i\}$, $\exp(\beta_h - \beta_i) = \mu_h/\mu_i$. With $\beta_0 = 0$ for identifiability, the model has the form $\log \boldsymbol{\mu} = \boldsymbol{X\beta}$ with

$$\boldsymbol{\mu} = \begin{pmatrix} \mu_1 \mathbf{1}_{n_1} \\ \mu_2 \mathbf{1}_{n_2} \\ \vdots \\ \mu_c \mathbf{1}_{n_c} \end{pmatrix}, \quad \boldsymbol{X\beta} = \begin{pmatrix} \mathbf{1}_{n_1} & \mathbf{0}_{n_1} & \cdots & \mathbf{0}_{n_1} \\ \mathbf{0}_{n_2} & \mathbf{1}_{n_2} & \cdots & \mathbf{0}_{n_2} \\ \vdots & \vdots & \ddots & \vdots \\ \mathbf{0}_{n_c} & \mathbf{0}_{n_c} & \cdots & \mathbf{1}_{n_c} \end{pmatrix} \begin{pmatrix} \beta_1 \\ \beta_2 \\ \vdots \\ \beta_c \end{pmatrix}.$$

The likelihood equation induced by parameter β_i for group i is

$$\sum_{j=1}^{n_i} (y_{ij} - \mu_i) = 0,$$

so $\hat{\mu}_i = \bar{y}_i = (\sum_j y_{ij})/n_i$ and $\hat{\beta}_i = \log \bar{y}_i$. In fact, the same likelihood equations and fitted means occur with any link function, or if we use baseline-category constraints (as we would for higher-way layouts) such as $\beta_1 = 0$. For the log link, $\widehat{\boldsymbol{W}}$ has the sample means on the main diagonal. For the model matrix shown above, $\widehat{\text{var}}(\hat{\boldsymbol{\beta}}) = (\boldsymbol{X}^{\text{T}}\widehat{\boldsymbol{W}}\boldsymbol{X})^{-1}$ is a diagonal matrix with $\widehat{\text{var}}(\hat{\beta}_i) = 1/n_i\bar{y}_i$. It follows that for large $\{n_i\mu_i\}$ a Wald 95% confidence interval for μ_h/μ_i is

$$\exp[(\hat{\beta}_h - \hat{\beta}_i) \pm 1.96\sqrt{(n_h\bar{y}_h)^{-1} + (n_i\bar{y}_i)^{-1}}].$$

Analogous to the one-way ANOVA for a normal response, we can test H_0: $\mu_1 = \cdots = \mu_c$. By direct construction or by applying the result in Section 4.4.3 about using the difference of deviances to compare the null model with the model for the one-way layout, we can construct the likelihood-ratio statistic. It equals

$$2\sum_{i=1}^{c} n_i\bar{y}_i \log(\bar{y}_i/\bar{y}),$$

where $\bar{y}$ is the grand mean of all $n = \sum_i n_i$ observations. As $\{n_i\}$ grow for fixed c, $\{\bar{y}_i\}$ have approximate normal distributions, and this statistic has null distribution converging to chi-squared with $df = (c-1)$.

These inferences assume validity of the Poisson model. They are not robust to violation of the Poisson assumption. If the data have greater than Poisson variability, the large-sample $\text{var}(\hat{\beta}_i)$ will exceed $1/n_i\mu_i$. It is sensible to compare results with those for analogous inferences using a model that permits greater dispersion, such as the negative binomial model introduced in Section 7.3.

The deviance (7.2) and the Pearson statistic for the Poisson model for the one-way layout simplify to

$$G^2 = 2\sum_{i=1}^{c}\sum_{j=1}^{n_i} y_{ij} \log\left(\frac{y_{ij}}{\bar{y}_i}\right), \quad X^2 = \sum_{i=1}^{c}\sum_{j=1}^{n_i} \frac{(y_{ij} - \bar{y}_i)^2}{\bar{y}_i}.$$

For testing goodness of fit of this model with relatively large $\{\bar{y}_i\}$, G^2 and X^2 have approximate chi-squared distributions with $df = \sum_i(n_i - 1)$ (Fisher 1970, p. 58). For a single group, Cochran (1954) referred to $[\sum_j(y_{1j} - \bar{y}_1)^2]/\bar{y}_1$ as the *variance test* for the fit of a Poisson distribution, since it compares the sample variance of the data with the estimated Poisson variance $\bar{y}_1$. This asymptotic theory applies, however, as $\{\mu_i\}$ grow for fixed $\{n_i\}$, which is not realistic in most applications. For checking fit, chi-squared asymptotics usually apply better for comparing the model with a more complex model and for comparing the model with the null model as just described.

7.1.6 Modeling Rates: Including an Offset in the Model

Often the expected value of a response count y_i is proportional to an index t_i. For instance, t_i might be an amount of time and/or a population size, such as in modeling crime counts for various cities. Or, it might be a spatial area, such as in modeling counts of a particular animal or plant species. Then the sample rate is y_i/t_i, with expected value μ_i/t_i. With explanatory variables, a loglinear model for the expected rate has the form

$$\log(\mu_i/t_i) = \sum_{j=1}^{p} \beta_j x_{ij}.$$

Because $\log(\mu_i/t_i) = \log \mu_i - \log t_i$, the model makes the adjustment $-\log t_i$ to the log link of the mean. This adjustment term is called an *offset*. The fit corresponds to using $\log t_i$ as an explanatory variable in the linear predictor for $\log(\mu_i)$ and forcing its coefficient to equal 1.

For this model, the expected response count satisfies

$$\mu_i = t_i \exp\left(\sum_{j=1}^{p} \beta_j x_{ij}\right).$$

The mean has a proportionality constant for t_i that depends on the values of the explanatory variables. The identity link is also occasionally useful, such as with a sole qualitative explanatory variable. The model with identity link is

$$\mu_i/t_i = \sum_{j=1}^{p} \beta_j x_{ij}, \quad \text{or} \quad \mu_i = \sum_{j=1}^{p} \beta_j x_{ij} t_i.$$

This corresponds to an ordinary Poisson GLM using the identity link with no intercept and with explanatory variables $x_{i1}t_i, \ldots, x_{ip}t_i$. It provides additive, rather than multiplicative, effects of explanatory variables.

7.1.7 Example: Lung Cancer Survival

Table 7.1, from Holford (1980), shows survival and death for 539 males diagnosed with lung cancer. The prognostic factors are histology and stage of disease, with

Table 7.1 Number of Deaths from Lung Cancer, by Histology, Stage of Disease, and Follow-up Time Interval[a]

		Histology								
Follow-up Time Interval (months)	Disease	I			II			III		
	Stage:	1	2	3	1	2	3	1	2	3
0–2		9	12	42	5	4	28	1	1	19
		(157	134	212	77	71	130	21	22	101)
2–4		2	7	26	2	3	19	1	1	11
		(139	110	136	68	63	72	17	18	63)
4–6		9	5	12	3	5	10	1	3	7
		(126	96	90	63	58	42	14	14	43)
6–8		10	10	10	2	4	5	1	1	6
		(102	86	64	55	42	21	12	10	32)
8–10		1	4	5	2	2	0	0	0	3
		(88	66	47	50	35	14	10	8	21)
10–12		3	3	4	2	1	3	1	0	3
		(82	59	39	45	32	13	8	8	14)
12+		1	4	1	2	4	2	0	2	3
		(76	51	29	42	28	7	6	6	10)

[a]Values in parentheses represent total follow-up months at risk.
Source: Reprinted from Holford (1980) with permission of John Wiley & Sons, Inc.

observations grouped into 2-month intervals of follow-up after the diagnosis. For each cell specifying a particular length of follow-up, histology, and stage of disease, the table shows the number of deaths and the number of months of observations of subjects still alive during that follow-up interval. We treat[2] the death counts in the table as independent Poisson variates.

Let μ_{ijk} denote the expected number of deaths and t_{ijk} the total time at risk for histology i and stage of disease j, in follow-up time interval k. The Poisson GLM for the death rate,

$$\log(\mu_{ijk}/t_{ijk}) = \beta_0 + \beta_i^{\mathrm{H}} + \beta_j^{\mathrm{S}} + \beta_k^{\mathrm{T}},$$

treats each explanatory variable as a qualitative factor, where the superscript notation shows the classification labels. It has residual deviance $G^2 = 43.92$ ($df = 52$). Models that assume a lack of interaction between follow-up interval and either prognostic factor are called *proportional hazards* models. They have the same effects of histology and stage of disease in each time interval. Then a ratio of hazards for two groups is the same at all times. Further investigation reveals that, although the stage of disease is an important prognostic factor, histology did not contribute significant additional

[2]This corresponds to a survival modeling approach that assumes piecewise exponential densities for survival times, yielding a constant hazard function in each two-month interval.

information. Adding interaction terms between stage and time does not significantly improve the fit (change in deviance = 14.86 with $df = 12$).

```
--------------------------------------------------------------------
> Cancer # file Cancer.dat at www.stat.ufl.edu/~aa/glm/data
   time histology stage count risktime
1     1          1     1     9      157
2     1          2     1     5       77
...
63    7          3     3     3       10
> attach(Cancer)
> logrisktime = log(risktime)
> fit <- glm(count ~ factor(histology) + factor(stage) + factor(time),
+         family = poisson(link = log), offset = logrisktime)
> summary(fit)
                   Estimate  Std. Error  z value  Pr(>|z|)
(Intercept)         -3.0093     0.1665   -18.073    <2e-16
factor(histology)2   0.1624     0.1219     1.332    0.1828
factor(histology)3   0.1075     0.1474     0.729    0.4658
factor(stage)2       0.4700     0.1744     2.694    0.0070
factor(stage)3       1.3243     0.1520     8.709    <2e-16
factor(time)2       -0.1274     0.1491    -0.855    0.3926
...
factor(time)7       -0.1752     0.2498    -0.701    0.4832
---
    Null deviance: 175.718  on 62  degrees of freedom
Residual deviance:  43.923  on 52  degrees of freedom
--------------------------------------------------------------------
```

The estimated stage-of-disease effects show the progressively worsening death rate as the stage advances. The estimated death rate at the third stage of disease is $\exp(1.324) = 3.76$ times that at the first stage, adjusting for follow-up time and histology, with Wald 95% confidence interval $\exp[1.324 \pm 1.96(0.152)]$, or (2.79, 5.06).

7.2 POISSON/MULTINOMIAL MODELS FOR CONTINGENCY TABLES

Chapters 5 and 6 introduced binomial and multinomial models for categorical response variables. For *multivariate* categorical responses, we can apply such models marginally to each response, as Section 9.6 shows. Alternatively, we can formulate multinomial models for their joint distribution, to investigate potential independence, association, and interaction structure. Many such multinomial models are equivalent to models for independent Poisson counts in cells of a contingency table. The Poisson model generates the multinomial model after we condition on an overall sample size. We illustrate in this section.

7.2.1 Connection Between Poisson and Multinomial Distributions

For independent Poisson random variables $(y_1, \ldots, y_c)$ with means $(\mu_1, \ldots, \mu_c)$, the joint probability mass function for $\{y_i\}$ is the product of the mass functions of form (4.5). The total $n = \sum_i y_i$ also has a Poisson distribution, with parameter $\sum_i \mu_i$. Conditional on n, $\{y_i\}$ no longer have Poisson distributions, because each y_i cannot exceed n, and $\{y_i\}$ are also no longer independent, because the value of one affects the possible range for the others.

The conditional probability of a set of counts $\{y_i\}$ satisfying $\sum_j y_j = n$ is

$$P\left[(y_1 = n_1, y_2 = n_2, \ldots, y_c = n_c) \mid \sum_{j=1}^{c} y_j = n\right]$$

$$= \frac{P(y_1 = n_1, y_2 = n_2, \ldots, y_c = n_c)}{P\left(\sum_j y_j = n\right)}$$

$$= \frac{\prod_i (e^{-\mu_i} \mu_i^{n_i}/n_i!)}{\exp(-\sum_j \mu_j)(\sum_j \mu_j)^n/n!} = \left(\frac{n!}{\prod_i n_i!}\right) \prod_{i=1}^{c} \pi_i^{n_i},$$

where $\{\pi_i = \mu_i/(\sum_j \mu_j)\}$. This is the multinomial distribution characterized by the sample size n and the probabilities $\{\pi_i\}$.

Because of this relation, many Poisson models for independent counts in c fixed categories have corresponding multinomial models that treat the total count as fixed. In the multinomial model, the sample size is the total count and the category probabilities are proportional to the Poisson means.

7.2.2 GLM of Independence in Two-Way Contingency Tables

To illustrate Poisson loglinear models for counts in contingency tables, we first consider $r \times c$ tables that cross-classify two categorical response variables, which we denote by A and B. Suppose $\{y_{ij}\}$ are independent counts having Poisson distributions with means $\{\mu_{ij}\}$ that satisfy

$$\mu_{ij} = \mu \phi_i \psi_j,$$

where $\{\phi_i\}$ and $\{\psi_j\}$ are positive constants satisfying $\sum_i \phi_i = \sum_j \psi_j = 1$. This model is multiplicative, but the log link yields a GLM for $\{\mu_{ij}\}$ whose linear predictor has the structure

$$\log \mu_{ij} = \beta_0 + \beta_i^A + \beta_j^B. \tag{7.3}$$

This Poisson loglinear model has additive main effects of the two classifications but no interaction. Identifiability requires a constraint on $\{\beta_i^A\}$ and on $\{\beta_j^B\}$.

Because $\{y_{ij}\}$ are independent, the total sample size $\sum_i \sum_j y_{ij}$ has a Poisson distribution with mean $\sum_i \sum_j \mu_{ij} = \mu$. Conditional on $\sum_i \sum_j y_{ij} = n$, the cell counts have a multinomial distribution with joint cell probabilities $\{\pi_{ij} = \mu_{ij}/\mu = \phi_i \psi_j\}$. Because $\sum_i \phi_i = 1$ and $\sum_j \psi_j = 1$, we have[3] $\phi_i = \pi_{i+}$, $\psi_j = \pi_{+j}$, and $\{\pi_{ij} = \pi_{i+}\pi_{+j}\}$. This is the expression of the multinomial joint distribution for *independence* between the categorical response variables. When we express the Poisson loglinear model (7.3) in multiplicative multinomial form by exponentiating both sides and dividing by μ, the intercept parameter β_0 cancels. That is, the Poisson model has $[1 + (r-1) + (c-1)]$ parameters, whereas the multinomial model has $[(r-1) + (c-1)]$ parameters.

As in the two-way layout for a linear model with main effects only, the model matrix $\boldsymbol{X}$ for the Poisson loglinear model has a simple form containing indicator variables that are the coefficients of the parameters for the row and column factors. For example, for a 2×2 table with constraints $\beta_1^A = \beta_1^B = 0$, the model is

$$\log \boldsymbol{\mu} = \begin{bmatrix} \log \mu_{11} \\ \log \mu_{12} \\ \log \mu_{21} \\ \log \mu_{22} \end{bmatrix} = \begin{bmatrix} 1 & 0 & 0 \\ 1 & 0 & 1 \\ 1 & 1 & 0 \\ 1 & 1 & 1 \end{bmatrix} \begin{bmatrix} \beta_0 \\ \beta_2^A \\ \beta_2^B \end{bmatrix} = \boldsymbol{X\beta}.$$

From such a model matrix for the independence model, you can verify that the likelihood equations (7.1) simplify to $\hat{\mu}_{i+} = y_{i+}$ and $\hat{\mu}_{+j} = y_{+j}$, for all i and j. These equate the fitted and sample marginal distributions. Or we can easily derive these directly from the model. The joint Poisson probability of cell counts $\{y_{ij}\}$ is

$$\prod_{i=1}^{r} \prod_{j=1}^{c} \frac{e^{-\mu_{ij}} \mu_{ij}^{y_{ij}}}{y_{ij}!},$$

from which the kernel of the log-likelihood is

$$L(\boldsymbol{\mu}) = \sum_{i=1}^{r} \sum_{j=1}^{c} y_{ij} \log \mu_{ij} - \sum_{i=1}^{r} \sum_{j=1}^{c} \mu_{ij}.$$

Substituting the model formula (7.3) for $\log \mu_{ij}$, we have

$$L(\beta_0, \boldsymbol{\beta}^A, \boldsymbol{\beta}^B) = n\beta_0 + \sum_{i=1}^{r} y_{i+}\beta_i^A + \sum_{j=1}^{c} y_{+j}\beta_j^B - \sum_{i=1}^{r} \sum_{j=1}^{c} \exp\left(\beta_0 + \beta_i^A + \beta_j^B\right).$$

[3]A + subscript denotes summing over that index.

The log-likelihood derivatives

$$\frac{\partial L}{\partial \beta_i^A} = y_{i+} - \sum_{j=1}^{c} \exp\left(\beta_0 + \beta_i^A + \beta_j^B\right) = y_{i+} - \mu_{i+} \quad \text{and}$$

$$\frac{\partial L}{\partial \beta_j^B} = y_{+j} - \sum_{i=1}^{r} \exp\left(\beta_0 + \beta_i^A + \beta_j^B\right) = y_{+j} - \mu_{+j}$$

yield these likelihood equations, when equated to 0. The solution of these equations that satisfies the model is the set of maximum likelihood (ML) fitted values, $\{\hat{\mu}_{ij} = y_{i+}y_{+j}/n\}$. The same fit results if we condition on $n = \sum_i \sum_j y_{ij}$ and maximize the corresponding multinomial likelihood $\prod_i \prod_j \pi_{ij}^{y_{ij}}$, for which the kernel of the log-likelihood, $\sum_i \sum_j y_{ij} \log \pi_{ij}$, is the same as the Poisson kernel except for the intercept parameter. The fitted joint multinomial probability $\hat{\pi}_{ij}$ is the product of the sample marginal proportions, $\hat{\pi}_{i+} = y_{i+}/n$ and $\hat{\pi}_{+j} = y_{+j}/n$.

The Pearson statistic for testing the independence-model goodness of fit,

$$X^2 = \sum_{i=1}^{r} \sum_{j=1}^{c} \frac{(y_{ij} - \hat{\mu}_{ij})^2}{\hat{\mu}_{ij}},$$

was proposed by Karl Pearson in 1900. When the model holds, the large-sample distributions of X^2 and the corresponding deviance G^2 are chi-squared, the approximation being reasonably good if most cell means exceed about 5. The Poisson model has rc observations described by $[1 + (r-1) + (c-1)]$ parameters. Equivalently, the multinomial model has $rc - 1$ counts described by $(r-1) + (c-1)$ parameters. So the residual df for the chi-squared test are $df = rc - (r + c - 1) = (r-1)(c-1)$. Pearson mistakenly concluded that $df = rc - 1$, as would be the case if H_0 specified particular values for $\{\pi_{ij}\}$. The correct df were not proven until an article by R. A. Fisher in 1922. This correction engendered a lifelong enmity[4] in which each of these giants of the Statistics community treated the other disparagingly.

7.2.3 Loglinear Association Parameters Relate to Odds Ratios

To allow association between the two classification variables, we add a two-factor interaction term to loglinear model (7.3), yielding

$$\log \mu_{ij} = \beta_0 + \beta_i^A + \beta_j^B + \gamma_{ij}^{AB}.$$

We can specify the model so that $\{\gamma_{ij}^{AB}\}$ are coefficients of cross-products of $r - 1$ indicator variables for the rows with $c - 1$ indicator variables for the columns. With

[4]For details, see Agresti (2013, Section 17.2) and *R. A. Fisher: The Life of a Scientist* by Joan Fisher Box (Wiley 1978).

appropriate constraints for identifiability, such as $\gamma_{1j}^{AB} = \gamma_{i1}^{AB} = 0$ for all i and j, this adds an additional $(r-1)(c-1)$ parameters, so the model is saturated.

The $\{\gamma_{ij}^{AB}\}$ association parameters pertain to odds ratios. We illustrate for $r = c = 2$. For the multinomial $\{\pi_{ij}\}$ or the Poisson $\{\mu_{ij}\}$, the log odds ratio is

$$\log \frac{\pi_{11}\pi_{22}}{\pi_{12}\pi_{21}} = \log \frac{\mu_{11}\mu_{22}}{\mu_{12}\mu_{21}} = \log \mu_{11} + \log \mu_{22} - \log \mu_{12} - \log \mu_{21}$$

$$= \left(\beta_0 + \beta_1^A + \beta_1^B + \gamma_{11}^{AB}\right) + \left(\beta_0 + \beta_2^A + \beta_2^B + \gamma_{22}^{AB}\right)$$

$$- \left(\beta_0 + \beta_1^A + \beta_2^B + \gamma_{12}^{AB}\right) - \left(\beta_0 + \beta_2^A + \beta_1^B + \gamma_{21}^{AB}\right)$$

$$= \gamma_{11}^{AB} + \gamma_{22}^{AB} - \gamma_{12}^{AB} - \gamma_{21}^{AB}.$$

Under the constraints just stated, the odds ratio simplifies to $\exp(\gamma_{22}^{AB})$.

7.2.4 Poisson/Multinomial Loglinear Models for Multiway Contingency Tables

Loglinear models for multidimensional contingency tables describe independence, association, and interaction patterns. We illustrate for $r \times c \times \ell$ cross-classifications of three categorical response variables, which we denote by A, B, and C. The models apply to Poisson sampling with independent cell counts $\{y_{ijk}\}$ having means $\{\mu_{ijk}\}$. They also apply to a multinomial distribution with cell probabilities $\{\pi_{ijk}\}$ having $\sum_i \sum_j \sum_k \pi_{ijk} = 1.0$.

Mutual independence: Three categorical response variables are *mutually independent* when the cell probabilities satisfy, for all i, j, and k,

$$P(A = i, B = j, C = k) = P(A = i)P(B = j)P(C = k).$$

That is, all $\pi_{ijk} = \pi_{i++}\pi_{+j+}\pi_{++k}$. For expected frequencies $\{\mu_{ijk}\}$, mutual independence has the loglinear structural form

$$\log \mu_{ijk} = \beta_0 + \beta_i^A + \beta_j^B + \beta_k^C. \tag{7.4}$$

Joint independence: A is *jointly independent* of B and C when for all i, j, and k,

$$P(A = i, B = j, C = k) = P(A = i)P(B = j, C = k).$$

That is, all $\pi_{ijk} = \pi_{i++}\pi_{+jk}$. This is ordinary independence for the two-way contingency table that cross-classifies A with a variable composed of the $c\ell$ combinations of levels of B and C. The corresponding loglinear model is

$$\log \mu_{ijk} = \beta_0 + \beta_i^A + \beta_j^B + \beta_k^C + \gamma_{jk}^{BC}. \tag{7.5}$$

We use the hierarchical structure by which the presence of a two-factor term implies inclusion of the lower order (single-factor) terms.

Conditional independence: A and B are *conditionally independent, given C*, when for all i, j, and k,

$$P(A = i, B = j \mid C = k) = P(A = i \mid C = k)P(B = j \mid C = k).$$

That is, independence holds for the $r \times c$ partial table relating A and B at each fixed category of C. Equivalently, by expressing each conditional probability in terms of the joint probabilities $\{\pi_{ijk}\}$ and their marginals,

$$\pi_{ijk} = \pi_{i+k}\pi_{+jk}/\pi_{++k}.$$

Conditional independence of A and B, given C, has loglinear model form

$$\log \mu_{ijk} = \beta_0 + \beta_i^A + \beta_j^B + \beta_k^C + \gamma_{ik}^{AC} + \gamma_{jk}^{BC}. \tag{7.6}$$

A model that permits all three pairs of variables to be conditionally dependent is

$$\log \mu_{ijk} = \beta_0 + \beta_i^A + \beta_j^B + \beta_k^C + \gamma_{ij}^{AB} + \gamma_{ik}^{AC} + \gamma_{jk}^{BC}. \tag{7.7}$$

From exponentiating both sides, the cell probabilities have the form

$$\pi_{ijk} = \phi_{ij}\psi_{ik}\omega_{jk}.$$

No closed-form expression exists for the three components in terms of margins of $\{\pi_{ijk}\}$ except in certain special cases. All these loglinear models have constraints on parameters to satisfy identifiability. For example, for any conditional association term, we can take $\gamma_{1j} = \gamma_{i1} = 0$ for all i and j, as R software does by default.

Interpretations of loglinear model parameters use their highest-order terms. The two-factor terms describe conditional association as measured by log odds ratios. At a fixed level k of C, the *conditional association* between A and B is specified by $(r-1)(c-1)$ odds ratios, such as

$$\theta_{ij(k)} = \frac{\mu_{ijk}\mu_{rck}}{\mu_{ick}\mu_{rjk}}, \qquad 1 \le i \le r-1, \quad 1 \le j \le c-1.$$

For example, when $r = c = 2$, substituting model (7.7) into $\log\theta_{11(k)}$ yields

$$\log\theta_{11(k)} = \log\frac{\mu_{11k}\mu_{22k}}{\mu_{12k}\mu_{21k}} = \gamma_{11}^{AB} + \gamma_{22}^{AB} - \gamma_{12}^{AB} - \gamma_{21}^{AB}.$$

Thus, $\theta_{11(k)}$ simplifies to $\exp(\gamma_{22}^{AB})$ under constraints such as R software imposes. Analogous expressions occur with arbitrary r and c. In such expressions, because the right-hand side is the same for all k, an absence of three-factor interaction is equivalent to

$$\theta_{ij(1)} = \theta_{ij(2)} = \cdots = \theta_{ij(\ell)} \quad \text{for all } i \text{ and } j.$$

Because of this property, model (7.7) is called a loglinear model of *homogeneous association*. Any loglinear model not having the three-factor interaction term has a homogeneous conditional association for each pair of variables.

The general Poisson loglinear model for a three-way contingency table is

$$\log\mu_{ijk} = \beta_0 + \beta_i^A + \beta_j^B + \beta_k^C + \gamma_{ij}^{AB} + \gamma_{ik}^{AC} + \gamma_{jk}^{BC} + \gamma_{ijk}^{ABC}.$$

With indicator variables for each factor, γ_{ijk}^{ABC} is the coefficient of the product of the ith indicator variable for A, jth indicator variable for B, and kth indicator variable for C. The total number of nonredundant parameters is

$$1 + (r-1) + (c-1) + (\ell-1) + (r-1)(c-1) + (r-1)(\ell-1)$$
$$+ (c-1)(\ell-1) + (r-1)(c-1)(\ell-1) = rc\ell,$$

which is the total number of cell counts. This model has as many parameters as Poisson observations and is saturated. It describes all possible $\{\mu_{ijk} > 0\}$.

Table 7.2 summarizes unsaturated loglinear models for three-way contingency tables. For all such Poisson models, corresponding multinomial models have one fewer parameter (the β_0 intercept in the Poisson models) after conditioning on the total count. The common parameters contribute in the same way to Poisson or

Table 7.2 Loglinear Models for Three-Way Contingency Tables, for Poisson Means or Multinomial Probabilities $\{\pi_{ijk}\}$

Model Formula	Probability Form for π_{ijk}	Association Terms in Loglinear Model	Interpretation
(7.4)	$\pi_{i++}\pi_{+j+}\pi_{++k}$	None	A, B, C mutually independent
(7.5)	$\pi_{i++}\pi_{+jk}$	γ_{jk}^{BC}	A jointly independent of B and C
(7.6)	$\pi_{i+k}\pi_{+jk}/\pi_{++k}$	$\gamma_{ik}^{AC} + \gamma_{jk}^{BC}$	A, B conditionally indep., given C
(7.7)	$\phi_{ij}\psi_{ik}\omega_{jk}$	$\gamma_{ij}^{AB} + \gamma_{ik}^{AC} + \gamma_{jk}^{BC}$	Homogeneous association

multinomial likelihoods and have the same ML estimates and *SE* values. The fit and the X^2 and G^2 goodness-of-fit statistics are identical for the Poisson and multinomial formulations.

Because the model matrices for these loglinear models contain indicator variables and their products, the likelihood equations (7.1) take the simple form of equating the observed counts to the fitted values in the margins of the contingency table that correspond to the highest-order terms in the model. For example, the mutual independence model (7.4) has likelihood equations, for all i, j, and k,

$$y_{i++} = \hat{\mu}_{i++}, \quad y_{+j+} = \hat{\mu}_{+j+}, \quad y_{++k} = \hat{\mu}_{++k},$$

whereas the homogeneous association model (7.7) has likelihood equations

$$y_{ij+} = \hat{\mu}_{ij+}, \quad y_{i+k} = \hat{\mu}_{i+k}, \quad y_{+jk} = \hat{\mu}_{+jk}.$$

It is straightforward to derive these, much as we did for the independence model in Section 7.2.2. For many models having some independence structure, closed-form solutions exist. In all cases the Newton–Raphson method, which is equivalent to Fisher scoring for these canonical-link models, yields fitted values and corresponding model parameter estimates. When cell means mostly exceed about 5, X^2 and G^2 statistics have approximate chi-squared null distributions for testing the model goodness of fit. Standardized residuals can detect particular cells for which the fit is poor.

7.2.5 Connections Between Logistic and Loglinear Models

Loglinear models for contingency tables treat all categorical classifications symmetrically and regard the cell count as the response. They are useful for modeling the joint distribution of categorical variables. By contrast, logistic models distinguish between response and explanatory classifications. Although different in purpose, the two types of models are connected.

We illustrate with the homogeneous association loglinear model (7.7). Suppose we treat A as a response variable and B and C as explanatory, conditioning on $\{n_{+jk}\}$. For the binary case $r = 2$, we are then modeling $c\ell$ binomial distributions on A. When we construct the logit for each binomial distribution of A, we obtain

$$\begin{aligned}\log \frac{P(A=1 \mid B=j,\ C=k)}{P(A=2 \mid B=j,\ C=k)} &= \log \frac{\mu_{1jk}}{\mu_{2jk}} = \log \mu_{1jk} - \log \mu_{2jk} \\ &= \left(\beta_0 + \beta_1^A + \beta_j^B + \beta_k^C + \gamma_{1j}^{AB} + \gamma_{1k}^{AC} + \gamma_{jk}^{BC}\right) \\ &\quad - \left(\beta_0 + \beta_2^A + \beta_j^B + \beta_k^C + \gamma_{2j}^{AB} + \gamma_{2k}^{AC} + \gamma_{jk}^{BC}\right) \\ &= \left(\beta_1^A - \beta_2^A\right) + \left(\gamma_{1j}^{AB} - \gamma_{2j}^{AB}\right) + \left(\gamma_{1k}^{AC} - \gamma_{2k}^{AC}\right).\end{aligned}$$

The first parenthetical term is a constant, not depending on j or k. The second parenthetical term depends on the category j of B. The third parenthetical term depends on the category k of C. This logit has the additive form

$$\text{logit}[P(A = 1 \mid B = j,\ C = k)] = \lambda + \delta_j^B + \delta_k^C.$$

In fact, the Poisson loglinear model and the binomial logistic model have the same likelihood equations and the same fit. An analogous correspondence holds when A has several categories, using a multinomial baseline-category logit model for A in terms of additive factor effects for B and C.

The loglinear model that has the same fit as a logistic model with factors as explanatory variables contains a general interaction term for relations among those explanatory variables. The logistic model does not assume anything about relations among explanatory variables, so it allows an arbitrary interaction pattern for them. For example, for a main-effects logistic model that predicts A using factors B, C, and D, the corresponding loglinear model has pairwise associations between A and B, A and C, and A and D, as well as the BCD three-factor interaction term and all its lower-order relatives.

7.2.6 Example: Loglinear Models for Student Substance Use

Table 7.3 refers to a survey by Wright State University that asked 2276 students in their final year of high school in a rural area near Dayton, Ohio whether they had ever used alcohol, cigarettes, or marijuana. Denote the variables in this $2 \times 2 \times 2$ table by A, C, and M.

Table 7.4 shows results of testing fit for four loglinear models. Models that lack any association term fit poorly. The homogeneous association model fits well. It is suggested by other criteria also, such as minimizing AIC.

The following output shows some results from fitting the homogeneous association model. The AC fitted conditional odds ratios at each level of M equal $\exp(\hat{\gamma}_{11}^{AC} + \hat{\gamma}_{22}^{AC} - \hat{\gamma}_{12}^{AC} - \hat{\gamma}_{21}^{AC})$, which is $\exp(\hat{\gamma}_{22}^{AC}) = e^{2.0545} = 7.80$ for the R constraints. For those who have used cigarettes, the odds of having used alcohol are estimated to be 7.80 times the odds of having used alcohol for those who have not used cigarettes, and this applies

Table 7.3 Alcohol, Cigarette, and Marijuana Use Among High School Seniors

Alcohol Use (A)	Cigarette Use (C)	Marijuana Use (M) Yes	No
Yes	Yes	911	538
	No	44	456
No	Yes	3	43
	No	2	279

Source: Data courtesy of Harry Khamis, Wright State University.

Table 7.4 Goodness-of-Fit Tests for Loglinear Models Fitted to the Data in Table 7.3

Loglinear Associations	Deviance G^2	Pearson X^2	*df*	*P*-value[a]	AIC
$\gamma_{ij}^{AC} + \gamma_{ik}^{AM}$	497.37	443.76	2	< 0.001	558.4
$\gamma_{ij}^{AC} + \gamma_{jk}^{CM}$	92.02	80.81	2	< 0.001	153.1
$\gamma_{ik}^{AM} + \gamma_{jk}^{CM}$	187.75	177.61	2	< 0.001	248.8
$\gamma_{ij}^{AC} + \gamma_{ik}^{AM} + \gamma_{jk}^{CM}$	0.37	0.40	1	0.54	63.4

[a] *P*-value for G^2 statistic.

both for those who have used marijuana and those who have not. The corresponding Wald 95% confidence interval is exp[2.0545 ± 1.96(0.1741)], or (5.5, 11.0).

```
-----------------------------------------------------------------------
> Drugs  # file Drugs.dat at www.stat.ufl.edu/~aa/glm/data
    A   C   M count
1 yes yes yes   911
2 yes yes  no   538
...
8  no  no  no   279
> attach(Drugs)
> alc <- factor(A); cig <- factor(C); mar <- factor(M)
> mutual.indep <- glm(count ~ alc + cig + mar, family=poisson(link=log))
> homo.assoc <- update(mutual.indep, .~. + alc:cig + alc:mar + cig:mar)
> summary(homo.assoc)
             Estimate  Std. Error  z value  Pr(>|z|)
(Intercept)   6.8139     0.0331   205.699   < 2e-16
alc2         -5.5283     0.4522   -12.225   < 2e-16
cig2         -3.0157     0.1516   -19.891   < 2e-16
mar2         -0.5249     0.0543    -9.669   < 2e-16
alc2:cig2     2.0545     0.1741    11.803   < 2e-16 # odds ratio  7.8
alc2:mar2     2.9860     0.4647     6.426  1.31e-10 # odds ratio 19.8
cig2:mar2     2.8479     0.1638    17.382   < 2e-16 # odds ratio 17.3
---
Residual deviance:    0.3740  on 1  degrees of freedom
AIC: 63.417
-----------------------------------------------------------------------
```

For a loglinear model with residual $df = 1$, each standardized residual has the same absolute value and has square equal to the Pearson X^2 statistic for testing goodness of fit. The Pearson residuals are less appealing: they have eight separate values, even though $|y_{ijk} - \hat{\mu}_{ijk}|$ is identical for each cell (because the likelihood equations imply that the two-way observed and fitted marginal tables are identical) and the residual $df = 1$.

```
-----------------------------------------------------------------------
> pearson.resid <- resid(homo.assoc, type="pearson")
> std.resid <- rstandard(homo.assoc, type="pearson")
> sum(pearson.resid^2)  # Pearson chi-squared statistic
[1] 0.4011006
```

```
> cbind(count, fitted(homo.assoc), pearson.resid, std.resid)
  count  fitted(homo.assoc)  pearson.resid  std.resid
1   911             910.383          0.020      0.633
2   538             538.617         -0.027     -0.633
3    44              44.617         -0.092     -0.633
4   456             455.383          0.029      0.633
5     3               3.617         -0.324     -0.633
6    43              42.383          0.095      0.633
7     2               1.383          0.524      0.633
8   279             279.617         -0.037     -0.633
---------------------------------------------------------------------------
```

Using a logistic model, we find the same results for the association between marijuana use and each of alcohol use and cigarette use (i.e., estimated log odds ratios of 2.99 and 2.85). We model the logit of the probability of using marijuana with additive effects for alcohol use and cigarette use, treating the data as four binomial observations instead of eight Poisson observations.

```
---------------------------------------------------------------------------
> Drugs2  # file Drugs2.dat at www.stat.ufl.edu/~aa/glm/data
    A    C  M_yes  M_no     n  # data entered as 4 binomials
1 yes  yes    911   538  1449
2 yes   no     44   456   500
3  no  yes      3    43    46
4  no   no      2   279   281
> attach(Drugs2)
> alc <- factor(A); cig <- factor(C)
> fit.logistic <- glm(M_yes/n ~ alc + cig, weights=n, # specify weights
     family = binomial(link = logit)) # when enter proportion responses
> summary(fit.logistic)
            Estimate  Std. Error  z value  Pr(>|z|)
(Intercept)  -5.3090      0.4752  -11.172   < 2e-16
alcyes        2.9860      0.4647    6.426  1.31e-10 # odds ratio 19.8
cigyes        2.8479      0.1638   17.382   < 2e-16 # odds ratio 17.3
---
    Null deviance: 843.8266  on 3  degrees of freedom
Residual deviance:   0.3740  on 1  degrees of freedom
---------------------------------------------------------------------------
```

The null logistic model in this case is equivalent to the loglinear model by which marijuana use is jointly independent of alcohol use and cigarette use.

7.2.7 Graphical Loglinear Models: Portraying Conditional Independence Structure

Many loglinear models have graphical portrayals of the conditional independence structure among the responses. This representation also helps to reveal implications of models, such as when an association is unchanged when a variable is dropped from an analysis.

From graph theory, an undirected graph consists of a set of vertices and a set of edges connecting some vertices. In a probabilistic *conditional independence graph*, each vertex represents a variable, and the absence of an edge connecting two variables represents conditional independence between them. For instance, Figure 7.1 portrays the conditional independence graph for categorical response variables A, B, C, and D and the loglinear model that assumes independence between A and C and between A and D, conditional on the other two variables. The four variables form the vertices, and the four edges represent pairwise conditional associations. Edges do not connect A and C or connect A and D, the conditionally independent pairs.

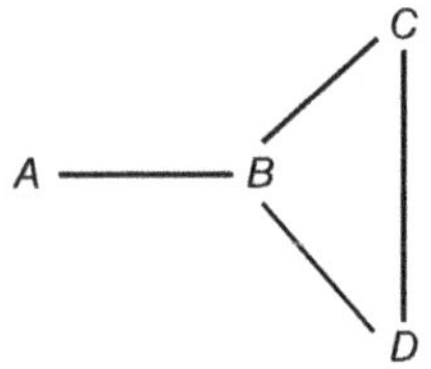

Figure 7.1 Conditional independence graph for the loglinear model that assumes conditional independence between A and C and between A and D.

Darroch et al. (1980) used undirected graphs to represent *graphical models*, which are essentially loglinear models for contingency tables that have a conditional independence structure. The graphical model corresponding to Figure 7.1 is the loglinear model having the three-factor BCD interaction and the two-factor AB association and their lower-order relatives, namely,

$$\log \mu_{hijk} = \beta_0 + \beta_h^A + \beta_i^B + \beta_j^C + \beta_k^D + \gamma_{hi}^{AB} + \gamma_{ij}^{BC} + \gamma_{ik}^{BD} + \gamma_{jk}^{CD} + \gamma_{ijk}^{BCD}.$$

A *path* in a conditional independence graph is a sequence of edges between one variable and another. Two variables A and C are said to be *separated* by a subset of variables if all paths connecting A and C intersect that subset. For instance, in Figure 7.1, B separates A and C. The subset $\{B, D\}$ also separates A and C. *Markov properties* that pertain to paths and separation allow us to deduce from the graph the conditional independence structure between variables and groups of variables. One such property, the *global Markov property*, states that two variables are conditionally independent given *any* subset of variables that separates them in the graph. Thus, in Figure 7.1, not only are A and C conditionally independent given B and D, but also given B alone. Similarly, A and D are conditionally independent given B alone. This property is equivalent to a *local Markov property*, according to which a variable is conditionally independent of all other variables, given the adjacent neighbors to which it is connected by an edge.

Conditional associations usually differ[5] from marginal associations. Under certain *collapsibility conditions*, however, they are the same. For loglinear and logistic models, the association parameters pertain to odds ratios, so such conditions relate to

[5]Recall, for example, Section 3.4.3 and Simpson's paradox.

equality of conditional and marginal odds ratios. Bishop et al. (1975, p. 47) provided a parametric collapsibility condition for multiway contingency tables:

> Suppose that a model for a multiway contingency table partitions variables into three mutually exclusive subsets, $\{S_1, S_2, S_3\}$, such that S_2 separates S_1 and S_3. After collapsing the table over the variables in S_3, parameters relating variables in S_1 and parameters relating variables in S_1 to variables in S_2 are unchanged.

For the graphical model corresponding to Figure 7.1, let $S_1 = \{A\}, S_2 = \{B\}$, and $S_3 = \{C, D\}$. Since the AC and AD terms do not appear in the model, all parameters linking set S_1 with set S_3 equal zero, and S_2 separates S_1 and S_3. If we collapse over C and D, the AB association is unchanged. Next, identify $S_1 = \{C, D\}$, $S_2 = \{B\}$, $S_3 = \{A\}$. Then conditional associations among B, C, and D remain the same after collapsing over A. By contrast, the homogeneous loglinear model that provides a good fit for the student substance use data of Table 7.3 does not satisfy collapsibility conditions. The fitted (AC, AM, CM) conditional odds ratios of (7.8, 19.8, 17.3) obtained by exponentiating the log odds ratio estimates from the R output differ from the corresponding two-way marginal odds ratios, (17.7, 61.9, 25.1).

7.3 NEGATIVE BINOMIAL GLMS

For the Poisson distribution, the variance equals the mean. In practice, count observations often exhibit variability exceeding that predicted by the Poisson. This phenomenon is called *overdispersion*.

7.3.1 Overdispersion for a Poisson GLM

A common reason for overdispersion is heterogeneity: at fixed levels of the explanatory variables, the mean varies according to values of unobserved variables. For example, for the horseshoe crab dataset introduced in Section 1.5.1 that we analyze further in Section 7.5, suppose that a female crab's carapace width, weight, color, and spine condition are the four explanatory variables that affect her number of male satellites. Suppose that y has a Poisson distribution at each fixed combination of those variables, but we use the model that has weight alone as an explanatory variable. Crabs having a certain weight are then a mixture of crabs of various widths, colors, and spine conditions. Thus, the population of crabs having that weight is a mixture of several Poisson populations, each having its own mean for the response. This heterogeneity results in an overall response distribution at that weight having greater variation than the Poisson. If the variance equals the mean when *all* relevant explanatory variables are included, it exceeds the mean when only *some* are included. Another severe limitation of Poisson GLMs is that, because the variance of y must equal the mean, at a fixed mean the variance cannot decrease as additional explanatory variables enter the model.

Overdispersion is not an issue in ordinary linear models that assume normally distributed y, because that distribution has a separate variance parameter to describe variability. For Poisson and binomial distributions, however, the variance is a function of the mean. Overdispersion is common in the modeling of counts. Suppose the model for the mean has the correct link function and linear predictor, but the true response distribution has more variability than the Poisson. Then the ML estimators of model parameters assuming a Poisson response are still consistent, converging in probability to the parameter values, but standard errors are too small. Extensions of the Poisson GLM that have an extra parameter account better for overdispersion. We present one such extension here, and Sections 7.4, 8.1, and 9.4 present others.

7.3.2 Negative Binomial as a Gamma Mixture of Poissons

A mixture model is a flexible way to account for overdispersion. At a fixed setting of the explanatory variables actually observed, given the mean λ, suppose the distribution of y is Poisson(λ), but λ itself varies because of unmeasured covariates. Let $\mu = E(\lambda)$. Then unconditionally,

$$E(y) = E[E(y \mid \lambda)] = E(\lambda) = \mu,$$

$$\text{var}(y) = E[\text{var}(y \mid \lambda)] + \text{var}[E(y \mid \lambda)] = E(\lambda) + \text{var}(\lambda) = \mu + \text{var}(\lambda) > \mu.$$

Here is an important example of a mixture model for count data: suppose that (1) given λ, y has a Poisson(λ) distribution, and (2) λ has the gamma distribution (4.29). Recall that the gamma distribution has $E(\lambda) = \mu$ and $\text{var}(\lambda) = \mu^2/k$ for a shape parameter $k > 0$, so the standard deviation is proportional to the mean. Marginally, the gamma mixture of the Poisson distributions yields the *negative binomial distribution* for y. Its probability mass function is

$$p(y; \mu, k) = \frac{\Gamma(y+k)}{\Gamma(k)\Gamma(y+1)} \left(\frac{\mu}{\mu+k}\right)^y \left(\frac{k}{\mu+k}\right)^k, \quad y = 0, 1, 2, \ldots . \tag{7.8}$$

With k fixed, this is a member of an exponential dispersion family appropriate for discrete variables (Exercise 7.23), with natural parameter $\log[\mu/(\mu + k)]$.

In the two-parameter negative binomial family, let $\gamma = 1/k$. Then y has

$$E(y) = \mu, \quad \text{var}(y) = \mu + \gamma\mu^2.$$

The index $\gamma > 0$ is a type of dispersion parameter. The greater the value of γ, the greater the overdispersion relative to the Poisson. As $\gamma \to 0$, $\text{var}(y) \to \mu$ and the negative binomial distribution converges[6] to the Poisson.

The negative binomial distribution has much greater scope than the Poisson. For example, the Poisson mode is the integer part of the mean and equals 0 only when $\mu < 1$. The negative binomial is also unimodal, but the mode is 0 when $\gamma \geq 1$ and otherwise it is the integer part of $\mu(1 - \gamma)$. The mode can be 0 for any μ.

[6]For a proof, see Cameron and Trivedi (2013, p. 85).

7.3.3 Negative Binomial GLMs

Negative binomial GLMs commonly use the log link, as in Poisson loglinear models, rather than the canonical link. For simplicity, we let the dispersion parameter γ be the same constant for all n observations but treat it as unknown, much like the variance in normal models. This corresponds to a constant coefficient of variation in the gamma mixing distribution, $\sqrt{\text{var}(\lambda)}/E(\lambda) = \sqrt{\gamma}$.

From Equation (7.8) expressed in terms of the dispersion parameter γ, the log-likelihood function for a negative binomial GLM with n independent observations is

$$L(\boldsymbol{\beta}, \gamma; \boldsymbol{y}) = \sum_{i=1}^{n} \left[\log \Gamma\left(y_i + \frac{1}{\gamma}\right) - \log \Gamma\left(\frac{1}{\gamma}\right) - \log \Gamma(y_i + 1) \right]$$
$$+ \sum_{i=1}^{n} \left[y_i \log\left(\frac{\gamma\mu_i}{1 + \gamma\mu_i}\right) - \left(\frac{1}{\gamma}\right) \log(1 + \gamma\mu_i) \right],$$

where μ_i is a function of $\boldsymbol{\beta}$ through $\eta_i = g(\mu_i) = \sum_j \beta_j x_{ij}$ with the link function g. The likelihood equations obtained by differentiating $L(\boldsymbol{\beta}, \gamma; \boldsymbol{y})$ with respect to $\boldsymbol{\beta}$ have the usual form (4.10) for a GLM,

$$\sum_{i=1}^{n} \frac{(y_i - \mu_i)x_{ij}}{\text{var}(y_i)} \left(\frac{\partial \mu_i}{\partial \eta_i}\right) = \sum_{i} \frac{(y_i - \mu_i)x_{ij}}{\mu_i + \gamma\mu_i^2} \left(\frac{\partial \mu_i}{\partial \eta_i}\right) = 0, \quad j = 1, 2, \ldots, p.$$

The log-likelihood yields a Hessian matrix that has

$$\frac{\partial^2 L(\boldsymbol{\beta}, \gamma; \boldsymbol{y})}{\partial \beta_j \partial \gamma} = -\sum_{i} \frac{(y_i - \mu_i)x_{ij}}{(1 + \gamma\mu_i)^2} \left(\frac{\partial \mu_i}{\partial \eta_i}\right).$$

Thus, $E(\partial^2 L/\partial \beta_j \partial \gamma) = 0$ for each j, and $\boldsymbol{\beta}$ and γ are orthogonal parameters (Recall Section 4.2.4). So $\hat{\boldsymbol{\beta}}$ and $\hat{\gamma}$ are asymptotically independent, and the large-sample *SE* for $\hat{\beta}_j$ is the same whether γ is known or estimated.

The iteratively reweighted least squares algorithm for Fisher scoring applies for ML model fitting. The estimated covariance matrix of $\hat{\boldsymbol{\beta}}$ is

$$\widehat{\text{var}}(\hat{\boldsymbol{\beta}}) = (\boldsymbol{X}^{\text{T}} \widehat{\boldsymbol{W}} \boldsymbol{X})^{-1},$$

where, with log link, $\boldsymbol{W}$ is the diagonal matrix with $w_i = (\partial \mu_i/\partial \eta_i)^2/\text{var}(y_i) = \mu_i/(1 + \gamma\mu_i)$. The deviance for a negative binomial GLM is

$$D(\boldsymbol{y}, \hat{\boldsymbol{\mu}}) = 2 \sum_{i} \left[y_i \log\left(\frac{y_i}{\hat{\mu}_i}\right) - \left(y_i + \frac{1}{\hat{\gamma}}\right) \log\left(\frac{1 + \hat{\gamma} y_i}{1 + \hat{\gamma}\hat{\mu}_i}\right) \right].$$

This is close to the Poisson GLM deviance (7.2) when $\hat{\gamma}$ is near 0.

7.3.4 Comparing Poisson and Negative Binomial GLMs

How can we compare Poisson and negative binomial GLMs that have the same explanatory variables, to determine whether the negative binomial model gives a better fit? An informal comparison can be based on AIC values. For a formal significance test, we can test H_0: $\gamma = 0$, because the Poisson is the limiting case of the negative binomial as $\gamma \downarrow 0$.

Since γ is positive, $\gamma = 0$ on the boundary of the parameter space. Thus, the likelihood-ratio statistic does not have an asymptotic null chi-squared distribution. Rather, it is an equal mixture of a single-point distribution at 0 (which occurs when $\hat{\gamma} = 0$) and chi-squared with $df = 1$. The P-value is half that from treating the statistic as chi-squared with $df = 1$ (Self and Liang 1987).

7.3.5 Negative Binomial Model with Variance Proportional to Mean

An alternative negative binomial parameterization results from writing the gamma density formula with $k\mu$ as the shape parameter,

$$f(\lambda; k, \mu) = \frac{k^{k\mu}}{\Gamma(k\mu)} \exp(-k\lambda)\lambda^{k\mu-1} \quad \lambda \geq 0,$$

so $E(\lambda) = \mu$ and $\text{var}(\lambda) = \mu/k$. For this parameterization, the gamma mixture of Poisson distributions yields a negative binomial distribution with

$$E(y) = \mu, \quad \text{var}(y) = \mu(1 + k)/k.$$

The variance is now linear rather than quadratic in μ. It corresponds to an inflation of the Poisson variance, converging to it as $k \to \infty$.

The two parameterizations of the negative binomial are sometimes denoted by NB1 (linear) and NB2 (quadratic). Only the NB2 falls within the traditional GLM framework, being expressible as an exponential dispersion family distribution, and it is much more commonly used. Unlike the NB2 model, for an NB1 model $\boldsymbol{\beta}$ and k are not orthogonal parameters, and $\hat{\boldsymbol{\beta}}$ is not a consistent estimator when the model for the mean holds but the true distribution is not negative binomial (Cameron and Trivedi 2013, Section 3.3). Lee and Nelder (1996) presented ML model fitting for NB1 models.

7.4 MODELS FOR ZERO-INFLATED DATA

In practice, the frequency of 0 outcomes is often larger than expected under standard discrete models. In particular, because the mode of a Poisson distribution is the integer part of the mean, a Poisson GLM is inadequate when means can be relatively large but the modal response is 0. Such data, which are *zero-inflated* relative to data expected for a Poisson GLM, are common when many subjects have a 0 response and many also have much larger responses, so the overall mean is not near 0. An example of a variable that might be zero-inflated is the number of times in the past week that individuals report exercising, such as by going to a gym. Some would do so frequently, some would do it occasionally but not in the past week (a random 0), and a substantial percentage would never do so, causing zero inflation. Other examples are counts of activities for which many subjects would necessarily report 0, such as the number of times during some period of having an alcoholic drink, or smoking marijuana, or having sexual intercourse.

Zero-inflation is less problematic for negative binomial GLMs, because that distribution can have a mode of 0 regardless of the value of the mean. However, a negative binomial model fits poorly when the data are strongly bimodal, with a mode at zero and a separate mode around some considerably higher value. This could occur for the frequency of an activity in which many subjects never participate but many others quite often do. Then a substantial fraction of the population necessarily has a zero outcome, and the remaining fraction follows some distribution that may have small probability of a zero outcome.

7.4.1 Zero-Inflated Poisson and Negative Binomial Models

The representation just mentioned, of one set of observations that necessarily are zero and another set that may be zero according to a random event, leads naturally to a mixture model in which two types of zeros can occur. The relevant distribution is a mixture of an ordinary count model such as the Poisson or negative binomial with one that places all its mass at zero.

The *zero-inflated Poisson (ZIP) model* (Lambert 1992) assumes that

$$y_i \sim \begin{cases} 0 & \text{with probability } 1 - \phi_i \\ \text{Poisson}(\lambda_i) & \text{with probability } \phi_i. \end{cases}$$

The unconditional probability distribution has

$$P(y_i = 0) = (1 - \phi_i) + \phi_i e^{-\lambda_i},$$

$$P(y_i = j) = \phi_i \frac{e^{-\lambda_i} \lambda_i^j}{j!}, \qquad j = 1, 2, \ldots .$$

Explanatory variables affecting ϕ_i need not be the same as those affecting λ_i. The parameters could be modeled by

$$\text{logit}(\phi_i) = \boldsymbol{x}_{1i}\boldsymbol{\beta}_1 \quad \text{and} \quad \log(\lambda_i) = \boldsymbol{x}_{2i}\boldsymbol{\beta}_2.$$

A latent class construction that yields this model posits an unobserved binary variable z_i. When $z_i = 0$, $y_i = 0$, and when $z_i = 1$, y_i is a Poisson(λ_i) variate. For this mixture distribution,

$$E(y_i) = E[E(y_i \mid z_i)] = (1 - \phi_i)0 + \phi_i(\lambda_i) = \phi_i \lambda_i.$$

Also, because $E[\text{var}(y_i \mid z_i)] = (1 - \phi_i)0 + \phi_i(\lambda_i) = \phi_i\lambda_i$ and $\text{var}[E(y_i \mid z_i)] = (1 - \phi_i)(0 - \phi_i\lambda_i)^2 + \phi_i(\lambda_i - \phi_i\lambda_i)^2 = \lambda_i^2\phi_i(1 - \phi_i)$,

$$\text{var}(y_i) = E[\text{var}(y_i \mid z_i)] + \text{var}[E(y_i \mid z_i)] = \phi_i\lambda_i[1 + (1 - \phi_i)\lambda_i].$$

Since $\text{var}(y_i) > E(y_i)$, overdispersion occurs relative to a Poisson model.

When λ_i and ϕ_i are not functionally related, the joint log-likelihood function for the two parts of the model is

$$L(\boldsymbol{\beta}_1, \boldsymbol{\beta}_2) = \sum_{y_i=0} \log[1 + e^{\boldsymbol{x}_{1i}\boldsymbol{\beta}_1} \exp(-e^{\boldsymbol{x}_{2i}\boldsymbol{\beta}_2})] - \sum_{i=1}^{n} \log(1 + e^{\boldsymbol{x}_{1i}\boldsymbol{\beta}_1})$$

$$+ \sum_{y_i>0} [\boldsymbol{x}_{1i}\boldsymbol{\beta}_1 + y_i\boldsymbol{x}_{2i}\boldsymbol{\beta}_2 - e^{\boldsymbol{x}_{2i}\boldsymbol{\beta}_2} - \log(y_i!)].$$

Lambert (1992) expressed the log-likelihood in terms of the latent variables $\{z_i\}$. She used the EM algorithm for ML fitting, treating each z_i as a missing value. Alternatively, the Newton–Raphson method can be used.

A disadvantage of the ZIP model is the larger number of parameters compared with ordinary Poisson or negative binomial models. Sometimes the explanatory variables in the two parts of the model are the same, and their effects have similar relative size. For such cases, Lambert proposed a simpler model in which $\boldsymbol{x}_{1i} = \boldsymbol{x}_{2i}$ and $\boldsymbol{\beta}_2 = \tau\boldsymbol{\beta}_1$ for a shape parameter τ. Another disadvantage of the general ZIP model is that the parameters do not directly describe the effects of explanatory variables on $E(y_i) = \phi_i\lambda_i$, because $\boldsymbol{\beta}_1$ pertains to effects on ϕ_i and $\boldsymbol{\beta}_2$ pertains to effects on λ_i. In addition, when $\boldsymbol{x}_{1i}$ and $\boldsymbol{x}_{2i}$ are the same or overlap substantially, the correlation between them could cause further problems with interpretation. A simpler alternative fits only an intercept in the model for ϕ_i. In that case $E(y_i)$ is proportional to λ_i.

In practice, overdispersion often occurs even when we condition on the response being positive or when we condition on $z_i = 1$ in the latent formulation of the ZIP model. The equality of mean and variance assumed by the ZIP model, conditional on $z_i = 1$, may not be realistic. When we use a ZIP model but there is overdispersion, standard error estimates can be badly biased downward. A zero-inflated negative binomial (ZINB) model is then more appropriate. For it, with probability $1 - \phi_i$, $y_i = 0$, and with probability ϕ_i, y_i has a negative binomial distribution with mean λ_i and dispersion parameter γ.

7.4.2 Hurdle Models: Handling Zeroes Separately

An alternative approach to modeling zero-inflation uses a two-part model called a *hurdle model*. One part is a binary model such as a logistic or probit model for whether the response outcome is zero or positive. If the outcome is positive, the "hurdle is crossed." Conditional on a positive outcome, to analyze its level, the second part uses a truncated model that modifies an ordinary distribution by conditioning on a positive outcome. The hurdle model can handle both zero inflation and zero deflation.

Suppose that the first part of the process is governed by probabilities $P(y_i > 0) = \pi_i$ and $P(y_i = 0) = 1 - \pi_i$ and that $\{y_i \mid y_i > 0\}$ follows a truncated-at-zero probability mass function $f(y_i; \mu_i)$, such as a truncated Poisson. The complete distribution is

$$P(y_i = 0) = 1 - \pi_i,$$

$$P(y_i = j) = \pi_i \frac{f(j; \mu_i)}{1 - f(0; \mu_i)}, \quad j = 1, 2, \ldots.$$

With explanatory variables, we could use a logistic regression model for π_i and a loglinear model for the mean μ_i of the untruncated f distribution,

$$\text{logit}(\pi_i) = \boldsymbol{x}_{1i}\boldsymbol{\beta}_1 \quad \text{and} \quad \log(\mu_i) = \boldsymbol{x}_{2i}\boldsymbol{\beta}_2.$$

The joint likelihood function for the two-part hurdle model is

$$\ell(\boldsymbol{\beta}_1, \boldsymbol{\beta}_2) = \prod_{i=1}^{n} (1 - \pi_i)^{I(y_i=0)} \left[\pi_i \frac{f(y_i; \mu_i)}{1 - f(0; \mu_i)} \right]^{1-I(y_i=0)},$$

where $I(\cdot)$ is the indicator function. If $(1 - \pi_i) > f(0; \mu_i)$ for every i, the model represents zero inflation. The log-likelihood separates into two terms, $L(\boldsymbol{\beta}_1, \boldsymbol{\beta}_2) = L_1(\boldsymbol{\beta}_1) + L_2(\boldsymbol{\beta}_2)$, where

$$\begin{aligned} L_1(\boldsymbol{\beta}_1) &= \sum_{y_i=0} \left[\log(1 - \pi_i)\right] + \sum_{y_i>0} \log(\pi_i) \\ &= \sum_{y_i>0} \boldsymbol{x}_{1i}\boldsymbol{\beta}_1 - \sum_{i=1}^{n} \log(1 + e^{\boldsymbol{x}_{1i}\boldsymbol{\beta}_1}) \end{aligned}$$

is the log-likelihood function for the binary process and

$$L_2(\boldsymbol{\beta}_2) = \sum_{y_i>0} \left\{ \log f(y_i; \exp(\boldsymbol{x}_{2i}\boldsymbol{\beta}_2)) - \log\left[1 - f(0; \exp(\boldsymbol{x}_{2i}\boldsymbol{\beta}_2))\right] \right\}$$

is the log-likelihood function for the truncated model. With a truncated Poisson model for the positive outcome,

$$L_2(\boldsymbol{\beta}_2) = \sum_{y_i>0} \{y_i \boldsymbol{x}_{2i}\boldsymbol{\beta}_2 - e^{\boldsymbol{x}_{2i}\boldsymbol{\beta}_2} - \log[1 - \exp(-e^{\boldsymbol{x}_{2i}\boldsymbol{\beta}_2})]\} - \sum_{y_i>0} \log(y_i!)$$

is the log-likelihood function for the truncated model. When overdispersion occurs, using a truncated negative binomial for the positive outcome performs better. We obtain ML estimates by separately maximizing L_1 and L_2.

Zero-inflated models are more natural than the hurdle model when the population is naturally regarded as a mixture, with one set of subjects that necessarily has a 0 response. However, the hurdle model is also suitable when, at some settings, the data have fewer zeros than are expected under standard distributional assumptions.

7.4.3 Truncated Discrete Models for Positive Count Data

The part of the hurdle model that applies to the positive counts uses a truncation of a discrete distribution. Such a truncated distribution is of use in its own right in applications in which a count of 0 is not possible. Examples of such response variables are the number of people in a household, the number of occupants of a car, and the number of days a patient admitted to a hospital stays there.

If y_i has a truncated Poisson distribution with parameter λ_i, then

$$E(y_i) = \frac{\lambda_i}{1 - e^{-\lambda_i}}, \qquad \text{var}(y_i) = \frac{\lambda_i}{1 - e^{-\lambda_i}} - \frac{\lambda_i^2 e^{-\lambda_i}}{(1 - e^{-\lambda_i})^2}.$$

Conditional on a Poisson variate being positive, the variance is smaller than the mean. When this is substantially violated, more flexibility is provided by the zero-truncated negative binomial distribution. It can be derived as a gamma mixture of zero-truncated Poisson distributions. Software is available for fitting zero-truncated distributions[7].

7.5 EXAMPLE: MODELING COUNT DATA

We illustrate models for discrete data using the horseshoe crab dataset introduced in Section 1.5.1 The response variable for the $n = 173$ mating female crabs is $y =$ number of "satellites"—male crabs that group around the female and may fertilize her eggs. Explanatory variables are the female crab's color, spine condition, weight, and carapace width.

7.5.1 Fits to Marginal Distribution of Satellite Counts

To illustrate the Poisson, negative binomial, ZIP, and ZINB distributions introduced in this chapter, we first investigate the marginal distribution of satellite counts. From Section 1.5.1, the mean of 2.919 and variance of 9.912 suggest overdispersion relative to the Poisson.

```
> attach(Crabs) # file Crabs.dat at www.stat.ufl.edu/~aa/glm/data
> hist(y, breaks=c(0:16)-0.5) # Histogram display with sufficient bins
```

[7] Examples are the `pospois` and `posnegbinom` functions in the `VGAM` package of R.

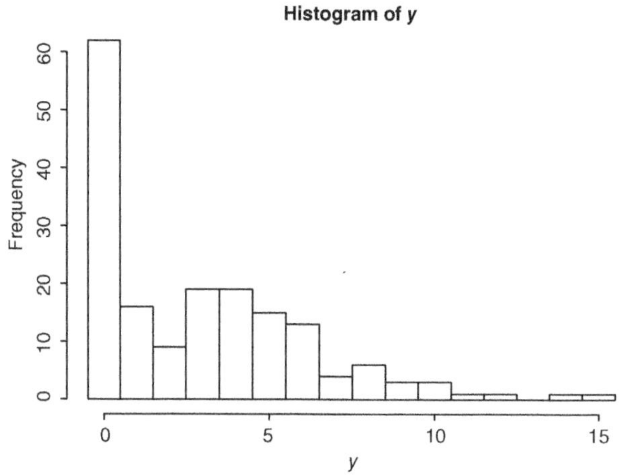

Figure 7.2 Histogram for sample distribution of y = number of horseshoe crab satellites.

The histogram (Figure 7.2) shows a strong mode at 0 but slightly elevated frequencies for satellite counts of 3 through 6 before decreasing substantially. Because the distribution may not be unimodal, the negative binomial may not fit as well as a zero-inflated distribution.

We fit the Poisson distribution and negative binomial distribution with quadratic variance (NB2) by fitting GLMs having only an intercept.

```
-------------------------------------------------------------------------
> summary(glm(y ~ 1, family=poisson, data=Crabs)) # default link is log
            Estimate  Std. Error  z value  Pr(>|z|)
(Intercept)  1.0713      0.0445    24.07  <2e-16 # exp(1.0713) = 2.919
---
> logLik(glm(y ~ 1, family=poisson, data=Crabs))
'log Lik.' -494.045

> library(MASS)
> summary(glm.nb(y ~ 1, data=Crabs)) # default link is log
            Estimate  Std. Error  z value  Pr(>|z|)
(Intercept)  1.0713      0.0980    10.93  <2e-16
---
Theta: 0.758, Std. Err.: 0.126
> logLik(glm.nb(y ~ 1, data=Crabs))
'log Lik.' -383.705
-------------------------------------------------------------------------
```

The estimated NB2 dispersion parameter[8] is $\hat{\gamma} = 1/0.758 = 1.32$. This estimate, the much larger *SE* (0.0980 vs. 0.0445) for the log mean estimate of log(2.919) =

[8]SAS (PROC GENMOD) reports $\hat{\gamma}$ as having $SE = 0.22$.

1.071, and the much larger log-likelihood also suggest that the Poisson distribution is inadequate.

Next, we consider zero-inflated models[9].

```
-----------------------------------------------------------------------
> library(pscl) # pscl package can fit zero-inflated distributions
> summary(zeroinfl(y ~ 1)) # uses log link
Count model coefficients (poisson with log link):
            Estimate  Std. Error  z value  Pr(>|z|)
(Intercept)  1.50385     0.04567    32.93    <2e-16

Zero-inflation model coefficients (binomial with logit link):
            Estimate  Std. Error  z value  Pr(>|z|)
(Intercept)  -0.6139      0.1619   -3.791   0.00015
---
Log-likelihood: -381.615 on 2 Df # 2 is model df, not residual df
-----------------------------------------------------------------------
```

The fitted ZIP distribution is a mixture with probability $e^{-0.6139}/[1 + e^{-0.6139}] = 0.351$ for the degenerate distribution at 0 and probability $1 - 0.351 = 0.649$ for a Poisson with mean $e^{1.50385} = 4.499$. The fitted value of $173[0.351 + 0.649e^{-4.499}] = 62.0$ for the 0 count reproduces the observed value of 62. The fitted value for the ordinary Poisson model is only $173e^{-2.919} = 9.3$. The log-likelihood increases substantially when we fit a zero-inflated negative binomial (ZINB) model.

```
-----------------------------------------------------------------------
> summary(zeroinfl(y ~ 1, dist="negbin")) # uses log link in pscl lib.
Count model coefficients (negbin with log link):
            Estimate  Std. Error  z value   Pr(>|z|)
(Intercept)  1.46527     0.06834   21.440   < 2e-16
Log(theta)   1.49525     0.34916    4.282  1.85e-05

Zero-inflation model coefficients (binomial with logit link):
            Estimate  Std. Error  z value  Pr(>|z|)
(Intercept)  -0.7279      0.1832   -3.973   7.1e-05
---
Theta = 4.4605     Log-likelihood: -369.352 on 3 Df
-----------------------------------------------------------------------
```

This distribution is a mixture with probability $e^{-0.7279}/[1 + e^{-0.7279}] = 0.326$ for the degenerate distribution at 0 and probability 0.674 for a negative binomial with mean $e^{1.465} = 4.33$ and dispersion parameter estimate $\hat{\gamma} = 1/4.4605 = 0.22$.

To further investigate lack of fit, we grouped the counts into 10 categories, using a separate category for each count from 0 to 8 and then combining counts of 9 and

[9]Such models can also be fitted with the `vglm` function in the VGAM package.

above into a single category. Comparing these with the ZINB fitted distribution of the 173 observations into these 10 categories, we obtained $X^2 = 7.7$ for $df = 10 - 3 = 7$ (since the model has three parameters), an adequate fit. For the other fits, $X^2 = 522.3$ for the Poisson model, 33.6 for the ordinary negative binomial model, and 31.3 for the ZIP model. Here are the fitted counts for the four models:

```
-----------------------------------------------------------------
count      observed    fit.p   fit.nb  fit.zip  fit.zinb
0              62       9.34    52.27    62.00     62.00
1              16      27.26    31.45     5.62     12.44
2               9      39.79    21.94    12.63     16.73
3              19      38.72    16.01    18.94     17.74
4              19      28.25    11.94    21.31     16.30
5              15      16.50     9.02    19.17     13.58
6              13       8.03     6.87    14.38     10.55
7               4       3.35     5.27     9.24      7.76
8               6       1.22     4.06     5.20      5.48
9 or more      10       0.55    14.16     4.51     10.43
-----------------------------------------------------------------
```

The ZIP model tends to be not dispersed enough, having fitted value that is too small for the counts of 1 and ≥ 9.

7.5.2 GLMs for Crab Satellite Numbers

We now consider zero-inflated negative binomial models with the explanatory variables from Table 1.3. Weight and carapace width have a correlation of 0.887, and we shall use only weight to avoid issues with collinearity. Darker-colored crabs tend to be older. Most crabs have both spines worn or broken (category 3). When we fit the ZINB main-effects model using weight, color, and spine condition for each component, with color and spine condition as qualitative factors, we find that weight is significant in each component but neither of color or spine condition are. Adding interaction terms does not yield an improved fit. Analyses using color in a quantitative manner with category scores $\{c_i = i\}$ gives relatively strong evidence that darker crabs tend to have more 0 counts. If we use weight w_i in both components of the model but quantitative color only in the zero-component, we obtain:

```
-----------------------------------------------------------------
> summary(zeroinfl(y ~ weight | weight + color, dist="negbin"))
Count model coefficients (negbin with log link):
             Estimate  Std. Error  z value  Pr(>|z|)
(Intercept)   0.8961      0.3070     2.919    0.0035
weight        0.2169      0.1125     1.928    0.0538
Log(theta)    1.5802      0.3574     4.422    9.79e-06
```

```
Zero-inflation model coefficients (binomial with logit link):
             Estimate  Std. Error  z value  Pr(>|z|)
(Intercept)    1.8662      1.2415    1.503     0.133
weight        -1.7531      0.4429   -3.958  7.55e-05
color          0.5985      0.2572    2.326     0.020
---
Theta = 4.8558     Log-likelihood: -349.865 on 6 Df
-------------------------------------------------------------------------
```

The fitted distribution is a mixture with probability $\hat{\phi}_i$ of a negative binomial having mean $\hat{\mu}_i$ satisfying

$$\log \hat{\mu}_i = 0.896 + 0.217 w_i$$

with dispersion parameter estimate $\hat{\gamma} = 1/4.8558 = 0.21$, and a probability mass $1 - \hat{\phi}_i$ at 0 satisfying

$$\text{logit}(1 - \hat{\phi}_i) = 1.866 - 1.753 w_i + 0.598 c_i.$$

The overall fitted mean response at a particular weight and color equals

$$\hat{E}(y_i) = \hat{\phi}_i \hat{E}(y_i \mid z_i = 1) = \left(\frac{1}{1 + e^{1.866 - 1.753 w_i + 0.598 c_i}} \right) e^{0.896 + 0.217 w_i}.$$

As weight increases for a particular color, the fitted probability mass at the 0 outcome decreases, and the fitted negative binomial mean increases. Figure 7.3 plots the overall fitted mean as a function of weight for the dark crabs (color 4) and as a function of color at the median weight of 2.35 kg.

If we drop color completely and exclude weight from the NB2 component of the model, the log-likelihood decreases to −354.7 but we obtain the simple expression for the overall fitted mean of $\exp(1.47094)/[1 + \exp(3.927 - 1.985 w_i)]$. This has a logistic shape for the increase in the fitted mean as a function of weight.

If we ignore the zero inflation and fit an ordinary NB2 model with weight and quantitative color predictors, we obtain:

```
-------------------------------------------------------------------------
> summary(glm.nb(y ~ weight + color))
Coefficients:
             Estimate  Std. Error  z value  Pr(>|z|)
(Intercept)   -0.3220      0.5540   -0.581     0.561
weight         0.7072      0.1612    4.387  1.15e-05
color         -0.1734      0.1199   -1.445     0.148
---
Theta: 0.956   2 x log-likelihood: -746.452 # L = -373.226
-------------------------------------------------------------------------
```

This describes the tendency of the overall mean response to increase with weight and decrease with color (but not significantly). In not having a separate component

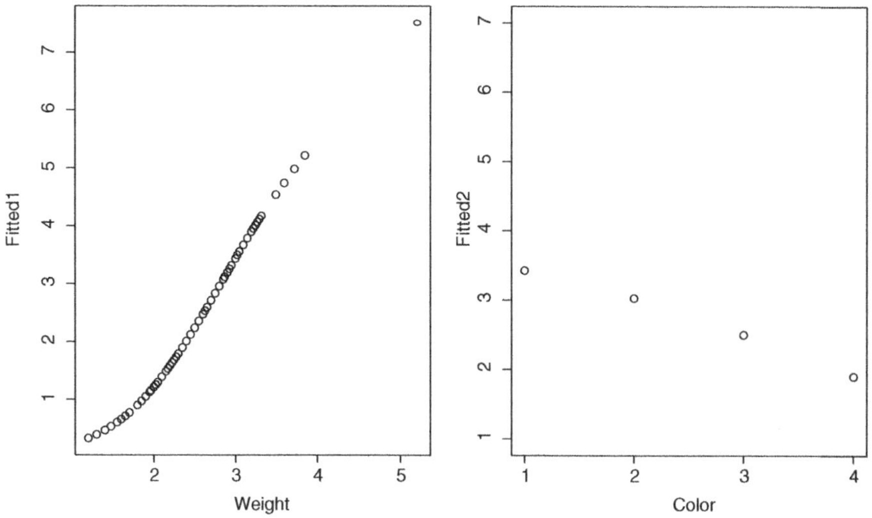

Figure 7.3 Fitted mean number of horseshoe crab satellites for zero-inflated negative binomial model, plotted as a function of weight for dark crabs and as a function of color for median-weight crabs.

to handle the zero count, the NB2 model has dispersion parameter estimate $\hat{\gamma} = 1/0.956 = 1.05$ that is much greater than $\hat{\gamma}$ for the NB2 component of ZINB models. The fit is similar to that of the geometric distribution, which is NB2 with $\gamma = 1$. But its log-likelihood of -373.2 is considerably worse than values obtained for ZINB models.

Unless previous research or theory suggests more-complex models, it seems adequate to use a zero-inflated NB2 model with weight as the primary predictor, adding color as a predictor of the mass at 0. In these analyses, however, we have ignored that the dataset contains an outlier—an exceptionally heavy crab weighing 5.2 kg of medium color that had 7 satellites. As exercise, you can fit models without that observation to investigate how the results change.

CHAPTER NOTES

Section 7.1: Poisson GLMs for Counts and Rates

7.1 **Poisson GLMs**: See Cameron and Trivedi (2013) for details about Poisson and other models for count data and an extensive bibliography.

Section 7.2: Poisson/Multinomial Models for Contingency Tables

7.2 **Loglinear models**: For more details about loglinear models for contingency tables, see Agresti (2013, Chapters 9, 10) and Bishop et al. (1975).

7.3 Graphical models: For more on conditional independence graphs, see Darroch et al. (1980), Lauritzen (1996), and Madigan and York (1995). More general probabilistic contexts include directed graphs, which are natural for hierarchical Bayesian models, and explanatory variables (e.g., Jordan 2004).

Section 7.3: Negative Binomial GLMs

7.4 NB modeling: Greenwood and Yule (1920) derived the negative binomial as a gamma mixture of Poisson distributions. Johnson et al. (2005, Chapter 5) summarized properties of the distribution. Cameron and Trivedi (2013, Section 3.3) discussed NB modeling and presented an asymptotic variance expression for $\hat{\gamma}$. They also presented moment estimators for γ and studied robustness properties (Section 3.3) and discussed analogs of R-squared for count data models (Section 5.3.3). See also Anscombe (1950), Hilbe (2011), Hinde and Demétrio (1998), and Lawless (1987). Alternatives to the gamma for mixing Poisson distributions include the log-normal and inverse-Gaussian distributions. See Cameron and Trivedi (2013, Section 4.2).

Section 7.4: Models for Zero-Inflated Data

7.5 Hurdle model, and ZIP versus ZINB: Mullahy (1986) proposed the hurdle model using the truncated Poisson or geometric distribution. Ridout et al. (2001) provided a score test of the ZIP model against the ZINB alternative. Estimators for the ZIP model can be unstable compared to the hurdle model (e.g., for estimating a predictor effect in the logit component of the model) when zero deflation occurs at some predictor settings. See Min and Agresti (2005) for discussion, more references, and extensions to handling repeated measurements with zero-inflated data. See also Cameron and Trivedi (2013, Chapter 4) and Hilbe (2011, Chapter 11) for zero-inflated models, hurdle models, truncated models, and other generalized count regression models.

7.6 Zero-truncated models: Models for zero-truncated data have a long history. See Amemiya (1984), Cameron and Trivedi (2013, Section 4.3), Johnson et al. (2005, Section 4.10, 5.11), and Meng (1997).

EXERCISES

7.1 Suppose $\{y_i\}$ are independent Poisson observations from a single group. Find the likelihood equation for estimating $\mu = E(y_i)$. Show that $\hat{\mu} = \bar{y}$ regardless of the link function.

7.2 Suppose $\{y_i\}$ are independent Poisson variates, with $\mu = E(y_i)$, $i = 1, \ldots, n$. For testing H_0: $\mu = \mu_0$, show that the likelihood-ratio statistic simplifies to

$$-2(L_0 - L_1) = 2[n(\mu_0 - \bar{y}) + n\bar{y}\log(\bar{y}/\mu_0)].$$

Explain how to use this to obtain a large-sample confidence interval for μ.

7.3 Refer to the previous exercise. Explain why, alternatively, for large samples you can test H_0 using the standard normal test statistic $z = \sqrt{n}(\bar{y} - \mu_0)/\sqrt{\mu_0}$. Explain how to invert this test to obtain a confidence interval. (These are the score test and score-test based confidence interval.)

7.4 When y_1 and y_2 are independent Poisson with means μ_1 and μ_2, find the likelihood-ratio statistic for testing H_0: $\mu_1 = \mu_2$. Specify its asymptotic null distribution, and describe the condition under which the asymptotics apply.

7.5 For the one-way layout for Poisson counts (Section 7.1.5), using the identity link function, show how to obtain a large-samples confidence interval for $\mu_h - \mu_i$. If there is overdispersion, explain why it is better to use a formula $(\bar{y}_h - \bar{y}_i) \pm z_{\alpha/2}\sqrt{(s_h^2/n_h) + (s_i^2/n_i)}$ based only on the central limit theorem.

7.6 For the one-way layout for Poisson counts, derive the likelihood-ratio statistic for testing H_0: $\mu_1 = \cdots = \mu_c$.

7.7 For the one-way layout for Poisson counts, derive a test of H_0: $\mu_1 = \cdots = \mu_c$ by applying a Pearson chi-squared goodness-of-fit test (with $df = c - 1$) for a multinomial distribution that compares sample proportions in c categories against H_0 values of multinomial probabilities, (**a**) when $n_1 = \cdots = n_c$, (**b**) for arbitrary $\{n_i\}$, with $n = \sum_i n_i$.

7.8 In a balanced two-way layout for a count response, let y_{ijk} be observation k at level i of factor A and level j of factor B, $k = 1, \ldots, n$. Formulate a Poisson loglinear main-effects model for $\{\mu_{ijk} = E(y_{ijk})\}$. Find the likelihood equations, and show that $\{\mu_{ij+} = \sum_k E(y_{ijk})\}$ have fitted values $\{\hat{\mu}_{ij+} = (y_{i++}y_{+j+})/y_{+++}\}$.

7.9 Refer to Note 1.5. For a Poisson loglinear model containing an intercept, show that the average estimated rate of change in the mean as a function of explanatory variable j satisfies $\frac{1}{n}\sum_i(\partial\hat{\mu}_i/\partial x_{ij}) = \hat{\beta}_j\bar{y}$.

7.10 A method for negative exponential modeling of survival times relates to the Poisson loglinear model for rates (Aitkin and Clayton 1980). Let T denote the time to some event, with pdf f and cdf F. For subject i, let $w_i = 1$ for death and 0 for censoring, and let $t = \sum_i t_i$ and $w = \sum_i w_i$.

a. Explain why the survival-time log-likelihood for n independent observations is[10]

$$L(\lambda) = \sum_i w_i \log[f(t_i)] + \sum_i (1 - w_i)\log[1 - F(t_i)].$$

[10]This actually applies only for *noninformative* censoring mechanisms.

Assuming $f(t) = \lambda \exp(-\lambda t)$, show that $\hat{\lambda} = w/t$. Conditional on t, explain why w has a Poisson distribution with mean $t\lambda$. Using the Poisson likelihood, show that $\hat{\lambda} = w/t$.

b. With λ replaced by $\lambda \exp(\boldsymbol{x\beta})$ and with $\mu_i = t_i \lambda \exp(\boldsymbol{x}_i \boldsymbol{\beta})$, show that L simplifies to

$$L(\lambda, \boldsymbol{\beta}) = \sum_i w_i \log \mu_i - \sum_i \mu_i - \sum_i w_i \log t_i.$$

Explain why maximizing $L(\lambda, \boldsymbol{\beta})$ is equivalent to maximizing the likelihood for the Poisson loglinear model

$$\log \mu_i - \log t_i = \log \lambda + \boldsymbol{x}_i \boldsymbol{\beta}$$

with offset $\log(t_i)$, using "observations" $\{w_i\}$.

c. When we sum terms in L for subjects having a common value of $\boldsymbol{x}$, explain why the observed data are the numbers of deaths ($\sum_i w_i$) at each setting of $\boldsymbol{x}$, and the offset is $\log(\sum_i t_i)$ at each setting.

7.11 Consider the loglinear model of conditional independence between A and B, given C, in a $r \times c \times \ell$ contingency table. Derive the likelihood equations, and interpret. Give the solution of fitted values that satisfies the model and the equations. (From Birch (1963), it follows that only one solution exists, namely the ML fit.) Explain the connection with the fitted values for the independence model for a two-way table. Find the residual *df* for testing fit.

7.12 Two balanced coins are flipped, independently. Let A = whether the first flip resulted in a head (yes, no), B = whether the second flip resulted in a head, and C = whether both flips had the same result. Using this example, show that marginal independence for each pair of three variables does not imply that the variables are mutually independent.

7.13 For three categorical variables A, B, and C:

a. When C is jointly independent of A and B, show that A and C are conditionally independent, given B.

b. Prove that mutual independence of A, B, and C implies that A and B are (i) marginally independent and (ii) conditionally independent, given C.

c. Suppose that A is independent of B and that B is independent of C. Does this imply that A is independent of C? Explain.

7.14 Express the loglinear model of mutual independence for a $2 \times 2 \times 2$ table in the form $\log \boldsymbol{\mu} = \boldsymbol{X\beta}$. Show that the likelihood equations equate $\{y_{ijk}\}$ and $\{\hat{\mu}_{ijk}\}$ in the one-dimensional margins, and their solution is $\{\hat{\mu}_{ijk} = y_{i++}y_{+j+}y_{++k}/n^2\}$.

7.15 For a $2 \times c \times \ell$ table, consider the loglinear model by which A is jointly independent of B and C. Treat A as a response variable and B and C as explanatory, conditioning on $\{n_{+jk}\}$. Construct the logit for the conditional distribution of A, and identify the corresponding logistic model.

7.16 For the homogeneous association loglinear model (7.7) for a $r \times c \times \ell$ contingency table, treating A as a response variable, find the equivalent baseline-category logit model.

7.17 For a four-way contingency table, consider the loglinear model having AB, BC, and CD two-factor terms and no three-factor interaction terms. Explain why A and D are independent given B alone or given C alone or given both B and C. When are A and C conditionally independent?

7.18 Suppose the loglinear model (7.7) of homogeneous association holds for a three-way contingency table. Find $\log \mu_{ij+}$ and explain why marginal associations need not equal conditional associations for this model.

7.19 Consider the loglinear model for a four-way table having AB, AC, and AD two-factor terms and no three-factor interaction term. What is the impact of collapsing over B on the other associations? Contrast that with what the collapsibility condition in Section 7.2.7 suggests, treating group $S_3 = \{B\}$, (**i**) if $S_1 = \{C\}$ and $S_2 = \{A, C\}$, (**ii**) if $S_1 = \{C, D\}$ and $S_2 = \{A\}$. This shows that different groupings for that condition can give different information.

7.20 A county's highway department keeps records of the number of automobile accidents reported each working day on a superhighway that runs through the county. Describe factors that are likely to cause the distribution of this count over time to show overdispersion relative to the Poisson distribution.

7.21 Show that a gamma mixture of Poisson distributions yields the negative binomial distribution.

7.22 Given u, y is Poisson with $E(y \mid u) = u\mu$, where u is a positive random variable with $E(u) = 1$ and $\text{var}(u) = \tau$. Show that $E(y) = \mu$ and $\text{var}(y) = \mu + \tau\mu^2$. Explain how you can formulate the negative binomial distribution and a negative binomial GLM using this construction.

7.23 For discrete distributions, Jørgensen (1987) showed that it is natural to define the exponential dispersion family as

$$f(y_i; \theta_i, \phi) = \exp[y_i\theta_i - b(\theta_i)/a(\phi) + c(y_i, \phi)].$$

a. For fixed k, show that the negative binomial distribution (7.8) has this form with $\theta_i = \log[\mu_i/(\mu_i + k)]$, $b(\theta_i) = -\log(1 - e^{\theta_i})$, and $a(\phi) = 1/k$.

b. For this version, show that $x_i = y_i a(\phi)$ has the usual exponential dispersion family form (4.1).

7.24 For a sequence of independent Bernoulli trials, let y = the number of successes before the kth failure. Show that y has the negative binomial distribution,

$$f(y; \pi, k) = \frac{\Gamma(y+k)}{\Gamma(k)\Gamma(y+1)} \pi^y (1-\pi)^k, \quad y = 0, 1, 2, \ldots .$$

(The *geometric distribution* is the special case $k = 1$.) Relate π to the parameters μ and k in the parameterization (7.8).

7.25 With independent negative binomial observations from a single group, find the likelihood equation and show that $\hat{\mu} = \bar{y}$. (ML estimation for γ requires iterative methods, as R. A. Fisher showed in an appendix to Bliss (1953). See also Anscombe (1950).) How does this generalize to the one-way layout?

7.26 For the ZIP null model (i.e., without explanatory variables), show from the likelihood equations that the ML-fitted 0 count equals the observed 0 count.

7.27 The text website contains an expanded version (file `Drugs3.dat`) of the student substance use data of Table 7.3 that also has each subject's G = gender (1 = female, 2 = male) and R = race (1 = white, 2 = other). It is sensible to treat G and R as explanatory variables. Explain why any loglinear model for the data should include the GR two-factor term. Use a model-building process to select a model for these data. Interpret the estimated conditional associations.

7.28 Other than a formal goodness-of-fit test, one analysis that provides a sense of whether a particular GLM is plausible is the following: Suppose the ML fitted equation were the true equation. At the observed $\boldsymbol{x}$ values for the n observations, randomly generate n variates with distributions specified by the fitted GLM. Construct scatterplots. Do they look like the scatterplots that were actually observed? Do this for a Poisson loglinear model for the horseshoe crab data, with y = number of satellites and x = width. Does the variability about the fit resemble that in the actual data, including a similar number of 0's and large values? Repeat this a few times to get a better sense of how the scatterplot observed differs from what you would observe if the Poisson GLM truly held.

7.29 Another model (Dobbie and Welsh 2001) for zero-inflated count data uses the *Neyman type A distribution*, which is a compound Poisson–Poisson mixture. For observation i, let z_i denote a Poisson variate with expected value λ_i. Conditional on z_i, let w_{ij} $(j = 1, \ldots, z_i)$ denote independent Poisson(ϕ_i)

observations. The model expresses y_i using the decomposition $y_i = \sum_{j=0}^{z_i} w_{ij}$, $i = 1, 2, \dots, n$. Find $E(y_i)$. Relating λ_i and ϕ_i to explanatory variables through $\log(\lambda_i) = \boldsymbol{x}_{1i}\boldsymbol{\beta}_1$ and $\log(\phi_i) = \boldsymbol{x}_{2i}\boldsymbol{\beta}_2$, show the model for $E(y_i)$ and interpret its parameters.

7.30 A headline in *The Gainesville Sun* (February 17, 2014) proclaimed a worrisome spike in shark attacks in the previous 2 years. The reported total number of shark attacks in Florida per year from 2001 to 2013 were 33, 29, 29, 12, 17, 21, 31, 28, 19, 14, 11, 26, 23. Are these counts consistent with a null Poisson model or a null negative binomial model? Test the Poisson model against the negative binomial alternative. Analyze the evidence of a positive linear trend over time.

7.31 Table 7.5, also available at `www.stat.ufl.edu/~aa/glm/data`, summarizes responses of 1308 subjects to the question: within the past 12 months, how many people have you known personally that were victims of homicide? The table shows responses by race, for those who identified their race as white or as black.

a. Let y_i denote the response for subject i and let $x_i = 1$ for blacks and $x_i = 0$ for whites. Fit the Poisson GLM $\log \mu_i = \beta_0 + \beta x_i$ and interpret $\hat{\beta}$.

b. Describe factors of heterogeneity such that a Poisson GLM may be inadequate. Fit the corresponding negative binomial GLM, and estimate how the variance depends on the mean. What evidence does this model fit provide that the Poisson GLM had overdispersion? (Table 7.5 also shows the fits for these two models.)

c. Show that the Wald 95% confidence interval for the ratio of means for blacks and whites is (4.2, 7.5) for the Poisson GLM but (3.5, 9.0) for the negative binomial GLM. Which do you think is more reliable? Why?

Table 7.5 Number of Victims of Murder Known in the Past Year, by Race, with Fit of Poisson and Negative Binomial Models

	Data		Poisson GLM		Negative Binomial GLM	
Response	Black	White	Black	White	Black	White
0	119	1070	94.3	1047.7	122.8	1064.9
1	16	60	49.2	96.7	17.9	67.5
2	12	14	12.9	4.5	7.8	12.7
3	7	4	2.2	0.1	4.1	2.9
4	3	0	0.3	0.0	2.4	0.7
5	2	0	0.0	0.0	1.4	0.2
6	0	1	0.0	0.0	0.9	0.1

Source: 1990 General Social Survey, file `Homicide.dat` at `www.stat.ufl.edu/~aa/glm/data`.

7.32 For the horseshoe crab data, the negative binomial modeling shown in the R output first treats color as nominal-scale and then in a quantitative manner, with the category numbers as scores. Interpret the result of the likelihood-ratio test comparing the two models. For the simpler model, interpret the color effect and interpret results of the likelihood-ratio test of the null hypothesis of no color effect.

```
-------------------------------------------------------------------
> fit.nb.color <- glm.nb(y ~ factor(color)) # Using Crabs.dat file
> summary(fit.nb.color)
              Estimate  Std. Error  z value  Pr(>|z|)
(Intercept)     1.4069      0.3526    3.990  6.61e-05
factor(color)2 -0.2146      0.3750   -0.572     0.567
factor(color)3 -0.6061      0.4036   -1.502     0.133
factor(color)4 -0.6913      0.4508   -1.533     0.125
---
> fit.nb.color2 <- glm.nb(y ~ color) # using color scores (1,2,3,4)
> summary(fit.nb.color2)
           Estimate  Std. Error  z value  Pr(>|z|)
(Intercept)  1.7045      0.3095    5.507  3.66e-08
color       -0.2689      0.1225   -2.194    0.0282
---
> anova(fit.nb.color2, fit.nb.color)
Likelihood ratio tests of Negative Binomial Models
Response: y
Model   theta  Res.df  2 x log-lik.   Test  df  LR stat. Pr(Chi)
1      0.7986     171   -762.6794
2      0.8019     169   -762.2960  1 vs. 2   2   0.3834   0.8256
---
> 1 - pchisq(767.409-762.679, df=172-171) # LR test vs. null model
[1] 0.0296
-------------------------------------------------------------------
```

7.33 For the horseshoe crab data, the following output shows a zero-inflated negative binomial model using quantitative color for the zero component. Interpret results, and compare with the NB2 model fitted in the previous exercise with quantitative color. Can you conduct a likelihood-ratio test comparing them? Why or why not?

```
--------------------------------------------------------------------
> summary(zeroinfl(y ~ 1 | color, dist = "negbin")) # Using Crabs.dat
Count model coefficients (negbin with log link):
            Estimate  Std. Error  z value  Pr(>|z|)
(Intercept)   1.4632      0.0689   21.231   < 2e-16
Log(theta)    1.4800      0.3511    4.215   2.5e-05

Zero-inflation model coefficients (binomial with logit link):
            Estimate  Std. Error  z value  Pr(>|z|)
(Intercept)  -2.7520      0.6658   -4.133  3.58e-05
color         0.8023      0.2389    3.358  0.000785
---
Theta = 4.3928    Log-likelihood:-362.997 on 4 Df
--------------------------------------------------------------------
```

7.34 Refer to Section 7.5.2. Redo the zero-inflated NB2 model building, deleting the outlier crab weighing 5.2 kg. Compare results against analyses that used this observation and summarize conclusions.

7.35 A question in a GSS asked subjects how many times they had sexual intercourse in the preceding month. The sample means were 5.9 for males and 4.3 for females; the sample variances were 54.8 and 34.4. The mode for each gender was 0. Specify a GLM that would be inappropriate for these data, explaining why. Specify a model that may be appropriate.

7.36 Table 7.6 is based on a study involving British doctors.

Table 7.6 Data for Exercise 7.36 on Coronary Death Rates

	Person-Years		Coronary Deaths	
Age	Nonsmokers	Smokers	Nonsmokers	Smokers
35–44	18,793	52,407	2	32
45–54	10,673	43,248	12	104
55–64	5710	28,612	28	206
65–74	2585	12,663	28	186
75–84	1462	5317	31	102

Source: Doll R. and A. Bradford Hill. 1966. *Natl. Cancer Inst. Monogr.* **19**: 205–268.

a. Fit a main-effects model for the log rates using age and smoking as factors. In discussing lack of fit, show that this model assumes a constant ratio of nonsmokers' to smokers' coronary death rates over age, and evaluate how the sample ratio depends on age.

b. Explain why it is sensible to add a quantitative interaction of age and smoking. For this model, show that the log ratio of coronary death rates changes linearly with age. Assign scores to age, fit the model, and interpret.

CHAPTER 8

Quasi-Likelihood Methods

For a GLM $\eta_i = g(\mu_i) = \sum_j \beta_j x_{ij}$, the likelihood equations

$$\sum_{i=1}^{n} \frac{(y_i - \mu_i)x_{ij}}{v(\mu_i)} \left(\frac{\partial \mu_i}{\partial \eta_i} \right) = 0, \quad j = 1, \ldots, p, \tag{8.1}$$

depend on the assumed probability distribution for y_i only through μ_i and the variance function, $v(\mu_i) = \text{var}(y_i)$. The choice of distribution for y_i determines the relation $v(\mu_i)$ between the variance and the mean. Higher moments such as the skewness can affect properties of the model, such as how fast $\hat{\boldsymbol{\beta}}$ converges to normality, but they have no impact on the value of $\hat{\boldsymbol{\beta}}$ and its large-sample covariance matrix.

An alternative approach, *quasi-likelihood estimation*, specifies a link function and linear predictor $g(\mu_i) = \sum_j \beta_j x_{ij}$ like a generalized linear model (GLM), but it does not assume a particular probability distribution for y_i. This approach estimates $\{\beta_j\}$ by solving equations that resemble the likelihood equations (8.1) for GLMs, but it assumes only a mean–variance relation for the distribution of y_i. The estimates are the solution of Equation (8.1) with $v(\mu_i)$ replaced by whatever variance function seems appropriate in a particular situation, with a corresponding adjustment for standard errors. To illustrate, a standard modeling approach for counts assumes that $\{y_i\}$ are independent Poisson variates, for which $v(\mu_i) = \mu_i$. However, in the previous chapter we noted that overdispersion often occurs, perhaps because of unmodeled heterogeneity among subjects. To allow for this, we could set $v(\mu_i) = \phi\mu_i$ for some unknown constant ϕ.

In Section 8.1, we present a simple quasi-likelihood (QL) approach for overdispersed Poisson and binomial response variables that merely assumes an inflation of the variance from a standard model. For binary data, Section 8.2 presents alternative approaches that imply overdispersion because of positively correlated Bernoulli trials or because the success probability satisfies a mixture model. In Section 8.3, we show

---

Foundations of Linear and Generalized Linear Models, First Edition. Alan Agresti.

how to adjust for misspecification of the variance function in finding standard errors of parameter estimates.

8.1 VARIANCE INFLATION FOR OVERDISPERSED POISSON AND BINOMIAL GLMS

This section introduces a simple quasi-likelihood way of adjusting for overdispersion in Poisson and binomial models. This method uses the same estimates as an ordinary GLM but inflates standard errors by taking into account the empirical variability.

8.1.1 Quasi-Likelihood Approach of Variance Inflation

Suppose a standard model specifies a function $v^*(\mu_i)$ for the variance as a function of the mean, but we believe that the actual variance may differ from $v^*(\mu_i)$. To allow for this, we might instead assume that

$$\text{var}(y_i) = \phi v^*(\mu_i)$$

for some constant ϕ. The value $\phi > 1$ represents overdispersion.

When we substitute $v(\mu_i) = \phi v^*(\mu_i)$ in Equation (8.1), ϕ drops out. The equations are identical to the likelihood equations for the GLM with variance function $v^*(\mu_i)$, and estimates of model parameters are also identical. With the generalized variance function,

$$w_i = (\partial \mu_i / \partial \eta_i)^2 / \text{var}(y_i) = (\partial \mu_i / \partial \eta_i)^2 / \phi v^*(\mu_i),$$

so the asymptotic $\text{var}(\hat{\boldsymbol{\beta}}) = (\boldsymbol{X}^{\text{T}}\boldsymbol{W}\boldsymbol{X})^{-1}$ is ϕ times that for the ordinary GLM.

When a variance function has the form $v(\mu_i) = \phi v^*(\mu_i)$, usually ϕ is also unknown. Let

$$X^2 = \sum_{i=1}^{n} \frac{(y_i - \hat{\mu}_i)^2}{v^*(\hat{\mu}_i)}$$

be the generalized Pearson statistic (4.17) for the simpler model with $\phi = 1$. When X^2/ϕ is approximately chi-squared, then with p parameters in the linear predictor, $E(X^2/\phi) \approx n - p$. Hence, $E[X^2/(n-p)] \approx \phi$. Using the motivation of estimation by matching moments, $\hat{\phi} = X^2/(n-p)$ is an estimated multiplier to apply to the ordinary estimated covariance matrix.

In summary, this quasi-likelihood approach is simple: fit the ordinary GLM and use its p maximum likelihood (ML) parameter estimates $\hat{\boldsymbol{\beta}}$. Multiply the ordinary standard error estimates by $\sqrt{X^2/(n-p)}$. This method is appropriate; however, only if the model chosen describes well the structural relation between $E(y_i)$ and the explanatory variables. If a large X^2 statistic is due to some other type of lack of fit, such as failing to include a relevant interaction term, adjusting for overdispersion will not address the inadequacy.

8.1.2 Overdispersed Poisson and Binomial GLMs

We illustrate the quasi-likelihood variance-inflation approach with the alternative to a Poisson GLM in which the mean–variance relation has the form

$$v(\mu_i) = \phi\mu_i.$$

The QL parameter estimates are identical to the ML estimates under the Poisson GLM assumption. With the canonical log link, the adjusted covariance matrix is $(\boldsymbol{X}^{\mathrm{T}}\boldsymbol{W}\boldsymbol{X})^{-1}$ with $w_i = (\partial\mu_i/\partial\eta_i)^2/\mathrm{var}(y_i) = (\mu_i)^2/\phi\mu_i = \mu_i/\phi$. Regardless of the link function, the Pearson statistic is

$$X^2 = \sum_i \frac{(y_i - \hat{\mu}_i)^2}{\hat{\mu}_i},$$

and $\hat{\phi} = X^2/(n-p)$ is the variance-inflation estimate.

An alternative approach uses a parametric model that permits extra variability, such as a negative binomial GLM (Section 7.3). An advantage of that approach is that it is an actual model with a likelihood function.

Overdispersion also can occur for counts from grouped binary data. Suppose y_i is the proportion of successes in n_i Bernoulli trials with parameter π_i for each trial, $i = 1, \ldots, n$. The $\{y_i\}$ may exhibit more variability than the binomial allows. This can happen in two common ways. One way involves heterogeneity, with observations at a particular setting of explanatory variables having success probabilities that vary according to values of unobserved variables. To deal with this, we could use a hierarchical mixture model that lets π_i itself have a distribution, such as a beta distribution. Alternatively, extra variability could occur because the Bernoulli trials at each i are positively correlated. We present models that reflect these possibilities in Section 8.2. Here we consider the simpler variance-inflation approach.

To adjust supposedly binomial sampling (i.e., independent, identical Bernoulli trials), the inflated-variance QL approach uses variance function

$$v(\pi_i) = \phi\pi_i(1-\pi_i)/n_i$$

for the proportion y_i. The QL estimates are the same as ML estimates for the binomial model, and the asymptotic covariance matrix multiplies by ϕ. The $X^2/(n-p)$ estimate of ϕ uses the X^2 fit statistic for the ordinary binomial model with p parameters, which from Equation (5.10) is

$$X^2 = \sum_i \frac{(y_i - \hat{\pi}_i)^2}{[\hat{\pi}_i(1-\hat{\pi}_i)]/n_i}.$$

Although this QL approach with $v(\pi_i) = \phi\pi_i(1-\pi_i)/n_i$ has the advantage of simplicity, it is inappropriate when $n_i = 1$: then $P(y_i = 1) = \pi_i = 1 - P(y_i = 0)$, and necessarily $E(y_i^2) = E(y_i) = \pi_i$ and $\text{var}(y_i) = \pi_i(1-\pi_i)$. For ungrouped binary data, necessarily $\text{var}(y_i) = \pi_i(1-\pi_i)$, and only $\phi = 1$ makes sense. This structural problem does not occur for mixture models or for a QL approach having variance function corresponding to a mixture model (Section 8.2).

8.1.3 Example: Quasi-Likelihood for Horseshoe Crab Counts

The horseshoe crab satellite counts analyzed in Section 7.5 display overdispersion for Poisson GLMs. For example, using the female crab's weight to predict the number of male satellites, the Poisson loglinear fit is $\log \hat{\mu}_i = -0.428 + 0.589x_i$, with SE = 0.065 for $\hat{\beta}_1 = 0.589$. Comparing the observed counts and fitted values for the $n = 173$ crabs, Pearson $X^2 = 535.9$ with $df = 173 - 2 = 171$. With the QL inflated-variance approach, $\hat{\phi} = X^2/(n-p) = 535.9/(173-2) = 3.13$. Thus, $SE = \sqrt{3.13}(0.065) = 0.115$ is a more plausible standard error for $\hat{\beta}_1$ in this prediction equation.

```
-------------------------------------------------------------------------
> attach(Crabs)
> fit.pois <- glm(y ~ weight, family=poisson) # ML Poisson loglinear
> summary(fit.pois)
             Estimate  Std. Error  z value  Pr(>|z|)
(Intercept)   -0.4284      0.1789   -2.394    0.0167
weight         0.5893      0.0650    9.064    <2e-16

> (X2 <- sum(residuals(fit.pois, type="pearson")^2))
[1] 535.90 # Pearson statistic is sum of squared Pearson residuals
> (phi <- X2/(173 - 2))
[1] 3.13
    # quasi family can use QL inflated Poisson variance directly:
> summary(glm(y ~ weight, family=quasi(link="log",variance="mu")))
             Estimate  Std. Error  t value  Pr(>|t|)
(Intercept)   -0.4284      0.3168   -1.352     0.178
weight         0.5893      0.1151    5.120  8.17e-07
    (Dispersion parameter for quasi family taken to be 3.134)
-------------------------------------------------------------------------
```

The QL approach with an inflated quadratic variance function yields larger $\hat{\beta}_1$ and SE values, similar to what we obtain with a negative binomial (NB2) model.

```
-------------------------------------------------------------------------
> summary(glm(y ~ weight, family=quasi(link="log",variance="mu^2")))
             Estimate  Std. Error  t value  Pr(>|t|)
(Intercept)   -1.0122      0.3863   -2.621   0.00957
weight         0.8184      0.1542    5.306  3.44e-07
    (Dispersion parameter for quasi family taken to be 1.362496)
> library(MASS)
```

```
> summary(glm.nb(y ~ weight)) # negative binomial (NB2) model
             Estimate  Std. Error  z value  Pr(>|z|)
(Intercept)  -0.8647      0.4048   -2.136    0.0327
weight        0.7603      0.1578    4.817  1.45e-06
---
Theta: 0.931  2 x log-likelihood: -748.644 # -916.164 for Poisson fit
----------------------------------------------------------------------
```

8.2 BETA-BINOMIAL MODELS AND QUASI-LIKELIHOOD ALTERNATIVES

We next describe ways to handle binomial overdispersion that are more satisfying than the variance-inflation approach. We first present a QL method based on correlated Bernoulli trials and then a mixture model that lets success probabilities vary according to values of unobserved variables.

8.2.1 Overdispersion Caused by Correlated Bernoulli Trials

Denote the n_i Bernoulli trials for observation i by $y_{i1}, y_{i2}, \dots, y_{in_i}$. That is, $P(y_{ij} = 1) = \pi_i = 1 - P(y_{ij} = 0)$, and $y_i = \sum_j y_{ij}/n_i$ is the sample proportion. For independent trials, $n_i y_i \sim \text{bin}(n_i, \pi_i)$, with $\text{var}(y_i) = v(\pi_i) = \pi_i(1 - \pi_i)/n_i$.

Instead of independent trials, suppose that y_{i1} is random but then $y_{ij} = y_{i1}$ for $j = 2, \dots, n_i$. For instance, in an election, perhaps in each household the head of the household decides how to vote, and then everyone else in the household votes the same way. Then the sample proportion in household i voting for a particular candidate has

$$P(y_i = 1) = \pi_i, \quad P(y_i = 0) = 1 - \pi_i.$$

That is, y_i can take only its extreme possible values. Then $\text{var}(y_i) = \pi_i(1 - \pi_i) > \pi_i(1 - \pi_i)/n_i$, and there is overdispersion relative to the binomial. By contrast, suppose that the observations occur sequentially, and

$$y_{ij} \mid y_{i1}, \dots, y_{i,j-1} \quad \text{equals} \quad 1 - y_{i,j-1}.$$

That is, trial j for observation i necessarily has the opposite result of trial $j - 1$. Then when n_i is an even number, $P(y_i = 1/2) = 1$, so $\text{var}(y_i) = 0$ and there is underdispersion.

In practice, a more likely scenario than one trial being completely dependent on another one is exchangeability of trials, with a common correlation ρ between each pair of $\{y_{i1}, y_{i2}, \dots, y_{in_i}\}$, as is often assumed in cluster sampling. When $\text{corr}(y_{is}, y_{it}) = \rho$ for $s \neq t$, then $\text{var}(y_{it}) = \pi_i(1 - \pi_i)$, $\text{cov}(y_{is}, y_{it}) = \rho\pi_i(1 - \pi_i)$, and

$$\text{var}(y_i) = \text{var}\left(\frac{\sum_{t=1}^{n_i} y_{it}}{n_i}\right) = \frac{1}{n_i^2}\left[\sum_{t=1}^{n_i} \text{var}(y_{it}) + 2\sum_{s<t}\sum \text{cov}(y_{is}, y_{it})\right]$$

$$= \frac{1}{n_i^2}\left[n_i\pi_i(1 - \pi_i) + n_i(n_i - 1)\rho\pi_i(1 - \pi_i)\right] = [1 + \rho(n_i - 1)]\frac{\pi_i(1 - \pi_i)}{n_i}.$$

The ordinary variance for a binomial sample proportion results when $\rho = 0$. Overdispersion occurs when $\rho > 0$.

The inflated binomial variance is not a special case of this variance function, unless all n_i are identical. When $n_i = 1$, it is not possible to have overdispersion or underdispersion, and this variance formula is still valid, unlike the inflated binomial variance.

8.2.2 QL with Variance Function for Correlated Bernoulli Trials

For binary count data, a quasi-likelihood approach can use a variance function motivated by the one just found with correlated Bernoulli trials,

$$v(\pi_i) = [1 + \rho(n_i - 1)]\pi_i(1 - \pi_i)/n_i$$

with $|\rho| \leq 1$. The estimates using it differ from ML estimates for an ordinary binomial model, because the multiple of the binomial variance does not drop out of the quasi-likelihood equations (8.1).

For this QL approach, Williams (1982) proposed an iterative routine for estimating $\boldsymbol{\beta}$ and the overdispersion parameter ρ. He let $\hat{\rho}$ be such that the resulting generalized Pearson X^2 statistic (4.17) equals the residual $df = (n - p)$ for the model. This requires an iterative two-step process of (1) solving the quasi-likelihood equations for $\boldsymbol{\beta}$ for a given $\hat{\rho}$, and then (2) using the updated $\hat{\boldsymbol{\beta}}$, solving for $\hat{\rho}$ in the equation that equates

$$X^2 = \sum_{i=1}^{n} \frac{(y_i - \hat{\pi}_i)^2}{[1 + \hat{\rho}(n_i - 1)]\hat{\pi}_i(1 - \hat{\pi}_i)/n_i} = n - p.$$

8.2.3 Models Using the Beta-Binomial Distribution

The *beta-binomial model* is a parametric mixture model that is an alternative to quasi-likelihood generalizations of binomial GLMs. As with other mixture models that assume a binomial distribution at a fixed parameter value, the marginal distribution permits more variation than the binomial. We will see that the variance function for the beta-binomial model has the same form as the one resulting from correlated Bernoulli trials.

The beta-binomial distribution results from a *beta distribution* mixture of binomials. Suppose that (1) given π, s has a binomial distribution, bin(n, π), and (2) π has a beta distribution. The beta pdf is

$$f(\pi; \alpha_1, \alpha_2) = \frac{\Gamma(\alpha_1 + \alpha_2)}{\Gamma(\alpha_1)\Gamma(\alpha_2)} \pi^{\alpha_1 - 1}(1 - \pi)^{\alpha_2 - 1}, \quad 0 \leq \pi \leq 1,$$

with parameters $\alpha_1 > 0$ and $\alpha_2 > 0$, and $\Gamma(\cdot)$ denotes the gamma function. The beta family provides a wide variety of pdf shapes over (0, 1), including uniform

($\alpha_1 = \alpha_2 = 1$), unimodal symmetric ($\alpha_1 = \alpha_2 > 1$), unimodal skewed left ($\alpha_1 > \alpha_2 > 1$) or skewed right ($\alpha_2 > \alpha_1 > 1$), and U-shaped ($\alpha_1 < 1, \alpha_2 < 1$). Let

$$\mu = \frac{\alpha_1}{\alpha_1 + \alpha_2}, \qquad \theta = 1/(\alpha_1 + \alpha_2).$$

The beta distribution for π has mean and variance

$$E(\pi) = \mu, \quad \text{var}(\pi) = \mu(1 - \mu)\theta/(1 + \theta).$$

Marginally, averaging over the beta distribution for π, s has the *beta-binomial distribution*. Its probability mass function is

$$p(s; n, \mu, \theta) = \binom{n}{s} \frac{\left[\prod_{k=0}^{s-1}(\mu + k\theta)\right]\left[\prod_{k=0}^{n-s-1}(1 - \mu + k\theta)\right]}{\prod_{k=0}^{n-1}(1 + k\theta)}, \qquad s = 0, 1, \ldots, n.$$

As $\theta \to 0$, $\text{var}(\pi) \to 0$, and the beta distribution for π converges to a degenerate distribution at μ. Then $\text{var}(s) \to n\mu(1 - \mu)$, and the beta-binomial distribution converges to the bin(n, μ). But the beta-binomial can look[1] quite different from the binomial. For example, when $\mu = 1/2$, it is uniform over the integers 0 to n when $\theta = 1/2$ (i.e., when $\alpha_1 = \alpha_2 = 1$ and the beta distribution is uniform), and it is bimodal at 0 and n when $\theta > 1/2$. For the beta-binomial proportion $y = s/n$,

$$E(y) = \mu, \quad \text{var}(y) = [1 + (n - 1)\theta/(1 + \theta)]\mu(1 - \mu)/n.$$

In fact, $\rho = \theta/(1 + \theta)$ is the correlation between each pair of the individual Bernoulli random variables that sum to s. The variance function in the beta-binomial and in the QL approach of Section 8.2.2 also results merely from assuming that π has a distribution with $\text{var}(\pi) = \rho\mu(1 - \mu)$.

Models using the beta-binomial distribution usually let θ be the same unknown constant for all observations. Models can use any link function for binary data, but the logit is most common. For observation i with n_i trials, assuming that $n_i y_i$ has a beta-binomial distribution with index n_i and parameters (μ_i, θ), the model links μ_i to explanatory variables by

$$\text{logit}(\mu_i) = \boldsymbol{x}_i \boldsymbol{\beta}, \quad i = 1, \ldots, n.$$

Model fitting can employ a variety of methods, including the Newton–Raphson method. See Note 8.4. The beta-binomial distribution is not in the exponential dispersion family, even for known θ. When the linear predictor is correct, the beta-binomial ML estimator $\hat{\boldsymbol{\beta}}$ is not consistent if the actual distribution is not beta-binomial. Quasi-likelihood methods have greater robustness (Liang and Hanfelt 1994).

[1] `distributome.org/V3/calc/BetaBinomialCalculator.html` displays shapes as a function of n, α_1, and α_2.

8.2.4 Example: Modeling Overdispersion in a Teratology Study

Teratology is the study of abnormalities of physiological development. Some teratology experiments investigate effects of dietary regimens or chemical agents on the fetal development of rats in a laboratory setting. Table 8.1 shows results from one such study. Female rats on iron-deficient diets were assigned to four groups. Rats in group 1 were given placebo injections, and rats in other groups were given injections of an iron supplement. This was done on days 7 and 10 in group 2, on days 0 and 7 in group 3, and weekly in group 4. The 58 rats were made pregnant, sacrificed after 3 weeks, and then the total number of dead fetuses was counted in each litter, as was the mother's hemoglobin level. The overall sample proportions of deaths for the four groups were = 0.758 (placebo), 0.102, 0.034, and 0.048. Because of unmeasured covariates and genetic variability, the probability of death may vary among litters within a particular treatment group and hemoglobin level.

Table 8.1 Response Proportions $y = s/n$ Dead in Teratology Study for n Fetuses in Rat Litter in Group i with Mother's Hemoglobin Level h

i	h	y	i	h	y	i	h	y	i	h	y
1	4.1	1/10	2	8.6	1/10	3	11.2	0/8	4	16.6	0/3
1	3.2	4/11	2	11.1	1/3	3	11.5	1/11	4	14.5	0/13
1	4.7	9/12	2	7.2	1/13	3	12.6	0/14	4	15.4	2/9
...											

Source: From Moore and Tsiatis (1991), reproduced with permission of John Wiley & Sons, Inc. Complete data for 58 rats are in the file `Rats.dat` at text website.

Let y_{ij} denote the dead proportion of the n_{ij} fetuses in litter j in treatment group i. Let π_{ij} denote the probability of death for a fetus in that litter. Moore and Tsiatis modeled π_{ij} using only the hemoglobin level or only group indicators as the explanatory variable. Here, we will use hemoglobin level and whether the litter is in the placebo group, to judge whether the death rate differs between the placebo group and the other groups after adjusting for the hemoglobin level.

Let z_i denote an indicator for the placebo group ($z_1 = 1$, $z_2 = z_3 = z_4 = 0$) and let h_{ij} denote the hemoglobin level for litter j in group i. We present four fits for the model

$$\text{logit}(\pi_{ij}) = \beta_0 + \beta_1 z_i + \beta_2 h_{ij}.$$

We first treat $n_{ij}y_{ij}$ as a bin(n_{ij}, π_{ij}) variate and find ML estimates.

```
-----------------------------------------------------------------------
> Rats # data in file Rats.dat at www.stat.ufl.edu/~aa/glm/data.html
  litter group   h   n  s  # s dead of n fetuses with hemoglobin h
1      1     1  4.1  10  1
2      2     1  3.2  11  4
...
```

```
58    58     4  12.4   17   0
> attach(Rats)
> placebo <- ifelse(group==1, 1, 0)
> fit.ML <- glm(s/n ~ placebo + h, weights=n, data=Rats, family=binomial)
> summary(fit.ML) # ML, assuming independent binomials
Coefficients:
             Estimate  Std. Error  z value  Pr(>|z|)
(Intercept)  -0.6239       0.7900   -0.790    0.4296
placebo       2.6509       0.4824    5.495   3.9e-08
h            -0.1871       0.0743   -2.519    0.0118
---
> logLik(fit.ML)
'log Lik.' -121.0219
------------------------------------------------------------------------
```

Summing the squared Pearson residuals for the $n = 58$ litters, we obtain $X^2 = 159.815$ with $df = 58 - 3 = 55$, considerable evidence of overdispersion. With the QL inflated-variance approach, $\hat{\phi} = 159.815/55 = 2.906$, so standard errors multiply by $\hat{\phi}^{1/2} = 1.70$. Even with this adjustment for overdispersion and for the hemoglobin level, strong evidence remains that the probability of death is substantially higher for the placebo group.

```
----------------------------------------------------------------------
> summary(glm(s/n ~ placebo + h, weights=n, data=Rats,
+              family=quasi(link = "logit", variance="mu(1-mu)")))
Coefficients: # QL inflated-variance approach
             Estimate  Std. Error  t value  Pr(>|t|)
(Intercept)  -0.6239       1.3466   -0.463   0.64495
placebo       2.6509       0.8223    3.224   0.00213
h            -0.1871       0.1266   -1.478   0.14514
---
(Dispersion parameter for quasi family taken to be 2.906)
----------------------------------------------------------------------
```

Because of unmeasured covariates, it is natural to permit the probability of death to vary among litters having particular values of z_i and h_{ij}. For the beta-binomial logistic model, $\hat{\rho} = \hat{\theta}/(1 + \hat{\theta}) = 0.237$, so the fit treats

$$\text{var}(y_{ij}) = [1 + 0.237(n_{ij} - 1)]\mu_{ij}(1 - \mu_{ij})/n_{ij}.$$

This corresponds roughly to a doubling of the variance relative to the binomial with a litter size of 5 and a tripling with $n_{ij} = 9$. The log-likelihood shows great improvement over the ordinary binomial GLM.

```
----------------------------------------------------------------------
> library(VGAM) # beta-binomial model is available in VGAM package
> fit.bb <- vglm(cbind(s, n-s) ~ placebo + h,
+            betabinomial(zero=2,irho=.2), data=Rats)
 # two parameters, mu and rho; zero=2 specifies 0 covariates for 2nd
```

```
 # parameter (rho); irho is initial guess for rho in beta-bin variance
Coefficients:
              Estimate  Std. Error  z value
(Intercept):1  -0.5009      1.1907  -0.4207
(Intercept):2  -1.1676      0.3251  -3.5918
placebo         2.5601      0.7642   3.3501
h              -0.1546      0.1085  -1.4243
Names of linear predictors: logit(mu), logit(rho)
Log-likelihood: -93.1849
> logit(-1.1676, inverse=T) # Inverse logit is a function in VGAM
[1] 0.2373                  # Estimate of rho in beta-binomial variance
----------------------------------------------------------------------
```

For the QL approach using beta-binomial-type variance, $\hat{\rho} = 0.1985$. It corresponds to using $v(\pi_{ij}) = [1 + 0.1985(n_{ij} - 1)]\mu_{ij}(1 - \mu_{ij})/n_{ij}$.

```
----------------------------------------------------------------------
> library(aod)
 # betabin fn. fits beta-bin., quasibin fn. fits QL with beta-bin. var.
>  quasibin(cbind(s, n-s) ~ placebo + h, data=Rats)
           Estimate  Std. Error  z value  Pr(>|z|)
(Intercept)  -0.7237      1.3785  -0.5250    0.5996
placebo       2.7573      0.8522   3.2355    0.0012
h            -0.1758      0.1284  -1.3692    0.1709
Overdispersion parameter: phi 0.1985  # estimate of rho in our notation
----------------------------------------------------------------------
```

Table 8.2 summarizes results for the four analyses. The QL approaches and the beta-binomial model have similar standard errors, quite different from those for the ordinary binomial ML estimates.

Liang and McCullagh (1993) showed several analyses using the inflated variance and beta-binomial-type variance. A plot of the standardized residuals for the ordinary binomial model against the indices $\{n_i\}$ can provide insight about which is more appropriate. When the residuals show an increasing trend in their spread as n_i increases, the beta-binomial-type variance function may be more appropriate.

Table 8.2 Parameter Estimates (with Standard Errors in Parentheses) for Four Fits of a Model with Logit Link to Table 8.1

	Type of Logistic Model Fit[a]			
Parameter	Binomial ML	QL(1)	QL(2)	Beta-Binomial ML
Intercept	0.62 (0.79)	0.62 (1.35)	0.72 (1.38)	0.50 (1.19)
Placebo	2.65 (0.48)	2.65 (0.82)	2.76 (0.85)	2.56 (0.76)
Hemoglobin	−0.19 (0.07)	−0.19 (0.13)	−0.18 (0.13)	−0.15 (0.11)
Overdispersion	None	$\hat{\phi} = 2.906$	$\hat{\rho} = 0.1985$	$\frac{\hat{\theta}}{1+\hat{\theta}} = 0.237$

[a]Quasi-likelihood (QL) has (1) inflated binomial variance, (2) beta-binomial-type variance.

Figure 8.1 plots these for the teratology data. The apparent increase in their variability as litter size increases suggests that the beta-binomial variance function is plausible.

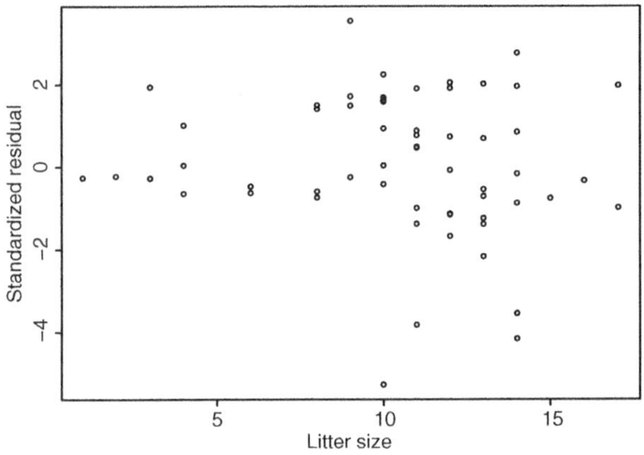

Figure 8.1 Standardized residuals and litter size for binomial logistic model fitted to Table 8.1.

8.3 QUASI-LIKELIHOOD AND MODEL MISSPECIFICATION

For the model $\eta_i = g(\mu_i) = \boldsymbol{x}_i\boldsymbol{\beta}$, the quasi-likelihood parameter estimates $\hat{\boldsymbol{\beta}}$ are the solutions of quasi-score equations

$$\boldsymbol{u}(\boldsymbol{\beta}) = \sum_{i=1}^{n} \left(\frac{\partial \mu_i}{\partial \boldsymbol{\beta}}\right)^{\mathrm{T}} \frac{(y_i - \mu_i)}{v(\mu_i)} = \mathbf{0}. \tag{8.2}$$

These equations are the same as the likelihood equations (8.1) when we substitute

$$\frac{\partial \mu_i}{\partial \beta_j} = \frac{\partial \mu_i}{\partial \eta_i}\frac{\partial \eta_i}{\partial \beta_j} = \frac{\partial \mu_i}{\partial \eta_i} x_{ij},$$

but it is more convenient to use Equation (8.2) for the covariance matrix expressions introduced in this section. With the assumption that $\{y_i\}$ has distribution in the exponential dispersion family, these equations are likelihood equations, in which case $v(\mu_i)$ characterizes the distribution. Our interest here, however, is still in variance functions that take us outside that family.

8.3.1 Estimating Equations and Properties of Quasi-Likelihood

The quasi-score equations (8.2) that determine $\hat{\boldsymbol{\beta}}$ in the QL method are called *estimating equations*. The quasi-score function $u_j(\boldsymbol{\beta})$ in Equation (8.2) is called an *unbiased estimating function*; this term refers to any function $h(\boldsymbol{y}; \boldsymbol{\beta})$ of $\boldsymbol{y}$ and $\boldsymbol{\beta}$ such that

$E[h(\mathbf{y}; \boldsymbol{\beta})] = 0$ for all $\boldsymbol{\beta}$. For an unbiased estimating function, the estimating equation $h(\mathbf{y}; \hat{\boldsymbol{\beta}}) = 0$ determines an estimator $\hat{\boldsymbol{\beta}}$ of $\boldsymbol{\beta}$. The maximum likelihood estimator is but one example of an estimator that results from estimating equations (namely, likelihood equations) with an unbiased estimating function.

The QL method treats the quasi-score function $\boldsymbol{u}(\boldsymbol{\beta})$ as the derivative of a *quasi-log-likelihood* function. Although this need not be a proper log-likelihood function, the QL estimators that maximize it have properties similar to those of ML estimators: under the correct specification of μ_i and $v(\mu_i)$, they are asymptotically efficient among estimators that are locally linear in $\{y_i\}$. This result generalizes the Gauss–Markov theorem, although in an asymptotic rather than exact manner. The QL estimators $\hat{\boldsymbol{\beta}}$ are asymptotically normal with a model-based covariance matrix approximated by

$$\boldsymbol{V} = \left[\sum_{i=1}^{n} \left(\frac{\partial \mu_i}{\partial \boldsymbol{\beta}} \right)^{\mathrm{T}} [v(\mu_i)]^{-1} \left(\frac{\partial \mu_i}{\partial \boldsymbol{\beta}} \right) \right]^{-1}. \tag{8.3}$$

This is equivalent to the formula for the large-sample covariance matrix of the ML estimator in a GLM, namely $(\boldsymbol{X}^{\mathrm{T}}\boldsymbol{W}\boldsymbol{X})^{-1}$ with $w_i = (\partial \mu_i/\partial \eta_i)^2/\mathrm{var}(y_i)$.

A key result is that the QL estimator $\hat{\boldsymbol{\beta}}$, like ML estimators for GLMs, is consistent for $\boldsymbol{\beta}$ even if $v(\mu_i)$ is misspecified, as long as the specification is correct for the link function and linear predictor. That is, assuming that the model form $g(\mu_i) = \sum_j \beta_j x_{ij}$ is correct, the consistency of $\hat{\boldsymbol{\beta}}$ holds even if the true variance function is not $v(\mu_i)$. Here is a heuristic explanation: when truly $\mu_i = g^{-1}(\sum_j \beta_j x_{ij})$, then from Equation (8.2), $E[u_j(\boldsymbol{\beta})] = 0$ for all j. Also from Equation (8.2), $\boldsymbol{u}(\boldsymbol{\beta})/n$ is a vector of sample means. By a law of large numbers, it converges in probability to its expected value of $\mathbf{0}$. But the solution $\hat{\boldsymbol{\beta}}$ of the quasi-likelihood equations is the value of $\boldsymbol{\beta}$ for which the sample mean is exactly equal to $\mathbf{0}$. Since $\hat{\boldsymbol{\beta}}$ is a continuous function of these sample means, it converges to $\boldsymbol{\beta}$ by the continuous mapping theorem.

8.3.2 Sandwich Covariance Adjustment for Variance Misspecification

In practice, when we assume a particular variance function $v(\mu_i)$, it is likely that the true $\mathrm{var}(y_i) \neq v(\mu_i)$. Then the asymptotic covariance matrix of the QL estimator $\hat{\boldsymbol{\beta}}$ is not $\boldsymbol{V}$ as given in Equation (8.3). To find the actual $\mathrm{var}(\hat{\boldsymbol{\beta}})$, we use a Taylor-series expansion for the quasi-score function in Equation (8.2),

$$\boldsymbol{u}(\hat{\boldsymbol{\beta}}) \approx \boldsymbol{u}(\boldsymbol{\beta}) + \frac{\partial \boldsymbol{u}(\boldsymbol{\beta})}{\partial \boldsymbol{\beta}}(\hat{\boldsymbol{\beta}} - \boldsymbol{\beta}).$$

Since $\boldsymbol{u}(\hat{\boldsymbol{\beta}}) = \mathbf{0}$,

$$(\hat{\boldsymbol{\beta}} - \boldsymbol{\beta}) \approx -\left(\frac{\partial \boldsymbol{u}(\boldsymbol{\beta})}{\partial \boldsymbol{\beta}} \right)^{-1} \boldsymbol{u}(\boldsymbol{\beta}),$$

$$\text{so that} \quad \mathrm{var}(\hat{\boldsymbol{\beta}}) \approx \left(\frac{\partial \boldsymbol{u}(\boldsymbol{\beta})}{\partial \boldsymbol{\beta}} \right)^{-1} \mathrm{var}[\boldsymbol{u}(\boldsymbol{\beta})] \left(\frac{\partial \boldsymbol{u}(\boldsymbol{\beta})}{\partial \boldsymbol{\beta}} \right)^{-1}.$$

But $[\partial \boldsymbol{u}(\boldsymbol{\beta})/\partial \boldsymbol{\beta}]$ is the Hessian matrix for the quasi-log-likelihood. So $-[\partial \boldsymbol{u}(\boldsymbol{\beta})/\partial \boldsymbol{\beta}]^{-1}$ is the analog of an inverse observed information matrix for the specified model and approximates the model-based covariance matrix $\boldsymbol{V}$. Also,

$$\text{var}[\boldsymbol{u}(\boldsymbol{\beta})] = \text{var}\left[\sum_{i=1}^{n} \left(\frac{\partial \mu_i}{\partial \boldsymbol{\beta}}\right)^{\text{T}} \frac{(y_i - \mu_i)}{v(\mu_i)}\right] = \sum_{i=1}^{n} \left(\frac{\partial \mu_i}{\partial \boldsymbol{\beta}}\right)^{\text{T}} \frac{\text{var}(y_i)}{[v(\mu_i)]^2} \left(\frac{\partial \mu_i}{\partial \boldsymbol{\beta}}\right).$$

In summary, the actual asymptotic covariance matrix of $\hat{\boldsymbol{\beta}}$ is

$$\text{var}(\hat{\boldsymbol{\beta}}) \approx \boldsymbol{V} \left[\sum_{i=1}^{n} \left(\frac{\partial \mu_i}{\partial \boldsymbol{\beta}}\right)^{\text{T}} \frac{\text{var}(y_i)}{[v(\mu_i)]^2} \left(\frac{\partial \mu_i}{\partial \boldsymbol{\beta}}\right)\right] \boldsymbol{V}. \tag{8.4}$$

This matrix simplifies to $\boldsymbol{V}$ if $\text{var}(y_i) = v(\mu_i)$.

In practice, the true variance function, $\text{var}(y_i)$, is unknown. With large n we can estimate the asymptotic covariance matrix (8.4) by a sample analog, replacing μ_i by $\hat{\mu}_i$ and $\text{var}(y_i)$ by $(y_i - \hat{\mu}_i)^2$. This estimator of the covariance matrix is valid regardless of whether the model-based variance specification $v(\mu_i)$ is correct, in the sense that n times this estimator converges in probability to the asymptotic covariance matrix of $\sqrt{n}(\hat{\boldsymbol{\beta}} - \boldsymbol{\beta})$. It is called a *sandwich estimator*, because the empirical evidence is sandwiched between the model-based covariance matrices.

The purpose of the sandwich estimator is to use the data's empirical evidence about variation to adjust the standard errors, in case the true variance function differs substantially from the variance function assumed in the modeling. Inference then uses the asymptotic normality of the estimator $\hat{\boldsymbol{\beta}}$ together with the sandwich-estimated covariance matrix.

8.3.3 Example: Robust Adjustment of Naïve Standard Errors

To illustrate, suppose $\{y_i\}$ are counts and we assume that $v(\mu_i) = \mu_i$, as in Poisson GLMs, but actually $\text{var}(y_i) = \mu_i^2$. Consider the null model, $\mu_i = \beta, i = 1, \ldots, n$. Since $\partial \mu_i / \partial \beta = 1$, from Equation (8.2),

$$u(\beta) = \sum_{i=1}^{n} \left(\frac{\partial \mu_i}{\partial \beta}\right) v(\mu_i)^{-1}(y_i - \mu_i) = \sum_{i=1}^{n} \frac{(y_i - \mu_i)}{\mu_i} = \sum_{i=1}^{n} \frac{(y_i - \beta)}{\beta}.$$

Setting this equal to 0 and solving, $\hat{\beta} = (\sum_i y_i)/n = \bar{y}$. The model-based variance (8.3) simplifies to

$$V = \left[\sum_{i=1}^{n} \left(\frac{\partial \mu_i}{\partial \beta}\right) [v(\mu_i)]^{-1} \left(\frac{\partial \mu_i}{\partial \beta}\right)\right]^{-1} = \left[\sum_{i=1}^{n} \mu_i^{-1}\right]^{-1} = \frac{\beta}{n}.$$

If we truly believe that $v(\mu_i) = \mu_i$, a sensible estimate of the variance of $\hat{\beta} = \bar{y}$ is $\hat{V} = \hat{\beta}/n = \bar{y}/n$.

The actual asymptotic variance (8.4) of $\hat{\beta}$, which incorporates the true variance function, $\text{var}(y_i) = \mu_i^2$, is

$$V\left[\sum_{i=1}^{n}\left(\frac{\partial\mu_i}{\partial\beta}\right)\frac{\text{var}(y_i)}{[v(\mu_i)]^2}\left(\frac{\partial\mu_i}{\partial\beta}\right)\right]V = \frac{\beta}{n}\left[\sum_{i=1}^{n}\frac{\mu_i^2}{(\mu_i)^2}\right]\frac{\beta}{n} = \frac{\beta^2}{n}.$$

This is considerably different from the naive model-based variance when β is not close to 1. In practice, not knowing the true variance function, we obtain a robust estimator of this actual asymptotic variance by replacing $\text{var}(y_i)$ in Equation (8.4) by $(y_i - \bar{y})^2$. Then the sandwich estimator simplifies (using $\mu_i = \beta$) to $\sum_i (y_i - \bar{y})^2/n^2$, which is a sensible estimator of $\text{var}(\bar{y})$ regardless of the model. Using this estimator instead of $\widehat{V} = \bar{y}/n$ protects against an incorrect choice of variance function.

In summary, even with an incorrect specification of the variance function, we can consistently estimate $\boldsymbol{\beta}$. We can also consistently estimate the asymptotic covariance matrix of $\hat{\boldsymbol{\beta}}$ by the sandwich estimator of Equation (8.4). However, we lose some efficiency in estimating $\boldsymbol{\beta}$ when the chosen variance function $v(\mu_i)$ is wildly inaccurate. Also, n needs to be large for the sample sandwich estimator of Equation (8.4) to work well; otherwise, the empirically based standard errors tend to underestimate the true ones (Kauermann and Carroll 2001). If the assumed variance function is only slightly wrong, the model-based standard errors are more reliable. Finally, in practice, keep in mind that, just as the chosen variance function only approximates the true one, the specification for the mean is also only approximate.

8.3.4 Example: Horseshoe Crabs Revisited

For the horseshoe crabs data, in Section 8.1.3 we used the variance-inflation approach to adjust standard errors for overdispersion from using a Poisson loglinear model to predict male satellite counts using the female crab weights. We obtain similar results from the empirical sandwich adjustment for the standard error. In the next chapter (Section 9.6), we will use a generalization of the sandwich covariance matrix as a way of dealing with correlated observations for a multivariate response. The method solves *generalized estimating equations* (GEE). Software for GEE can also perform the analysis described in this section, using empirical variability to find robust standard errors that adjust for variance misspecification. In the following R printout, the naive standard error comes from the ordinary ML fit of the Poisson model.

```
-----------------------------------------------------------------------
> library(gee) # sandwich adjustment for generalized estimating equa's
> obs <- c(1:173) # labeling of observations needed for GEE method
> summary(gee(y ~ weight, id=obs, family=poisson, scale.fix=TRUE))
               Estimate  Naive S.E.   Naive z  Robust S.E.  Robust z
(Intercept)     -0.4284      0.1789   -2.3942       0.3083   -1.3896
weight           0.5893      0.0650    9.0638       0.1103    5.3418
-----------------------------------------------------------------------
```

CHAPTER NOTES

Section 8.1: Variance Inflation for Overdispersed Poisson and Binomial GLMs

8.1 **QL approach**: Wedderburn (1974) proposed the quasi-likelihood approach with the Pearson moment adjustment for estimating ϕ in the variance inflation approach. Finney (1947) had proposed this for binomial overdispersion. More generally, the QL approach can model ϕ in terms of explanatory variables, thus simultaneously modeling the mean and the variability (McCullagh and Nelder 1989, Chapter 10; Lee et al. 2006, Chapter 3). McCullagh (1983) and Godambe and Heyde (1987) analyzed properties of QL estimators.

8.2 **Poisson overdispersion**: For other QL approaches for Poisson overdispersion, see Cameron and Trivedi (2013, Section 3.2) and Hinde and Demétrio (1998).

Section 8.2: Beta-Binomial Models and Quasi-Likelihood Alternatives

8.3 **Correlated trials**: For extensions of the binomial that permit correlated Bernoulli trials, see Altham (1978) and Ochi and Prentice (1984).

8.4 **Beta-binomial**: Skellam (1948) introduced the beta-binomial distribution. More general beta-binomial models let θ depend on covariates, such as by allowing a different θ for each group of interest (Prentice 1986). For other modeling using this distribution or related QL approaches, see Capanu and Presnell (2008), Crowder (1978), Hinde and Demétrio (1998), Lee et al. (2006), Liang and Hanfelt (1994), Liang and McCullagh (1993), Lindsey and Altham (1998), and Williams (1982).

8.5 **Dirichlet-multinomial**: The beta-binomial generalizes to a *Dirichlet-multinomial* (Mosimann 1962): conditional on the probabilities, the distribution is multinomial, and the probabilities themselves have a Dirichlet distribution. For modeling, see Guimarães and Lindrooth (2007).

Section 8.3: Quasi-Likelihood and Model Misspecification

8.6 **Estimating equations**: Extending Fisher's work, in 1960 Godambe showed that of solutions for unbiased estimating functions, ML estimators are optimal. See Godambe and Heyde (1987), who reviewed the theory of QL estimating equations, and McCullagh (1983).

8.7 **Sandwich**: The sandwich covariance matrix and related results about adjustments for model misspecification using moment-based models evolved from literature in statistics (Huber 1967; Fahrmeir 1990), econometrics (Gourieroux et al. 1984; Hansen 1982; White 1980, 1982), and biostatistics (Liang and Zeger 1986). Cameron and Trivedi (2013, Chapter 2) and Royall (1986) presented motivation and examples.

EXERCISES

8.1 Does the inflated-variance QL approach make sense as a way to generalize the ordinary normal linear model with $v(\mu_i) = \sigma^2$? Why or why not?

8.2 Using $E(y) = E[E(y|x)]$ and $\text{var}(y) = E[\text{var}(y|x)] + \text{var}[E(y|x)]$, derive the mean and variance of the beta-binomial distribution.

8.3 Let y_1 and y_2 be independent negative binomial variates with common dispersion parameter γ.

a. Show that $y_1 + y_2$ is negative binomial with dispersion parameter $\gamma/2$.

b. Conditional on $y_1 + y_2$, show that y_1 has a beta-binomial distribution.

c. State the multicategory extension of (**b**) that yields a Dirichlet-multinomial distribution. Explain the analogy with the Poisson-multinomial result in Section 7.2.1.

8.4 Altham (1978) introduced the discrete distribution

$$f(x; \pi, \theta) = c(\pi, \theta) \binom{n}{x} \pi^x (1-\pi)^{n-x} \theta^{x(n-x)}, \quad x = 0, 1, \ldots, n,$$

where $c(\pi, \theta)$ is a normalizing constant. Show that this is in the two-parameter exponential family and that the binomial occurs when $\theta = 1$. (Altham noted that overdispersion occurs when $\theta < 1$. Lindsey and Altham (1998) used this as the basis of an alternative model to the beta-binomial.)

8.5 Sometimes sample proportions are continuous rather than of the binomial form (number of successes)/(number of trials). Each observation is any real number between 0 and 1, such as the proportion of a tooth surface that is covered with plaque. For independent responses $\{y_i\}$, Bartlett (1937) modeled $\text{logit}(y_i) \sim N(\boldsymbol{x}_i\boldsymbol{\beta}, \sigma^2)$. Then y_i itself has a *logit-normal distribution.*

a. Expressing a $N(\boldsymbol{x}_i\boldsymbol{\beta}, \sigma^2)$ variate as $\boldsymbol{x}_i\boldsymbol{\beta} + \sigma z$, where z is a standard normal variate, show that $y_i = \exp(\boldsymbol{x}_i\boldsymbol{\beta} + \sigma z)/[1 + \exp(\boldsymbol{x}_i\boldsymbol{\beta} + \sigma z)]$ and for small σ,

$$y_i = \frac{e^{\boldsymbol{x}_i\boldsymbol{\beta}}}{1 + e^{\boldsymbol{x}_i\boldsymbol{\beta}}} + \frac{e^{\boldsymbol{x}_i\boldsymbol{\beta}}}{1 + e^{\boldsymbol{x}_i\boldsymbol{\beta}}} \frac{1}{1 + e^{\boldsymbol{x}_i\boldsymbol{\beta}}} \sigma z + \frac{e^{\boldsymbol{x}_i\boldsymbol{\beta}}(1 - e^{\boldsymbol{x}_i\boldsymbol{\beta}})}{2(1 + e^{\boldsymbol{x}_i\boldsymbol{\beta}})^3} \sigma^2 z^2 + \cdots.$$

b. Letting $\mu_i = e^{\boldsymbol{x}_i\boldsymbol{\beta}}/(1 + e^{\boldsymbol{x}_i\boldsymbol{\beta}})$, when σ is close to 0 show that

$$E(y_i) \approx \mu_i, \quad \text{var}(y_i) \approx [\mu_i(1 - \mu_i)]^2 \sigma^2.$$

c. The approximate moments for the logit-normal motivate a QL approach with $v(\mu_i) = \phi[\mu_i(1 - \mu_i)]^2$ for unknown ϕ. Explain why this approach provides similar results as fitting an ordinary linear model to the sample logits, assuming constant variance. (The QL approach has the advantage of not requiring adjustment of 0 or 1 observations, for which sample logits do not exist. Papke and Wooldridge (1996) proposed an alternative QL approach using a sandwich covariance adjustment.)

d. Wedderburn (1974) used QL to model the proportion of a leaf showing a type of blotch. Envision an approximation of binomial form based on cutting each leaf into a very large number of tiny regions of the same size and observing for each region whether it is covered with blotch. Explain why this suggests using $v(\mu_i) = \phi\mu_i(1-\mu_i)$. What violation of the binomial assumptions might make this questionable? (Recall that the parametric family of beta distributions has variance function of this form.)

8.6 Motivation for the quasi-score equations (8.2): suppose we replace $v(\mu_i)$ by known variance v_i. Show that the equations result from the weighted least squares approach of minimizing $\sum_i[(y_i - \mu_i)^2/v_i]$.

8.7 Before R. A. Fisher introduced the method of maximum likelihood in 1922, Karl Pearson had proposed the *method of moments* as a general-purpose method for statistical estimation[2]. Explain how this method can be formulated as having estimating equations with an unbiased estimating function.

8.8 Ordinary linear models assume that $v(\mu_i) = \sigma^2$ is constant. Suppose instead that actually $\text{var}(y_i) = \mu_i$. Using the QL approach for the null model $\mu_i = \beta$, $i = 1, \ldots, n$, show that $u(\beta) = (1/\sigma^2)\sum_i(y_i - \beta)$, so $\hat{\beta} = \bar{y}$ and $V = \sigma^2/n$. Find the model-based estimate of $\text{var}(\hat{\beta})$, the actual variance, and the robust estimate of that variance that adjusts for misspecification of the variance.

8.9 Suppose we assume $v(\mu_i) = \mu_i$ but actually $\text{var}(y_i) = \sigma^2$. For the null model $\mu_i = \beta$, find the model-based $\text{var}(\hat{\beta})$, the actual $\text{var}(\hat{\beta})$, and the robust estimate of that variance.

8.10 Suppose we assume $v(\mu_i) = \mu_i$ but actually $\text{var}(y_i) = v(\mu_i)$ for some unspecified function v. For the null model $\mu_i = \beta$, find the model-based $\text{var}(\hat{\beta})$, the actual $\text{var}(\hat{\beta})$, and the robust estimate of that variance.

8.11 Consider the null model $\mu_i = \beta$ when the observations are independent counts. Of the Poisson-model-based and robust estimators of the variance of $\hat{\beta} = \bar{y}$ presented in Section 8.3.3, which would you expect to be better (**a**) if the Poisson model truly holds, (**b**) if there is severe overdispersion? Explain your reasoning.

8.12 Let y_{ij} denote the response to a question about belief in life after death (1 = yes, 0 = no) for person j in household i, $j = 1, \ldots, n_i$, $i = 1, \ldots, n$. In modeling $P(y_{ij} = 1)$ with explanatory variables, describe a scenario in which you would expect binomial overdispersion. Specify your preferred method for dealing with it, presenting your reasoning for that choice.

8.13 Use QL methods to construct a model for the horseshoe crab satellite counts, using weight, color, and spine condition as explanatory variables. Compare results with those obtained with zero-inflated GLMs in Section 7.5.

[2] J. Aldrich in *Statistical Science* (**12**: 162–176, 1997) gave a historical overview.

8.14 Use QL methods to analyze Table 7.5 on counts of homicide victims. Interpret, and compare results with Poisson and negative binomial GLMs.

8.15 Refer to Exercise 7.35 on the frequency of sexual intercourse. Use QL methods to obtain a confidence interval for the (**a**) difference, (**b**) ratio of means for males and females.

8.16 For the teratology study analyzed in Section 8.2.4, analyze the data using only the group indicators as explanatory variables (i.e., ignoring hemoglobin). Interpret results. Is it sufficient to use the simpler model having only the placebo indicator for the explanatory variable?

8.17 Table 8.3 shows the three-point shooting, by game, of Ray Allen of the Boston Celtics during the 2010 NBA (basketball) playoffs (e.g., he made 0 of 4 shots in game 1). Commentators remarked that his shooting varied dramatically from game to game. In game i, suppose that $n_i y_i$ = number of three-point shots made out of n_i attempts is a bin(n_i, π_i) variate and the $\{y_i\}$ are independent.

Table 8.3 Data for Exercise 8.17 on Three-Point Shooting in Basketball

Game	y_i	Game	y_i	Game	y_i	Game	y_i	Game	y_i
1	0/4	6	2/7	11	0/5	16	1/3	21	0/4
2	7/9	7	3/7	12	2/5	17	3/7	22	0/4
3	4/11	8	0/1	13	0/5	18	0/2	23	2/5
4	3/6	9	1/8	14	2/4	19	8/11	24	2/7
5	5/6	10	6/9	15	5/7	20	0/8		

Source: http://boston.stats.com/nba. Data at file `Basketball.dat` at text website.

a. Fit the model, $\pi_i = \beta_0$. Find and interpret $\hat{\beta}_0$ and its standard error.

b. Describe a factor that could cause overdispersion. Adjust the standard error for overdispersion. Using the original *SE* and its correction, find and compare 95% confidence intervals for β_0. Interpret.

CHAPTER 9

Modeling Correlated Responses

Many studies have multivariate response variables. For example, a social survey might ask a subject's opinion about whether government spending should decrease, stay the same, or increase in each of several areas (defense, health, education, environment, ...). A clinical trial studying patients taking a new drug might measure whether each of several side effects (e.g., headaches, nausea) occurs, and its severity. Longitudinal[1] studies observe a response variable repeatedly for each subject, at several times. A clinical trial comparing treatments for some malady, for example, might randomize patients to take either a new drug or a placebo and then observe them after 1 month, 3 months, and 6 months to evaluate whether the treatment response is positive.

In this chapter, we present models for a d-dimensional response variable $\boldsymbol{y} = (y_1, y_2, \dots, y_d)$. Each subject has a *cluster* of d observations. In a longitudinal study, for example, a cluster consists of the observations over time for a particular subject. Often d varies by cluster, such as when some subjects drop out of the study and are missing some observations. For multivariate data, observations within a cluster are typically correlated, and models need to account for that correlation. Section 9.1 presents two primary types of models for multivariate responses. One type, a *marginal model*, simultaneously models only each marginal distribution, but takes into account the correlation structure in finding valid standard errors. The other type models the clusters of correlated responses, generating a multivariate distribution for $\boldsymbol{y}$ by including in the linear predictor an unobserved random variable for each cluster, called a *random effect*. The extension of the generalized linear model (GLM) to include random effects in addition to the usual fixed effects is called a *generalized linear mixed model* (GLMM).

We present normal linear mixed models in Section 9.2 and their fitting in Section 9.3. We introduce GLMMs in Section 9.4, focusing on binomial and Poisson cases, and Section 9.5 presents fitting methods. Section 9.6 then summarizes the

[1] *Panel data* is an alternate name for longitudinal data.

Foundations of Linear and Generalized Linear Models, First Edition. Alan Agresti.

marginal modeling approach. For non-normal responses, it uses a multivariate extension of quasi-likelihood methods. Section 9.7 presents an example illustrating and comparing the two types of models.

9.1 MARGINAL MODELS AND MODELS WITH RANDOM EFFECTS

For cluster i, denote the observations on the d components of the multivariate response vector by $\boldsymbol{y}_i = (y_{i1}, \ldots, y_{id})^{\mathrm{T}}$, $i = 1, \ldots, n$. Let $\boldsymbol{x}_{ij}$ denote the row vector of p explanatory variable values for observation y_{ij}, with $\mu_{ij} = E(y_{ij})$. Values of the explanatory variables may vary for the observations in a cluster, as would happen for variables such as weight, blood pressure readings, and total cholesterol level in a longitudinal study about heart disease.

9.1.1 Effect of Correlation on Within-Subject and Between-Subject Effects

Models for multivariate response data can analyze effects of two types. For an explanatory variable with constant value for observations in a cluster, we can compare clusters that differ in values of the variable. This effect is "between-cluster," also known as *between-subject* when each cluster is an individual. An example is a demographic variable, such as the gender of an individual, where the effect is a comparison of females with males. For an explanatory variable that varies among observations in a cluster, we can analyze the effect of change in its value within a cluster. This effect is "within-cluster" (*within-subject*). For example, in a longitudinal study that regularly observes the weights of anorexic girls, we might analyze the within-subject effect of daily calorie intake.

Analyses that ignore the correlations among observations within a cluster have invalid standard errors. To illustrate, consider a 2×2 design with treatments A and B having independent sets of n individuals who are observed at $d = 2$ times. Treatment is a between-subjects factor and the time of observation is a within-subjects factor. For $i = 1, \ldots, n$, let (y_{i1}^A, y_{i2}^A) be the observations at the two times for subject i in treatment A, and let (y_{i1}^B, y_{i2}^B) be the observations at the two times for subject i in treatment B. Suppose that $\mathrm{corr}(y_{i1}^A, y_{i2}^A) = \mathrm{corr}(y_{i1}^B, y_{i2}^B) = \rho$ for all i and $\mathrm{corr}(y_{it}^A, y_{ju}^B) = 0$ for all i and j and for $t, u = 1, 2$, with all $\mathrm{var}(y_{it}^A) = \mathrm{var}(y_{it}^B) = \sigma^2$. Let

$$\bar{y}_t^A = \left(\sum_{i=1}^n y_{it}^A\right) \Big/ n \quad \text{and} \quad \bar{y}_t^B = \left(\sum_{i=1}^n y_{it}^B\right) \Big/ n, \qquad t = 1, 2.$$

For a linear model that assumes an absence of interaction between the treatment and time factors, the estimated between-subjects effect is

$$b = \left[\left(\bar{y}_1^A + \bar{y}_2^A\right)/2\right] - \left[\left(\bar{y}_1^B + \bar{y}_2^B\right)/2\right],$$

which compares the treatment means. The estimated within-subjects effect is

$$w = \left[\left(\bar{y}_1^A + \bar{y}_1^B\right)/2\right] - \left[\left(\bar{y}_2^A + \bar{y}_2^B\right)/2\right],$$

which compares the means at the two times. You can verify that

$$\text{var}(b) = \frac{\sigma^2(1+\rho)}{n}, \quad \text{var}(w) = \frac{\sigma^2(1-\rho)}{n}. \tag{9.1}$$

For each effect, if all observations had been independent, we would have variance σ^2/n for comparing two groups of $2n$ observations each. But the correlation between observations within a cluster is typically positive. So the variance is then smaller for inference about within-cluster effects but larger for inference about between-cluster effects. If we mistakenly treat observations within a cluster as independent, the *SE* values we report will be too large for within-cluster effects and too small for between-cluster effects.

9.1.2 Two Types of Multivariate Models

Marginal model: A *marginal model* for $\boldsymbol{y}$, with link function g, has the form

$$g(\mu_{ij}) = \boldsymbol{x}_{ij}\boldsymbol{\beta}, \quad i = 1, \ldots, n, \quad j = 1, \ldots, d. \tag{9.2}$$

The model refers to the marginal distribution at each j rather than the joint distribution. For example, in a battery of achievement exams, let y_{ij} denote the score on exam j for student i, who has grade-point average (GPA) x_i. For the model $\mu_{ij} = \beta_{0j} + \beta_{1j}x_i$, each exam score has a separate linear relation with GPA. This has the form (9.2) with $\boldsymbol{\beta} = (\beta_{01}, \beta_{11}, \ldots, \beta_{0d}, \beta_{1d})^{\mathrm{T}}$ and $\boldsymbol{x}_{ij} = (0, 0, \ldots, 1, x_i, \ldots 0, 0)$ having zero entries except for the coefficients of β_{0j} and β_{1j}.

A marginal model has the usual GLM structure for each component in the multivariate vector. To complete the model, we assume a parametric joint distribution for $(y_{i1}, \ldots, y_{id})$. Then, with independent observations for $i = 1, \ldots, n$, we can fit the model by maximum likelihood (ML). In an important case, $\boldsymbol{y}_i$ has a multivariate normal distribution, and g is the identity link function (Section 9.6.1). Discrete responses, however, do not have multivariate distributions for $\boldsymbol{y}_i$ that account in a simple manner for correlation among responses. Because of this, ML is often not viable for marginal models. In Section 9.6.4 we fit the model by assuming a parametric distribution only for each marginal component, using a quasi-likelihood method to produce estimates and a sandwich covariance matrix to find valid standard errors.

GLMM: A generalized linear mixed model (GLMM) for $\boldsymbol{y}$ has the form

$$g[E(y_{ij} \mid \boldsymbol{u}_i)] = \boldsymbol{x}_{ij}\boldsymbol{\beta} + \boldsymbol{z}_{ij}\boldsymbol{u}_i, \quad i = 1, \ldots, n, \quad j = 1, \ldots, d, \tag{9.3}$$

where the parameters $\boldsymbol{\beta}$ are *fixed effects* of the explanatory variables and $\{\boldsymbol{u}_i\}$ are *random effects* assumed to have a particular probability distribution.

Here $\boldsymbol{z}_{ij}$, like $\boldsymbol{x}_{ij}$, is a row vector of known values of explanatory variables. The random effects $\{\boldsymbol{u}_i\}$ are usually assumed to be independent from a $N(\boldsymbol{0}, \boldsymbol{\Sigma}_u)$ distribution specified by unknown variance and correlation parameters. Their shared common value for all j connects the d dimensions of the model. When we assume a conditional distribution for $(y_{i1}, \ldots, y_{id})$ given $\boldsymbol{u}_i$, model (9.3) determines a multivariate distribution for $\boldsymbol{y}$.

The adjective *mixed* in *generalized linear mixed model* refers to the presence of both fixed effects (i.e., parameters) and random effects (i.e., random variables) in the linear predictor. GLMs extend ordinary regression by allowing non-normal responses and a link function of the mean. The GLMM is the further extension that permits random effects $\boldsymbol{u}_i$ as well as fixed effects $\boldsymbol{\beta}$ in the linear predictor. Fixed effects apply to *all* values or categories of interest for a variable, such as genders, education levels, or age groups. By contrast, random effects usually apply to a *sample*, with $\boldsymbol{u}_i$ referring to cluster i in the sample. With large n, if we treated $\{\boldsymbol{u}_i\}$ as fixed effects, including them in the model would increase greatly the number of parameters. With a random effects approach, we instead treat these as a random sample from the population of interest. We then have only the additional variance and correlation parameters for the $N(\boldsymbol{0}, \boldsymbol{\Sigma}_u)$ distribution of $\{\boldsymbol{u}_i\}$ in that population. We will find that estimation of effects in any particular cluster is strengthened by "borrowing from the whole," using data from all the clusters.

Often, the random effect in a GLMM is one-dimensional with coefficient 1, and $\boldsymbol{z}_{ij}\boldsymbol{u}_i = u_i$ merely adds a random term to the linear predictor for each subject. For modeling y_{ij} = score for student i on achievement exam j with x_i = GPA, a possible GLMM is

$$E(y_{ij} \mid u_i) = \beta_{0j} + \beta_{1j}x_i + u_i = (\beta_{0j} + u_i) + \beta_1 x_j.$$

Here, u_i is unobserved and perhaps summarizes characteristics such as ability, achievement motivation, and parental encouragement for student i. For students of a given GPA, those with relatively higher u_i tend to perform better on any particular exam. The second expression for the linear predictor shows that the model has a separate intercept for each subject. This type of GLMM is called a *random-intercept model*.

The variability of $\{\boldsymbol{u}_i\}$ in a GLMM might represent that different subjects at common explanatory variable values are heterogeneous in their distributions on the response variable, perhaps because the linear predictor did not include some relevant explanatory variables. In a random-intercept model with u_i replaced by $u_i^*\sigma_u$, where

$\{u_i^*\}$ are $N(0, 1)$, the linear predictor $\boldsymbol{x}_{ij}\boldsymbol{\beta} + u_i^*\sigma_u$ has the form of one for an ordinary GLM with unobserved values $\{u_i^*\}$ of a covariate. Random effects also sometimes represent random measurement error in the explanatory variables. In either case, they provide a mechanism for handling overdispersion relative to a standard model. We will also find that random effects having greater variability induce stronger correlation between pairs of responses within clusters of observations.

9.1.3 GLMMs Imply Marginal Models

For the GLMM, by inverting the link function,

$$E(y_{ij} \mid \boldsymbol{u}_i) = g^{-1}(\boldsymbol{x}_{ij}\boldsymbol{\beta} + \boldsymbol{z}_{ij}\boldsymbol{u}_i).$$

Marginally, averaging over the random effects, the mean is

$$\mu_{ij} = E(y_{ij}) = E[E(y_{ij} \mid \boldsymbol{u}_i)] = \int g^{-1}(\boldsymbol{x}_{ij}\boldsymbol{\beta} + \boldsymbol{z}_{ij}\boldsymbol{u}_i) f(\boldsymbol{u}_i; \boldsymbol{\Sigma}_u) d\boldsymbol{u}_i,$$

where $f(\boldsymbol{u}; \boldsymbol{\Sigma}_u)$ is the $N(\boldsymbol{0}, \boldsymbol{\Sigma}_u)$ density function for the random effects. This is the marginal model implied by the GLMM.

For the identity link function,

$$\mu_{ij} = \int (\boldsymbol{x}_{ij}\boldsymbol{\beta} + \boldsymbol{z}_{ij}\boldsymbol{u}_i) f(\boldsymbol{u}_i; \boldsymbol{\Sigma}_u) d\boldsymbol{u}_i = \boldsymbol{x}_{ij}\boldsymbol{\beta}.$$

The marginal model has the same link function and fixed effects $\boldsymbol{\beta}$ as the GLMM. This is not generally true for other link functions. The following example and Section 9.4 illustrate.

9.1.4 Example: Bivariate Models for Binary Matched-Pairs Data

To illustrate the distinction between marginal models and GLMMs, we present two simple models for binary matched-pairs data. Let (y_{i1}, y_{i2}) denote the pair of observations for subject (matched pair) i, where 1 = success and 0 = failure. For example, in a crossover study for comparing two drugs on a chronic condition (such as migraine headaches), y_{ij} may refer to whether subject i has a successful outcome with drug j, $j = 1, 2$. To compare drugs, we compare $P(y_{i1} = 1)$ with $P(y_{i2} = 1)$. Cross-classifying y_{i1} by y_{i2} for the n observations yields a 2×2 contingency table with cell counts $\{n_{ab}\}$ as shown in Table 9.1.

A possible marginal model for the paired binary responses is

$$\text{logit}[P(y_{ij} = 1)] = \beta_0 + \beta_1 x_j, \tag{9.4}$$

Table 9.1 Contingency Table Representation of n Observations for Binary Matched Pairs

	Observation 2 (y_{i2})		
Observation 1 (y_{i1})	Success	Failure	Total
Success	n_{11}	n_{12}	n_{1+}
Failure	n_{21}	n_{22}	n_{2+}
Total	n_{+1}	n_{+2}	n

where $x_1 = 0$ and $x_2 = 1$. Then $\text{logit}[P(y_{i1} = 1)] = \beta_0$ and $\text{logit}[P(y_{i2} = 1)] = \beta_0 + \beta_1$, so β_1 is a log odds ratio describing the difference between the column and row marginal distributions for the population analog of Table 9.1.

By contrast, the GLMM approach focuses on a 2×2 table for each subject, as shown in Table 9.2. The random-intercept model

$$\text{logit}[P(y_{ij} = 1 \mid u_i)] = \beta_0 + \beta_1 x_j + u_i \tag{9.5}$$

for these n tables, where $\{u_i\}$ are independent from a $N(0, \sigma_u^2)$ distribution, allows success probabilities to vary by subject.

Table 9.2 Representation of Pair i of Observations for Binary Matched-Pairs Data

	Response		
Observation	Success	Failure	Total
1	y_{i1}	$1 - y_{i1}$	1
2	y_{i2}	$1 - y_{i2}$	1

The sample has n "partial tables" of this form, one for each subject.

In the random-intercept model (9.5), the effect β_1 pertains at the cluster level and is called *subject-specific*. Defined conditional on the subject, it describes conditional association for the $2 \times 2 \times n$ table with subject-specific partial tables of the form of Table 9.2. The model[2] implies that the odds ratio for the underlying probabilities in each of the n partial tables equals $\exp(\beta_1)$. By contrast, the effect in marginal model (9.4) is *population-averaged*, because it results from averaging over the entire population. Its sample version is the odds ratio for the single 2×2 table obtained by adding together the n partial tables. That summary table has rows that are the margins of Table 9.1.

[2]More generally, Section 9.2.4 shows that a GLMM can also have a random slope, so this effect itself varies among clusters.

If we replace the logit link by the identity link function in these models, the population-averaged and subject-specific effects are identical. For instance, for the random-intercept model

$$P(y_{ij} = 1 \mid u_i) = \beta_0 + \beta_1 x_j + u_i,$$

the effect $\beta_1 = P(y_{i2} = 1 \mid u_i) - P(y_{i1} = 1 \mid u_i)$ for all i. Averaging this over subjects in the population equates β_1 to the parameter in the marginal model using an identity link, namely $P(y_{ij} = 1) = \beta_0 + \beta_1 x_j$. For nonlinear link functions, however, the effects differ. If we begin with the logistic random-intercept model (9.5), equivalently

$$P(y_{ij} = 1 \mid u_i) = \exp(\beta_0 + \beta_1 x_j + u_i)/[1 + \exp(\beta_0 + \beta_1 x_j + u_i)],$$

and average over the distribution of u_i to obtain the implied marginal model for the population, that model does not have the logistic form

$$P(y_{ij} = 1) = \exp(\beta_0 + \beta_1 x_j)/[1 + \exp(\beta_0 + \beta_1 x_j)]$$

corresponding to the marginal model (9.4). So the two models describe different effects. It is beyond our scope to derive this here, but the ML estimate of β_1 for the marginal model and, when the sample log odds ratio in Table 9.1 is nonnegative, the ML estimate of β_1 for the random-intercept model are

$$\text{Marginal model}: \hat{\beta}_1 = \log \frac{(n_{+1}/n_{+2})}{(n_{1+}/n_{2+})}, \quad \text{GLMM}: \hat{\beta}_1 = \log \frac{n_{21}}{n_{12}}.$$

The estimate for the marginal model is the log odds ratio of the marginal counts. The two estimates can be quite different in magnitude. We discuss the difference further in Section 9.4.1.

Incidentally, if we instead treat $\{u_i\}$ in logistic model (9.5) as fixed effects, the ML estimator of β_1 is poor because the number of $\{u_i\}$ equals the sample size n. The analysis then violates the regularity condition for asymptotic optimality of ML that the number of parameters is fixed as n increases. ML estimators need not be consistent[3] when the number of parameters grows with the sample size, and in fact, $\hat{\beta}_1$ then converges to $2\beta_1$ as $n \to \infty$ (Exercise 9.20). The remedy of conditional ML treats $\{u_i\}$ as nuisance parameters and maximizes the likelihood function for a conditional distribution that eliminates them (Exercise 9.19). This yields $\hat{\beta}_1 = \log(n_{21}/n_{12})$, the same estimate that usually occurs with the GLMM approach (Lindsay et al. 1991).

With the random-intercept model, averaged over the unobserved $\{u_i\}$, Section 9.4.1 shows that the responses are nonnegatively correlated. A subject with a large positive u_i has a relatively high $P(y_{ij} = 1 \mid u_i)$ for each j and is likely to have a success at any particular time; a subject with a large negative u_i has low $P(y_{ij} = 1 \mid u_i)$

[3]This result is often referred to as the *Neyman–Scott phenomenon*, recognizing a 1948 *Econometrica* article by Jerzy Neyman and Elizabeth Scott.

for each j and is likely to have a failure at any particular time. The greater the variability in $\{u_i\}$, the greater the overall positive association between responses. The positive association reflects the shared value of u_i for each observation in a cluster.

9.1.5 Choice of Marginal Model versus GLMM

As explained in Section 9.1.1, clustered data have two types of effects, between-cluster and within-cluster. GLMMs explicitly include the cluster in the model, so they naturally describe within-cluster effects. By contrast, effects in marginal models are averaged over clusters (i.e., population-averaged), so those effects do not refer to a comparison at a fixed value of a random effect. The GLMM approach is preferable when we want to estimate cluster-specific effects, estimate their variability, specify a mechanism for generating nonnegative association among clustered observations, or model the joint distribution. Latent variable constructions that motivate model forms (e.g., for binary data, the threshold model of Section 5.1.2) apply more naturally at the cluster level than at the marginal level.

Many surveys and epidemiological studies have the goal of comparing distinct groups, such as smokers and non-smokers, on a mean response or the relative frequency of some outcome. Then quantities of primary interest include between-group comparisons of marginal means or probabilities for the groups. That is, the effects of interest are between-cluster rather than within-cluster. When between-cluster effects are the main focus, it can be simpler to model them directly using marginal models. A between-cluster fixed effect for two groups in a GLMM applies only when the random effect takes the same value in each group, such as a smoker and a non-smoker with the same random effect values, adjusting for the other explanatory variables in the model. In this sense, GLMMs are cluster-specific models, as both within-cluster and between-cluster effects apply conditional on the random effect. Modeling the joint distribution that generates those marginal effects, as a GLMM does, provides greater opportunity for misspecification, and the subsequent estimates of between-cluster effects can be more sensitive to violations of assumptions.

Although a GLMM does not naturally describe between-cluster effects, we can recover information about such effects by integrating out the random effects, as we showed in Section 9.1.3. But often this integration does not generate a closed form for the implied marginal model. By contrast, although marginal models naturally describe between-subjects effects, they are less general than GLMMs, in that a marginal model does not imply[4] a GLMM.

9.1.6 Transition Models and Other Multivariate Models

Marginal models and GLMMs are the primary types of model for multivariate responses. However, they are not the only models.

[4]However, Diggle et al. (2002, Section 11.3) showed that marginal model structure can be imbedded in GLMMs.

For contingency tables that cross-classify several categorical response variables, Poisson loglinear models and the corresponding multinomial models (Section 7.2) are multivariate models. Rather than modeling marginal distributions, such models focus on the joint distribution, such as to analyze whether a certain pair of variables is conditionally independent or has homogeneous conditional association or a more complex interaction structure. By contrast, marginal models regard the joint distribution as a nuisance, and use it merely to find valid *SE* values for estimates of model parameters.

A quite different type of multivariate model, called a *transition model*, has the form

$$g[E(y_{ij} \mid y_{i,j-1}, y_{i,j-2}, \ldots)] = \boldsymbol{x}_{ij}\boldsymbol{\beta} + \gamma_1 y_{i,j-1} + \gamma_2 y_{i,j-2} + \cdots .$$

Unlike other models, this model takes into account the *sequence* of the observations in a cluster, rather than treating them as exchangeable. With time series data, we can use transition models to predict a response at the next time using past observations as well as explanatory variables. In this model, $\boldsymbol{\beta}$ describes effects of explanatory variables after adjusting for past responses. A linear predictor that contains $y_{i,j-1}$ but not earlier observations treats y_{ij} as conditionally independent of the earlier observations, given $y_{i,j-1}$. This is a first-order *Markov model*. The Markov model and other transition models are beyond the scope of this text.

9.2 NORMAL LINEAR MIXED MODELS

The *linear mixed model* for y_{ij} is

$$E(y_{ij} \mid \boldsymbol{u}_i) = \boldsymbol{x}_{ij}\boldsymbol{\beta} + \boldsymbol{z}_{ij}\boldsymbol{u}_i, \quad \text{or} \quad y_{ij} = \boldsymbol{x}_{ij}\boldsymbol{\beta} + \boldsymbol{z}_{ij}\boldsymbol{u}_i + \epsilon_{ij},$$

where $\boldsymbol{\beta}$ is a $p \times 1$ vector of fixed effects and $\boldsymbol{u}_i \sim N(\boldsymbol{0}, \boldsymbol{\Sigma}_u)$ is a $q \times 1$ vector of random effects. Usually we assume that $\epsilon_{ij} \sim N(0, \sigma_\epsilon^2)$, yielding the *normal linear mixed model*. The basic model assumes that $\{\boldsymbol{u}_i\}$ and $\{\epsilon_{ij}\}$ are independent between clusters (i.e., over i) and of each other. To begin, we also assume that $\{\epsilon_{ij}\}$ are independent within clusters (i.e., over j for each i).

The model for y_{ij} decomposes into a term $\boldsymbol{x}_{ij}\boldsymbol{\beta}$ for the mean, a term $\boldsymbol{z}_{ij}\boldsymbol{u}_i$ for between-cluster variability, and a term ϵ_{ij} for within-cluster variability. For $\boldsymbol{y}_i = (y_{i1}, \ldots, y_{id})^{\mathrm{T}}$, the model has the form

$$\boldsymbol{y}_i = \boldsymbol{X}_i\boldsymbol{\beta} + \boldsymbol{Z}_i\boldsymbol{u}_i + \boldsymbol{\epsilon}_i \tag{9.6}$$

(Laird and Ware 1982), where $\boldsymbol{X}_i$ is the $d \times p$ model matrix for observation i that has $\boldsymbol{x}_{ij}$ in row j, $\boldsymbol{Z}_i$ is a $d \times q$ model matrix for the random effects that has $\boldsymbol{z}_{ij}$ in row j, and $\boldsymbol{\epsilon}_i \sim N(\boldsymbol{0}, \sigma_\epsilon^2\boldsymbol{I})$. Conditional on the random effects, the model $E(\boldsymbol{y}_i \mid \boldsymbol{u}_i) = \boldsymbol{X}_i\boldsymbol{\beta} + \boldsymbol{Z}_i\boldsymbol{u}_i$

looks like an ordinary linear model with $\boldsymbol{Z}_i\boldsymbol{u}_i$ as an offset term. With both sources of random variability, marginally

$$\text{var}(\boldsymbol{y}_i) = \boldsymbol{Z}_i\boldsymbol{\Sigma}_u\boldsymbol{Z}_i^{\text{T}} + \sigma_\epsilon^2\boldsymbol{I}. \tag{9.7}$$

Here, $\boldsymbol{Z}_i\boldsymbol{\Sigma}_u\boldsymbol{Z}_i^{\text{T}}$ describes the between-cluster variability and $\sigma_\epsilon^2\boldsymbol{I}$ describes the within-cluster variability.

9.2.1 The Random-Intercept Linear Mixed Model

An important special case of a linear mixed model has $\boldsymbol{u}_i = u_i$, $\boldsymbol{Z}_i = \boldsymbol{1}$, and $\text{var}(u_i) = \sigma_u^2$, that is,

$$\boldsymbol{y}_i = \boldsymbol{X}_i\boldsymbol{\beta} + u_i\boldsymbol{1} + \boldsymbol{\epsilon}_i. \tag{9.8}$$

For this random-intercept model, marginally

$$\text{var}(\boldsymbol{y}_i) = \sigma_u^2\boldsymbol{1}\boldsymbol{1}^{\text{T}} + \sigma_\epsilon^2\boldsymbol{I}.$$

The two variances in this expression are referred to as *variance components*.

This model has the exchangeable correlation structure, for $j \neq k$,

$$\text{corr}(y_{ij}, y_{ik}) = \frac{\sigma_u^2}{\sigma_u^2 + \sigma_\epsilon^2},$$

called *compound symmetry*. Having u_i in the model, shared among clusters, implies that $\text{corr}(y_{ij}, y_{ik}) \geq 0$ marginally. Greater within-cluster correlation occurs as σ_u^2 increases. This type of correlation, using within-cluster association to summarize the effect of clustering, is referred to as an *intraclass* (or *intracluster*) *correlation*. The correlation summarizes the proportion of the response variability due to the clustering.

Compound symmetry structure also occurs in traditional repeated-measures ANOVA methods for comparing groups with repeated-measurement data (Diggle et al. 2002, Section 6.4). Analyses based on a linear mixed model have advantages compared with those ANOVA methods. Especially important is that the linear mixed modeling analysis can use subjects who have different numbers of observations, such as when some observations are missing. For instance, in the clinical trials example in Section 9.2.4 below, complete results for all five times were available for only 211 of the 627 subjects. Using only the available data in linear mixed modeling does not introduce bias as long as the data are *missing at random*; that is, what caused the data to be missing can depend on the observed data but not on the data that are missing. This is true, for example, if whether someone drops out of the study may depend on values observed prior to the drop-out but not on the later unobserved values. Other advantages of linear mixed models over repeated-measures ANOVA are that the models generalize to accommodate common characteristics of longitudinal studies, such as irregularly spaced observations, time-varying explanatory variables, increased variability over time, and correlated errors within clusters.

9.2.2 Hierarchical Modeling: Multilevel Models

In some research studies the data structure is hierarchical, with sampled units nested in clusters that are themselves nested in other clusters. Hierarchical models are natural when subjects and institutions form clusters, such as educational studies of students within schools or medical studies of doctors within hospitals. GLMMs for hierarchically structured data are called *multilevel models*. Such models enable us to study the effects of the relevant explanatory variables at each level. Also, the total error variability decomposes into variance components attributable to each level. Standard error estimators can be badly biased if we ignore the clustering and the consequent within-cluster correlations.

Suppose that a study of characteristics that affect student performance on a battery of exams samples students from a sample of schools. The model should take into account the student and the school (or school district). Just as observations on the same students tend to be more alike than observations on different students, students in the same school tend to be more alike than students from different schools. Random effects can enter the model at each level of the hierarchy, for students at level 1 and for schools at level 2. Let y_{ist} denote the score for student i in school s on test t in the battery of exams. A multilevel model with fixed effects $\boldsymbol{\beta}$ for explanatory variables and random effects $\{u_s\}$ for schools and $\{v_{is}\}$ for students has the form

$$y_{ist} = \boldsymbol{x}_{ist}\boldsymbol{\beta} + u_s + v_{is} + \epsilon_{ist}.$$

The explanatory variables $\boldsymbol{x}$ might include student demographic characteristics and past performance such as scores from other exams, as well as school-level variables. We assume that the random effects u_s and v_{is} and the errors ϵ_{ist} are independent with distributions $N(0, \sigma_u^2)$, $N(0, \sigma_v^2)$, and $N(0, \sigma_\epsilon^2)$ having unknown variances. The level 1 random effects $\{v_{is}\}$ account for variability among students in characteristics that are not fully captured in $\boldsymbol{x}$, such as perhaps their ability and achievement motivation. The level 2 random effects $\{u_s\}$ account for variability among schools from unmeasured variables, such as perhaps the quality of the teachers. The model could even have additional levels, such as if classrooms are selected within schools or if sampled schools come from a sample of counties.

Although here the random effects enter at two levels, the linear predictor shows that the model actually has three levels: a particular observation is affected (beyond the influence of the explanatory variables) by random variability among schools, among students within the school, and among exams taken by the student. The total variability, having three variance components, is $\sigma_u^2 + \sigma_v^2 + \sigma_\epsilon^2$. You can verify that the intraclass correlation between scores on different exams for a student and the intraclass correlation between scores on a particular exam for pairs of students in the same school are

$$\text{corr}(y_{ist}, y_{ist'}) = \frac{\sigma_u^2 + \sigma_v^2}{\sigma_u^2 + \sigma_v^2 + \sigma_\epsilon^2}, \quad \text{corr}(y_{ist}, y_{i'st}) = \frac{\sigma_u^2}{\sigma_u^2 + \sigma_v^2 + \sigma_\epsilon^2}.$$

The correlations increase as the variability σ_u^2 among schools increases. If this variability is much less than the variability σ_v^2 among students, then the within-student correlation is much larger than the within-school correlation.

9.2.3 Example: Smoking Prevention and Cessation Study

Hedeker and Gibbons (2006, p. 9) analyzed data from a study[5] of the efficacy of two programs for discouraging young people from starting or continuing to smoke. The study compared four groups, defined by a 2x2 factorial design according to whether a student was exposed to a school-based curriculum (SC; 1 = yes, 0 = no) and a television-based prevention program (TV; 1 = yes, 0 = no). The subjects were 1600 seventh-grade students from 135 classrooms in 28 Los Angeles schools. The schools were randomly assigned to the four intervention conditions. The response variable was a tobacco and health knowledge (THK) scale, measured at the end of the study. This variable was also observed at the beginning of the study, and that measure (PTHK = Pre-THK) was used as a covariate. THK took values between 0 and 7, with $\bar{y} = 2.66$ and $s_y = 1.38$. The data, shown partly in Table 9.3, are available in the file `Smoking.dat` at the text website.

Table 9.3 Part of Smoking Prevention and Cessation Data File

School	Class	SC	TV	PTHK	THK
403	403101	1	0	2	3
403	403101	1	0	4	4
...					
515	515113	0	0	3	3

Complete data (file `Smoking.dat`), courtesy of Don Hedeker, are at `www.stat.ufl.edu/~aa/glm/data`.

Let y_{ics} denote the follow-up THK score for student i within classroom c in school s. We fitted the multilevel model

$$y_{ics} = \beta_0 + \beta_1 \text{PTHK}_{ics} + \beta_2 \text{SC}_{ics} + \beta_3 \text{TV}_{ics} + u_s + v_{cs} + \epsilon_{ics},$$

where $u_s \sim N(0, \sigma_u^2)$, $v_{cs} \sim N(0, \sigma_v^2)$, and $\epsilon_{ics} \sim N(0, \sigma_\epsilon^2)$. The estimated fixed effects do not exhibit a significant TV effect. The SC effect (0.47) is highly statistically

[5]See also the articles by D. Hedeker et al. (1994) *J. Consult. Clin. Psychol.* **62**: 757–765, and by B. R. Flay et al. (1995), *Prev. Med.* **24**: 29–40.

significant but not large in practical terms. Adding an interaction between SC and TV does not improve the fit.

```
------------------------------------------------------------------------
> library(lme4) # Doug Bates's linear mixed models package
> attach(Smoking)
> Smoking # data in file Smoking.dat at www.stat.ufl.edu/~aa/glm/data
   school   class SC TV PTHK  y
1     403  403101  1  0    2  3
2     403  403101  1  0    4  4
...
1600  515  515113  0  0    3  3

> fit <- lmer(y ~ PTHK + SC + TV + (1|school) + (1|class))
> summary(fit) # school and classroom random intercepts
Random effects: # These are "REML" variance estimates; see Sec. 9.3.3
 Groups   Name        Variance  Std.Dev.
 class    (Intercept)  0.0685    0.2618
 school   (Intercept)  0.0393    0.1981
 Residual              1.6011    1.2653
Number of obs: 1600, groups: class, 135; school, 28
Fixed effects:
             Estimate  Std. Error  t value
(Intercept)   1.7849       0.1129   15.803
PTHK          0.3052       0.0259   11.786
SC            0.4715       0.1133    4.161
TV            0.0196       0.1133    0.173
# use ranef(fit) to predict random effects at levels of school, class
# use predict(fit) to get predicted values for the observations
------------------------------------------------------------------------
```

The residual standard deviation $\hat{\sigma}_\epsilon = 1.265$ is not much less than the marginal response standard deviation (1.383), partly reflecting that the correlation is not strong (0.289) between pre-THK and the follow-up THK. The variance component estimates indicate more variability among classrooms within schools than among schools. The estimated intraclass correlation between responses of two students in the same classroom,

$$\frac{\hat{\sigma}_u^2 + \hat{\sigma}_v^2}{\hat{\sigma}_u^2 + \hat{\sigma}_v^2 + \hat{\sigma}_\epsilon^2} = \frac{0.039 + 0.069}{0.039 + 0.069 + 1.601} = 0.063,$$

and the estimated intraclass correlation between responses of two students in the same school but different classrooms,

$$\frac{\hat{\sigma}_u^2}{\hat{\sigma}_u^2 + \hat{\sigma}_v^2 + \hat{\sigma}_\epsilon^2} = \frac{0.039}{0.039 + 0.069 + 1.601} = 0.023$$

are quite modest.

Suppose we ignored the clustering of observations in classrooms and schools and treated the 1600 observations as independent by fitting the ordinary normal linear model

$$y_{ics} = \beta_0 + \beta_1 \text{PTHK}_{ics} + \beta_2 \text{SC}_{ics} + \beta_3 \text{TV}_{ics} + \epsilon_{ics}.$$

The estimated fixed effects are similar to those in the multilevel model, but the *SE* values are quite dramatically underestimated for the between-subjects effects (SC and TV). This might seem surprising, given the small within-classroom and within-school correlations. However, the relative sizes of the *SE* values are also affected by the cluster sizes, with the difference tending to increase as the cluster sizes increase.

```
-------------------------------------------------------------------------
> summary(lm(y ~ PTHK + SC + TV))
              Estimate  Std. Error  t value  Pr(>|t|)
(Intercept)    1.7373      0.0787   22.088    < 2e-16
PTHK           0.3252      0.0259   12.561    < 2e-16
SC             0.4799      0.0653    7.350   3.15e-13
TV             0.0453      0.0652    0.696      0.487
---
Residual standard error: 1.303 on 1596 degrees of freedom # This is s
-------------------------------------------------------------------------
```

9.2.4 Linear Models with Random Intercept and Random Slope

Random-intercept models generalize to allow slopes also to be random. Longitudinal studies often use such models to describe trajectories of change in a response over time.

To illustrate, we describe a simplified version of models from Gueorguieva and Krystal (2004), who analyzed data from a clinical trial for the effect of using a drug (naltrexone) instead of placebo in treating 627 veterans suffering from chronic alcohol dependence. The response variable was a financial satisfaction score, observed initially and then after 4, 26, 52, and 78 weeks. Let y_{ij} be the response of subject i at observation time j, and let x_i be a treatment indicator of whether the veteran receives the drug (1 = yes, 0 = no). Graphical inspection showed approximately a positive linear trend of the response when plotted against $t_j = \log(\text{week number} + 1)$, for each group. A possible model is

$$y_{ij} = (\beta_0 + u_{i1}) + (\beta_1 + u_{i2})t_j + \beta_2 x_i + \beta_3 t_j x_i + \epsilon_{ij}.$$

For each i, $E(y_{ij} \mid \boldsymbol{u}_i)$ has a linear trend in the time metric. Considered for all i, the intercepts vary around a mean intercept of β_0 for the placebo group and $(\beta_0 + \beta_2)$ for the drug group. The slopes of the linear trends vary around a mean of β_1 for the placebo group and $(\beta_1 + \beta_3)$ for the drug group. A bivariate normal distribution for (u_{i1}, u_{i2}) can permit u_{i1} and u_{i2} to be correlated, with possibly different variances.

The correlation could be negative, for example, if subjects who tend to have a high response at the initial observation tend to increase more slowly. Figure 9.1 shows a plot portraying the model.

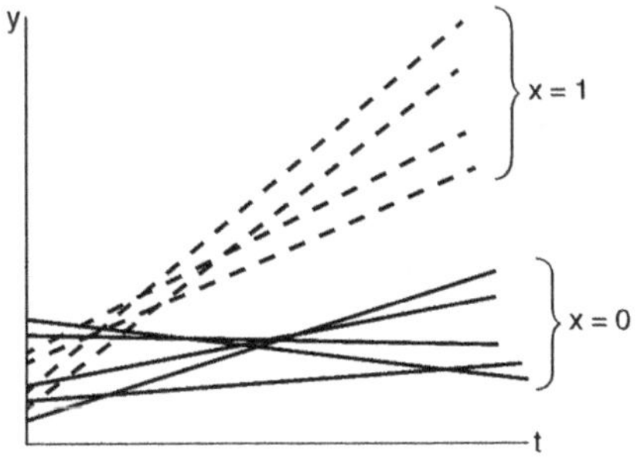

Figure 9.1 Portrayal of linear growth curve model, with subject-specific intercepts and slopes for each treatment group.

9.2.5 Models Generalize to Allow Correlated Errors

The normal linear mixed model $\boldsymbol{y}_i = \boldsymbol{X}_i\boldsymbol{\beta} + \boldsymbol{Z}_i\boldsymbol{u}_i + \boldsymbol{\epsilon}_i$ with $\boldsymbol{\epsilon}_i \sim N(\boldsymbol{0}, \sigma_\epsilon^2\boldsymbol{I})$, assumes that $(y_{i1}, \dots, y_{id})$ are conditionally independent, given $\boldsymbol{u}_i$. Linear mixed models can allow more general structures for correlated observations. When subjects are observed spatially or over time in a longitudinal study, it is often more realistic to permit $(\epsilon_{i1}, \epsilon_{i2}, \dots, \epsilon_{id})$ to be correlated, replacing the covariance matrix $\sigma_\epsilon^2\boldsymbol{I}$ by a matrix $\boldsymbol{R}$ that is non-diagonal. Then,

$$\text{var}(\boldsymbol{y}_i) = \boldsymbol{Z}_i\boldsymbol{\Sigma}_u\boldsymbol{Z}_i^{\text{T}} + \boldsymbol{R}.$$

In the clinical trial analysis just mentioned, the authors fitted models that take $(\epsilon_{i1}, \dots, \epsilon_{i5})$ to have an *autoregressive* correlation structure[6]: $\text{corr}(\epsilon_{ij}, \epsilon_{ik}) = \rho^{|j-k|}$. This is often sensible for observations over time, when one expects the correlation to diminish with increasing time distance between observations. Substantial correlation between errors within a cluster, given the random effect, affects *SE* values for within-cluster effects.

To illustrate, we show two model fits for a dataset from a pharmaceutical clinical trial analyzed by Littell et al. (2000). In the study, 24 patients were randomly assigned to each of three treatment groups (drug A, drug B, placebo) and compared on a measure of respiratory ability, FEV = forced expiratory volume in 1 second (in liters). The study observed FEV for a baseline measurement and then for each of 8 hours

[6]One can replace the index j by an actual time t_j in the autoregressive structure.

Table 9.4 Part of FEV Clinical Trial Data File

Observation	Patient	Baseline	Drug	Hour	fev
1	01	2.46	a	1	2.68
2	02	3.50	a	1	3.95
3	03	1.96	a	1	2.28
...					
25	01	2.46	a	2	2.76
...					
576	72	2.88	p	8	2.76

Complete data (file `FEV2.dat`), courtesy of Ramon Littell, are at `www.stat.ufl.edu/~aa/glm/data`.

after the drug was administered. The data file has the form shown in Table 9.4, with eight response observations per patient.

We show results here for a simple model with a linear trend effect of time after administering the drug and no interaction between time and the drug; you can develop better models in Exercise 9.34. For the FEV observation at hour j for subject i, we assume that

$$y_{ij} = \beta_0 + \beta_1 \text{baseline}_i + \beta_2 b_i + \beta_3 p_i + \beta_4 j + u_i + \epsilon_{ij},$$

where $b_i = 1$ for drug B and 0 otherwise and $p_i = 1$ for placebo and 0 otherwise. We fit a model with independent within-patient errors and a model with autoregressive error structure. The models have similar estimated fixed effects, but the *SE* values differ for the within-patient explanatory variable (hour). Within-cluster *SE* values tend to be too small under compound-symmetry assumptions (and likewise under more general *sphericity* assumptions in repeated-measures ANOVA) when the assumption is badly violated. The log-likelihood and AIC are substantially better for the model permitting autocorrelated errors.

```
-----------------------------------------------------------------------
> library(nlme) # has lme function that can have error structure
> attach(FEV2)
> summary(lme(fev ~ baseline + factor(drug) + hour, random = ~1|patient))
       AIC      BIC    logLik
    388.91   419.35   -187.46
Random effects:
       (Intercept)   Residual
StdDev:     0.4527     0.2717 # random intercept estimated std.dev.=0.4527
Fixed effects: fev ~ baseline + factor(drug) + hour
                 Value  Std.Error  DF  t-value  p-value
(Intercept)     1.0492     0.2922 503   3.5911   0.0004
baseline        0.9029     0.1033  68   8.7417   0.0000
factor(drug)b   0.2259     0.1336  68   1.6907   0.0955 # relative to
```

```
factor(drug)p -0.2815      0.1336   68  -2.1065    0.0389 # drug a
hour          -0.0746      0.0049  503 -15.0950    0.0000

> summary(lme(fev ~ baseline + factor(drug) + hour,
          random =~1|patient, correlation = corAR1(form = ~1|patient)))
       AIC      BIC  logLik
    243.35   278.13 -113.67
Random effects:
       (Intercept)  Residual
StdDev:     0.4075    0.3355 # random intercept estimated std.dev.=0.4075
Correlation Structure: AR(1)
0.6481 # autoregressive correlation estimate for 8 within-patient errors
Fixed effects: fev ~ baseline + factor(drug) + hour
                    Value  Std.Error  DF  t-value    p-value
(Intercept)        1.0723     0.2914 503    3.680     0.0003
baseline           0.8918     0.1026  68    8.694     0.0000
factor(drug)b      0.2130     0.1327  68    1.605     0.1131
factor(drug)p     -0.3142     0.1327  68   -2.367     0.0208
hour              -0.0691     0.0077 503   -8.979     0.0000 # larger SE
-----------------------------------------------------------------------
```

Other correlation structures besides autoregressive are possible in $\text{var}(\epsilon_i)$. Most generally, an *unstructured* correlation form has $\binom{d}{2}$ unspecified correlations. When d is large, however, this has many additional parameters. Yet, with increasing time separation, correlations sometimes do not die out as quickly as the autoregressive structure implies. An alternative, between these two in generality, is the *Toeplitz* structure. It assumes $\text{corr}(\epsilon_{ij}, \epsilon_{ik}) = \rho_{|j-k|}$, permitting $(d-1)$ separate correlations.

Just as it can be helpful to allow more complex correlation structure than independent within-cluster errors, it also can be helpful to allow different groups or times to have different variances. Exercise 9.35 shows an example.

For a particular dataset, how does one decide whether a certain covariance structure $\boldsymbol{R}$ for $\text{var}(\boldsymbol{\epsilon}_i)$ is better than $\sigma_\epsilon^2\boldsymbol{I}$? As usual, summary fit measures such as AIC are useful for comparing models. It is also sensible to compare the model-fitted variances and $\text{corr}(y_{ij}, y_{ik})$ for all (j, k) with the sample values. See Diggle et al. (2002, Chapter 5), Littell et al. (2000), and Verbeke and Molenberghs (2000, Chapter 9, 10) for examples.

9.3 FITTING AND PREDICTION FOR NORMAL LINEAR MIXED MODELS

In this section we present model fitting and inference for normal linear mixed models. Unless otherwise stated, we use the general error covariance structure $\boldsymbol{\epsilon}_i \sim N(\boldsymbol{0}, \boldsymbol{R})$. More generally, we could let $\boldsymbol{R}$ depend on i, to accommodate structure such as missing data or autoregressive correlations with subjects observed at different times.

9.3.1 Maximum Likelihood Model Fitting

We combine the n vectors of observations $\boldsymbol{y}_i$ into an $nd \times 1$ vector $\boldsymbol{y}$ and express the linear mixed model (9.6) simultaneously for all n observations as

$$\boldsymbol{y} = \boldsymbol{X}\boldsymbol{\beta} + \boldsymbol{Z}\boldsymbol{u} + \boldsymbol{\epsilon},$$

where

$$\boldsymbol{X} = \begin{pmatrix} \boldsymbol{X}_1 \\ \vdots \\ \boldsymbol{X}_n \end{pmatrix}, \quad \boldsymbol{Z} = \begin{pmatrix} \boldsymbol{Z}_1 & \boldsymbol{0} & \cdots & \boldsymbol{0} \\ \boldsymbol{0} & \boldsymbol{Z}_2 & \cdots & \boldsymbol{0} \\ \boldsymbol{0} & \boldsymbol{0} & \ddots & \boldsymbol{0} \\ \boldsymbol{0} & \boldsymbol{0} & \cdots & \boldsymbol{Z}_n \end{pmatrix}, \quad \boldsymbol{u} = \begin{pmatrix} \boldsymbol{u}_1 \\ \vdots \\ \boldsymbol{u}_n \end{pmatrix}, \quad \boldsymbol{\epsilon} = \begin{pmatrix} \boldsymbol{\epsilon}_1 \\ \vdots \\ \boldsymbol{\epsilon}_n \end{pmatrix}$$

and

$$\boldsymbol{\Sigma_u} = \mathrm{var}(\boldsymbol{u}) = \begin{pmatrix} \boldsymbol{\Sigma}_u & \boldsymbol{0} & \cdots & \boldsymbol{0} \\ \boldsymbol{0} & \boldsymbol{\Sigma}_u & \cdots & \boldsymbol{0} \\ \boldsymbol{0} & \boldsymbol{0} & \ddots & \boldsymbol{0} \\ \boldsymbol{0} & \boldsymbol{0} & \cdots & \boldsymbol{\Sigma}_u \end{pmatrix}, \quad \boldsymbol{R}_\epsilon = \mathrm{var}(\boldsymbol{\epsilon}) = \begin{pmatrix} \boldsymbol{R} & \boldsymbol{0} & \cdots & \boldsymbol{0} \\ \boldsymbol{0} & \boldsymbol{R} & \cdots & \boldsymbol{0} \\ \boldsymbol{0} & \boldsymbol{0} & \ddots & \boldsymbol{0} \\ \boldsymbol{0} & \boldsymbol{0} & \cdots & \boldsymbol{R} \end{pmatrix}.$$

Marginally, assuming independent normal errors and random effects with $\boldsymbol{0}$ means,

$$\boldsymbol{y} \sim N(\boldsymbol{X\beta}, \boldsymbol{Z\Sigma_u Z}^{\mathrm{T}} + \boldsymbol{R}_\epsilon). \tag{9.9}$$

Let $\boldsymbol{V} = \boldsymbol{Z\Sigma_u Z}^{\mathrm{T}} + \boldsymbol{R}_\epsilon$. Based on the multivariate normal pdf shown in Section 3.1.1 and ignoring the constant term, the log-likelihood function for the model is

$$L(\boldsymbol{\beta}, \boldsymbol{V}) = -\frac{1}{2}\log|\boldsymbol{V}| - \frac{1}{2}(\boldsymbol{y} - \boldsymbol{X\beta})^{\mathrm{T}}\boldsymbol{V}^{-1}(\boldsymbol{y} - \boldsymbol{X\beta}).$$

If $\boldsymbol{V}$ is known, then maximizing $L(\boldsymbol{\beta}, \boldsymbol{V})$ with respect to $\boldsymbol{\beta}$ yields the generalized least squares solution (2.7.2),

$$\tilde{\boldsymbol{\beta}} = \tilde{\boldsymbol{\beta}}(\boldsymbol{V}) = \left(\boldsymbol{X}^{\mathrm{T}}\boldsymbol{V}^{-1}\boldsymbol{X}\right)^{-1}\boldsymbol{X}^{\mathrm{T}}\boldsymbol{V}^{-1}\boldsymbol{y}. \tag{9.10}$$

Now $\boldsymbol{V}$ has block-diagonal form with $d \times d$ blocks $\boldsymbol{V}_i = \boldsymbol{Z}_i\boldsymbol{\Sigma}_u\boldsymbol{Z}_i^{\mathrm{T}} + \boldsymbol{R}$, so this esimator has the form

$$\tilde{\boldsymbol{\beta}} = \left(\sum_{i=1}^{n} \boldsymbol{X}_i^{\mathrm{T}}\boldsymbol{V}_i^{-1}\boldsymbol{X}_i\right)^{-1} \sum_{i=1}^{n} \boldsymbol{X}_i^{\mathrm{T}}\boldsymbol{V}_i^{-1}\boldsymbol{y}_i.$$

Since $E(y_i) = X_i\beta$, this estimator is unbiased. Since it is a linear function of y, it has a normal distribution with

$$\text{var}(\tilde{\beta}) = \left(\sum_{i=1}^{n} X_i^{\mathrm{T}} V_i^{-1} X_i \right)^{-1}.$$

In practice, V is rarely known. We discuss its estimation in Section 9.3.3. Substituting an estimate $\hat{V}$ in $\tilde{\beta}$ yields

$$\hat{\beta} = \tilde{\beta}(\hat{V}) = \left(\sum_{i=1}^{n} X_i^{\mathrm{T}} \hat{V}_i^{-1} X_i \right)^{-1} \sum_{i=1}^{n} X_i^{\mathrm{T}} \hat{V}_i^{-1} y_i.$$

Under regularity conditions, its asymptotic distribution is the same as the normal distribution that applies when V is known. Inference about fixed effects can use the usual methods, such as likelihood-ratio tests and Wald confidence intervals.

9.3.2 Best Linear Unbiased Prediction of Random Effects

After we have obtained $\hat{\beta}$ and $\hat{V}$, we can predict values of the random effects $\{u_i\}$. We say "predict" rather than "estimate" because u_i is a random effect rather than a parameter.

Prediction of random effects is useful in various sorts of applications. In multilevel models of performance, they are used to rank institutions such as schools and hospitals. For example, a study might predict school random effects $\{u_s\}$ to evaluate whether some schools are unusually high or unusually low in student achievement exam scores, adjusted for the fixed covariates[7]. Prediction of random effects is also useful in *small-area estimation*, which involves estimating characteristics for many geographical areas when each has relatively few observations. Examples are county-specific estimates of mean family income and of the proportion of unemployed adult residents. In a national or statewide survey, many counties may have few, if any, observations. Then sample means and sample proportions may poorly estimate the true countywide values. Models with random effects that treat each county as a cluster can provide improved estimates, because those estimates borrow from the whole, using all the data rather than only the county-specific data.

How do we characterize a predictor $\tilde{u}_i$ of a random effect u_i? A predictor $\tilde{u}_i$ is called the *best linear unbiased predictor* (BLUP) of u_i if $\tilde{u}_i$ is linear in y, $E(\tilde{u}_i) = 0$ (the value for $E(u_i)$), and for any linear combination $a^{\mathrm{T}} u_i$ of the random effects, $E(a^{\mathrm{T}} \tilde{u}_i - a^{\mathrm{T}} u_i)^2$ is minimized, among all such linear unbiased predictors.

[7]Intervals are more informative than predictions, which may suggest differences between institutions that merely reflect random variability. See Goldstein (2014) and Goldstein and Spiegelhalter (1996) for issues in constructing and interpreting "league tables."

For the normal linear mixed model $y = X\beta + Zu + \epsilon$ with $\text{var}(u) = \Sigma_u$ and $\text{var}(\epsilon) = R_\epsilon$, $\text{cov}(y, u) = \text{cov}(Zu + \epsilon, u) = Z\Sigma_u$, so

$$\begin{pmatrix} y \\ u \end{pmatrix} \sim N\left[\begin{pmatrix} X\beta \\ 0 \end{pmatrix}, \begin{pmatrix} Z\Sigma_u Z^T + R_\epsilon & Z\Sigma_u \\ \Sigma_u Z^T & \Sigma_u \end{pmatrix}\right].$$

To maximize the joint normal density of y and u with respect to β and u, with known variances, we differentiate the log-density with respect to β and u to obtain normal-like equations

$$\begin{pmatrix} X^T R_\epsilon^{-1} X & X^T R_\epsilon^{-1} Z \\ Z^T R_\epsilon^{-1} X & \Sigma_u^{-1} + Z^T R_\epsilon^{-1} Z \end{pmatrix} \begin{pmatrix} \tilde{\beta} \\ \tilde{u} \end{pmatrix} = \begin{pmatrix} X^T R_\epsilon^{-1} y \\ Z^T R_\epsilon^{-1} y \end{pmatrix}.$$

These equations are often referred to as *Henderson's mixed-model equations*, because they are due to the statistician Charles Henderson, who developed them and BLUP for applications in animal breeding (see Henderson 1975). The solution $\tilde{u}$ to these equations is the BLUP of u and is the best linear unbiased estimator of $E(u \mid y)$, the *posterior mean* given the data. These results are true even without the normality assumption. The solution $\tilde{\beta}$ is identical to the generalized least squares solution (9.10).

Applying the expression in Section 3.1.1 for conditional distributions of multivariate normal variables to the expression just given for the joint distribution of y and u,

$$E(u \mid y) = \Sigma_u Z^T V^{-1}(y - X\beta)$$

for $V = Z\Sigma_u Z^T + R_\epsilon$. With known variances, the BLUP $\tilde{u}$ of u is therefore

$$\tilde{u} = \Sigma_u Z^T V^{-1}(y - X\tilde{\beta}) = \Sigma_u Z^T V^{-1}[I - X(X^T V^{-1} X)^{-1} X^T V^{-1}]y.$$

The fitting process uses data from all the clusters to estimate characteristics in any given one. Bayesian approaches naturally do this, through the impact of the prior distribution on Bayes estimates, as shown in an example in Section 10.4.3. The prediction $\tilde{u}$ is a weighted combination of $\mathbf{0}$ and the generalized least squares estimate based on treating u as a fixed effect[8].

We illustrate with the model for the balanced one-way layout,

$$y_{ij} = \beta_0 + u_i + \epsilon_{ij}, \quad i = 1, \ldots, c, \quad j = 1, \ldots, n,$$

[8]See Laird and Ware (1982) and Robinson (1991), and the following example for the one-way layout.

treating the factor as random instead of fixed, and with $\text{var}(\boldsymbol{\epsilon}_i) = \sigma_\epsilon^2 \boldsymbol{I}_n$. This model has the form $\boldsymbol{y} = \boldsymbol{X\beta} + \boldsymbol{Zu} + \boldsymbol{\epsilon}$ with

$$\boldsymbol{X} = \begin{pmatrix} \mathbf{1}_n \\ \vdots \\ \mathbf{1}_n \end{pmatrix}, \quad \boldsymbol{\beta} = \beta_0, \quad \boldsymbol{Z} = \begin{pmatrix} \mathbf{1}_n & \mathbf{0} & \cdots & \mathbf{0} \\ \mathbf{0} & \mathbf{1}_n & \cdots & \mathbf{0} \\ \mathbf{0} & \mathbf{0} & \ddots & \mathbf{0} \\ \mathbf{0} & \mathbf{0} & \cdots & \mathbf{1}_n \end{pmatrix}, \quad \boldsymbol{u} = \begin{pmatrix} u_1 \\ \vdots \\ u_c \end{pmatrix},$$

with $\boldsymbol{\Sigma_u} = \text{var}(\boldsymbol{u}) = \sigma_u^2 \boldsymbol{I}_c$ and $\boldsymbol{R}_\epsilon = \text{var}(\boldsymbol{\epsilon}) = \sigma_\epsilon^2 \boldsymbol{I}_{nc}$. For this structure, you can verify that $\hat{\beta}_0 = \bar{y}$ and

$$\tilde{u}_i = \frac{\sigma_u^2}{\sigma_u^2 + \sigma_\epsilon^2/n}(\bar{y}_i - \bar{y}), \quad \hat{\beta}_0 + \tilde{u}_i = \left(\frac{\sigma_u^2}{\sigma_u^2 + \sigma_\epsilon^2/n}\right)\bar{y}_i + \left(\frac{\sigma_\epsilon^2/n}{\sigma_u^2 + \sigma_\epsilon^2/n}\right)\bar{y}.$$

The prediction $\tilde{u}_i$ is a weighted average of 0 and the least squares estimate $(\bar{y}_i - \bar{y})$ from treating $\boldsymbol{u}$ as a fixed effect, with the weight $\sigma_u^2/(\sigma_u^2 + \sigma_\epsilon^2/n)$ for $(\bar{y}_i - \bar{y})$ increasing as n increases. The estimated mean $\hat{\beta}_0 + \tilde{u}_i$ for group i is a weighted average of $\bar{y}_i$ and $\bar{y}$, with greater weight for $\bar{y}_i$ as n increases. The solution has a form we will obtain with the Bayesian approach in Sections 10.2.3 and 10.4.2.

From Henderson (1975), $\tilde{\boldsymbol{\beta}}$ and $\tilde{\boldsymbol{u}}$ are uncorrelated, and when $\boldsymbol{X}$ has full rank,

$$\text{var}\begin{pmatrix} \tilde{\boldsymbol{\beta}} - \boldsymbol{\beta} \\ \tilde{\boldsymbol{u}} - \boldsymbol{u} \end{pmatrix} = \begin{pmatrix} \boldsymbol{X}^{\mathrm{T}}\boldsymbol{R}_\epsilon^{-1}\boldsymbol{X} & \boldsymbol{X}^{\mathrm{T}}\boldsymbol{R}_\epsilon^{-1}\boldsymbol{Z} \\ \boldsymbol{Z}^{\mathrm{T}}\boldsymbol{R}_\epsilon^{-1}\boldsymbol{X} & \boldsymbol{\Sigma}_{\boldsymbol{u}}^{-1} + \boldsymbol{Z}^{\mathrm{T}}\boldsymbol{R}_\epsilon^{-1}\boldsymbol{Z} \end{pmatrix}^{-1}.$$

In practice, we must estimate $\tilde{\boldsymbol{u}}$ and $\text{var}(\tilde{\boldsymbol{u}})$ by substituting estimates for the unknown variances, providing an *empirical BLUP*. We make similar substitutions to predict linear combinations of fixed and random effects and to estimate standard errors of the predictions.

9.3.3 Estimating Variance Components: REML

An alternative to ordinary ML for estimating covariance matrices and variance components of random effects adjusts for estimating $\boldsymbol{\beta}$ while estimating $\boldsymbol{V}$. This yields a *residual ML* (sometimes called *restricted ML*) estimate, abbreviated REML. Consider the marginal representation (9.9) for the normal linear mixed model, for which $E(\boldsymbol{y}) = \boldsymbol{X\beta}$. The REML approach uses a linear transformation $\boldsymbol{Ly}$ of the data satisfying $\boldsymbol{LX} = \mathbf{0}$ and so having $E(\boldsymbol{Ly}) = \mathbf{0}$. Then $\boldsymbol{Ly} = \boldsymbol{L\epsilon}^*$ with $\boldsymbol{\epsilon}^* = (\boldsymbol{Zu} + \boldsymbol{\epsilon})$ are *error contrasts* whose distribution does not depend on the fixed effects $\boldsymbol{\beta}$. The REML estimates, which are obtained from maximizing the likelihood for the distribution of $\boldsymbol{Ly}$, do not depend on the value of $\boldsymbol{\beta}$.

With projection matrix $\boldsymbol{P}_X$ for the model space $C(\boldsymbol{X})$, $\boldsymbol{L} = \boldsymbol{I} - \boldsymbol{P}_X$ satisfies $\boldsymbol{LX} = \mathbf{0}$. Then the error contrasts are $\boldsymbol{Ly} = (\boldsymbol{I} - \boldsymbol{P}_X)\boldsymbol{y} = \boldsymbol{y} - \hat{\boldsymbol{\mu}}$, the residuals. This is the reason for the name "residual ML."

We illustrate with the null model having n independent observations from $N(\mu, \sigma^2)$. Then $\boldsymbol{L}\boldsymbol{y} = (\boldsymbol{y} - \bar{\boldsymbol{y}})$ has a multivariate normal distribution with mean $\boldsymbol{0}$ and covariance matrix depending on σ^2 but not μ. For this distribution treated as a likelihood function, the estimate of σ^2 that maximizes it is $s^2 = [\sum_i (y_i - \bar{y})^2]/(n-1)$. More generally, for a normal linear model with projection matrix $\boldsymbol{P}_X$ of rank r, the REML estimator of σ^2 is the unbiased estimator

$$s^2 = \frac{\boldsymbol{y}^{\mathrm{T}}(\boldsymbol{I} - \boldsymbol{P}_X)\boldsymbol{y}}{n-r} = \frac{\sum_{i=1}^{n}(y_i - \hat{\mu}_i)^2}{n-r}$$

found in Section 2.4.1. The ordinary ML estimator has denominator n and is biased.

The REML estimates for more complex models are solutions of likelihood-like equations, found using methods such as Newton–Raphson or Fisher scoring. See Harville (1977) for details.

9.4 BINOMIAL AND POISSON GLMMS

To illustrate GLMMs with nonlinear link functions, we now present models for clustered binary data and for clustered count data.

9.4.1 Logistic-Normal Models for a Binary Response

The binary matched-pairs model (9.5) with normally distributed random intercept and logit link function is an example of a class of models for binary data called *logistic-normal models*. The model form is

$$\text{logit}[P(y_{ij} = 1 \mid \boldsymbol{u}_i)] = \boldsymbol{x}_{ij}\boldsymbol{\beta} + \boldsymbol{z}_{ij}\boldsymbol{u}_i, \tag{9.11}$$

where $\{\boldsymbol{u}_i\}$ are independent $N(\boldsymbol{0}, \boldsymbol{\Sigma}_u)$ variates.

To illustrate, the random-intercept model (9.5) for binary matched pairs extends to a model for $d > 2$ observations in each cluster,

$$\text{logit}[P(y_{ij} = 1 \mid u_i)] = \beta_0 + \beta_j + u_i,$$

with an identifiability constraint such as $\beta_1 = 0$. Early applications of this GLMM were in psychometrics, for describing (correct, incorrect) outcomes for d questions on an examination. The probability $P(y_{ij} = 1 \mid u_i)$ that subject i makes the correct response on question j depends on the overall ability of subject i, characterized by u_i, and on the easiness of question j, characterized by β_j. Such models are called *item-response models*. This particular model with logit link is called the *Rasch model* (Rasch 1961). In estimating $\{\beta_j\}$, Rasch treated $\{u_i\}$ as fixed effects and used conditional ML, as outlined in Exercise 9.19 for matched pairs. Later authors used the normal random effects approach and often the probit link (Bock and Aitkin 1981).

With that approach, the model assumes a latent variable u such that for each possible sequence $(a_1, \ldots, a_d)$ of response outcomes and each value u^* of u,

$$P(y_{i1} = a_1, \ldots, y_{id} = a_d \mid u_i = u^*) = P(y_{i1} = a_1 \mid u_i = u^*) \cdots P(y_{id} = a_d \mid u_i = u^*).$$

This model structure also applies with other types of latent variable. With a categorical latent variable, the model is called a *latent class model* (Lazarsfeld and Henry 1968). This model treats a 2^d contingency table as a finite mixture of unobserved tables generated under a conditional independence structure.

For the general binary random-intercept model in which the inverse link function is an arbitrary cdf F, model-fitting usually treats y_{ij} and y_{ik} as conditionally independent, given u_i. Marginally, for $j \neq k$,

$$\begin{aligned}\text{cov}(y_{ij}, y_{ik}) &= E[\text{cov}(y_{ij}, y_{ik} \mid u_i)] + \text{cov}[E(y_{ij} \mid u_i), E(y_{ik} \mid u_i)] \\ &= 0 + \text{cov}[F(\boldsymbol{x}_{ij}\boldsymbol{\beta} + u_i), F(\boldsymbol{x}_{ik}\boldsymbol{\beta} + u_i)].\end{aligned}$$

The functions in the last covariance term are both monotone increasing in u_i, and hence are non-negatively correlated. Thus, the model implies that $\text{corr}(y_{ij}, y_{ik}) \geq 0$.

Effects in binary random-effects models tend to be larger than those in corresponding marginal models. To show this, we first consider the probit analog of the random-intercept version of the logistic-normal model (9.11),

$$\Phi^{-1}[P(y_{ij} = 1 \mid u_i)] = \boldsymbol{x}_{ij}\boldsymbol{\beta} + u_i.$$

Let z denote a standard normal variate, and let $f(u; \sigma_u^2)$ denote the $N(0, \sigma_u^2)$ pdf of u_i. The corresponding marginal model satisfies

$$P(y_{ij} = 1) = \int P(y_{ij} = 1 \mid u_i) f\left(u_i; \sigma_u^2\right) du_i = \int P(z - u_i \leq \boldsymbol{x}_{ij}\boldsymbol{\beta}) f\left(u_i; \sigma_u^2\right) du_i.$$

Since $z - u_i \sim N(0, 1 + \sigma_u^2)$, we have $(z - u_i)/\sqrt{1 + \sigma_u^2} \sim N(0, 1)$, and

$$P(y_{ij} = 1) = \Phi\left(\boldsymbol{x}_{ij}\boldsymbol{\beta}/\sqrt{1 + \sigma_u^2}\right).$$

The implied marginal model is also a probit model, but the effects equal those from the GLMM divided by $\sqrt{1 + \sigma_u^2}$. The discrepancy increases as σ_u increases. With the logistic-normal model, the implied marginal model is not exactly of logistic form. An approximate relation exists, based on the similarity of the normal and logistic cdfs. The marginal model is approximately a logistic model with effects $\boldsymbol{\beta}/\sqrt{1 + (\sigma_u/c)^2}$ for $c \approx 1.7$.

Figure 9.2 illustrates why the marginal effect is smaller than the subject-specific effect in the binary GLMM. For a single explanatory variable x, the figure shows

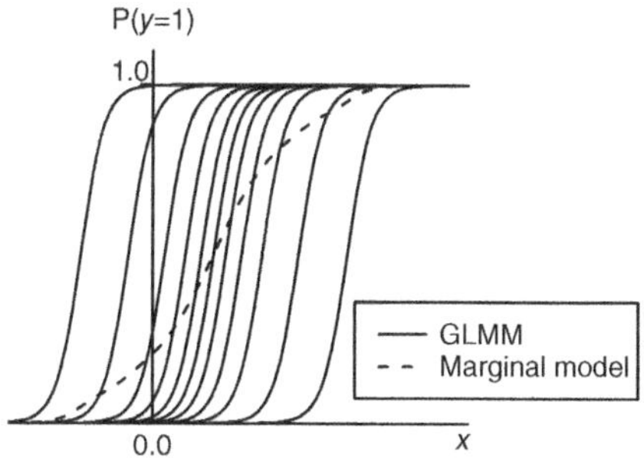

Figure 9.2 Logistic random-intercept GLMM, showing its subject-specific curves and the population-averaged marginal curve obtained at each x by averaging the subject-specific probabilities.

subject-specific curves for $P(y_{ij} = 1 \mid u_i)$ for several subjects when considerable heterogeneity exists. This corresponds to a relatively large var(u_i). At any fixed value of x, variability occurs in the conditional means, $E(y_{ij} \mid u_i) = P(y_{ij} = 1 \mid u_i)$. The average of these is the marginal mean, $E(y_{ij})$. These averages for various x values yield the superimposed dashed curve. It has a shallower slope. An example in Section 9.7 illustrates that effects in a GLMM can be very different from effects in the corresponding marginal model.

9.4.2 Poisson GLMM for Correlated Count Data

With count data, we have seen that mixture models provide a flexible way to account for overdispersion. Mixing the Poisson distribution using the gamma distribution for its mean yields the negative binomial distribution marginally. An alternative approach mixes the Poisson log mean with a normal random effect. The GLMM structure with the log link is

$$\log[E(y_{ij} \mid \boldsymbol{u}_i)] = \boldsymbol{x}_{ij}\boldsymbol{\beta} + \boldsymbol{z}_{ij}\boldsymbol{u}_i, \tag{9.12}$$

where $\{\boldsymbol{u}_i\}$ are independent $N(\boldsymbol{0}, \boldsymbol{\Sigma}_u)$. Conditional on $\boldsymbol{u}_i$, y_{ij} has a Poisson distribution. The identity link is also possible but has a structural problem: for a random intercept with $\sigma_u > 0$, a positive probability exists that the linear predictor is negative. The negative binomial model results when instead $\exp(u_i)$ has a gamma distribution. The Poisson GLMM with normal random effects has the advantage, relative to the negative binomial GLM, of easily permitting multivariate random effects and multilevel models.

The random-intercept version of Poisson GLMM (9.12) implies that the corresponding marginal model has

$$E(y_{ij}) = E[E(y_{ij} \mid u_i)] = E[e^{\boldsymbol{x}_{ij}\boldsymbol{\beta}+u_i}] = e^{\boldsymbol{x}_{ij}\boldsymbol{\beta}+\sigma_u^2/2}.$$

Here, $E[\exp(u_i)] = \exp(\sigma_u^2/2)$ because a $N(0, \sigma_u^2)$ variate u_i has moment generating function (mgf) $E[\exp(tu_i)] = \exp(t^2\sigma_u^2/2)$. So the log of the mean conditionally equals $\boldsymbol{x}_{ij}\boldsymbol{\beta} + u_i$ and marginally equals $\boldsymbol{x}_{ij}\boldsymbol{\beta} + \sigma_u^2/2$. A loglinear model still applies, and the marginal effects of the explanatory variables are the same as the subject-specific effects. Thus, the *ratio* of means at two settings of $\boldsymbol{x}_{ij}$ is the same conditionally and marginally. The variance of the marginal distribution is

$$\begin{aligned}\mathrm{var}(y_{ij}) &= E[\mathrm{var}(y_{ij} \mid u_i)] + \mathrm{var}[E(y_{ij} \mid u_i)] = E[e^{\boldsymbol{x}_{ij}\boldsymbol{\beta}+u_i}] + e^{2\boldsymbol{x}_{ij}\boldsymbol{\beta}}\mathrm{var}(e^{u_i}) \\ &= e^{\boldsymbol{x}_{ij}\boldsymbol{\beta}+\sigma_u^2/2} + e^{2\boldsymbol{x}_{ij}\boldsymbol{\beta}}\left(e^{2\sigma_u^2} - e^{\sigma_u^2}\right) = E(y_{ij}) + [E(y_{ij})]^2\left(e^{\sigma_u^2} - 1\right).\end{aligned}$$

Here $\mathrm{var}(e^{u_i}) = E(e^{2u_i}) - [E(e^{u_i})]^2 = e^{2\sigma_u^2} - e^{\sigma_u^2}$ by evaluating the mgf at $t = 2$ and $t = 1$. As in the negative binomial model, the marginal variance is a quadratic function of the marginal mean. The ordinary Poisson model results when $\sigma_u = 0$. When $\sigma_u > 0$, the marginal distribution is not Poisson, and the extent to which the variance exceeds the mean increases as σ_u increases. Marginally, observations within a cluster are non-negatively correlated, as in the binary GLMM (Exercise 9.24).

9.4.3 Multilevel Modeling with a Discrete Response

Multilevel models can have discrete responses. For example, for the battery-of-exams scenario introduced in Section 9.2.2, suppose the response y_{ist} for student i in school s on test t is binary, with 1 = pass and 0 = fail. A multilevel model with random effects $\{v_{is}\}$ for students and $\{u_s\}$ for schools and fixed effects for explanatory variables has the form

$$\mathrm{logit}[P(y_{ist} = 1 \mid u_s, v_{is})] = \boldsymbol{x}_{ist}\boldsymbol{\beta} + u_s + v_{is}. \tag{9.13}$$

As in Section 5.1.2, a latent variable model for a continuous response implies this model. Let y_{ist}^* denote the latent observation for student i in school s on exam t, such that we observe $y_{ist} = 1$ if y_{ist}^* falls above some threshold. The latent variable model is

$$y_{ist}^* = \boldsymbol{x}_{ist}\boldsymbol{\beta} + u_s + v_{is} + \epsilon_{ist}.$$

The assumption that $\{\epsilon_{ist}\}$ come from a standard logistic distribution, for which the inverse cdf is the logit link function, implies the logistic random-effects model. For it, conditional on u_s and v_{is}, the observed response satisfies the logistic model (9.13). The assumption that ϵ_{ist} comes from a standard normal distribution implies a corresponding probit random-effects model.

For the binary multilevel model, the total unexplained variability in this latent variable model is $(\sigma_u^2 + \sigma_v^2 + \sigma_\epsilon^2)$. Of these, $\sigma_\epsilon^2 = \pi^2/3 = 3.29$ for the logistic model and 1.0 for the probit model, which are the variances of the standard logistic and standard normal distributions. A large within-student correlation between scores on different exams again corresponds to a relatively large var(v_{is}), and hence relatively large ratio of $(\sigma_u^2 + \sigma_v^2)$ to $(\sigma_u^2 + \sigma_v^2 + \sigma_\epsilon^2)$.

9.4.4 Binary and Count Data Models with More Complex Correlation Structure

In Section 9.2.5 we noted that linear mixed models that assume conditionally independent responses in a cluster, given a random effect, are sometimes inadequate, especially for longitudinal or spatial data. We can instead permit the error term $\boldsymbol{\epsilon}_i$ for a cluster to have correlated components.

GLMMs for binary data and count data are expressed in terms of $E(y_{ij} \mid \boldsymbol{u}_i)$ and do not have a separate error term, unless we work with the latent variable version of the model. However, the model of conditional independence, given $\boldsymbol{u}_i$, generalizes by replacing $\boldsymbol{u}_i$ by $\boldsymbol{u}_{ij}$ with correlated components. For example, the simple logistic-normal random-intercept model, $\text{logit}[P(y_{ij} = 1 \mid u_i)] = \boldsymbol{x}_{ij}\boldsymbol{\beta} + u_i$, generalizes to

$$\text{logit}[P(y_{ij} = 1 \mid u_{ij})] = \boldsymbol{x}_{ij}\boldsymbol{\beta} + u_{ij},$$

where $(u_{i1}, \dots, u_{id})$ have a multivariate normal distribution with a certain correlation structure (Coull and Agresti 2000). The autoregressive structure $\text{cov}(u_{ij}, u_{ik}) = \rho^{|j-k|}\sigma^2$ has only a single extra parameter. The simpler random-intercept model is the special case $\rho = 1$, implying a common random effect u_i for each component in cluster i. With large cluster sizes, the generalized model is challenging to fit by ML.

9.5 GLMM FITTING, INFERENCE, AND PREDICTION

Model fitting is not simple for GLMMs, because the likelihood function does not have a closed form. Numerical methods for approximating it can be computationally intensive for models with multivariate random effects. In this section we present methods for ML fitting and inference for GLMMs.

9.5.1 Marginal Likelihood and Maximum Likelihood Fitting

The GLMM is a two-stage model. At the first stage, conditional on the random effects $\{\boldsymbol{u}_i\}$, observations are assumed to follow a GLM. That is, the observations are independent, with y_{ij} in cluster i having distribution in the exponential dispersion family with expected value linked to a linear predictor, $g[E(y_{ij} \mid \boldsymbol{u}_i)] = \boldsymbol{x}_{ij}\boldsymbol{\beta} + \boldsymbol{z}_{ij}\boldsymbol{u}_i$. At that first stage, $\boldsymbol{z}_{ij}\boldsymbol{u}_i$ is a known offset. Then at the second stage, $\{\boldsymbol{u}_i\}$ are assumed independent from a $N(\boldsymbol{0}, \boldsymbol{\Sigma}_u)$ distribution.

For observations $\boldsymbol{y}$ and random effects $\boldsymbol{u}$, let $f(\boldsymbol{y} \mid \boldsymbol{u}; \boldsymbol{\beta})$ denote the conditional probability mass or density function of $\boldsymbol{y}$, given $\boldsymbol{u}$. Let $f(\boldsymbol{u}; \boldsymbol{\Sigma}_u)$ denote the normal pdf for $\boldsymbol{u}$. The likelihood function $\ell(\boldsymbol{\beta}, \boldsymbol{\Sigma_u}; \boldsymbol{y})$ for a GLMM refers to the marginal distribution of $\boldsymbol{y}$ after integrating out the random effects,

$$\ell(\boldsymbol{\beta}, \boldsymbol{\Sigma_u}; \boldsymbol{y}) = f(\boldsymbol{y}; \boldsymbol{\beta}, \boldsymbol{\Sigma_u}) = \int f(\boldsymbol{y} \mid \boldsymbol{u}; \boldsymbol{\beta}) f(\boldsymbol{u}; \boldsymbol{\Sigma_u})\, d\boldsymbol{u}. \tag{9.14}$$

It is called a *marginal likelihood*. For example, the marginal likelihood function $\ell(\boldsymbol{\beta}, \sigma_u^2; \boldsymbol{y})$ for the logistic-normal random-intercept model is

$$\prod_{i=1}^{n} \left\{ \int_{-\infty}^{\infty} \prod_{j=1}^{d} \left[\frac{\exp(\boldsymbol{x}_{ij}\boldsymbol{\beta} + u_i)}{1 + \exp(\boldsymbol{x}_{ij}\boldsymbol{\beta} + u_i)} \right]^{y_{ij}} \left[\frac{1}{1 + \exp(\boldsymbol{x}_{ij}\boldsymbol{\beta} + u_i)} \right]^{1-y_{ij}} f\left(u_i; \sigma_u^2\right) du_i \right\}.$$

Many methods can approximate $\ell(\boldsymbol{\beta}, \boldsymbol{\Sigma_u}; \boldsymbol{y})$ numerically. We next briefly describe a few of them.

9.5.2 Gauss–Hermite Quadrature Methods for ML Fitting

When the dimension of $\boldsymbol{u}_i$ is small, as in the one-dimensional integral just shown for the logistic-normal random-intercept model, standard numerical integration methods can approximate $\ell(\boldsymbol{\beta}, \boldsymbol{\Sigma_u}; \boldsymbol{y})$ well. *Gauss–Hermite quadrature* approximates the integral of a function $h(\cdot)$ multiplied by a scaled normal density function. The approximation is a finite weighted sum that evaluates the function at certain points, essentially approximating the area under a curve by a discrete histogram. For univariate normal random effects, the approximation has the form

$$\int_{-\infty}^{\infty} h(u) \exp(-u^2) du \approx \sum_{k=1}^{q} c_k h(s_k),$$

for tabulated *weights* $\{c_k\}$ and *quadrature points* $\{s_k\}$ that are the roots of *Hermite polynomials*. The specified $\{c_k\}$ and $\{s_k\}$ make the approximation exact for polynomials of degree $2q - 1$ or less. The approximation improves as q increases. The approximated likelihood can be maximized with algorithms such as Newton–Raphson. Inverting an approximation for the observed information matrix provides standard errors. For complex models, second partial derivatives for the Hessian can be computed numerically rather than analytically.

When the function h to be integrated is not centered at 0, many quadrature points may fall outside the region where the variation in the function is concentrated. An adaptive version of Gauss–Hermite quadrature (Liu and Pierce 1994) centers the quadrature points with respect to the mode of the function being integrated and scales them according to the estimated curvature at the mode. This improves efficiency, dramatically reducing the number of quadrature points needed to approximate the integral effectively.

9.5.3 Other Fitting Methods: Monte Carlo, Laplace Approximation

With Gauss–Hermite quadrature, adequate approximation becomes computationally more difficult as the dimension of $\boldsymbol{u}_i$ increases much beyond bivariate, because of the "curse of dimensionality." Then Monte Carlo methods are more feasible. Various approaches, including Markov chain Monte Carlo (MCMC), Monte Carlo in combination with Newton–Raphson, Monte Carlo in combination with the EM algorithm, and simulation, estimate the likelihood directly.

These likelihood approximations yield parameter estimates that converge to the ML estimates as the number of quadrature points increases for Gauss–Hermite integration and as the Monte Carlo sample size increases for MC methods. This contrasts with other methods, such as *Laplace approximations* and *penalized quasi-likelihood*, that maximize an analytical approximation of the likelihood function but do not yield exact ML estimates. Using the exponential family representation of each component of the joint distribution of $\boldsymbol{y}$ and $\boldsymbol{u}$, the integrand of (9.14) is an exponential function of $\boldsymbol{u}$. The Laplace approximation for that function uses a second-order Taylor series expansion of its exponent around a point $\tilde{\boldsymbol{u}}$ at which the first-order term equals 0. That point is $\tilde{\boldsymbol{u}} \approx E(\boldsymbol{u} \mid \boldsymbol{y})$. The approximating function for the integrand is then exponential with quadratic exponent in $(\boldsymbol{u} - \tilde{\boldsymbol{u}})$ and has the form of a constant multiple of a multivariate normal density. Thus, its integral has closed form.

9.5.4 Inference for GLMM Parameters and Prediction of Random Effects

After fitting the model, inference about fixed effects proceeds in the usual way. For instance, likelihood-ratio tests can compare nested models. Inference about random effects (e.g., their variance components) is more complex. Often one model is a special case of another in which a variance component equals 0. The simpler model then falls on the boundary of the parameter space relative to the more complex model, so as we observed in comparing Poisson and negative binomial models in Section 7.3.4, ordinary likelihood-based inference does not apply. For the most common situation, testing H_0: $\sigma_u^2 = 0$ against H_1: $\sigma_u^2 > 0$ for a model containing a random intercept, the null asymptotic distribution of the likelihood-ratio statistic is an equal mixture of degenerate at 0 (which occurs when $\hat{\sigma}_u = 0$) and χ_1^2 random variables. When $\hat{\sigma}_u > 0$ and the observed test statistic equals t, the P-value for this large-sample test is $\frac{1}{2}P(\chi_1^2 > t)$.

Some effects, such as subject-specific means, involve linear combinations of fixed and random effects. Given the data, the conditional distribution of $(\boldsymbol{u} \mid \boldsymbol{y})$ contains the information about $\boldsymbol{u}$. As in Section 9.3.2 for the linear mixed model, the prediction $\tilde{\boldsymbol{u}}_i$ for $\boldsymbol{u}_i$ estimates $E(\boldsymbol{u}_i \mid \boldsymbol{y})$, and the standard error of $\tilde{\boldsymbol{u}}_i$ is the standard deviation of the distribution of $(\boldsymbol{u}_i \mid \boldsymbol{y})$. Estimated effects, like those from an *empirical Bayes* approach (Section 10.4), exhibit shrinkage relative to estimates using only data in the specific cluster. Shrinkage estimators can be far superior to sample values when the sample size for estimating each effect is small, when there are many effects to estimate, or when the true effect values are roughly equal. Estimation of $E(\boldsymbol{u}_i \mid \boldsymbol{y})$ also uses numerical integration or Monte Carlo approximation. The expectation depends on $\boldsymbol{\beta}$ and $\boldsymbol{\Sigma}_u$, so in practice we substitute $\hat{\boldsymbol{\beta}}$ and $\hat{\boldsymbol{\Sigma}}_u$ in the approximation.

9.5.5 Misspecification of Random Effects Distribution

In spite of its popularity and attractive features, the normality assumption for random effects in ordinary GLMMs can rarely be closely checked. Distributions of predicted values are highly dependent on their assumed distribution and are not reliable indicators of the true random effects distribution. An obvious concern of this or any parametric assumption for the random effects is possibly harmful effects of misspecification. However, choosing an incorrect random effects distribution does not tend to seriously bias estimators of those effects. Assuming different distributions for the random effects can yield quite different predicted values yet have similar performance in terms of overall accuracy of prediction (McCulloch and Neuhaus 2011).

Likewise, different assumptions for the random effects distribution usually provide similar results for estimating the fixed effects, with similar efficiency. However, when the variance of the random effects is assumed constant but actually depends strongly on values of covariates, between-cluster effects may be more sensitive to correct specification of the random effects distribution than within-cluster effects, because of the attenuation in the effect when we integrate out the random effect to estimate the between-cluster effect. Bias sensitivity to the random effects assumption is greater for estimating fixed effects in GLMMs than estimating their counterparts in corresponding marginal models. This is an advantage of using marginal models to estimate between-cluster effects. See Heagerty and Zeger (2000) for discussion. For the linear mixed model, a sandwich correction can better estimate standard errors under misspecification, especially for variance components (Verbeke and Lesaffre 1998).

9.6 MARGINAL MODELING AND GENERALIZED ESTIMATING EQUATIONS

In presenting the marginal modeling approach, we first show a multivariate extension of the normal regression model. We illustrate with a model for the multivariate one-way layout, which leads to a special case of multivariate ANOVA. When we allow non-normal responses, ML fitting is often not feasible. We can then fit the model with a multivariate quasi-likelihood method, for which the model parameter estimates are solutions of *generalized estimating equations* (GEE) and standard errors come from an estimated sandwich covariance matrix.

9.6.1 Multivariate Normal Regression Model

With an identity link function, the normal linear model for the marginal responses in $\boldsymbol{y}_i = (y_{i1}, \ldots, y_{id})^{\mathrm{T}}$, $i = 1, \ldots, n$, is

$$y_{ij} = \boldsymbol{x}_{ij}\boldsymbol{\beta} + \epsilon_{ij}, \quad j = 1, \ldots, d,$$

where $\epsilon_{ij} \sim N(0, \sigma_\epsilon^2)$. Let $\boldsymbol{\epsilon}_i = (\epsilon_{i1}, \dots, \epsilon_{id})^{\mathrm{T}}$. The basic model takes $\{\boldsymbol{\epsilon}_i\}$, and thus the multivariate observations, as independent. However, we allow components of $\boldsymbol{\epsilon}_i$ to be correlated, since in most applications we expect $\mathrm{corr}(y_{ij}, y_{ik}) \neq 0$. Let $\mathrm{var}(\boldsymbol{\epsilon}_i) = \boldsymbol{V}_i$. As in Section 9.2.5, we can increase parsimony by providing structure for $\boldsymbol{V}_i$.

Forming a $d \times p$ matrix $\boldsymbol{X}_i$ of coefficients of $\boldsymbol{\beta}$, we express the model as $\boldsymbol{y}_i = \boldsymbol{X}_i\boldsymbol{\beta} + \boldsymbol{\epsilon}_i$. To have a single equation for all n observations, we append $\{\boldsymbol{X}_i\}$ into a $dn \times p$ matrix $\boldsymbol{X}$ and also stack the n vectors of observations $\boldsymbol{y}_i$ and errors $\boldsymbol{\epsilon}_i$, yielding

$$\boldsymbol{y} = \boldsymbol{X}\boldsymbol{\beta} + \boldsymbol{\epsilon}.$$

Let $\boldsymbol{V}$ be the block-diagonal covariance matrix for $\boldsymbol{y}$, with block i being the covariance matrix $\boldsymbol{V}_i$ for $\boldsymbol{y}_i$. Then $\boldsymbol{y} \sim N(\boldsymbol{X}\boldsymbol{\beta}, \boldsymbol{V})$.

For this formulation, the ML estimator of $\boldsymbol{\beta}$ is the generalized least squares estimator,

$$\hat{\boldsymbol{\beta}} = \left(\boldsymbol{X}^{\mathrm{T}}\boldsymbol{V}^{-1}\boldsymbol{X}\right)^{-1}\boldsymbol{X}^{\mathrm{T}}\boldsymbol{V}^{-1}\boldsymbol{y}.$$

In practice, $\boldsymbol{V}$ is unknown, so ML fitting simultaneously estimates it (using ML or REML) and $\hat{\boldsymbol{\beta}}$, taking into account any assumed structure for $\boldsymbol{V}$. See Diggle et al. (2002, Sections 4.4, 4.5, 5.3.2) for details.

9.6.2 Multivariate Normal Linear Model for One-Way Layout

An extension of the one-way layout (Sections 1.3.3, 2.3.2, and 3.2.1) to multivariate data provides structure for comparing the mean of a multivariate response $\boldsymbol{y}$ for c groups. Let y_{gij} denote the response for subject i in group g on response variable j, for $i = 1, \dots, n_g$ and $j = 1, \dots, d$. Let $\boldsymbol{y}_{gi} = (y_{gi1}, \dots, y_{gid})^{\mathrm{T}}$ and $\boldsymbol{\mu}_{gi} = E(\boldsymbol{y}_{gi})$. A marginal model with an identity link function for this setting assumes that $\boldsymbol{y}_{gi} \sim N(\boldsymbol{\mu}_{gi}, \boldsymbol{V})$, with

$$\mu_{gij} = \beta_{0j} + \beta_{gj}$$

for all g, i, and j. For identifiability, we need constraints such as $\beta_{1j} = 0$ for each j. For fixed j, this resembles the ordinary model for a one-way layout.

For this model, $\boldsymbol{\mu}_{gi} = \boldsymbol{\mu}_g$, all subjects in the same group having the same vector of means. The hypothesis of identical response distributions for the c groups is H_0: $\boldsymbol{\mu}_1 = \cdots = \boldsymbol{\mu}_c$. In the linear model, it corresponds to H_0: $\beta_{1j} = \cdots = \beta_{cj}$ for $j = 1, \dots, d$, which is equivalent to the null model, $\mu_{gij} = \beta_{0j}$ for all g and j. The test of H_0 is an example of a *multivariate analysis of variance*, abbreviated *MANOVA*. The technical details are beyond our scope, but just as one-way ANOVA partitions the total variability into between-group variability and within-group variability, MANOVA does something analogous with covariance matrices. The likelihood-ratio test, due to S. Wilks, is often referred to as *Wilks' lambda*.

With $c = 2$ groups, the test statistic simplifies considerably. Let $\bar{y}_g = (\sum_{i=1}^{n_g} y_{gi})/n_g$ for $g = 1, 2$. Let S denote the pooled covariance matrix,

$$S = \frac{\sum_{i=1}^{n_1}(y_{1i} - \bar{y}_1)(y_{1i} - \bar{y}_1)^{\mathrm{T}} + \sum_{i=1}^{n_2}(y_{2i} - \bar{y}_2)(y_{2i} - \bar{y}_2)^{\mathrm{T}}}{n_1 + n_2 - 2}.$$

Then testing H_0: $\boldsymbol{\mu}_1 = \boldsymbol{\mu}_2$ using a likelihood-ratio statistic is equivalent to using *Hotelling's* T^2 statistic (Hotelling 1931), which is

$$T^2 = \frac{n_1 n_2}{n_1 + n_2}(\bar{y}_1 - \bar{y}_2)^{\mathrm{T}} S^{-1}(\bar{y}_1 - \bar{y}_2).$$

Under H_0, the transformation $[(n_1 + n_2 - d - 1)/(n_1 + n_2 - 2)d]T^2$ has an F distribution with $df_1 = d$ and $df_2 = n_1 + n_2 - d - 1$. See Johnson and Wichern (2007, Section 6.3) for details. With only $d = 1$ response variable, T^2 is the square of the t statistic with $df = n_1 + n_2 - 2$ for comparing two groups under the assumption of common variance (Exercise 3.9).

Over time, MANOVA methods have been losing popularity, relative to analyses using normal linear mixed models and the method discussed next, because of their restrictive structure for the data. MANOVA methods do not easily handle missing data, and modeling the covariance structure has advantages of parsimony.

9.6.3 Method of Generalized Estimating Equations (GEE)

Extensions of the multivariate linear marginal model just considered to more complex settings, such as factorial designs with interaction terms, are easily handled for multivariate normal distributions. Indeed many books are devoted entirely to analyzing multivariate normal responses (e.g., Anderson 2003). For non-normal responses such as binary data and count data, however, specifying multivariate GLMs that focus on marginal modeling is more difficult. For discrete data, there is a lack of multivariate families of distributions that can exhibit simple correlation structures, thus serving as discrete analogs of the multivariate normal.

As with a univariate response, the quasi-likelihood method states a model for $\mu_{ij} = E(y_{ij})$ and specifies a variance function $v(\mu_{ij})$. Now, though, that model applies to the marginal distribution for each y_{ij}, but without the necessity of specifying a full multivariate distribution. The method of *generalized estimating equations* (GEE) instead merely specifies a pairwise "working correlation" pattern for $(y_{i1}, y_{i2}, \ldots, y_{id})$. Common patterns are exchangeable $(\text{corr}(y_{ij}, y_{ik}) = \alpha)$, autoregressive $(\text{corr}(y_{ij}, y_{ik}) = \alpha^{|j-k|})$, independence $(\text{corr}(y_{ij}, y_{ik}) = 0)$, and unstructured $(\text{corr}(y_{ij}, y_{ik}) = \alpha_{jk})$. The choice for the working correlation matrix determines the GEE estimates of $\boldsymbol{\beta}$ and their model-based standard errors. For example, under the independence structure, the estimates are identical to the ML estimates obtained by treating all observations within and between clusters as independent.

When the chosen link function and linear predictor truly describe how $E(y_{ij})$ depends on the explanatory variables, GEE estimators of $\boldsymbol{\beta}$ are consistent even if the correlation structure is misspecified. In practice, a chosen model is never exactly correct. This consistency result is useful, however, for suggesting that a misspecified correlation structure need not adversely affect estimation of effects of interest, for whatever model we use. Although the estimates of $\boldsymbol{\beta}$ are usually fine whatever working correlation structure we choose, their model-based standard errors are not. More-appropriate standard errors result from an adjustment the GEE method makes using the empirical covariation, generalizing the robust sandwich covariance matrix for univariate responses presented in Section 8.3.2.

9.6.4 GEE and Sandwich Covariance Matrix

For cluster i with $\boldsymbol{y}_i = (y_{i1}, \dots, y_{id})^{\mathrm{T}}$ and $\boldsymbol{\mu}_i = (\mu_{i1}, \dots, \mu_{id})^{\mathrm{T}}$, the marginal model with link function g is $g(\mu_{ij}) = \boldsymbol{x}_{ij}\boldsymbol{\beta}$. Let $\boldsymbol{V}_i$ denote the working covariance matrix for $\boldsymbol{y}_i$, depending on a parameter or parameters $\boldsymbol{\alpha}$ that determine the working correlation matrix $\boldsymbol{R}(\boldsymbol{\alpha})$. If $\boldsymbol{R}(\boldsymbol{\alpha})$ is the true correlation matrix for $\boldsymbol{y}_i$, then $\boldsymbol{V}_i = \mathrm{var}(\boldsymbol{y}_i)$. Let $\boldsymbol{D}_i = \partial\boldsymbol{\mu}_i/\partial\boldsymbol{\beta}$ be the $d \times p$ matrix with jk element $\partial\mu_{ij}/\partial\beta_k$. From Equation (8.2), for univariate GLMs ($d = 1$) the quasi-likelihood estimating equations have the form

$$\sum_{i=1}^{n} (\partial\mu_i/\partial\boldsymbol{\beta})^{\mathrm{T}} v(\mu_i)^{-1}(y_i - \mu_i) = \boldsymbol{0},$$

where $\mu_i = g^{-1}(\boldsymbol{x}_i\boldsymbol{\beta})$. The analog of this for a multivariate response is the set of *generalized estimating equations*

$$\sum_{i=1}^{n} \boldsymbol{D}_i^{\mathrm{T}} \boldsymbol{V}_i^{-1}(\boldsymbol{y}_i - \boldsymbol{\mu}_i) = \boldsymbol{0}. \tag{9.15}$$

The GEE estimator $\hat{\boldsymbol{\beta}}$ is the solution of these equations.

The GEE estimates are computed by iterating between estimating $\boldsymbol{\beta}$, given current estimates of $\boldsymbol{\alpha}$ and any dispersion parameter or scaling factor ϕ for the variance of the marginal distributions, and moment estimation of $\boldsymbol{\alpha}$ and ϕ, given a current estimate of $\boldsymbol{\beta}$ (Liang and Zeger 1986). The estimation of $\boldsymbol{\beta}$ uses a modified Fisher scoring algorithm for solving the generalized estimating equations. Estimation of ϕ equates a Pearson statistic to the nominal *df* value, as in Section 8.1.1. Estimation of $\boldsymbol{\alpha}$ combines information from the pairwise empirical correlations. Under certain regularity conditions including appropriate consistency for estimates of $\boldsymbol{\alpha}$ and ϕ, Liang and Zeger showed that as n increases,

$$\sqrt{n}(\hat{\boldsymbol{\beta}} - \boldsymbol{\beta}) \xrightarrow{d} N(\boldsymbol{0}, \boldsymbol{V}_G).$$

Here, generalizing the heuristic argument that motivated the formula (8.4) used in quasi-likelihood to adjust for misspecified variance functions, they proved that $\boldsymbol{V}_G/n$ is approximately

$$\left[\sum_{i=1}^{n} \boldsymbol{D}_i^{\mathrm{T}} \boldsymbol{V}_i^{-1} \boldsymbol{D}_i\right]^{-1} \left[\sum_{i=1}^{n} \boldsymbol{D}_i^{\mathrm{T}} \boldsymbol{V}_i^{-1} \mathrm{var}(\boldsymbol{y}_i) \boldsymbol{V}_i^{-1} \boldsymbol{D}_i\right] \left[\sum_{i=1}^{n} \boldsymbol{D}_i^{\mathrm{T}} \boldsymbol{V}_i^{-1} \boldsymbol{D}_i\right]^{-1}. \quad (9.16)$$

The estimated sandwich covariance matrix $\widehat{\boldsymbol{V}}_G/n$ of $\hat{\boldsymbol{\beta}}$ estimates this expression by replacing $\boldsymbol{\beta}$ with $\hat{\boldsymbol{\beta}}$, ϕ with $\hat{\phi}$, $\boldsymbol{\alpha}$ with $\hat{\boldsymbol{\alpha}}$, and $\mathrm{var}(\boldsymbol{y}_i)$ with $(\boldsymbol{y}_i - \hat{\boldsymbol{\mu}}_i)(\boldsymbol{y}_i - \hat{\boldsymbol{\mu}}_i)^{\mathrm{T}}$.

When the working covariance structure is the true one, so $\mathrm{var}(\boldsymbol{y}_i) = \boldsymbol{V}_i$, the approximation $(1/n)\boldsymbol{V}_G$ for the asymptotic covariance matrix simplifies to the model-based covariance matrix, $\left(\sum_i \boldsymbol{D}_i^{\mathrm{T}} \boldsymbol{V}_i^{-1} \boldsymbol{D}_i\right)^{-1}$. This is the relevant covariance matrix if we put complete faith in our choice of that structure.

Advantages of the GEE approach include its computational simplicity compared with ML, its not requiring specification of a joint distribution for $(y_{i1}, y_{i2}, \dots, y_{id})$, and the consistency of estimation even with misspecified correlation structure. However, it has limitations. Since the GEE approach does not completely specify the joint distribution, it does not have a likelihood function. Likelihood-based methods are not available for testing fit, comparing models, and conducting inference about parameters. Also, with categorical responses, the correlation is not the most natural way to characterize within-cluster association. An alternative approach uses the *odds ratio* to characterize pairwise associations, and then uses the odds ratios together with the marginal probabilities to generate working correlation matrices (Note 9.9). Finally, when data are missing, the GEE method requires a stronger assumption about the missing data than ML does in order for estimators to be consistent. For GEE, the data must be "missing completely at random," which means that the probability an observation is missing is independent of that observation's value and the values of other variables in the entire data file. For ML, the data need only be "missing at random," with what caused the data to be missing not depending on their values.

9.6.5 Why Does ML Have Limited Feasibility for Marginal Models?

For fitting marginal models, why not use ML itself instead of quasi-likelihood methods? Sometimes this is possible, but once we move away from multivariate normal models, it is limited. A difficulty is the lack of multivariate families (like the normal) for which we can easily characterize the joint distribution in terms of a small set of correlations and for which we can express the parameters of that distribution in terms of parameters for marginal models.

To illustrate the difficulty, consider a d-dimensional binary response. Then $\boldsymbol{y}_i$ is a multinomial observation that falls in one of 2^d cells of a d-dimensional contingency table. For n observations, the multinomial likelihood results from the product of n multinomial trials, each defined over the 2^d cells at that setting for the explanatory variables. Let $z_{ij_1\cdots j_d} = 1$ if $(y_{i1} = j_1, \ldots, y_{id} = j_d)$ and $z_{ij_1\cdots j_d} = 0$ otherwise. The multinomial likelihood function is

$$\ell(\boldsymbol{\pi}) = \prod_{i=1}^{n} \left[\prod_{j_1=1}^{2} \cdots \prod_{j_d=1}^{2} P(y_{i1} = j_1, \ldots, y_{id} = j_d)^{z_{ij_1\cdots j_d}} \right].$$

This pertains to probabilities in the joint distribution, whereas a marginal model

$$\text{logit}[P(y_{ij} = 1)] = \boldsymbol{x}_{ij}\boldsymbol{\beta}, \quad i = 1, \ldots, n, \quad j = 1, \ldots, d,$$

describes the marginal probabilities. We cannot substitute the marginal model formula into the multinomial likelihood function to obtain the function to maximize in terms of $\boldsymbol{\beta}$ to fit the model.

Methods exist for fitting marginal models with ML, such as applying a method[9] for maximizing a function subject to constraints (Lang and Agresti 1994). However, such methods are infeasible when a study has many explanatory variables, especially if some are continuous and have possibly different numbers of observations for different clusters.

9.7 EXAMPLE: MODELING CORRELATED SURVEY RESPONSES

We illustrate marginal models and GLMMs for binary data using Table 9.5. The respondents in a General Social Survey indicated whether they supported legalizing abortion in each of $d = 3$ situations. Table 9.5 also classifies the subjects by gender. Let y_{ij} denote the response for subject i in situation j, with $y_{ij} = 1$ representing support of legalization.

We first fit the random-intercept GLMM

$$\text{logit}[P(y_{ij} = 1 \mid u_i)] = \beta_0 + \beta_j + \gamma x_i + u_i, \tag{9.17}$$

Table 9.5 Support for Legalized Abortion in Three Situations, by Gender

	Sequence of Responses (1 = Yes, 0 = No) in Three Situations							
Gender	(1,1,1)	(1,1,0)	(0,1,1)	(0,1,0)	(1,0,1)	(1,0,0)	(0,0,1)	(0,0,0)
Male	342	26	6	21	11	32	19	356
Female	440	25	14	18	14	47	22	457

Source: General Social Survey, available at `http://sda.berkeley.edu/GSS`. Situations are (1) if the family has a very low income and cannot afford any more children, (2) when the woman is not married and does not want to marry the man, and (3) when the woman wants it for any reason. Subject-specific data table is in the file `Abortion2.dat` at text website.

[9]For some models, this is available with the hmmm R package. See `cran.r-project.org/web/packages/hmmm/hmmm.pdf` and a 2014 article by Colombi et al. in *J. Statist. Software*.

where $x_i = 1$ for females and 0 for males and $\{u_i\}$ are independent $N(0, \sigma_u^2)$. The situation effects $\{\beta_j\}$ satisfy a constraint, $\beta_3 = 0$ in the R output shown below. Here, the gender effect γ is assumed to be identical for each situation. We need a large number of quadrature points to achieve adequate approximation of the log-likelihood function and its curvature.

```
-----------------------------------------------------------------------
> Abortion # data file Abortion.dat at www.stat.ufl.edu/~aa/glm/data
      case  gender  situation  response
1        1       1          1         1
2        1       1          2         1
3        1       1          3         1
4        2       1          1         1
...
5550  1850       0          3         0
> z1 <- ifelse(Abortion$situation==1,1,0)
> z2 <- ifelse(Abortion$situation==2,1,0)
> library(glmmML) # ML fitting of GLMMs
  # Alternative: glmer function in lme4 package, nAGQ quadrature pts.
> fit.glmm <- glmmML(response ~ gender + z1 + z2,
+     cluster=Abortion$case, family=binomial, data=Abortion,
+     method = "ghq", n.points=70, start.sigma=9) # uses adaptive GHQ
> summary(fit.glmm)
                 coef  se(coef)        z  Pr(>|z|)
(Intercept)  -0.6187    0.3777  -1.6384  1.01e-01
gender        0.0126    0.4888   0.0257  9.79e-01
z1            0.8347    0.1601   5.2135  1.85e-07
z2            0.2924    0.1567   1.8662  6.20e-02
Scale parameter in mixing distribution:  8.74 gaussian
Std. Error:                              0.542
        LR p-value for H_0: sigma = 0: 0
-----------------------------------------------------------------------
```

The estimates of fixed effects have log-odds-ratio interpretations, within-subject for situation effects and between-subject for the gender effect. For a given subject of either gender, for instance, the estimated odds of supporting legalized abortion in situation 1 equals exp(0.835) = 2.30 times the estimated odds in situation 3. Since $\hat{\gamma} = 0.013$, for each situation the estimated probability of supporting legalized abortion is similar for females and males having the same random effect values.

The random effects have $\hat{\sigma}_u = 8.74$, so strong associations exist among responses for the three situations. This is reflected by 1595 of the 1850 subjects making the same response in all three situations. For this application, a normality assumption for u_i is suspect. A polarized population might suggest a bimodal distribution. However, maximizing the likelihood with a two-point distribution for u_i yields similar results for the fixed effects estimates (Agresti et al. 2000).

Finding cell fitted values requires integrating over the estimated random effects distribution to obtain estimated marginal probabilities of any particular sequence of responses. For the ML parameter estimates, the probability of a particular sequence of responses (y_{i1}, y_{i2}, y_{i3}) for a given u_i is the appropriate product of estimated conditional probabilities, $\prod_j \hat{P}(y_{ij} \mid u_i)$. Integrating this product probability with respect to u_i for the $N(0, \hat{\sigma}_u^2)$ distribution (or simulating that integral) estimates the marginal probability for a given cell. Multiplying this estimated marginal probability of a given sequence by the gender sample size for that multinomial gives a fitted value. For instance, of the 1037 females, 440 indicated support under all three circumstances (457 under none of the three), and the fitted value was 436.5 (459.3). Overall chi-squared statistics comparing the 16 observed and fitted counts are $G^2 = 23.2$ and $X^2 = 27.8$ ($df = 9$). Here $df = 9$ since we are modeling 14 multinomial parameters ($8 - 1 = 7$ for each gender) using five GLMM parameters $(\beta_0, \beta_1, \beta_2, \gamma, \sigma_u)$. These statistics reflect some lack of fit. An analysis of residuals (not shown here) indicates the lack of fit mainly reflects, for each gender, a tendency for fewer observed (1,0,1) response sequences and more (0,0,1) response sequences than the model fit has. An extended model that allows interaction between gender and situation has different $\{\beta_j\}$ for men and women, but it does not fit better. The likelihood-ratio statistic comparing the models equals 1.0 ($df = 2$).

A marginal model analysis of the data focuses on the marginal distributions for the three situations for each gender, treating the dependence as a nuisance. A marginal model analog of (9.17) is

$$\text{logit}[P(y_{ij} = 1)] = \beta_0 + \beta_j + \gamma x_i. \tag{9.18}$$

For the exchangeable working correlation structure, the GEE analysis estimates a common correlation of 0.817 between pairs of responses. We also show results with the independence working correlation structure. For it, the estimates are very similar, but the model-based ("naive") *SE* values are badly biased relative to the sandwich-covariance-matrix-based ("robust") *SE* values. The *SE* values are underestimated for gender (the between-subject effect) and overestimated for the situations (the within-subject effects).

```
-------------------------------------------------------------------------
> library(gee)
> fit.gee <- gee(response ~ gender + z1 + z2, id=case, family=binomial,
+               corstr="exchangeable", data=Abortion)
> summary(fit.gee)
              Estimate  Naive S.E. Naive z Robust S.E. Robust z
(Intercept)   -0.1253     0.0678  -1.8478     0.0676   -1.8544
gender         0.0034     0.0879   0.0391     0.0878    0.0391
z1             0.1493     0.0281   5.3066     0.0297    5.0220
z2             0.0520     0.0281   1.8478     0.0270    1.9232
Working Correlation
       [,1]   [,2]   [,3]
[1,] 1.0000 0.8173 0.8173
[2,] 0.8173 1.0000 0.8173
[3,] 0.8173 0.8173 1.0000

> fit.gee2 <- gee(response ~ gender + z1 + z2, id=case, family=binomial,
+         corstr="independence", data=Abortion)
```

```
> summary(fit.gee2) # same estimates as ML, assuming independence
             Estimate  Naive S.E.  Naive z  Robust S.E.  Robust z
(Intercept)   -0.1254      0.0556  -2.2547       0.0676   -1.8556
gender         0.0036      0.0542   0.0661       0.0878    0.0408
z1             0.1493      0.0658   2.2680       0.0297    5.0220
z2             0.0520      0.0659   0.7897       0.0270    1.9232
-----------------------------------------------------------------------
```

Table 9.6 compares the estimates from the GLMM with the GEE estimates from the corresponding marginal model. It also shows the ML estimates for the marginal model, that is, ML that accounts for the actual dependence by treating each observation as a multinomial trial over eight cells, rather than naive ML from the GEE approach with independence estimating equations that treat the observation as three independent binomial trials. That marginal model fits well, with $G^2 = 1.10$. Its $df = 2$, because the model describes six marginal probabilities (three for each gender) using four parameters. The population-averaged $\{\hat{\beta}_j\}$ from the marginal model fit are much smaller than the subject-specific $\{\hat{\beta}_j\}$ from the GLMM fit. This reflects the very large GLMM heterogeneity ($\hat{\sigma}_u = 8.74$) and the corresponding strong correlations among the three responses. Although the GLMM $\{\hat{\beta}_j\}$ are about five to six times the marginal model $\{\hat{\beta}_j\}$, so are the standard errors. The two approaches provide similar substantive interpretations and conclusions. The contrasts of $\{\hat{\beta}_j\}$ indicate greater support for legalized abortion in situation 1 than in the other two situations.

Table 9.6 For Table 9.5, ML Estimates for GLMM (9.17) and ML and GEE Estimates for Marginal Model (9.18)

		GLMM ML		Marginal GEE		Marginal ML	
Effect	Parameter	Estimate	*SE*	Estimate	*SE*	Estimate	*SE*
Abortion	$\beta_1 - \beta_3$	0.835	0.160	0.149	0.030	0.148	0.030
	$\beta_1 - \beta_2$	0.542	0.157	0.097	0.028	0.098	0.027
	$\beta_2 - \beta_3$	0.292	0.157	0.052	0.027	0.049	0.027
Gender	γ	0.013	0.489	0.003	0.088	0.005	0.088
$\sqrt{\text{var}(u_i)}$	σ_u	8.74	0.54				

CHAPTER NOTES

Section 9.1: Marginal Models and Models with Random Effects

9.1 Longitudinal data: For overviews of GLMMs and marginal models, with emphasis on modeling longitudinal data, see Diggle et al. (2002), Fitzmaurice et al. (2011), and Hedeker and Gibbons (2006). Skrondal and Rabe-Hesketh (2004) presented latent variable models. For discrete data, see Cameron and Trivedi (2013, Chapter 9) and Molenberghs and Verbeke (2005). Regression models with time series data receive much attention in texts having econometrics emphasis, such as Cameron and Trivedi

(2013, Chapter 7) and Greene (2011). Fahrmeir and Tutz (2001) surveyed multivariate GLMs, including models for time series data (Chapter 8). For transition models, see Azzalini (1994) and Diggle et al. (2002, Chapter 10). Survival data are another form of longitudinal data, models for which are beyond our scope. See McCullagh and Nelder (1989, Chapter 13). For GLMMs for spatial data, see Fahrmeir and Tutz (2001, Section 8.5) and Stroup (2012, Chapter 15).

Section 9.2: Normal Linear Mixed Models

9.2 **Covariance structures**: For linear mixed models and related covariance structures, see Demidenko (2013), Diggle et al. (2002, Chapter 5), Fahrmeir et al. (2013, Chapter 7), Laird and Ware (1982), Littell et al. (2000), and Verbeke and Molenberghs (2000). For diagnostics, such as the influence of individual observations, see Verbeke and Molenberghs (2000, Chapter 11).

9.3 **Multilevel models**: For multilevel modeling, see Aitkin et al. (1981), Gelman and Hill (2006), Goldstein (2010), Hedeker and Gibbons (2006, Chapter 13), Scott et al. (2013), Skrondal and Rabe-Hesketh (2004), and Stroup (2012, Chapter 8).

Section 9.3: Fitting and Prediction for Normal Linear Mixed Models

9.4 **BLUP**: Robinson (1991) surveyed BLUP. See also the discussion of that article by T. Speed. A Bayesian approach with uniform improper prior for $\boldsymbol{\beta}$ and $N(\mathbf{0}, \boldsymbol{\Sigma}_u)$ prior for $\boldsymbol{u}$ results in a posterior density having mode that is the BLUP (Lindley and Smith 1972; Robinson 1991). Morris (1983b) discussed prediction of random effects. Rao (2003) presented methods for small-area estimation.

9.5 **REML**: For more about REML, see Harville (1977) and Patterson and Thompson (1971). Laird and Ware (1982) gave a Bayesian interpretation, in which REML results from the posterior for $\boldsymbol{V}$ after using a flat prior for $\boldsymbol{\beta}$.

Section 9.4: Binomial and Poisson GLMMs

9.6 **Discrete GLMM**: For GLMMs for discrete data, see Agresti (2013, Chapter 13, 14), Agresti et al. (2000), Cameron and Trivedi (2013, Chapter 9), Demidenko (2013), Diggle et al. (2002), Hedeker and Gibbons (2006, Chapter 12), McCulloch et al. (2008), Molenberghs and Verbeke (2005), Skrondal and Rabe-Hesketh (2004), Stroup (2012), and Zeger and Karim (1991). For negative binomial loglinear models with random effects, see Booth et al. (2003).

9.7 **Multinomial GLMM**: For nominal-response models with random effects, see Hartzel et al. (2001) and Hedeker and Gibbons (2006, Chapter 11). For ordinal-response models with random effects, see Hartzel et al. (2001), Hedeker and Gibbons (2006, Chapter 10), Skrondal and Rabe-Hesketh (2004), and Tutz and Hennevogl (1996).

Section 9.5: GLMM Fitting, Inference, and Prediction

9.8 **PQL, Laplace, h-likelihood, REML**: Breslow and Clayton (1993) presented a penalized quasi-likelihood (PQL) method for fitting GLMs. Later literature dealt with reducing the large bias that method can have when variance components are large. For an overview of the Laplace approximation, see Davison (2003, Section 11.3.1).

The *h-likelihood* approach essentially treats random effects like parameters and is also computationally simpler (Lee and Nelder 1996; Lee et al. 2006), but its justification is unclear (Meng 2009). For REML methods for GLMMs, see Schall (1991).

Section 9.6: Marginal Modeling and Generalized Estimating Equations

9.9 GEE methods: Liang and Zeger (1986) proposed the GEE method, generalizing moment-based methods for model misspecification in the econometrics literature, such as Gourieroux (1984), Hansen[10] (1982), and White (1982). For categorical responses, GEE methods can characterize pairwise associations using odds ratios. For binary data, see Fitzmaurice et al. (1993), Lipsitz et al. (1991), and Carey et al. (1993). For nominal and ordinal responses, see Touloumis et al. (2013).

9.10 Missing data: For ways of dealing with missing data in GLMs and GLMMs, see Diggle et al. (2002, Chapter 13), Hedeker and Gibbons (2006, Chapter 14), Ibrahim et al. (2005), Little and Rubin (2002), and Molenberghs and Verbeke (2005, Chapters 26–32).

9.11 Marginal models, copulas, and composite likelihood: For ML for marginal and related models for contingency tables, see Bartolucci et al. (2007), Bergsma et al. (2009), and Lang (2004, 2005). An alternative approach specifies a joint distribution from the marginal distributions using a *copula* function. A parametric copula is merely a formula, indexed by association parameters, that generates a multivariate distribution from specified marginal distributions. Its theory is based on *Sklar's theorem*, which states that any multivariate cdf can be expressed as a function of the marginal cdfs. See Trivedi and Zimmer (2007). Another approach, *composite likelihood*, is based on contributions to the likelihood function for all pairs of observations. For an overview, see Varin et al. (2011).

EXERCISES

9.1 Verify formula (9.1) for the effects of correlation on between-cluster and within-cluster effects.

9.2 How does positive correlation affect the *SE* for between-cluster effects with binary data? Let $y_{11}, \dots, y_{1d}$ be Bernoulli trials with $E(y_{1j}) = \pi_1$ and let $y_{21}, \dots, y_{2d}$ be Bernoulli trials with $E(y_{2j}) = \pi_2$. Suppose $\text{corr}(y_{ij}, y_{ik}) = \rho$ for $i = 1, 2$ and $\text{corr}(y_{1j}, y_{2k}) = 0$ for all j and k. Find the *SE* of $\hat{\pi}_1 - \hat{\pi}_2$. Show it is larger when $\rho > 0$ than when $\rho = 0$.

9.3 Suppose $y_1, \dots y_n$ have $E(y_i) = \mu$, $\text{var}(y_i) = \sigma^2$, and $\text{corr}(y_i, y_j) = \rho$ for $i \neq j$. Show that $E(s^2) = \sigma^2(1 - \rho)$.

9.4 Formulate a normal random-effects model that generates the within-cluster and between-cluster effects described in Section 9.1.1.

[10]In 2013 Lars Peter Hansen won the Nobel Prize in Economic Sciences, partly for this work.

9.5 In the *analysis of covariance* model, observation j in group i has $\mu_{ij} = \beta_0 + \beta_1 x_{ij} + \gamma_i$ for a quantitative variable x_{ij} and qualitative $\{\gamma_1, \dots, \gamma_c\}$. Describe an application in which the qualitative variable would naturally be treated as a random effect. Show then how to express the model as a normal linear mixed model (9.6).

9.6 A crossover study comparing $d = 2$ drugs observes a continuous response (y_{i1}, y_{i2}) for each subject for each drug. Let $\mu_1 = E(y_{i1})$ and $\mu_2 = E(y_{i2})$ and consider H_0: $\mu_1 = \mu_2$.

a. Construct the normal linear mixed model that generates a paired-difference t test (with test statistic $t = \sqrt{n}\bar{d}/s$, using mean and standard deviation of the differences $\{d_i = y_{i2} - y_{i1}\}$) and the corresponding confidence interval for $\mu_1 - \mu_2$.

b. Show the effect of the relative sizes of the variances of the random error and random effect on corr(y_{i1}, y_{i2}). Based on this, to compare two means, explain why it can be more efficient to use a design with dependent samples than with independent samples.

9.7 For the normal linear mixed model (9.6), derive expression (9.7) for var$(\boldsymbol{y}_i)$.

9.8 For the extension of the random-intercept linear mixed model (9.8) that assumes $\text{cov}(\epsilon_{ij}, \epsilon_{ik}) = \sigma_\epsilon^2 \rho^{|j-k|}$, show that

$$\text{corr}(y_{ij}, y_{ik}) = \left(\sigma_u^2 + \rho^{|j-k|}\sigma_\epsilon^2\right) / \left(\sigma_u^2 + \sigma_\epsilon^2\right).$$

9.9 Consider the model discussed in Section 9.2.4 having a random intercept and a random slope. Is the fit for subject i any different than using least squares to fit a line using only the data for subject i? Explain.

9.10 For the linear mixed model, show that $\tilde{\boldsymbol{\beta}}$ (for known $\{\boldsymbol{V}_i\}$) is unbiased, and derive its variance. Show how $\tilde{\boldsymbol{\beta}}$ and var$(\tilde{\boldsymbol{\beta}})$ simplify when the model does not contain random effects.

9.11 When $\boldsymbol{X}_i$ and $\boldsymbol{V}_i$ in the linear mixed model are the same for each subject, show that the generalized least squares solution (9.10) can be expressed in terms of $\bar{\boldsymbol{y}} = (1/n)\sum_i \boldsymbol{y}_i$.

9.12 For the balanced random-intercept linear mixed model (9.8) based on conditional independence given the random effect (Section 9.2.1), show that $\tilde{\boldsymbol{\beta}} = \left(\boldsymbol{X}^{\mathrm{T}}\boldsymbol{V}^{-1}\boldsymbol{X}\right)^{-1}\boldsymbol{X}^{\mathrm{T}}\boldsymbol{V}^{-1}\boldsymbol{y}$ simplifies to the ordinary least squares estimator, $\left(\boldsymbol{X}^{\mathrm{T}}\boldsymbol{X}\right)^{-1}\boldsymbol{X}^{\mathrm{T}}\boldsymbol{y}$; i.e., with compound symmetry, observations do not need weights. (Diggle et al. (2002, p. 63) showed that ordinary least squares can have poor efficiency; however, when a model instead has autoregressive error structure with a time-varying explanatory variable.)

9.13 Using the joint normal density of $\boldsymbol{y}$ and $\boldsymbol{u}$, derive Henderson's mixed-model equations.

9.14 When $\boldsymbol{R} = \sigma_\epsilon^2 \boldsymbol{I}$, show that as $\boldsymbol{\Sigma}_u^{-1}$ tends to the zero matrix, the mixed model equations for a linear mixed model tend to the ordinary normal equations that treat both $\boldsymbol{\beta}$ and $\boldsymbol{u}$ as fixed effects.

9.15 For the random-effects one-way layout model (Section 9.3.2), show that $\hat{\beta}_0$ and $\tilde{u}_i$ are as stated there (i.e., with the known variances). Show that $\tilde{u}_i$ is a weighted average of 0 and the least squares estimate based on treating $\boldsymbol{u}$ as fixed effects. Give an application for which you would prefer $\hat{\beta}_0 + \tilde{u}_i$ to the fixed-effect estimate $\bar{y}_i$.

9.16 BLUPs are unbiased. Explain why this does not imply that $E(\tilde{\boldsymbol{u}} \mid \boldsymbol{u}) = \boldsymbol{u}$. Illustrate by a simulation using a simple linear mixed model. For your simulation, show how $\tilde{\boldsymbol{u}}$ tends to shrink toward $\boldsymbol{0}$ relative to $\boldsymbol{u}$.

9.17 Show how fitting and prediction results of Section 9.3 for linear mixed models simplify when $\text{var}(\epsilon_i) = \sigma_\epsilon^2 \boldsymbol{I}$ instead of $\boldsymbol{R}$.

9.18 For the REML approach for the normal null model described in Section 9.3.3, find $\boldsymbol{L}$, derive the distribution of $\boldsymbol{L}\boldsymbol{y}$, and find the REML estimator of σ^2.

9.19 For the binary matched-pairs model (9.5), consider a strictly fixed effects approach, replacing $\beta_0 + u_i$ in the model by β_{0i}. Assume independence of responses between and within subjects.

a. Show that the joint probability mass function is proportional to

$$\exp\left[\sum_{i=1}^{n} \beta_{0i}(y_{i1} + y_{i2}) + \beta_1\left(\sum_{i=1}^{n} y_{i2}\right)\right].$$

b. To eliminate $\{\beta_{0i}\}$, explain why we can condition on $\{s_i = y_{i1} + y_{i2}\}$ (Recall Section 5.3.4). Find the conditional distribution.

c. Let $\{n_{ab}\}$ denote the counts for the four possible sequences, as in Table 9.1. For subjects having $s_i = 1$, explain why $\sum_i y_{i1} = n_{12}$ and $\sum_i y_{i2} = n_{21}$ and $\sum_i s_i = n^* = n_{12} + n_{21}$. Explain why the conditional distribution of n_{21} is $\text{bin}(n^*, \exp(\beta_1)/[1 + \exp(\beta_1)])$. Show that the conditional ML estimator is

$$\hat{\beta}_1 = \log\left(\frac{n_{21}}{n_{12}}\right), \quad \text{with} \quad SE = \sqrt{\frac{1}{n_{21}} + \frac{1}{n_{12}}}.$$

d. For testing marginal homogeneity, the binomial parameter equals $\frac{1}{2}$. Explain why the normal approximation to the binomial yields the test statistic

$$z = \frac{n_{21} - n_{12}}{\sqrt{n_{12} + n_{21}}}.$$

(The chi-squared test using z^2 is referred to as *McNemar's test*. Note that pairs in which $y_{i1} = y_{i2}$ are irrelevant to inference about β_1.)

9.20 Refer to the previous exercise. Unlike the conditional ML estimator of β_1, the unconditional ML estimator is inconsistent (Andersen 1980, pp. 244–245).

a. Averaging over the population, explain why

$$\pi_{21} = E\left[\frac{1}{1+\exp(\beta_{0i})}\frac{\exp(\beta_{0i}+\beta_1)}{1+\exp(\beta_{0i}+\beta_1)}\right],$$

where the expectation refers to the distribution for $\{\beta_{0i}\}$ and $\{\pi_{ab}\}$ are the probabilities for the population analog of Table 9.1. Similarly, state π_{12}. For a random sample of size $n \to \infty$, explain why $n_{21}/n_{12} \xrightarrow{p} \exp(\beta_1)$.

b. Find the log-likelihood. Show that the likelihood equations are $y_{+j} = \sum_i P(y_{ij} = 1)$ and $y_{i+} = \sum_j P(y_{ij} = 1)$. Substituting $\exp(\beta_{0i})/[1 + \exp(\beta_{0i})] + \exp(\beta_{0i} + \beta_1)/[1 + \exp(\beta_{0i} + \beta_1)]$ in the second likelihood equation, show that $\hat{\beta}_{0i} = -\infty$ for the n_{22} subjects with $y_{i+} = 0$, $\hat{\beta}_{0i} = \infty$ for the n_{11} subjects with $y_{i+} = 2$, and $\hat{\beta}_{0i} = -\hat{\beta}_1/2$ for the $n_{21} + n_{12}$ subjects with $y_{i+} = 1$.

c. By breaking $\sum_i P(y_{ij} = 1)$ into components for the sets of subjects having $y_{i+} = 0$, $y_{i+} = 2$, and $y_{i+} = 1$, show that the first likelihood equation is, for $j = 1$, $y_{+1} = n_{22}(0) + n_{11}(1) + (n_{21} + n_{12})\exp(-\hat{\beta}_1/2)/[1 + \exp(-\hat{\beta}_1/2)]$. Explain why $y_{+1} = n_{11} + n_{12}$, and solve the first likelihood equation to show that $\hat{\beta}_1 = 2\log(n_{21}/n_{12})$. Hence, as a result of (**a**), $\hat{\beta}_1 \xrightarrow{p} 2\beta_1$.

9.21 A binary response $y_{ij} = 1$ or 0 for observation j on subject i, $i = 1, \ldots, n$, $j = 1, \ldots, d$. Let $\bar{y}_{.j} = \sum_i y_{ij}/n$, $\bar{y}_{i.} = \sum_j y_{ij}/d$, and $\bar{y} = \sum_i \sum_j y_{ij}/nd$. Regard $\{y_{i+}\}$ as fixed, and suppose each way to allocate the y_{i+} "successes" to the d observations is equally likely. Show that $E(y_{ij}) = \bar{y}_{i.}$, $\text{var}(y_{ij}) = \bar{y}_{i.}(1 - \bar{y}_{i.})$, and $\text{cov}(y_{ij}, y_{ik}) = -\bar{y}_{i.}(1 - \bar{y}_{i.})/(d-1)$ for $j \neq k$. For large n with independent subjects, explain why $(\bar{y}_{.1}, \ldots, \bar{y}_{.d})$ is approximately multivariate normal with pairwise correlation $\rho = -1/(d-1)$. Conclude that *Cochran's Q statistic* (Cochran 1950)

$$Q = \frac{n^2(d-1)\sum_{j=1}^{d}(\bar{y}_{.j} - \bar{y})^2}{d\sum_{i=1}^{n}\bar{y}_{i.}(1 - \bar{y}_{i.})}$$

has an approximate chi-squared distribution with $df = (d-1)$ for testing homogeneity of the d marginal distributions. Show that Q is unaffected by deleting all observations for which $y_{i1} = \cdots = y_{id}$.

9.22 For the GLMM for binary data using probit link function,

$$\Phi^{-1}[P(y_{ij} = 1 \mid \boldsymbol{u}_i)] = \boldsymbol{x}_{ij}\boldsymbol{\beta} + \boldsymbol{z}_{ij}\boldsymbol{u}_i$$

show that the corresponding marginal model is $\Phi^{-1}[P(y_{ij} = 1] = \boldsymbol{x}_{ij}\boldsymbol{\beta}[1 + \boldsymbol{z}_{ij}\boldsymbol{\Sigma}_u\boldsymbol{z}_{ij}^{\mathrm{T}}]^{-1/2}$. Compare effects in the GLMM and marginal model.

9.23 From Exercise 7.25, with any link and a factor predictor in the one-way layout, the negative binomial ML fitted means equal the sample means. Show this is not true for the Poisson GLMM.

9.24 For the Poisson GLMM (9.12) with random intercept, use the normal mgf to show that for $j \neq k$,

$$\mathrm{cov}(y_{ij}, y_{ik}) = \exp[(\boldsymbol{x}_{ij} + \boldsymbol{x}_{ik})\boldsymbol{\beta}]\left[\exp\left(\sigma_u^2\right)\left(\exp\left(\sigma_u^2\right) - 1\right)\right]$$

Find $\mathrm{corr}(y_{ij}, y_{ik})$. Explain why, as in binary GLMMs, $\mathrm{corr}(y_{ij}, y_{ik}) \geq 0$.

9.25 For recent US Presidential elections, in each state wealthier voters tend to be more likely to vote Republican, yet states that are wealthier in an aggregate sense are more likely to have more Democrat than Republican votes (Gelman and Hill 2007, Section 14.2.). Sketch a plot that illustrates how this instance of Simpson's paradox could occur. Specify a GLMM with random effects for states that could be used to analyze data for a sample of voters using their state of residence, their household income, and their vote in an election. Explain how the model could be generalized to allow the income effect to vary by state, to reflect that Republican-leaning states tend to have stronger associations between income and vote.

9.26 Construct a marginal model that is a multivariate analog of the normal linear model for a balanced two-way layout, assuming an absence of interaction. Show how to express the two main effect hypotheses for that model in terms of model parameters.

9.27 For subject i, $\boldsymbol{y}_i = (y_{i1}, \ldots, y_{id})^{\mathrm{T}}$ are d repeated measures on a response variable. Consider a model by which $\boldsymbol{y}_i \sim N(\boldsymbol{\mu}, \boldsymbol{\Sigma}_\epsilon)$ with $\sigma_1 = \cdots = \sigma_d$ and $\rho = \mathrm{corr}(y_{ij}, y_{ik})$ for all $j \neq k$. The hypothesis of interest is H_0: $\mu_1 = \cdots = \mu_d$.

a. Express this scenario as a marginal model for a multivariate response, and show how to express H_0 in terms of the model parameters.

b. Construct a random-effects model for this scenario that would have the specified correlation structure.

c. For $d = 2$, find a test statistic formula for H_0: $\mu_1 = \mu_2$ (e.g., the *paired-difference t test* of Exercise 9.6).

9.28 When $\boldsymbol{y} \sim N(\mathbf{0}, \boldsymbol{\Sigma}_\epsilon)$, in terms of elements k_{ij} from the *concentration matrix* $\boldsymbol{K} = \boldsymbol{\Sigma}_\epsilon^{-1}$, show that the pdf f for $\boldsymbol{y}$ has the form

$$\log f(y) = \text{constant} - \frac{1}{2} \sum_i k_{ii} y_i^2 - \sum_i \sum_j k_{ij} y_i y_j,$$

and the elements of $\boldsymbol{K}$ are natural parameters in the exponential family representation. Explain why y_i and y_j are conditionally independent, given the other elements of $\boldsymbol{y}$, if $k_{ij} = 0$. (As in Section 7.2.7, one can construct a graphical representation. A *Gaussian graphical model*, also called a *covariance selection model*, portrays the conditionally independent pairs by the absence of an edge. See Dempster (1972) and Lauritzen (1996).)

9.29 For the GEE (9.15) with $\boldsymbol{R}(\boldsymbol{\alpha}) = \boldsymbol{I}$, show that the equations simplify to

$$(1/\phi) \sum_{i=1}^{n} \boldsymbol{X}_i^{\mathrm{T}} \boldsymbol{\Delta}_i (\boldsymbol{y}_i - \boldsymbol{\mu}_i) = \mathbf{0},$$

where $\boldsymbol{\Delta}_i$ is the diagonal matrix with elements $\partial\theta_{ij}/\partial\eta_{ij}$ on the main diagonal for $j = 1, \ldots, d$, for natural parameter θ_{ij} and linear predictor η_{ij}. Show that $\hat{\boldsymbol{\beta}}$ is then the same as the ordinary ML estimator for a GLM with the chosen link function and variance function, treating $(y_{i1}, \ldots, y_{id})$ as independent observations.

9.30 Generalizing the heuristic argument in Section 8.3.2, justify why formula (9.16) is valid for the sandwich covariance matrix (Liang and Zeger 1986, Appendix).

9.31 For both linear mixed models and marginal models, the generalized least squares estimator $\hat{\boldsymbol{\beta}} = \left(\boldsymbol{X}^{\mathrm{T}} \boldsymbol{V}^{-1} \boldsymbol{X}\right)^{-1} \boldsymbol{X}^{\mathrm{T}} \boldsymbol{V}^{-1} \boldsymbol{y}$ is useful. Suppose $\boldsymbol{V}$ is not the true var($\boldsymbol{y}$) but we can consistently estimate var($\boldsymbol{y}$) empirically by $\boldsymbol{S}$. Describe a way to do this. Explain why it is then sensible to estimate var($\hat{\boldsymbol{\beta}}$) by

$$\left(\boldsymbol{X}^{\mathrm{T}} \hat{\boldsymbol{V}}^{-1} \boldsymbol{X}\right)^{-1} \boldsymbol{X}^{\mathrm{T}} \hat{\boldsymbol{V}}^{-1} \boldsymbol{S} \hat{\boldsymbol{V}}^{-1} \boldsymbol{X} \left(\boldsymbol{X}^{\mathrm{T}} \hat{\boldsymbol{V}}^{-1} \boldsymbol{X}\right)^{-1}.$$

9.32 For the smoking prevention and cessation study (Section 9.2.3), fit multilevel models to analyze whether it helps to add any interaction terms. Interpret fixed and random effects for the model that has a SC × TV interaction.

9.33 Using the R output shown for the simple analyses of the FEV data in Section 9.2.5, show that the estimated values of corr(y_{i1}, y_{i2}) and corr(y_{i1}, y_{i8}) are 0.74 for the random intercept model and 0.86 and 0.62 for the model that also permits autoregressive within-patient errors.

9.34 Refer to Exercise 1.21 and the longitudinal analysis in Section 9.2.5. Analyze the data in file `FEV2.dat` at `www.stat.ufl.edu/~aa/glm/data` , investigating the correlation structure for the eight FEV responses and modeling how FEV depends on the hour and the drug, adjusting for the baseline observation. Take into account whether to treat hour as qualitative or quantitative, whether you need interaction terms, whether to have random slopes or only random intercepts, and whether to treat within-patient errors as correlated. Interpret results for your final chosen model. (You may want to read Littell et al. (2000). The book *SAS for Mixed Models*, 2nd ed., by Littell et al. (2006, SAS Institute), uses SAS to fit various models to these data.)

9.35 A field study[11] analyzed associations between a food web response measure and plant invasion and tidal restriction in salt marsh habitats. The response, observed in a species of small marsh fish (*Fundulus heteroclitus*) that swim in and out of the salt marshes with the tides, was a stable carbon isotope measure[12] ($\delta^{13}C$). It is used as a tracer in food studies because its uptake by plants and animals follows predictable patterns, and so it quantifies the carbon basis in a diet. A salt marsh is called "tidally restricted" if the tidal flow is restricted because of factors such as roads and dikes built across the marsh for residential developments. Such marshes tend to be invaded by non-native vegetation. A salt marsh is called "tidally restored" if it had been tidally restricted but is now in a restored state, with less non-native vegetation. It is called a "reference" marsh if it had never been restricted. The study analyzed the extent to which a particular invasive plant (*Phragmites australis*) contributes carbon to the food web in restricted marshes, relative to reference marshes, and how that changes as the cover of the native plants increases during restoration. The study's experimental design used four tidally restored and four tidally restricted marshes and eight control marshes downstream from them. This resulted in 16 marshes and four marsh types: 1 = tidally restricted marsh, 2 = tidally restored marsh, 3 = control marsh near a tidally restricted marsh, 4 = control marsh near a tidally restored marsh. Half the marshes of each type were in each of two locations—Long Island Sound and the Gulf of Maine. Each marsh used three fixed locations for stations. These 16 marshes with 48 stations were visited in summer and fall of 2010 and 2011. The researchers collected 10 fish at each station on every sample date and used

[11]Thanks to Penelope Pooler and Kimberly Dibble for these data. For details, see the article by K. Dibble and L. A. Meyerson in *Estuaries and Coasts*, August 2013, pp. 1–15.
[12]For its definition, see en.wikipedia.org/wiki/Δ13C.

Table 9.7 Part of Data File from Salt Marsh Habitat Study

obs	d13C	Time	Treatment	Region	Fishlength
1	−13.05	Fall 2010	control-restored	LIS	69.20
2	−12.52	Fall 2010	control-restored	LIS	68.90
...					
192	−21.23	Summer 2011	tide-restricted	GOM	65.00

Complete data (file `Dibble.xlsx`) are at `www.stat.ufl.edu/~aa/glm/data`.

the composited sample to measure the response variable. The data file has the form shown in Table 9.7.

a. For the stable carbon isotope response (variable *d13C* in the data file) at time t for station i, consider the model

$$y_{it} = \beta_0 + \beta_1 x_{it1} + \beta_2 x_{it2} + \cdots + \beta_8 x_{it8} + u_i + \epsilon_{it},$$

where $(x_{it1}, x_{it2}, x_{it3})$ are indicators for 3 of the 4 marsh types, x_{it4} is an indicator for the region, $(x_{it5}, x_{it6}, x_{it7})$ are indicators for 3 of the 4 time periods, and x_{it8} is the mean length (in *mm*) from the 10 fish samples for that observation. For this model and application, specify the components and their dimensions in the model expression $\boldsymbol{y}_i = \boldsymbol{X}_i\boldsymbol{\beta} + \boldsymbol{Z}_i\boldsymbol{u}_i + \boldsymbol{\epsilon}_i$.

b. Fit the model, assuming $\text{var}(\boldsymbol{\epsilon}_i) = \sigma_\epsilon^2\boldsymbol{I}$. Interpret results.

c. Conditional on a random-effect value u_i, the authors expected observations at the same location to be correlated because of ecological processes, but less so with increasing time between them. So they used an autoregressive correlation structure in $\text{var}(\boldsymbol{\epsilon}_i)$. Fit this model, analyze whether the fit is much better than assuming $\text{var}(\boldsymbol{\epsilon}_i) = \sigma_\epsilon^2\boldsymbol{I}$.

d. Because the data variability itself varies substantially between summer and fall and from year to year, the authors also permitted heterogeneous variances. Show that

$$\text{cov}(y_{ij}, y_{ik}) = \sigma_u^2 + \sigma_{\epsilon(j)}\sigma_{\epsilon(k)}\rho^{|j-k|}.$$

(The researchers also fitted more-complex models with interactions among treatment, region, and time.)

9.36 For Table 7.5 on counts of victims of homicide, specify and fit a Poisson GLMM. Interpret estimates. Show that the deviance decreases by 116.6 compared with the Poisson GLM, and interpret.

9.37 The data file `Maculatum.dat` at the text website is from a study[13] of salamander embryo development. These data refer to the spotted salamander

[13]Data courtesy of Rebecca Hale, University of North Carolina Asheville.

(*Ambystoma maculatum*). One purpose of the study was to compare four rearing environments (very humid air, and water with low, medium, and saturated dissolved oxygen) on the age at hatching, for each of the embryos that survived to hatching. In the experiment, embryos from the same family were divided into four groups for the four treatments, and each group was reared together in the same jar. For the embryos that survived to hatching, use multilevel modeling to compare the mean ages for the four treatments. Compare estimates and *SE* values to ones you would obtain with an ordinary linear model with treatment as the explanatory variable, ignoring the dependence due to embryos in the same jar being from the same family and due to the same family of embryos being in four jars.

9.38 Refer to the previous exercise. For all the embryos, use a logistic GLMM to model the probability of hatching in terms of the treatment. Interpret results, and compare to those obtained with an ordinary logistic GLM that ignores the clustering.

9.39 Download the file `Rats.dat` at the text website for the teratology study in Table 8.1.

a. Use the GEE approach to fit the logistic model, assuming an exchangeable working correlation structure for observations within a litter. Show how the empirical sandwich adjustment increases the *SE* values compared with naive binomial ML. Report the estimated within-litter correlation between the binary responses, and compare with the value of 0.192 that yields the quasi-likelihood results with beta-binomial variance function.

b. Fit the GLMM that adds a normal random intercept u_i for litter i to the binomial logistic model. Interpret the estimated effects, and explain why they are larger than with the GEE approach.

9.40 Refer to Exercise 9.34. Analyze the FEV data with marginal models, and compare results to those obtained with linear mixed models.

9.41 A crossover study analyzed by B. Jones and M. Kenward (1987, *Stat. Med.* **6**: 555–564) compared three drugs on a binary outcome (success = 1, failure = 0). Counts for the eight possible response patterns for drugs (A, B, C) were 6 for 000, 9 for 001, 4 for 010, 45 for 011, 3 for 100, 7 for 101, 4 for 110, and 8 for 111. Compare the drugs using (**a**) a GLMM, (**b**) a marginal model. In each case, state all assumptions, and interpret results.

CHAPTER 10

Bayesian Linear and Generalized Linear Modeling

This book has used the traditional, often referred to as *frequentist*, approach to statistical inference. That approach regards parameter values as fixed effects rather than random variables. Probability statements apply to possible values for the data, given the parameter values. Increasingly popular is the *Bayesian* alternative, which applies probability distributions to parameters as well as to data. This yields inferences in the form of probability statements about the parameters, given the data. For example, after observing the data in a clinical trial, a researcher might evaluate and report the probability that the population mean of the response variable is higher for the active drug than for the placebo.

In this chapter we first review the Bayesian approach to statistical inference. Section 10.2 presents a Bayesian analog of the normal linear model, and Section 10.3 presents Bayesian generalized linear models (GLMs). In each section we show how to obtain essentially the same substantive results as with a frequentist approach, but with Bayesian interpretations that make probability statements about the parameters. The final section presents *empirical Bayes* and *hierarchical Bayes* approaches, which make weaker assumptions about prior probability distributions for the parameters.

10.1 THE BAYESIAN APPROACH TO STATISTICAL INFERENCE

Let $\boldsymbol{\theta}$ be a generic symbol for the parameters in a particular model, such as the $\boldsymbol{\beta}$ effects in the linear predictor and variance components. Parametric models assume a particular distribution for the data $\boldsymbol{y}$, described by a probability density (or mass) function $f(\boldsymbol{y}; \boldsymbol{\theta})$. We now express that function as $f(\boldsymbol{y} \mid \boldsymbol{\theta})$ to emphasize that it specifies the distribution of the data, given a particular value for the parameters.

Foundations of Linear and Generalized Linear Models, First Edition. Alan Agresti.

10.1.1 Prior and Posterior Distributions

The Bayesian approach involves two distributions for $\boldsymbol{\theta}$. As their names indicate, the *prior* distribution describes knowledge about $\boldsymbol{\theta}$ before we see $\boldsymbol{y}$, whereas the *posterior* distribution combines that prior information with $\boldsymbol{y}$ to update our knowledge.

The *prior distribution* for $\boldsymbol{\theta}$ is characterized by a probability density function (pdf) $h(\boldsymbol{\theta})$ specified over the space $\boldsymbol{\Theta}$ of possible $\boldsymbol{\theta}$ values. This probability distribution may reflect subjective prior beliefs, perhaps based on results of other studies. Or it may be relatively uninformative, so that inferential results are based almost entirely on the data $\boldsymbol{y}$. When h is a member of some parametric family, $h(\boldsymbol{\theta} \mid \boldsymbol{\lambda})$ is specified by its own parameters $\boldsymbol{\lambda}$, referred to as *hyperparameters*.

The information that $\boldsymbol{y}$ provides combines with the prior distribution to generate a *posterior distribution* for $\boldsymbol{\theta}$. By Bayes' theorem, the posterior pdf h of $\boldsymbol{\theta}$, given $\boldsymbol{y}$, relates to the assumed pdf f for $\boldsymbol{y}$, given $\boldsymbol{\theta}$, and the prior pdf $h(\boldsymbol{\theta})$, by

$$h(\boldsymbol{\theta} \mid \boldsymbol{y}) = \frac{f(\boldsymbol{y} \mid \boldsymbol{\theta})h(\boldsymbol{\theta})}{f(\boldsymbol{y})} = \frac{f(\boldsymbol{y} \mid \boldsymbol{\theta})h(\boldsymbol{\theta})}{\int_{\boldsymbol{\Theta}} f(\boldsymbol{y} \mid \boldsymbol{\zeta})h(\boldsymbol{\zeta})d\boldsymbol{\zeta}}. \tag{10.1}$$

When we observe $\boldsymbol{y}$ and view $f(\boldsymbol{y} \mid \boldsymbol{\theta})$ as a function of $\boldsymbol{\theta}$, $\ell(\boldsymbol{\theta}) = f(\boldsymbol{y} \mid \boldsymbol{\theta})$ is the likelihood function. The denominator $f(\boldsymbol{y})$ in (10.1) is the marginal pdf for $\boldsymbol{y}$. It is constant with respect to $\boldsymbol{\theta}$ and merely causes the posterior density to integrate to 1. It is irrelevant for inference comparing different values for $\boldsymbol{\theta}$. So the prior pdf for $\boldsymbol{\theta}$ multiplied by the likelihood function determines the posterior pdf $h(\boldsymbol{\theta} \mid \boldsymbol{y})$. That is, for $\boldsymbol{\theta}$, the key part of Bayes' theorem is the numerator,

$$h(\boldsymbol{\theta} \mid \boldsymbol{y}) \propto f(\boldsymbol{y} \mid \boldsymbol{\theta})h(\boldsymbol{\theta}) = \ell(\boldsymbol{\theta})h(\boldsymbol{\theta}).$$

Bayesian inference depends on the data only through the likelihood function. Statistical inference for which this applies, such as maximum likelihood (ML) estimation as well as Bayesian inference, is said to satisfy the *likelihood principle*.

For a future observation y^*, the posterior *predictive distribution* is the conditional distribution of $(y^* \mid y_1, \dots, y_n)$. We find this from

$$f(y^* \mid \boldsymbol{y}) = \int f(y^* \mid \boldsymbol{y}, \boldsymbol{\theta})h(\boldsymbol{\theta} \mid \boldsymbol{y})d\boldsymbol{\theta},$$

that is, taking the distribution for y^* as if we know $\boldsymbol{\theta}$ and then integrating out with respect to its posterior pdf. This is a sort of mean distribution for y^*, averaged over the posterior distribution.

10.1.2 Types of Prior Distributions

A *subjective prior distribution* reflects the researcher's beliefs about the value for $\boldsymbol{\theta}$. For GLM parameters, it may be unclear to a researcher how to formulate this, and a danger is being overly optimistic, making the prior distribution too narrow. It is more common to use an *objective prior*, which is relatively uninformative, having very little influence on the posterior results; that is, inferential statements depend almost entirely on the data, through the likelihood function for the assumed model.

An objective prior is flat relative to the likelihood function. The uniform density over the parameter space satisfies this, but this depends on the parameterization. For a binomial parameter π, for example, uniformity for π over (0,1) differs from uniformity over real-line values for logit(π). And uniformity is not a proper distribution when the parameter space has an infinite range, as it then integrates to ∞ rather than 1. Such prior distributions are called *improper*. With an improper prior, the posterior distribution need not be proper. Not as extreme is a *diffuse* proper prior spread out over a large region of the parameter space, such as a normal prior with relatively large σ. Results for certain parameters, however, may depend critically on just how diffuse the prior is, especially when the number of parameters p is large. One way to implement this selects the prior distribution so that its impact on the posterior is comparable to that of a single observation (Kass and Wasserman 1995).

The *conjugate prior distribution* is the family of probability distributions such that, when combined with the likelihood function, the posterior distribution falls in the same family. For example, when we combine a normal prior distribution for the mean of a normal distribution with a normal likelihood function, the posterior distribution is normal (Section 10.2). When we combine a beta prior distribution for a binomial parameter with a binomial likelihood function, the posterior distribution is beta (Section 10.1.3). When we combine a gamma prior distribution for a Poisson mean parameter with a Poisson likelihood function, the posterior distribution is gamma (Exercise 10.20). Conjugate prior distributions have the advantage of computational simplicity for finding the posterior distribution, but they exist for relatively few models beyond simple ones such as for a single mean or the one-way layout.

The *Jeffreys prior distribution* is proportional to the square root of the determinant of the Fisher information matrix for the parameters of interest. It has the advantage of being invariant to the parameterization. The prior distributions for different functions of a parameter are equivalent (e.g., for the binomial parameter or its logit). For single-parameter analyses, this is a reasonable way to construct an objective prior in a straightforward way, but for models with a large p it can be cumbersome and unappealing (Berger et al. 2013).

Another possibility is hierarchical in nature (Section 10.4): We assume a probability distribution for the prior hyperparameters $\boldsymbol{\lambda}$ instead of assigning them fixed values. The prior distribution is then obtained by integrating $h(\boldsymbol{\theta} \mid \boldsymbol{\lambda})$ with respect to that distribution for $\boldsymbol{\lambda}$. Objectivity is enhanced by taking that second-stage prior to be diffuse.

Just as we never completely believe any model, likewise we should be skeptical about any particular choice of prior distribution. It is informative to check how posterior results vary according to the choice. As n increases relative to the number of parameters, the likelihood function has an increasingly dominant influence in Bayes' theorem, and results are less sensitive to the choice.

10.1.3 Binomial Parameter: Beta Prior and Posterior Distributions

We illustrate prior distributions for a binomial parameter π. The simplest Bayesian inference uses a beta distribution as the prior distribution. The beta(α_1, α_2) family (Section 8.2.3) has hyperparameter values $\alpha_1 > 0$ and $\alpha_2 > 0$ that provide a wide variety of pdf shapes over (0, 1). The Jeffreys prior is the symmetric U-shaped beta with $\alpha_1 = \alpha_2 = 0.5$.

For iid Bernoulli trials $y_1, \dots, y_n$ with parameter π, let $y = (\sum_i y_i)/n$, so that $(ny \mid \pi) \sim \text{bin}(n, \pi)$. The beta distribution is the conjugate prior distribution for inference about π. When we combine a beta(α_1, α_2) prior distribution with a binomial likelihood function, the posterior density $h(\pi \mid y) \propto f(y \mid \pi)h(\pi)$, or

$$h(\pi \mid y) \propto \pi^{ny}(1-\pi)^{n-ny}\pi^{\alpha_1 - 1}(1-\pi)^{\alpha_2 - 1} = \pi^{ny+\alpha_1-1}(1-\pi)^{n-ny+\alpha_2-1}$$

over (0, 1). So the posterior distribution is a beta($ny + \alpha_1, n - ny + \alpha_2$) distribution. The posterior mean

$$\frac{ny + \alpha_1}{n + \alpha_1 + \alpha_2} = \left(\frac{n}{n + \alpha_1 + \alpha_2}\right) y + \left(\frac{\alpha_1 + \alpha_2}{n + \alpha_1 + \alpha_2}\right) \frac{\alpha_1}{\alpha_1 + \alpha_2}$$

is a weighted average of the sample proportion y and the prior mean, y receiving greater weight as n increases. We can interpret $\alpha_1 + \alpha_2$ as the effective sample size represented by the prior distribution, corresponding to α_1 "successes" and α_2 "failures." The posterior predictive distribution has

$$P(y^* = 1 \mid y) = \int_0^1 P(y^* = 1 \mid y, \pi)h(\pi \mid y)d\pi = \int_0^1 \pi h(\pi \mid y)d\pi,$$

which is the posterior mean.

10.1.4 Markov Chain Monte Carlo (MCMC) Methods for Finding Posterior

For GLMs, usually the posterior distribution has no closed-form expression. The difficulty is in evaluating the denominator integral that determines $f(\mathbf{y})$, that is, determining the appropriate constant so that the posterior integrates to 1. Simulation methods can approximate the posterior distribution. The primary method for doing this is *Markov chain Monte Carlo* (MCMC). It is beyond our scope to present detailed descriptions of MCMC algorithms (Note 10.2 has references). In brief, a stochastic process of $\boldsymbol{\theta}$ values having Markov-chain form is constructed so that its long-run stationary distribution is the posterior distribution of $\boldsymbol{\theta}$. One or more such long Markov chains provide a very large number of simulated values from the posterior distribution, and the distribution of the simulated values approximates the posterior distribution.

Gelfand and Smith (1990) showed how to use MCMC methods to determine Bayesian posterior distributions, and the literature on implementing Bayesian methods exploded soon after that. The two primary MCMC methods are *Gibbs sampling* and the *Metropolis–Hastings algorithm*. These each require only that we know a function that is proportional to the posterior distribution, which for Bayesian inference is true once we multiply the prior distribution by the likelihood function.

The Metropolis–Hastings algorithm randomly generates a potential new value $\boldsymbol{\theta}^*$ in the chain from a "proposal density," conditional on the current value $\boldsymbol{\theta}^{(t)}$ of the chain. For symmetric proposal densities for which the density of $\boldsymbol{\theta}^*$ given $\boldsymbol{\theta}^{(t)}$ is identical to the density of $\boldsymbol{\theta}^{(t)}$ given $\boldsymbol{\theta}^*$, the process is quite simple: The next iterate $\boldsymbol{\theta}^{(t+1)}$ equals $\boldsymbol{\theta}^*$ with probability

$$h(\boldsymbol{\theta}^* \mid \mathbf{y})/h(\boldsymbol{\theta}^{(t)} \mid \mathbf{y}),$$

taking the probability to be 1 if this ratio exceeds 1, and otherwise $\boldsymbol{\theta}^{(t+1)} = \boldsymbol{\theta}^{(t)}$. In this process, although we do not know $h(\cdot)$, we can calculate this ratio because the unknown normalizing constant $f(\mathbf{y})$ appears both in the numerator and denominator, and hence cancels. Software that uses this algorithm incorporates a proposal density from which it is easy to generate random numbers, acceptance probabilities are relatively high, and successive $\boldsymbol{\theta}^{(t)}$ have a relatively low autocorrelation.

The Gibbs sampling scheme approximates the posterior distribution for $\boldsymbol{\theta}$ by iteratively sampling each element of $\boldsymbol{\theta}$ from its full conditional distribution, given the other elements of $\boldsymbol{\theta}$. That is, the new iterate $\theta_j^{(t+1)}$ is randomly generated from the density

$$h(\theta_j \mid \theta_1^{(t+1)}, \ldots, \theta_{j-1}^{(t+1)}, \theta_{j+1}^{(t)}, \ldots, \theta_p^{(t)}, \mathbf{y}),\ j = 1, 2, \ldots, p.$$

In many cases this conditional density is quite simple, with the unknown normalizing constant canceling in the ratio of probabilities specified by this conditional density. The samples generated of $(\theta_1^{(t)}, \ldots, \theta_p^{(t)})$ for large t approximate the joint posterior distribution of $\boldsymbol{\theta}$. See Casella and George (1992) for details and insight for why this works. Gibbs sampling is a special case of the Metropolis–Hastings algorithm in which the conditional densities form the proposal density at each step of an iteration.

Both Gibbs sampling and the Metropolis–Hastings algorithm are designed to ensure eventual convergence to the stationary distribution. Convergence can be slow, however, because MCMC samplers yield dependent draws. When the process stops, a Monte Carlo standard error indicates how close the final values are likely to be to the actual ML estimates.

10.1.5 de Finetti's Theorem: Independence and Exchangeability

Throughout this book, in univariate response modeling, a standard assumption is that $y_1, \ldots, y_n$ are independent. But if we observe y_1, that tells us something about the model parameters, so does not that also give us information about what to expect for $y_2, \ldots, y_n$? In classical frequentist statistics, no, because the assumed independence of $\{y_1, \ldots, y_n\}$ is conditional on the parameter values. For instance, in the normal linear model for the one-way layout, $y_{i1}, \ldots, y_{in_i}$ are assumed to be independent, conditional on the parameters (μ_i, σ^2) of their distribution.

In Bayesian inference, because parameters are random variables, it is more natural to treat observations at any particular setting of the explanatory variables as *exchangeable* than as independent. That is, their probability distribution is identical if we permute the observations in any way whatever. A generalization of a result due to the Italian statistician Bruno de Finetti in 1937 states that the observations are independent, conditional on the parameter value for a mixture distribution. This motivates the way Bayesian models are constructed: For a particular parametric family, observations are treated as independent, conditional on the parameter values; but assuming a prior probability distribution for those parameters implies that marginally the observations may be correlated but (at a fixed setting for explanatory variables) are exchangeable. We represent the pdf for $\boldsymbol{y}$ as

$$f(\boldsymbol{y}) = \int_{\boldsymbol{\Theta}} f(\boldsymbol{y} \mid \boldsymbol{\theta}) h(\boldsymbol{\theta}) d\boldsymbol{\theta} = \int_{\boldsymbol{\Theta}} \left[\prod_{i=1}^{n} f(y_i \mid \boldsymbol{\theta}) \right] h(\boldsymbol{\theta}) d\boldsymbol{\theta}.$$

Bayesian hierarchical models also apply exchangeability to sets of parameter values. As we'll observe, Bayesian methods then "borrow from the whole" in using all the data to estimate a parameter for any individual group.

10.1.6 Parallels Between Bayesian and Frequentist Inferences

Bayesian methods of statistical inference using the posterior distribution parallel those for frequentist inference, with analogs of point estimates, confidence intervals, and significance tests. The usual Bayes estimate of $\boldsymbol{\theta}$ is the mean of its posterior distribution. If we use a uniform prior $h(\boldsymbol{\theta})$ (possibly improper), then $h(\boldsymbol{\theta} \mid \boldsymbol{y})$ is a constant multiple of the likelihood function. That is, the posterior distribution (when it exists) is a scaling of the likelihood function so that it integrates to 1. The mode of $h(\boldsymbol{\theta} \mid \boldsymbol{y})$ is then the ML estimate $\hat{\boldsymbol{\theta}}$. With a proper prior density, when n is small or the posterior distribution is quite skewed, the posterior mean can be quite different from the posterior mode and thus from the ML estimate. It is then characteristic of Bayes estimates that they *shrink* the ML estimate toward the prior mean. This chapter shows numerous examples.

Analogous to the frequentist confidence interval is a *posterior interval*, also often called a *credible interval*. We can construct this for θ_j using percentiles of $h(\theta_j \mid \boldsymbol{y})$, with equal probabilities in the two tails. The 95% equal-tail posterior interval for θ_j is the region between the 2.5 percentile and 97.5 percentile of $h(\theta_j \mid \boldsymbol{y})$. An alternative *highest posterior density* (HPD) region has higher posterior density for every value inside the region than for every value outside it, subject to the posterior probability over the region equaling the desired confidence level. For unimodal posteriors this method produces an interval, the shortest possible one with the given confidence level.

For Bayesian posterior intervals, the coverage probability applies *after* observing the data, whereas for frequentist inferences it applies *before* (i.e., based on the distribution for the random data). For a Bayesian 95% posterior interval, $P(\theta_j \in \text{interval} \mid \boldsymbol{y}) = 0.95$ based on $h(\theta_j \mid \boldsymbol{y})$; by contrast, for a frequentist 95% confidence interval, $P(\text{interval contains } \theta_j \mid \boldsymbol{\theta}) = 0.95$ based on random endpoints for that interval constructed using $\boldsymbol{y}$.

For significance tests about θ_j, a frequentist P-value is a tail probability of "more extreme" results for the data, for a given θ_j value. For a Bayesian approach, in lieu of P-values, useful summaries are posterior tail probabilities about θ_j, given the data. Information about the direction of an effect is contained in $P(\theta_j > 0 \mid \boldsymbol{y})$ and $P(\theta_j < 0 \mid \boldsymbol{y})$. With a prior distribution that is flat relative to the likelihood function, $P(\theta_j < 0 \mid \boldsymbol{y})$ takes similar value as a frequentist P-value for a one-sided test with $H_1: \theta_j > 0$.

The Bayesian analog of a two-sided test of H_0: $\theta_j = 0$ against H_1: $\theta_j \neq 0$ is not so obvious. For GLM parameters that take values over the entire real line or some interval subset of it, it is common to use continuous prior distributions. Then the prior $P(\theta_j = 0) = 0$, as is the posterior probability. To evaluate whether H_0 is plausible, we can evaluate the posterior probability of the region of values having smaller density than at $\theta_j = 0$. Or we could base this judgment on whether the posterior interval for θ_j of desired confidence level contains 0. Analogous approaches apply for testing the general linear hypothesis H_0: $\boldsymbol{\Lambda\beta} = \boldsymbol{0}$.

For a continuous y, we can use the posterior predictive distribution $f(y^* \mid \boldsymbol{y})$ to find a prediction interval for a future response value. The interpretation for such an interval is more natural than the rather tortuous one presented in Section 3.3.5 for a frequentist prediction interval, because the posterior distribution naturally applies conditional on $\boldsymbol{y}$. We can state that a Bayesian 95% prediction interval has probability 0.95 of containing y^* (assuming, of course, that the model is correct).

10.1.7 Bayesian Model Checking

Many Bayesian model-checking methods also parallel frequentist methods. For example, sensitivity analyses investigate how posterior inferences change when alternative reasonable models are used. Case-deletion diagnostics summarize the influence of individual observations. If the model is adequate, new datasets generated randomly from the model using the predictive distribution should look like the observed data. Analogs of test statistics compare summaries of the observed data with the corresponding summaries of predictive simulations based on the model. Analogs of P-values find the probability that replicated data are more extreme, in some sense, than the observed data. See Gelman et al. (2013, Chapters 6 and 7) and Ntzoufras (2009, Chapter 10) for details.

The *Bayes factor* comparing two models, which need not be nested, is the ratio of their marginal densities, $f_1(\boldsymbol{y})/f_2(\boldsymbol{y})$. This can, however, be sensitive to the choice of the prior distribution and may provide quite different results than frequentist likelihood-ratio methods (O'Hagan and Forster 2004, pp. 177–183). A Bayesian analog of the Akaike information criterion (AIC) is the *Bayesian information criterion*

(BIC) introduced in Section 4.6.3 (Schwarz 1978), which adjusts the maximum log-likelihood L for a model with p parameters:

$$\text{BIC} = -2L + [\log(n)]p.$$

This does not depend on a prior distribution, but with large n, the difference between the BIC values for two models is approximately twice the log of the Bayes factor. For a set of reasonable candidate models, the model with the highest posterior probability tends to be the one that minimizes BIC.

Rather than selecting a model and then ignoring the model uncertainty in making inferences, a *Bayesian model averaging* approach provides a mechanism for accounting for model uncertainty. See Raftery et al. (1997) and Hoeting et al. (1999) for details. Note 10.4 lists additional references about model checking.

10.2 BAYESIAN LINEAR MODELS

Lindley and Smith (1972) presented the following Bayesian model for a normal linear model with known covariance matrices:

Bayesian normal linear model:

For positive-definite $\boldsymbol{\Sigma}_1$ and $\boldsymbol{\Sigma}_2$, suppose

$$(\boldsymbol{y} \mid \boldsymbol{\beta}_1, \boldsymbol{\Sigma}_1) \sim N(\boldsymbol{X}_1\boldsymbol{\beta}_1, \boldsymbol{\Sigma}_1), \quad \text{where } (\boldsymbol{\beta}_1 \mid \boldsymbol{\beta}_2, \boldsymbol{\Sigma}_2) \sim N(\boldsymbol{X}_2\boldsymbol{\beta}_2, \boldsymbol{\Sigma}_2). \quad (10.2)$$

Then the posterior distribution of $\boldsymbol{\beta}_1$ is $N(\tilde{\boldsymbol{\mu}}, \tilde{\boldsymbol{\Sigma}})$, where

$$\tilde{\boldsymbol{\mu}} = \tilde{\boldsymbol{\Sigma}}\left[\boldsymbol{X}_1^{\mathrm{T}}\boldsymbol{\Sigma}_1^{-1}\boldsymbol{y} + \boldsymbol{\Sigma}_2^{-1}\boldsymbol{X}_2\boldsymbol{\beta}_2\right], \quad \tilde{\boldsymbol{\Sigma}} = \left(\boldsymbol{X}_1^{\mathrm{T}}\boldsymbol{\Sigma}_1^{-1}\boldsymbol{X}_1 + \boldsymbol{\Sigma}_2^{-1}\right)^{-1}. \quad (10.3)$$

This result treats $\boldsymbol{\beta}_2$, $\boldsymbol{\Sigma}_1$, and $\boldsymbol{\Sigma}_2$ as known. In practice, a noninformative approach for the prior hyperparameters takes $\boldsymbol{\Sigma}_2$ to be a diagonal matrix with very large elements and takes $\boldsymbol{\beta}_2 = \boldsymbol{0}$. It is unrealistic that we would know $\boldsymbol{\Sigma}_1$, but this assumption is sufficient for our illustration of basic ideas of Bayesian linear modeling. For the ordinary linear model, $\boldsymbol{\Sigma}_1 = \sigma^2\boldsymbol{I}$, and we extend the model to include a prior distribution for σ^2 in Section 10.2.4. At any particular setting of the explanatory variables, the prior and posterior distributions for $\boldsymbol{\beta}_1$ induce a corresponding prior and posterior distribution for the expected response $\boldsymbol{X}_1\boldsymbol{\beta}_1$.

10.2.1 Bayesian Estimation of a Normal Mean

To illustrate the Bayesian normal linear model, we first consider the simplest case: The data is an observation $y \sim N(\mu, \sigma_1^2)$ and the goal is to estimate μ, which has a $N(\lambda, \sigma_2^2)$ prior distribution. Slightly more generally, we could let y be the sample mean of n observations and take σ_1^2 to be the variance of that sample mean.

The product of the likelihood function and the prior pdf is proportional to

$$e^{-\frac{(y-\mu)^2}{2\sigma_1^2}} e^{-\frac{(\mu-\lambda)^2}{2\sigma_2^2}}.$$

Combining terms in a common exponent, writing that term as quadratic in μ and then completing the square, we find that this product is proportional to

$$e^{-\frac{1}{2\tilde{\sigma}^2}(\mu-\tilde{\mu})^2} \quad \text{with} \quad \tilde{\mu} = \frac{\sigma_1^2 y + \sigma_2^2 \lambda}{\sigma_1^2 + \sigma_2^2} \quad \text{and} \quad \tilde{\sigma}^2 = \left(\sigma_1^{-2} + \sigma_2^{-2}\right)^{-1}.$$

This exponential function is a constant multiple of the $N(\tilde{\mu}, \tilde{\sigma}^2)$ pdf, so that is the posterior distribution of μ. The posterior mean $\tilde{\mu}$ is a weighted average of y and the prior mean λ, shrinking the observation toward λ. This is the same shrinkage behavior as found for the binomial model in Section 10.1.3. In fact, substituting in the Lindley and Smith formulation of the Bayesian normal linear model $\boldsymbol{y} = y$, $\boldsymbol{\beta}_1 = \mu$, $\boldsymbol{X}_1 = \boldsymbol{X}_2 = 1$, $\boldsymbol{\beta}_2 = \lambda$, $\boldsymbol{\Sigma}_1 = \sigma_1^2$, and $\boldsymbol{\Sigma}_2 = \sigma_2^2$ yields this result for the posterior distribution.

The reciprocal of a variance is a measure of information referred to as the *precision*. We see that the precision of the posterior estimator of the mean equals the sum of the precision of the sample observation and the precision of the prior mean.

10.2.2 A Bayesian Analog of the Normal Linear Model

From Chapter 3, the ordinary normal linear model for n independent observations $\boldsymbol{y} = (y_1, \ldots, y_n)^{\mathrm{T}}$, with $\boldsymbol{\mu} = (\mu_1, \ldots, \mu_n)^{\mathrm{T}}$ for $\mu_i = E(y_i)$, states that

$$\boldsymbol{y} \sim N(\boldsymbol{X}\boldsymbol{\beta}, \sigma^2 \boldsymbol{I}).$$

For a Bayesian analog, in the general Bayesian normal linear model (10.2), we take

$$(\boldsymbol{y} \mid \boldsymbol{\beta}, \sigma^2) \sim N(\boldsymbol{X}\boldsymbol{\beta}, \sigma^2 \boldsymbol{I}), \quad (\boldsymbol{\beta} \mid \lambda, \tau^2) \sim N(\lambda \boldsymbol{1}, \tau^2 \boldsymbol{I}),$$

that is, independent observations for which the elements of $\boldsymbol{\beta}$ have a common prior mean λ (usually taken to be 0) and a common prior variance τ^2. For the common prior variance to be sensible in this exchangeable treatment of the effects, we standardize the explanatory variables so that the effects are comparable in their prior magnitude.

From (10.3), the posterior distribution of $\boldsymbol{\beta}$ is $N(\tilde{\boldsymbol{\beta}}, \tilde{\boldsymbol{\Sigma}})$, with

$$\tilde{\boldsymbol{\beta}} = \tilde{\boldsymbol{\Sigma}}[\boldsymbol{X}^{\mathrm{T}}\sigma^{-2}\boldsymbol{y} + \tau^{-2}\lambda\boldsymbol{1}], \quad \tilde{\boldsymbol{\Sigma}} = \left(\sigma^{-2}\boldsymbol{X}^{\mathrm{T}}\boldsymbol{X} + \tau^{-2}\boldsymbol{I}\right)^{-1}. \quad (10.4)$$

The posterior mean $\tilde{\boldsymbol{\beta}}$ is

$$\tilde{\boldsymbol{\beta}} = \left(\sigma^{-2}\boldsymbol{X}^{\mathrm{T}}\boldsymbol{X} + \tau^{-2}\boldsymbol{I}\right)^{-1}[\sigma^{-2}(\boldsymbol{X}^{\mathrm{T}}\boldsymbol{X})\hat{\boldsymbol{\beta}} + \tau^{-2}\lambda\mathbf{1}],$$

where $\hat{\boldsymbol{\beta}} = \left(\boldsymbol{X}^{\mathrm{T}}\boldsymbol{X}\right)^{-1}\boldsymbol{X}^{\mathrm{T}}\boldsymbol{y}$ is the ordinary least squares estimate. The coefficients of $\hat{\boldsymbol{\beta}}$ and $\lambda\mathbf{1}$ sum to $\boldsymbol{I}$, so $\tilde{\boldsymbol{\beta}}$ is a weighted average of $\hat{\boldsymbol{\beta}}$ and the prior mean. The additivity of precision mentioned above for Bayesian inference about a single mean generalizes: The posterior $\tilde{\boldsymbol{\Sigma}}^{-1} = (\sigma^{-2}\boldsymbol{X}^{\mathrm{T}}\boldsymbol{X} + \tau^{-2}\boldsymbol{I})$ is a sum of the inverse covariance matrix for the least squares estimator and the inverse of the prior covariance matrix. As τ grows unboundedly, so that the prior distribution is more diffuse, $\tilde{\boldsymbol{\beta}}$ converges to $\hat{\boldsymbol{\beta}}$ and $\tilde{\boldsymbol{\Sigma}}$ converges to $\sigma^2(\boldsymbol{X}^{\mathrm{T}}\boldsymbol{X})^{-1}$, the covariance matrix of $\hat{\boldsymbol{\beta}}$.

Let us see how to obtain the posterior (10.4). By Bayes' theorem,

$$h(\boldsymbol{\beta} \mid \boldsymbol{y}) \propto \exp\left\{-\frac{1}{2}\left[\sigma^{-2}(\boldsymbol{y} - \boldsymbol{X}\boldsymbol{\beta})^{\mathrm{T}}(\boldsymbol{y} - \boldsymbol{X}\boldsymbol{\beta}) + \tau^{-2}(\boldsymbol{\beta} - \lambda\mathbf{1})^{\mathrm{T}}(\boldsymbol{\beta} - \lambda\mathbf{1})\right]\right\}.$$

We express this as $e^{-\frac{1}{2}Q}$, where, collecting terms together that are quadratic and linear in $\boldsymbol{\beta}$,

$$Q = \boldsymbol{\beta}^{\mathrm{T}}\left(\sigma^{-2}\boldsymbol{X}^{\mathrm{T}}\boldsymbol{X} + \tau^{-2}\boldsymbol{I}\right)\boldsymbol{\beta} - 2\left(\sigma^{-2}\boldsymbol{X}^{\mathrm{T}}\boldsymbol{y} + \tau^{-2}\lambda\mathbf{1}\right)^{\mathrm{T}}\boldsymbol{\beta} + c,$$

with c constant with respect to $\boldsymbol{\beta}$. Completing the quadratic form in $\boldsymbol{\beta}$ and taking $\tilde{\boldsymbol{\Sigma}}$ to be the expression in (10.4),

$$Q = \left[\boldsymbol{\beta} - \tilde{\boldsymbol{\Sigma}}\left(\sigma^{-2}\boldsymbol{X}^{\mathrm{T}}\boldsymbol{y} + \tau^{-2}\lambda\mathbf{1}\right)\right]^{\mathrm{T}}\tilde{\boldsymbol{\Sigma}}^{-1}\left[\boldsymbol{\beta} - \tilde{\boldsymbol{\Sigma}}\left(\sigma^{-2}\boldsymbol{X}^{\mathrm{T}}\boldsymbol{y} + \tau^{-2}\lambda\mathbf{1}\right)\right] + c',$$

where c' is another constant with respect to $\boldsymbol{\beta}$. But, as a function of $\boldsymbol{\beta}$, taking $\tilde{\boldsymbol{\beta}}$ to be the expression in (10.4) and ignoring the constant,

$$e^{-\frac{1}{2}Q} = e^{-\frac{1}{2}\left[(\boldsymbol{\beta}-\tilde{\boldsymbol{\beta}})\tilde{\boldsymbol{\Sigma}}^{-1}(\boldsymbol{\beta}-\tilde{\boldsymbol{\beta}})\right]}.$$

As a function of $\boldsymbol{\beta}$, this is proportional to the $N(\tilde{\boldsymbol{\beta}}, \tilde{\boldsymbol{\Sigma}})$ density.

10.2.3 Bayesian Approach to Normal One-Way Layout

We illustrate the Bayesian normal linear model for the one-way layout, with n_i observations for group i having $E(y_{ij}) = \mu_i, i = 1, \ldots, c$. We use the structure just given in which $\mathrm{var}(\boldsymbol{y}) = \sigma^2\boldsymbol{I}$ and the prior distribution has $E(\boldsymbol{\beta}) = \lambda\mathbf{1}$ and $\mathrm{var}(\boldsymbol{\beta}) = \tau^2\boldsymbol{I}$. We parameterize the model in $\boldsymbol{X}$ such that $\beta_i = \mu_i$ by setting the usual intercept $= 0$. In summary, then,

$$\left(y_{ij} \mid \mu_i, \sigma^2\right) \sim N(\mu_i, \sigma^2), \text{ independently for } j = 1, \ldots, n_i,$$

$$\text{with } (\mu_i \mid \lambda, \tau^2) \sim N(\lambda, \tau^2), \text{ independently for } i = 1, \ldots c.$$

The $N(\mu_i, \sigma^2)$ distribution describes the within-group variability, and the $N(\lambda, \tau^2)$ distribution describes the between-group variability.

With this structure, the posterior mean $\tilde{\mu}$ has elements

$$\tilde{\mu}_i = w\bar{y}_i + (1 - w)\lambda$$

with $w = \tau^2/\left[\frac{\sigma^2}{n_i} + \tau^2\right]$. This is a weighted average of $\bar{y}_i$ and the prior mean, with $\bar{y}_i$ receiving greater weight as n_i increases. The Bayes estimator shrinks the ML (and least squares) estimate $\bar{y}_i$ toward λ. With fixed n_i, less shrinkage occurs as τ^2 increases. The ML sample mean estimates emerge in the limit as τ^2 increases unboundedly, so that the prior distribution for $\{\mu_i\}$ moves toward the improper uniform prior over $\mathbb{R}^c$. The posterior $\tilde{\Sigma}^{-1}$ is a $c \times c$ diagonal matrix with precision entries $\{\frac{n_i}{\sigma^2} + \frac{1}{\tau^2}\}$.

Using the posterior normal distributions, we can construct posterior intervals for the $\mu_i - \mu_j$. Such differences themselves have a normal distribution, so it is straightforward to find useful summaries such as $P(\mu_i > \mu_j \mid y) = P(\mu_i - \mu_j > 0 \mid y)$.

10.2.4 Unknown Variance in Normal Linear Model

In practice, the error variance σ^2 in the ordinary normal linear model is unknown and requires its own prior distribution. In the frequentist approach, with a full-rank $n \times p$ model matrix $\boldsymbol{X}$ and error mean square

$$s^2 = (\boldsymbol{y} - \boldsymbol{X\beta})^{\mathrm{T}}(\boldsymbol{y} - \boldsymbol{X\beta})/(n - p),$$

by Cochran's theorem, $(n - p)s^2/\sigma^2$ has a chi-squared distribution with $df = n - p$. That approach treats s^2 as the random variable, for fixed σ^2. This suggests that in a prior distribution for σ^2, we could choose hyperparameters $\nu_0 > 0$ and σ_0^2 such that $\nu_0\sigma_0^2/\sigma^2$ has a chi-squared distribution with $df = \nu_0$. Here σ_0^2 is a prior guess for the value of σ^2, and ν_0 is a measure of the number of observations to which that information corresponds. Then σ^2 itself has an *inverse chi-squared distribution*. Its prior pdf

$$h(\sigma^2) \propto (\sigma^2)^{-(\nu_0/2+1)} \exp\left(\frac{-\nu_0\sigma_0^2}{2\sigma^2}\right), \quad \sigma^2 > 0. \tag{10.5}$$

This distribution, which is skewed to the right, is a special case of the *inverse-gamma* distribution; that is, $1/\sigma^2$ has the ordinary gamma distribution (4.29) with $(k, k/\mu) = (\nu_0/2, \nu_0\sigma_0^2/2)$. A limiting version of the inverse chi-squared prior for σ^2 that results from letting $\nu_0 \downarrow 0$ is improper. For that limiting case, $f(\sigma^2) \propto 1/\sigma^2$. This corresponds to an improper uniform prior distribution over the entire real line for $\log(\sigma^2)$.

We continue to assume that $\mathrm{var}(\boldsymbol{y}) = \sigma^2\boldsymbol{I}$ and that $\boldsymbol{\beta}$ has prior distribution $N(\lambda\mathbf{1}, \tau^2\boldsymbol{I})$ and is independent of σ^2. The likelihood function for the normal linear model is

$$f(\boldsymbol{y} \mid \boldsymbol{\beta}, \sigma^2) = \left(\frac{1}{\sqrt{2\pi}\sigma}\right)^n \exp\left[-\frac{1}{2\sigma^2}(\boldsymbol{y} - \boldsymbol{X\beta})^{\mathrm{T}}(\boldsymbol{y} - \boldsymbol{X\beta})\right].$$

Taking $(\boldsymbol{y} - \boldsymbol{X}\boldsymbol{\beta}) = (\boldsymbol{y} - \boldsymbol{X}\hat{\boldsymbol{\beta}}) + (\boldsymbol{X}\hat{\boldsymbol{\beta}} - \boldsymbol{X}\boldsymbol{\beta})]$ for the least squares estimate $\hat{\boldsymbol{\beta}}$, we can express the likelihood as

$$f(\boldsymbol{y} \mid \boldsymbol{\beta}, \sigma^2) = \left(\frac{1}{\sqrt{2\pi}\sigma}\right)^n \exp\left[-\frac{1}{2\sigma^2}[(n-p)s^2 + (\boldsymbol{\beta} - \hat{\boldsymbol{\beta}})^{\mathrm{T}}(\boldsymbol{X}^{\mathrm{T}}\boldsymbol{X})(\boldsymbol{\beta} - \hat{\boldsymbol{\beta}})\right].$$

Multiplying together the normal prior of $\boldsymbol{\beta}$, the inverse chi-squared prior of σ^2, and the likelihood function, and then completing the square, yields

$$h(\boldsymbol{\beta}, \sigma^2 \mid \boldsymbol{y}) \propto (\sigma^2)^{-[(\nu_0+n)/2+1]} \exp\left(-\frac{(\nu_0 + n)\sigma_n^2}{2\sigma^2}\right)$$
$$\times \sigma^{-p} \exp\left[-\frac{1}{2}(\boldsymbol{\beta} - \tilde{\boldsymbol{\beta}})^{\mathrm{T}}\tilde{\boldsymbol{\Sigma}}^{-1}(\boldsymbol{\beta} - \tilde{\boldsymbol{\beta}})\right]$$

where $\tilde{\boldsymbol{\beta}}$ and $\tilde{\boldsymbol{\Sigma}}$ are as in (10.4) and

$$(\nu_0 + n)\sigma_n^2 = \nu_0\sigma_0^2 + (n-p)s^2 + (\lambda\mathbf{1} - \tilde{\boldsymbol{\beta}})^{\mathrm{T}}(\lambda\mathbf{1} - \tilde{\boldsymbol{\beta}})(\sigma^2/\tau^2) + (\hat{\boldsymbol{\beta}} - \tilde{\boldsymbol{\beta}})^{\mathrm{T}}\boldsymbol{X}^{\mathrm{T}}\boldsymbol{X}(\hat{\boldsymbol{\beta}} - \tilde{\boldsymbol{\beta}}).$$

This posterior pdf factors as

$$h(\boldsymbol{\beta}, \sigma^2 \mid \boldsymbol{y}) = h(\sigma^2 \mid s^2)h(\boldsymbol{\beta} \mid \hat{\boldsymbol{\beta}}, \sigma^2),$$

where $h(\sigma^2 \mid s^2)$ has inverse chi-squared form and $h(\boldsymbol{\beta} \mid \hat{\boldsymbol{\beta}}, \sigma^2)$ is the same as the normal distribution obtained in Section 10.2.2 by treating σ^2 as known. Integrating σ^2 from $h(\boldsymbol{\beta}, \sigma^2 \mid \boldsymbol{y})$ yields the marginal posterior $h(\boldsymbol{\beta} \mid \boldsymbol{y})$, which has the form of a multivariate t distribution (Seber and Lee 2003, pp. 74–76). We next present an important special case of this.

10.2.5 Improper Priors and Equivalent Frequentist Inferences

When we use improper priors, with $h(\sigma^2) \propto 1/\sigma^2$ and an improper uniform prior distribution for $\boldsymbol{\beta}$ by taking $\tau^2 \to \infty$, we obtain posterior results that relate very closely to standard frequentist inference. Then $h(\boldsymbol{\beta} \mid \hat{\boldsymbol{\beta}}, \sigma^2)$ is a $N[\hat{\boldsymbol{\beta}}, (\boldsymbol{X}^{\mathrm{T}}\boldsymbol{X})^{-1}\sigma^2)]$ distribution. Recall that the frequentist distribution of $\hat{\boldsymbol{\beta}}$ is $N[\boldsymbol{\beta}, (\boldsymbol{X}^{\mathrm{T}}\boldsymbol{X})^{-1}\sigma^2]$, so the two inferential approaches then merely interchange the roles of $\boldsymbol{\beta}$ and $\hat{\boldsymbol{\beta}}$. Also, $h(\sigma^2 \mid s^2)$ is inverse chi-squared, such that $(n-p)s^2/\sigma^2$ has a chi-squared distribution with $df = n - p$. This is identical to the distribution in the frequentist case, except here it is the distribution of σ^2 given s^2 instead of s^2 given σ^2.

After integrating out σ^2, Box and Tiao (1973, Section 2.7.2) showed that

$$h(\boldsymbol{\beta} \mid \boldsymbol{y}) \propto \left[1 + \frac{(\boldsymbol{\beta} - \hat{\boldsymbol{\beta}})^{\mathrm{T}}(\boldsymbol{X}^{\mathrm{T}}\boldsymbol{X})(\boldsymbol{\beta} - \hat{\boldsymbol{\beta}})}{(n-p)s^2}\right]^{-n/2}.$$

This distribution for $\boldsymbol{\beta}$ is the multivariate t distribution with parameters $df = n - p$, $\hat{\boldsymbol{\beta}}$ (the mean of the distribution when $df > 1$), and $s^2(\boldsymbol{X}^{\mathrm{T}}\boldsymbol{X})^{-1}$ (which is the covariance matrix times $(df - 2)/df$ when $df > 2$). Let $(\boldsymbol{X}^{\mathrm{T}}\boldsymbol{X})^{-1}_{jj}$ denote the element from row j and column j of $(\boldsymbol{X}^{\mathrm{T}}\boldsymbol{X})^{-1}$. Then marginally for $(\beta_j \mid \boldsymbol{y})$,

$$t = \frac{\beta_j - \hat{\beta}_j}{s(\boldsymbol{X}^{\mathrm{T}}\boldsymbol{X})^{-1}_{jj}}$$

has the univariate t distribution with $df = n - p$. Setting $\beta_j = 0$ yields test statistic having the same absolute value as the test statistic used in a frequentist significance test of H_0: $\beta_j = 0$. Likewise, a posterior interval for β_j is identical to the corresponding frequentist confidence interval. In the Bayesian interpretation, the probability that the posterior interval contains β_j equals 0.95. With the frequentist approach, the "95% confidence" relates to long-run performance of the method; in repeated hypothetical sampling, 95% of the intervals contain β_j.

By integrating out the parameters using their posterior distribution, we can construct a posterior predictive distribution for a future observation y^* at a particular value $\boldsymbol{x}_0$ for the explanatory variables. For the improper prior structure, a prediction interval for y^* has the same formula as (3.3) presented for the frequentist approach with the normal linear model.

10.2.6 Example: Normal Linear Model for Record Running Times

Section 2.6 introduced data on record times for Scottish hill races (in minutes), with the distance of the race (in miles) and the cumulative climb (in thousands of feet) as explanatory variables. There we found that a frequentist linear model with an interaction term has good predictive power.

```
-------------------------------------------------------------------------------
> attach(ScotsRaces)
> summary(lm(time ~ climb + distance + climb:distance))
Coefficients: # ordinary least squares for linear  model
                 Estimate  Std. Error  t value  Pr(>|t|)
(Intercept)       -0.7672      3.9058   -0.196   0.84556
climb              3.7133      2.3647    1.570   0.12650
distance           4.9623      0.4742   10.464  1.07e-11
climb:distance     0.6598      0.1743    3.786   0.00066
---
Residual standard error: 7.338 on 31 degrees of freedom
Multiple R-squared:  0.9807,    Adjusted R-squared:  0.9788
-------------------------------------------------------------------------------
```

We next show two Bayesian analyses in R that give essentially the same estimates as the least squares estimates. The `MCMCregress` function in the `MCMCpack` package can select improper uniform priors for $\boldsymbol{\beta}$ by taking the normal prior to have precision

$1/\tau^2 = 0$. It does not allow an improper chi-squared prior for $1/\sigma^2$, but the impact of that prior can be approximated well by taking tiny values for the two parameters of a gamma prior distribution ($c0$ and $d0$). With a huge number of MCMC iterations for the Metropolis–Hastings algorithm (here 5,000,000), the standard errors for the posterior mean estimates are very small.

```
-------------------------------------------------------------------------------
> library(MCMCpack)
> fit.bayes <- MCMCregress(time ~ climb + distance + climb:distance,
+mcmc=5000000, b0=0, B0=0, c0=10^(-10), d0=10^(-10))
> summary(fit.bayes) # normal prior mean = b0, variance = 1/B0
1. Empirical mean and standard deviation for each variable,
   plus standard error of the mean:
                    Mean       SD   Naive SE
(Intercept)      -0.7662   4.0376  1.806e-03
climb             3.7133   2.4443  1.093e-03
distance          4.9622   0.4902  2.192e-04
climb:distance    0.6598   0.1802  8.057e-05
sigma2           57.5774  15.6825  7.013e-03
2. Quantiles for each variable:
                    2.5%      25%      50%      75%   97.5%
(Intercept)      -8.7292  -3.4310  -0.7672   1.8996   7.196
climb            -1.1110   2.0999   3.7135   5.3262   8.536
distance          3.9958   4.6384   4.9619   5.2859   5.928
climb:distance    0.3045   0.5409   0.6599   0.7788   1.015
sigma2           34.6010  46.5164  55.0384  65.7743  95.216 # error var.
-------------------------------------------------------------------------------
```

The `bayesglm` function in the `arm` package, based on Gelman et al. (2008), uses t distribution priors. It provides normal priors by taking $df = \infty$ for the t distribution, and that prior becomes flat when the prior scale parameter is infinite. For model fitting, rather than employing Gibbs sampling or the Metropolis–Hastings algorithm, it provides a very fast calculation that approximates a posterior mode and *SE* by incorporating an EM algorithm into the iteratively reweighted least squares algorithm.

```
-------------------------------------------------------------------------------
> library(arm)
> fit.bayes <- bayesglm(time ~ climb + distance + climb:distance,
+         family=gaussian, prior.mean=0, prior.scale=Inf, prior.df=Inf)
> summary(fit.bayes)
Coefficients:
               Estimate  Std. Error  t value  Pr(>|t|)
(Intercept)     -0.7672      3.6758   -0.209    0.8359
climb            3.7133      2.2254    1.669    0.1041
distance         4.9623      0.4463   11.118  4.93e-13
climb:distance   0.6598      0.1640    4.023    0.0003
(Dispersion parameter for gaussian family taken to be 47.6967)
-------------------------------------------------------------------------------
```

The Bayesian estimates are essentially the same as the least squares estimates. Interpretations can use Bayesian posterior probability statements. For example, the probability is 0.95 that the interaction parameter falls between 0.30 and 1.01.

10.3 BAYESIAN GENERALIZED LINEAR MODELS

As in Bayesian normal linear models, Bayesian GLMs provide point estimators that shrink ML estimators toward means of prior distributions. A simple nearly noninformative prior distribution takes $\boldsymbol{\beta} \sim N(\mathbf{0}, \sigma^2 \boldsymbol{I})$ with a very large σ. Again, standardizing the explanatory variables makes the effects comparable. Alternatively, as a default prior for GLMs, Gelman et al. (2008) proposed the t distribution, in particular a Cauchy[1] (the t with $df = 1$) as a conservative special case, with default location $= 0$ and a fixed scale value. This prior is less informative than the normal, being heavier in the tails.

We illustrate Bayesian inference for GLMs in this section by focusing on models for binary data. For Bayesian modeling of multinomial data and count data, see Notes 10.7 and 10.8.

10.3.1 Prior Specifications for Binary GLMs

For GLMs for binary data, such as logistic and probit regression, it is usually not advisable to use an improper uniform prior for $\boldsymbol{\beta}$. It does not sufficiently shrink ML estimates in cases that are problematic for ML estimation, such as when the data have complete or quasi-complete separation in the space of $\boldsymbol{x}$ values and at least one ML estimate is infinite.

When we use independent $N(\mu, \sigma^2)$ priors, the posterior Bayes estimate is well defined, even if the data have complete separation (Gelman et al. 2008). With the logit link, the corresponding normal prior for the linear predictor $\boldsymbol{x}_i\boldsymbol{\beta}$ induces a prior distribution for

$$P(y_i = 1) = \exp(\boldsymbol{x}_i\boldsymbol{\beta})/[1 + \exp(\boldsymbol{x}_i\boldsymbol{\beta})]$$

that is a special case of the *logit-normal* (also called *logistic-normal*) distribution. This two-parameter family of distributions over (0, 1) is an alternative to the beta family. When $\mu = 0$, the logit-normal density is symmetric[2]. It is then unimodal when $\text{var}(\boldsymbol{x}_i\boldsymbol{\beta}) \leq 2$ and bimodal otherwise. The modes are closer to 0 and 1 as σ increases. Relatively noninformative (large σ) priors imply priors on the probability scale that are highly U-shaped, with about half the probability very close to 0 and half very close to 1. Although seemingly rather informative, such priors usually have little

[1]The Cauchy pdf with location 0 and scale γ is $h(\beta) = 1/\pi\gamma[1 + (\beta/\gamma)^2]$.

[2]See `logitnorm.r-forge.r-project.org` and the "Logit-normal distribution" entry in wikipedia.org for figures illustrating the shapes described here.

influence, and the posterior distribution has much the same shape as the likelihood function.

Data analysts who use a subjective approach may find it is easier to formulate prior beliefs about probabilities than about logistic or probit $\boldsymbol{\beta}$ that pertain to a nonlinear function of the probabilities. When one constructs prior distributions for $P(y = 1)$ (such as the beta) at various settings of $\boldsymbol{x}$, those prior distributions induce a corresponding prior distribution for $\boldsymbol{\beta}$. MCMC methods can then generate posterior distributions. See Christensen et al. (2010, Section 8.4) for details.

10.3.2 Example: Risk Factors Revisited for Endometrial Cancer

Section 5.7.1 analyzed data from a study of y = histology of 79 cases of endometrial cancer (0 = low grade, 1 = high grade) with three risk factors: x_1 = neovasculation (1 = present, 0 = absent), x_2 = pulsatility index of arteria uterina, and x_3 = endometrium height. The data exhibit quasi-complete separation on x_1, and the ML fit of the main effects model

$$\text{logit}[P(y_i = 1)] = \beta_0 + \beta_1 x_{i1} + \beta_2 x_{i2} + \beta_3 x_{i3}$$

has $\hat{\beta}_1 = \infty$.

```
---------------------------------------------------------------------------
> Endometrial # file Endometrial.dat at www.stat.ufl.edu/~aa/glm/data
   NV PI   EH HG
1   0 13 1.64  0
2   0 16 2.26  0
...
> attach(Endometrial)
> PI2 <- (PI-mean(PI))/sd(PI); EH2 <- (EH-mean(EH))/sd(EH); NV2 <- NV-0.5
  # standardize quantitative explanatory var's, center indicator at 0
> fitML <- glm(HG ~ NV2 + PI2 + EH2, family=binomial)
> summary(fitML)
Coefficients:  # actual ML estimate for NV2 effect is infinite
             Estimate  Std. Error  z value  Pr(>|z|)
(Intercept)    7.8411    857.8755    0.009    0.9927
NV2           18.1856   1715.7509    0.011    0.9915
PI2           -0.4217      0.4432   -0.952    0.3413
EH2           -1.9219      0.5599   -3.433    0.0006
---------------------------------------------------------------------------
```

For Bayesian analyses, we use independent $N(0, \sigma^2)$ prior distributions for $\{\beta_j\}$, with standardized versions of x_2 and x_3. Instead of the usual (0, 1) coding for the indicator variable x_1, we assign values −0.5 and 0.5. The prior distribution is then symmetric in the sense that the logits for each group have the same prior variability as well as the same prior means, yet β_1 retains its usual interpretation as a conditional log odds ratio. For these data, because the log likelihood is relatively flat in the β_1 dimension, posterior means for β_1 can be highly sensitive to the choice of prior distribution. To reflect a lack of information about the sizes of the effects, we first used quite-diffuse prior distributions, with $\sigma = 10$.

```
> library(MCMCpack) # b0 = prior mean, B0 = prior precision = 1/variance
> fitBayes <- MCMClogit(HG ~ NV2+PI2+EH2, mcmc=10000000, b0=0, B0=0.01)
> summary(fitBayes)
1. Empirical mean and standard deviation:
              Mean     SD
(Intercept)   3.214  2.560
NV2           9.118  5.096
PI2          -0.473  0.454
EH2          -2.138  0.593
2. Quantiles for each variable:
               2.5%     25%     50%     75%   97.5%
(Intercept)  -0.343   1.270   2.721   4.686   9.344
NV2           2.107   5.233   8.126  12.047  21.336
PI2          -1.413  -0.767  -0.455  -0.159   0.366
EH2          -3.402  -2.515  -2.101  -1.722  -1.082
# Alternatively, bayesglm function in R arm package uses t priors
```

Table 10.1 shows posterior means and standard deviations and the 95% equal-tail posterior interval for β_1, based on an MCMC process with 10,000,000 iterations. The table also shows the ML results, for comparison. With such a long process, the Monte Carlo standard errors for the approximations to the Bayes estimates were negligible—about 0.005 for the neovasculation effect and much less for the others. The results yield the inference that $\beta_1 > 0$, and the effect seems to be large. The estimated size of the effect is imprecise, because of the flat log likelihood and the relatively diffuse priors. Inferences about the model parameters were substantively similar to those using the ML frequentist analysis.

Corresponding to the frequentist P-value for H_1: $\beta_1 > 0$, the Bayesian approach provides $P(\beta_1 < 0 \mid \mathbf{y})$. This is 0.002; that is, 0.0 is the 0.002 quantile of the posterior distribution. For this relatively flat prior distribution, this posterior tail probability is similar to the P-value of 0.001 for the one-sided frequentist likelihood-ratio test of H_0: $\beta_1 = 0$ against H_1: $\beta_1 > 0$. Each has very strong evidence that $\beta_1 > 0$.

For comparison, we used a highly informative prior distribution. To reflect a prior belief that the effects are not strong, we took $\sigma = 1.0$. Then nearly all the prior probability for the conditional odds ratio $\exp(\beta_1)$ falls between $\exp(-3.0) = 0.05$ and

Table 10.1 Results of Bayesian and Frequentist Fitting of Models to the Endometrial Cancer Dataset of Table 5.3

Analysis	$\hat{\beta}_1$ (*SD*)	Interval[a]	$\hat{\beta}_2$ (*SD*)	$\hat{\beta}_3$ (*SD*)
ML	∞ (—)	$(1.3, \infty)$	−0.42 (0.44)	−1.92 (0.56)
Bayes, $\sigma = 10$	9.12 (5.10)	(2.1, 21.3)	−0.47 (0.45)	−2.14 (0.59)
Bayes, $\sigma = 1$	1.65 (0.69)	(0.3, 3.0)	−0.22 (0.33)	−1.77 (0.43)

[a]Profile-likelihood interval for ML and equal-tail posterior interval for Bayes.

$\exp(3.0) = 20$. As Table 10.1 shows, results were quite different from the ML frequentist analysis or the Bayesian analysis with $\sigma = 10$. Because $y_i = 1$ for all 13 patients having $x_{i1} = 1$, the frequentist approach tells us we cannot rule out any very large value for β_1. By contrast, if we had strong prior beliefs that $|\beta_1| < 3$, then even with these sample results the Bayesian posterior inference has an upper bound of about 3 for β_1.

10.3.3 Bayesian Fitting for Probit Models

For Bayesian model fitting of binary GLMs with normal priors, a simple analysis is possible for the probit link. The probit model is simpler to handle than the logistic model, because results apply directly from Bayesian normal linear models. Albert and Chib (1993) exploited the normal-threshold latent variable model presented in Section 5.1.2, assuming multivariate normal prior distributions for $\boldsymbol{\beta}$ and the independent normal latent variables. Then the posterior distribution of $\boldsymbol{\beta}$, conditional on $\boldsymbol{y}$ and the latent variables, is multivariate normal. Implementation of MCMC methods is relatively simple because the Monte Carlo sampling is from normal distributions.

Consider the data in ungrouped form, so that all $n_i = 1$. For subject i, a latent variable y_i^* is assumed to relate to y_i by $y_i = 1$ if $y_i^* > 0$ and $y_i = 0$ if $y_i^* \leq 0$. Assuming that y_i^* has a $N(\boldsymbol{x}_i\boldsymbol{\beta}, 1)$ distribution,

$$P(y_i = 1) = P\left(y_i^* > 0\right) = P(\boldsymbol{x}_i\boldsymbol{\beta} + \epsilon_i > 0),$$

where ϵ_i is a $N(0, 1)$ random variable. The corresponding probit model is

$$\Phi^{-1}[P(y_i = 1)] = \boldsymbol{x}_i\boldsymbol{\beta}.$$

If $\{y_i^*\}$ were observed and a multivariate normal prior were chosen for $\boldsymbol{\beta}$, then the posterior distribution for $\boldsymbol{\beta}$ would result from ordinary normal linear model results. Given the $\{y_i\}$ actually observed, however, $\{y_i^*\}$ are left- or right-truncated at 0. Thus, their distributions are truncated normal. Nonetheless, it is still possible to use MCMC (here, Gibbs sampling) to simulate the exact posterior distribution. The likelihood function is expressed in terms of the model for y_i^*. If y_i^* were observed, the contribution to the likelihood function would be $\phi(y_i^* - \boldsymbol{x}_i\boldsymbol{\beta})$. With y_i^* unknown except for its sign, the contribution to the likelihood function is

$$\left[I\left(y_i^* > 0\right)^{y_i} I\left(y_i^* \leq 0\right)^{1-y_i}\right]\phi\left(y_i^* - \boldsymbol{x}_i\boldsymbol{\beta}\right).$$

For n independent observations, the likelihood function is proportional to the product of n such terms. Then for prior density function $h(\boldsymbol{\beta})$, the joint posterior density of $\boldsymbol{\beta}$ and of $\{y_i^*\}$ given the data $\{y_i\}$ is proportional to

$$h(\boldsymbol{\beta})\prod_{i=1}^{n}\left[I\left(y_i^* > 0\right)^{y_i} I\left(y_i^* \leq 0\right)^{1-y_i}\right]\phi\left(y_i^* - \boldsymbol{x}_i\boldsymbol{\beta}\right).$$

With the ML estimates as initial values, Albert and Chib used a Gibbs sampling scheme that successively sampled from the density of $\boldsymbol{y}^* = (y_1^*, \ldots, y_n^*)^{\mathrm{T}}$ given $\boldsymbol{\beta}$ and of $\boldsymbol{\beta}$ given $\boldsymbol{y}^*$. With the conjugate normal prior, they noted that the posterior density of $\boldsymbol{\beta}$ given $\boldsymbol{y}^*$ is normal. Specifically, suppose that the prior distribution of $\boldsymbol{\beta}$ is $N(\boldsymbol{0}, \tau^2\boldsymbol{I})$, and let $\boldsymbol{X}$ be the matrix with ith row $\boldsymbol{x}_i$, so the latent variable model is $\boldsymbol{y}^* = \boldsymbol{X}\boldsymbol{\beta} + \boldsymbol{\epsilon}$. Conditional on $\boldsymbol{y}^*$, from (10.4) the distribution of $\boldsymbol{\beta}$ is $N(\tilde{\boldsymbol{\beta}}, \tilde{\boldsymbol{\Sigma}})$ with

$$\tilde{\boldsymbol{\beta}} = (\tau^{-2}\boldsymbol{I} + \boldsymbol{X}^{\mathrm{T}}\boldsymbol{X})^{-1}\boldsymbol{X}^{\mathrm{T}}\boldsymbol{y}^*, \quad \tilde{\boldsymbol{\Sigma}} = (\tau^{-2}\boldsymbol{I} + \boldsymbol{X}^{\mathrm{T}}\boldsymbol{X})^{-1}.$$

Conditional on $\boldsymbol{\beta}$, the elements of $\boldsymbol{y}^*$ are independent with the density of y_i^* being $N(\boldsymbol{x}_i\boldsymbol{\beta}, 1)$ truncated at the left by 0 if $y_i = 1$ and truncated at the right by 0 if $y_i = 0$.

Albert and Chib (1993) also proposed using a link function corresponding to the *cdf* of a t distribution, to investigate the sensitivity of results to the choice of link function. This approach yields the Cauchy link when $df = 1$ and the probit link as $df \to \infty$. It also can provide close approximations to results for corresponding logistic models, because a t variate with $df = 8$ divided by 0.63 is approximately a standard logistic variate.

10.3.4 Extensions to Models for a Multivariate Response

The Bayesian approach is also possible for the models for a multivariate response introduced in Chapter 9. Here we only briefly outline the normal case.

Models assuming multivariate normality with correlated components require a parametric family for prior distributions for the covariance matrix. The *Wishart distribution* is a multivariate generalization of the gamma distribution that generates the inverse chi-squared distribution for a variance. If $z_1, \ldots, z_\nu$ are iid from a $N(\boldsymbol{0}, \boldsymbol{\Sigma}_0)$ distribution, then $\boldsymbol{W} = \sum_i z_i z_i^{\mathrm{T}}$ has a Wishart distribution with parameters $(\nu, \boldsymbol{\Sigma}_0)$. The mean of this distribution is $\nu\boldsymbol{\Sigma}_0$. A Wishart distribution often serves as the prior distribution for the precision matrix $\boldsymbol{\Sigma}^{-1}$ in multivariate normal linear models. For a prior value $\boldsymbol{\Sigma}_0$ for the covariance matrix, one takes $\boldsymbol{\Sigma}^{-1}$ to have a Wishart $(\nu, \boldsymbol{\Sigma}_0^{-1})$ distribution. The parameter ν is a sort of prior sample size that determines the strength of prior beliefs, with smaller values being less informative. For details, see Hoff (2009, Chapter 11).

10.4 EMPIRICAL BAYES AND HIERARCHICAL BAYES MODELING

Some methodologists who otherwise like the Bayesian approach find it unappealing to have to select values for the hyperparameters in prior distributions. We next present two alternative approaches: (1) an *empirical Bayes* approach uses the data to estimate the hyperparameters; (2) in a hierarchical approach, the hyperparameters themselves have prior distributions.

10.4.1 Empirical Bayes Approach Estimates Hyperparameters

When the prior distribution has hyperparameters λ, as it usually does, the marginal pdf of y in the denominator of Bayes' theorem (10.1) is actually

$$f(y \mid \lambda) = \int_{\Theta} f(y \mid \theta) h(\theta \mid \lambda) d\theta,$$

itself depending on λ. The empirical Bayes approach uses the value of λ that maximizes $f(y \mid \lambda)$, given the observed data. That is, it uses marginal ML estimators of the hyperparameters, the values under which the data would have highest density. Inference then uses the posterior distribution, $h(\theta \mid y, \hat{\lambda})$, generated by that empirically estimated prior distribution.

The name "empirical Bayes" refers to this method's use of the data to estimate the hyperparameters. *Parametric empirical Bayes* uses a parametric form for the prior distribution. The *nonparametric empirical Bayes* approach makes no assumption about its form.

10.4.2 Parametric Empirical Bayes for One-Way Layout Models

In Section 10.2.3 we posed a Bayesian normal linear model for the one-way layout, with $N(\mu_i, \sigma^2)$ distributions having $\mu_i \sim N(\lambda, \tau^2)$. For simplicity here in illustrating the basic ideas, suppose σ^2 is known and $n_1 = \cdots = n_c$. Let n denote that common value. We can then summarize the data by independent observations $\bar{y}_1, \ldots, \bar{y}_c$ with common variance σ^2/n. From Section 10.2.3, the posterior mean is

$$\tilde{\mu}_i = \left(\frac{\tau^2}{\tau^2 + \sigma^2/n}\right)\bar{y}_i + \left(\frac{\sigma^2/n}{\tau^2 + \sigma^2/n}\right)\lambda. \tag{10.6}$$

We now adapt this estimate with the empirical Bayes approach for estimating the hyperparameters λ and τ.

For this normal/normal conjugate model, the marginal density of $\bar{y}_i$ is

$$f(\bar{y}_i \mid \lambda, \tau) \propto \int e^{-n(\bar{y}_i - \mu_i)^2/2\sigma^2} e^{-(\mu_i - \lambda)^2/2\tau^2} d\mu_i.$$

After combining the terms in the exponent involving μ_i and completing the square, we find that the terms in the integrand are a constant (relative to μ_i) times

$$\exp\left[-\frac{1}{2(\tau^{-2} + \sigma^{-2}n)^{-1}}(\mu_i - \tilde{\mu}_i)^2\right].$$

This integrand has the form of an integral of a normal density with respect to μ_i, and equals 1.0 times another constant that (as a function of $\bar{y}_i$) has the form of the

$N(\lambda, \tau^2 + \sigma^2/n)$ density. That is, the marginal distribution for each $\bar{y}_i$ is $N(\lambda, \tau^2 + \sigma^2/n)$. Since $\{\bar{y}_i \mid \lambda, \tau\}$ are independent, the joint marginal density of $\bar{\mathbf{y}}$ is

$$f(\bar{\mathbf{y}} \mid \lambda, \tau) = [2\pi(\tau^2 + \sigma^2/n)]^{-c/2} \exp\left\{ -\left[\sum_{i=1}^{c}(\bar{y}_i - \lambda)^2\right] \Big/ 2(\tau^2 + \sigma^2/n)\right\}.$$

Maximizing this with respect to λ, we obtain the marginal ML estimate $\hat{\lambda} = \bar{y} = (\sum_i \bar{y}_i)/c$. So when we specify τ^2, the empirical Bayes estimate of μ_i is (10.6) with λ replaced by $\bar{y}$. This estimate shrinks each sample mean toward the overall sample mean. It is the same as obtained with a frequentist random-effects model in Section 9.3.2. The shrinkage factor decreases as the common sample size n increases.

When we also treat τ^2 as unknown, the ML estimate of the marginal variance $(\tau^2 + \sigma^2/n)$ is $\frac{1}{c}\sum_i(\bar{y}_i - \bar{y})^2$. We are treating σ^2 as known, so the corresponding marginal ML estimate of τ^2 is $\hat{\tau}^2 = \frac{1}{c}\sum_i(\bar{y}_i - \bar{y})^2 - \sigma^2/n$, unless this difference is negative, in which case the estimate is 0. The estimated shrinkage proportion applied to $\bar{y}_i$ is then $\hat{\tau}^2/(\hat{\tau}^2 + \sigma^2/n)$. In practice, with σ^2 also unknown, we could substitute the unbiased estimate $s^2 = \sum_{i=1}^{c}\sum_{j=1}^{n}(y_{ij} - \bar{y}_i)^2/[c(n-1)]$ from one-way ANOVA.

With binary data in the one-way layout, with a beta prior (the conjugate) for the binomial parameters $\{\pi_i\}$, the marginal distribution of $\mathbf{y}$ resembles a beta-binomial likelihood (Section 8.2). For fixed beta hyperparameters α_1 and α_2, $\tilde{\pi}_i$ is a weighted average of the sample proportion and the mean of the beta prior distribution (Exercise 10.14). When we estimate (α_1, α_2) using the marginal beta-binomial distribution, we obtain empirical Bayes estimates that shrink the sample proportions toward the overall sample proportion.

With count data in the one-way layout, we could start with a Poisson model in which $\{\mu_i\}$ have a gamma distribution prior (the conjugate). Then the marginal distribution of $\mathbf{y}$ resembles a negative binomial likelihood (Section 7.3.2). For fixed gamma hyperparameters, $\tilde{\mu}_i$ is a weighted average of $\bar{y}_i$ and the mean of the gamma prior distribution. When we estimate those hyperparameters using the marginal negative binomial distribution, we obtain empirical Bayes estimates that shrink $\{\bar{y}_i\}$ toward the overall $\bar{y}$.

Regardless of the distribution for y_{ij}, the empirical Bayesian approach borrows from the whole to estimate any one mean. The analysis need not use conjugate priors, as the next example illustrates.

10.4.3 Example: Smoothing Election Poll Results

This example uses a simulated sample of 2000 people to mimic a poll taken before the 2012 US presidential election, in which Barack Obama faced Mitt Romney. For n_i observations in state i ($i = 1, \ldots, 51$, where $i = 51$ is District of Columbia), let y_i be the sample proportion favoring Obama in that election. Let π_i be the corresponding population proportion. Here, we take n_i proportional to the number of people in state i who voted in that election, subject to $\sum_i n_i = 2000$. Table 10.2, available in the file `Election.dat` at the text website, shows $\{n_i\}$, $\{\pi_i\}$, and $\{y_i\}$.

Table 10.2 Empirical Bayes Estimates $\{\hat{\pi}_i\}$ and Hierarchical Bayes Estimates $\{\hat{\pi}_i^h\}$ of Proportions of Vote $\{\pi_i\}$ for Obama in 2012 US Presidential Election, for Sample Size n_i in State i with Sample Proportion y_i

State	n_i	π_i	y_i	$\hat{\pi}_i$	$\hat{\pi}_i^h$	State	n_i	π_i	y_i	$\hat{\pi}_i$	$\hat{\pi}_i^h$
AK	5	0.408	0.200	0.450	0.448	MT	7	0.417	0.714	0.538	0.540
AL	32	0.384	0.531	0.514	0.514	NC	66	0.483	0.288	0.351	0.347
AR	17	0.369	0.470	0.485	0.485	ND	5	0.387	0.200	0.450	0.447
AZ	35	0.444	0.514	0.505	0.506	NE	12	0.379	0.583	0.521	0.521
CA	207	0.603	0.652	0.633	0.637	NH	11	0.520	0.545	0.509	0.509
CO	37	0.515	0.486	0.490	0.490	NJ	59	0.582	0.593	0.561	0.563
CT	25	0.581	0.640	0.562	0.565	NM	13	0.529	0.462	0.484	0.483
DC	4	0.836	0.750	0.526	0.528	NV	15	0.524	0.267	0.416	0.413
DE	6	0.586	0.500	0.495	0.495	NY	114	0.626	0.667	0.632	0.636
FL	128	0.499	0.430	0.441	0.440	OH	87	0.507	0.448	0.460	0.459
GA	60	0.455	0.517	0.509	0.510	OK	22	0.332	0.409	0.457	0.456
HI	7	0.706	0.714	0.538	0.540	OR	28	0.542	0.464	0.479	0.479
IA	23	0.520	0.565	0.526	0.527	PA	92	0.520	0.576	0.557	0.558
ID	10	0.324	0.300	0.444	0.441	RI	7	0.627	0.428	0.481	0.480
IL	84	0.576	0.607	0.579	0.581	SC	29	0.441	0.448	0.471	0.470
IN	42	0.439	0.619	0.569	0.570	SD	6	0.399	0.167	0.437	0.434
KS	19	0.380	0.263	0.402	0.398	TN	40	0.391	0.475	0.483	0.482
KY	28	0.378	0.321	0.409	0.406	TX	123	0.414	0.382	0.403	0.401
LA	30	0.406	0.267	0.378	0.374	UT	15	0.247	0.267	0.416	0.413
MA	47	0.606	0.638	0.584	0.586	VA	57	0.512	0.509	0.504	0.504
MD	40	0.620	0.650	0.585	0.588	VT	5	0.666	0.200	0.450	0.447
ME	11	0.563	0.545	0.509	0.509	WA	46	0.562	0.522	0.511	0.512
MI	75	0.541	0.613	0.581	0.583	WI	45	0.528	0.600	0.559	0.560
MN	44	0.526	0.568	0.539	0.540	WV	11	0.355	0.454	0.483	0.483
MO	45	0.444	0.333	0.396	0.393	WY	4	0.278	0.500	0.495	0.494
MS	20	0.438	0.400	0.455	0.454						

Let y_{ij} be the binary observation for subject j in state i, where "1" indicates a preference for Obama. We consider the model for a one-way layout,

$$\text{logit}[P(y_{ij} = 1)] = \beta_i, \quad i = 1, \ldots, 51. \tag{10.7}$$

In the frequentist approach, the ML estimator of π_i is the sample proportion y_i, and $\hat{\beta}_i$ is the sample logit, $\log[y_i/(1 - y_i)]$. Many states have small n_i, and better estimators borrow from the whole.

A simple Bayesian approach treats $\{\beta_i\}$ as independent $N(\mu, \sigma^2)$ variates. When we combine the 51 independent binomial likelihood functions with the normal prior distributions, the resulting Bayes estimates shrink the sample logits toward μ (Exercise 10.24). We instead use an empirical Bayes approach, treating μ and σ^2 as unknown. The marginal distribution that we maximize to find $\hat{\mu}$ and $\hat{\sigma}^2$ is the same as the marginal likelihood in the frequentist approach for the random-effects model for the one-way layout,

$$\text{logit}[P(y_{ij} = 1)] = u_i, \quad i = 1, \ldots, 51, \tag{10.8}$$

for independent $u_i \sim N(\mu, \sigma^2)$.

The marginal ML analysis provides hyperparameter estimates $\hat{\mu} = -0.023$ and $\hat{\sigma} = 0.376$. The posterior means $\{\tilde{\beta}_i\}$ of $\{\beta_i\}$ transform to the proportion estimates $\{\hat{\pi}_i = \exp(\tilde{\beta}_i)/[1 + \exp(\tilde{\beta}_i)]\}$ shown[3] in Table 10.2. Because $\{n_i\}$ are mostly small and $\hat{\sigma}$ is relatively small, these estimates shrink the sample proportions considerably. The empirical Bayes model estimates tend to be closer than the sample proportions to the true values. The root-mean-square error about the true proportions (weighted by the per-state sample sizes) is 0.064 for the model-based estimates and 0.091 for the sample proportions.

States with relatively few observations and sample proportions rather far from the overall proportion, such as Alaska and DC, have empirical Bayes estimates that tend to shrink more toward the overall proportion. In such cases, the posterior interval benefits strongly from borrowing from the whole, being much shorter than a confidence interval based on that state alone. For Alaska, for instance, the 95% frequentist likelihood-ratio test-based confidence interval for π_1 based on 1 Obama supporter in $n_1 = 5$ voters sampled is (0.01, 0.63), which is similar to a Bayesian posterior interval using very flat priors for $\{\beta_i\}$. By contrast, the empirical Bayesian equal-tail posterior interval[4] of $(-0.68, 0.40)$ for β_1 translates to (0.34, 0.60) for π_1. Figure 10.1 compares this shrinkage effect with the effect for a state with a much larger sample, California. It has frequentist interval (0.59, 0.71) for π_5 shrinking only to the empirical Bayes interval of (0.58, 0.68).

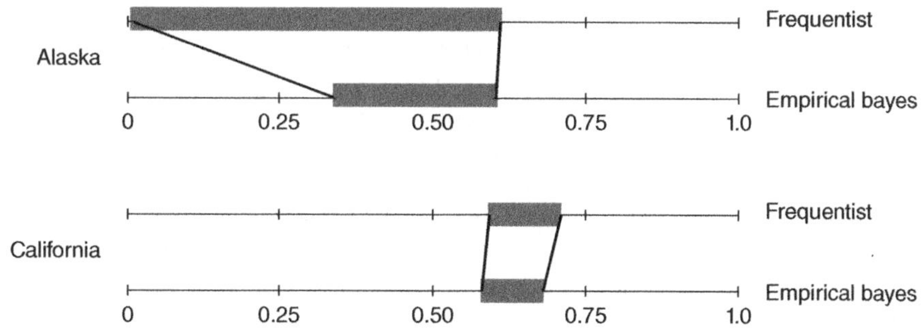

Figure 10.1 Comparison of frequentist confidence intervals and empirical Bayesian posterior intervals, for Alaska and California.

The empirical Bayes estimates of $\{\beta_i\}$ in model (10.7) and the corresponding best linear unbiased predictors (BLUPs) of $\{u_i\}$ in model (10.8) both exhibit strong

[3]These are not identical to the posterior means of $\{\pi_i\}$, because this is a nonlinear transformation.
[4]Obtained using `MCMClogit` in R, this is only approximate because of Monte Carlo error.

shrinkage. With a standard Bayesian analysis, rather than use noninformative normal priors with very large σ, one could select a relatively small σ based on historical information. For example, suppose past elections suggest that all (or nearly all) states tend to have π_i between about 0.25 and 0.75, for which logit(π_i) falls between -1.1 and $+1.1$. Then we could take the normal prior to have $\mu \pm 3\sigma$ equal to ± 1.1, or $\sigma \approx 1.1/3 = 0.37$. This results in similar Bayesian estimates for $\{\beta_i\}$ as obtained with the empirical Bayesian and frequentist random-effects approaches.

10.4.4 Hierarchical Bayes Has Priors for Hyperparameters

A disadvantage of the empirical Bayesian approach is not accounting for the variability caused by substituting estimates for prior hyperparameters $\boldsymbol{\lambda}$. This method uses $\boldsymbol{y}$ both to obtain the likelihood function and to estimate the prior distribution. So posterior intervals tend to have actual coverage probabilities below the nominal level. Supplementary methods, beyond our scope, can inflate the naive *SE* values for parameter estimates and predicted values to account for estimating $\boldsymbol{\lambda}$ (Carlin and Louis 2009, Section. 5.4).

An alternative approach quantifies the uncertainty about the hyperparameters $\boldsymbol{\lambda}$ by using a hierarchical approach in which those hyperparameters have their own prior distribution, $h(\boldsymbol{\lambda})$. Integration with respect to that second-stage prior distribution then yields the overall prior,

$$h(\boldsymbol{\theta}) = \int_{\boldsymbol{\lambda}} h(\boldsymbol{\theta} \mid \boldsymbol{\lambda}) h(\boldsymbol{\lambda}) d\boldsymbol{\lambda},$$

which is used in the ordinary way for the Bayesian analysis. Using noninformative distributions for the hyperparameters reduces the impact of the choice of prior distribution. The analysis may then be more robust, with the subjectivity reduced because posterior results are averaged over a family of prior distributions. When it is unclear how to select a prior distribution in a Bayesian analysis, it is sensible to specify the prior with a hierarchical structure, because the tail behavior of the prior has a large impact on robustness and parameter shrinkage (Polson and Scott 2010).

To illustrate the hierarchical approach, we return to the Bayesian analysis in Section 10.2.4 for the normal one-way layout. The observations from group i are independent $N(\mu_i, \sigma^2)$, where $\{\mu_i\}$ are, a priori, independent $N(\lambda, \tau^2)$, and σ^2 has an inverse chi-squared prior distribution with hyperparameters ν_0 and σ_0^2. The hierarchical approach assumes prior distributions for λ and τ^2. The usual ones are a $N(\lambda_0, \gamma_0^2)$ distribution for λ and an inverse chi-squared distribution for τ^2 with hyperparameters η_0 and τ_0^2. We could select λ_0, σ_0^2, and τ_0^2 to reflect our prior beliefs about the mean of the means, the variability of observations within groups, and the variability of the means. To make the prior information relatively diffuse, we would choose a relatively large value for γ_0 and relatively small values for ν_0 and η_0, such as 1 each. Hoff (2009, Section 8.3) described a Gibbs sampling scheme for approximating the joint posterior distribution for $(\{\mu_i\}, \lambda, \sigma^2, \tau^2)$, by iteratively sampling each parameter from its full conditional distribution, given the other parameters. As usual, the Bayesian estimate of μ_i employs less shrinkage of $\bar{y}_i$ as n_i increases.

Another example is the logistic model logit$[P(y_{ij} = 1)] = \beta_i$ for the election poll results in Section 10.4.3. There, we used an empirical Bayes approach for estimating the hyperparameters in the $N(\mu, \sigma^2)$ prior distribution. With the hierarchical approach, we use a diffuse inverse chi-squared prior distribution for σ^2. As Table 10.2 shows, we obtain very similar posterior mean predictions with this approach, within the limits of MCMC error.

```
-------------------------------------------------------------------------
> Election <- read.table("Election2.dat", header = TRUE)
      y   state  # individual-level data file at www.stat.ufl.edu/~aa/glm
1     1       1
2     0       1
...
2000  0      51

> library(MCMCglmm)
> prior<-list(R=list(V=1, fix=1), G=list(G1=list(V=1, nu=0.002)))
  # In G list, V and nu are sigma^2 and nu_0 for inverse chi-squared
> fit.h <- MCMCglmm(fixed = y ~ 1, random = ~ state, family = "categorical",
+ prior=prior, data=Election, pr=TRUE, slice=TRUE, nitt=400000)
> fitted <- predict(fit.h, marginal=fit.h$random)
> fitted
        [,1]
1      0.448
2      0.448
...
2000   0.494
-------------------------------------------------------------------------
```

CHAPTER NOTES

Section 10.1: The Bayesian Approach to Statistical Inference

10.1 **Priors**: Conjugate priors became popular following their introduction by Raiffa and Schlaifer (1961), computational simplicity being crucial 50 years ago. Kass and Wasserman (1996) reviewed the Jeffreys prior and rules for selecting priors that are relatively noninformative. Much literature deals with defining classes of prior distributions for objective Bayes analyses, such as Berger (2006) and Berger et al. (2013). Ferguson (1973) introduced a class of *Dirichlet process priors* for a nonparametric Bayesian approach. See Hjort et al. (2010) and Christensen et al. (2010, Chapter 13) for recent advances.

10.2 **MCMC**: For introductory surveys of Bayesian computation, see Craiu and Rosenthal (2014) and Davison (2003, Section 11.3). Gelfand and Smith (1990) and Casella and George (1992) focused on Gibbs sampling. Hastings (1970) generalized the Metropolis–Hastings algorithm for nonsymmetric proposal densities and justified the appropriate convergence.

10.3 **Exchangeability**: Draper et al. (1993) discussed the role of exchangeability in data analysis. Davison (2003, Chapter 11) provided an informative overview of this and other key topics for Bayesian inference, such as the likelihood principle and controversies over the choice of the prior.

10.4 **Bayesian model checking**: For Bayesian model checking and model selection, see Carlin and Louis (2009, Chapter 4), Christensen et al. (2010, Section 8.3), Draper (1995), Gelman et al. (2013, Chapter 6), Hoff (2009, Section 9.3), and Spiegelhalter et al. (2002) and references therein. Kass and Raftery (1995) surveyed Bayes factors and applications of their use. For comparing models, Spiegelhalter et al. (2002) proposed a *deviance information criterion* (DIC), which adds double the effective number of parameters to a *mean posterior deviance* for checking fit. See also Spiegelhalter et al. (2014).

10.5 **Experimental design**: The Bayesian approach applies naturally to design issues. Carlin and Louis (2009, Chapter 6) and Chaloner and Verdinelli (1995) presented reviews, the latter focusing on issues of optimal design.

Section 10.2: Bayesian Linear Models

10.6 **Modeling and variable selection**: For Bayesian linear modeling, see Box and Tiao (1973, Section 2.7, Chapters 5–7), Carlin and Louis (2009, Sections 2.4.1 and 4.1.1), Gelman et al. (2013, Chapters 14 and 15), Hoff (2009, Chapter 9), Ntzoufras (2009, Chapter 5), and O'Hagan and Forster (2004, Chapter 11). For variable selection, see Brown et al. (1998), Carlin and Louis (2008, Chapter 4), George (2000), George and McCulloch (1997), O'Hagan and Forster (2004, pp. 320–322), and Rǒcková and George (2014).

Section 10.3: Bayesian Generalized Linear Models

10.7 **Binary and multinomial GLMs**: For Bayesian modeling of binary data, see Christensen et al. (2010, Chapter 8), Hosmer et al. (2013, Section 10.6), Gelman et al. (2013, Chapter 16), Ntzoufras (2009, Sections 7.5 and 9.3), O'Hagan and Forster (2004, Chapter 12), and Zellner and Rossi (1984). For Bayesian modeling of multinomial data including contingency tables, see Albert and Chib (1993), Congdon (2005, Chapter 6), Leonard and Hsu (1994), and O'Hagan and Forster (2004, Chapter 12). For ordinal models such as the cumulative logit, see Albert and Chib (1993) and Hoff (2009, Chapter 12).

10.8 **Count data GLMs**: For Bayesian modeling of count data, see Cameron and Trivedi (2013, Chapter 12), Christensen et al. (2010, Chapter 10), Gelman et al. (2013, Chapter 16), and Ntzoufras (2009, Sections 7.4, 8.3, and 9.3).

10.9 **Multivariate GLMs**: For extensions of normal linear models to multivariate responses, see Box and Tiao (1973, Chapter 8), Brown et al. (1998), Christensen et al. (2010, Chapter 10), and Hoff (2009, Chapters 7 and 11). For various GLMs and multivariate generalizations of GLMs, see Dey et al. (2000) and Zeger and Karim (1991). For Bayesian graphical models, see Madigan and York (1995).

Section 10.4: Empirical Bayes and Hierarchical Bayes Modeling

10.10 **Empirical Bayes**: For parametric empirical Bayes methodology, see Efron and Morris (1975), Morris (1983b), and Carlin and Louis (2008, Chapter 5).

10.11 **Hierarchical Bayes**: For the hierarchical Bayesian approach, see Carlin and Louis (2008, Sections 2.4 and 4.1), Christensen et al. (2010, Section 4.12), Gelman et al. (2013, Chapters 5, 15), Gelman and Hill (2006), Good (1965), Hoff (2009, Chapter 8), and Lindley and Smith (1972).

EXERCISES

10.1 Suppose $y_1, \ldots, y_c$ are independent from a Poisson distribution with mean μ. Conditional on $\sum y_i = n$, are $y_1, \ldots, y_c$ exchangeable? Independent? Explain.

10.2 Independent observations $\boldsymbol{y} = (y_1, \ldots, y_n)$ come from the $N(\mu, \sigma^2)$ distribution, with σ^2 known, and μ has a $N(\lambda, \tau^2)$ prior. Show that the posterior predictive distribution for a future y^* is normal with mean equal to the posterior mean of μ and variance equal to the posterior variance plus σ^2. For large n, show that this is approximately a $N(\bar{y}, \sigma^2)$ distribution.

10.3 Find the posterior mean and variance for μ in the null model with a $N(\mu, \sigma^2)$ response for unknown σ^2, using the improper-priors approach of Section 10.2.5.

10.4 Suppose $\boldsymbol{y}_1, \ldots, \boldsymbol{y}_n$ are independent from a $N(\boldsymbol{\mu}, \boldsymbol{\Sigma})$ distribution, and $\boldsymbol{\mu}$ has a $N(\boldsymbol{\mu}_0, \boldsymbol{\Sigma}_0)$ prior distribution. With $\boldsymbol{\Sigma}$ known, derive the posterior distribution. Explain how the posterior mean is a weighted average of the prior mean and the sample mean and the posterior precision is the sum of the prior precision and the data precision. Discuss their behavior as n increases.

10.5 For the Bayesian ordinary normal linear model, using a flat improper prior for $\boldsymbol{\beta}$ but a proper inverse-gamma distribution (10.5) for σ^2, find the posterior Bayes estimate of σ^2 and express it as a weighted average of s^2 and the prior mean of σ^2.

10.6 Suppose we assume the Bayesian normal linear model for a one-way layout, but the actual conditional distribution of y is highly skewed to the right (e.g., y = annual income). For large $\{n_i\}$, would you expect Bayesian inference about $\{\mu_i\}$ to be relatively robust? Would you expect Bayesian prediction intervals based on the posterior predictive distribution to be robust? Explain.

10.7 You regard m potential models, $M_1, \ldots, M_m$, to be (a priori) equally likely. Use Bayes' theorem to conduct Bayesian model averaging by finding an expression for the posterior $P(M_i \mid \boldsymbol{y})$ in terms of the marginal $\{P(\boldsymbol{y} \mid M_j)\}$ for the models (integrating out the parameters). Extend the result to possibly unequal prior probabilities $\{P(M_i)\}$.

10.8 With a beta(α_1, α_2) prior for the binomial parameter π and sample proportion y, if n is large relative to $\alpha_1 + \alpha_2$, show that the posterior distribution of π has approximate mean y and approximate variance $y(1 - y)/n$. Interpret.

10.9 With a beta prior for the binomial parameter π having $\mu = \alpha_1/(\alpha_1 + \alpha_2)$ and letting $n^* = \alpha_1 + \alpha_2$, find $E(\tilde{\pi} - \pi)^2$ and express it as a weighted average of $[\pi(1 - \pi)]/n$ and $(\mu - \pi)^2$. Compare this with $E(\hat{\pi} - \pi)^2$, for $\hat{\pi} = y$. Evaluate when the Bayes estimator is better and when the ML estimator is better.

10.10 A beta prior for the binomial parameter π has $\alpha_1 = \alpha_2 = \alpha$.

a. When $y = 0$, for what values of α is the posterior density of π monotone decreasing and hence the HPD posterior interval of the form $(0, U)$? For such α, report U as the quantile of a beta distribution.

b. Show that the improper case $\alpha = 0$ corresponds to an improper uniform prior for $\log[\pi/(1 - \pi)]$. With it, when $y = 0$ or 1, show that the posterior distribution of π is also improper.

c. Show that the uniform prior for π corresponds to a logistic distribution prior for the logit.

10.11 In Exercise 4.11 for the binomial probability π of being a vegetarian, the proportion $y = 0$ of $n = 25$ students were vegetarians.

a. Report the ML estimate. Find the 95% confidence interval based on inverting the likelihood-ratio chi-squared test.

b. Using a uniform prior distribution for π, find the posterior mean, the posterior 95% equal-tail interval, and the 95% HPD interval.

10.12 This exercise is based on an example in the keynote lecture by Carl Morris (see `www.youtube.com/watch?v=JOovvj_SKOg`) at a symposium held in his honor in October 2012. Before a Presidential election, polls are taken in two states that are usually swing states. In State A, the proportion $y_1 = 0.590$ of $n_1 = 100$ sampled state a preference for the Democratic candidate. In State B, the proportion $y_2 = 0.525$ of $n_2 = 1000$ sampled state a preference for the Democratic candidate. Treat these as independent binomial samples with population proportions π_1 and π_2. In which state is there greater evidence supporting a Democratic victory (i.e., $\pi_i > 0.50$)?

a. With a frequentist approach, show that the binomial P-values for testing H_0: $\pi_i = 0.50$ against H_1: $\pi_i > 0.50$ are 0.044 for π_1 and 0.061 for π_2.

b. A Bayesian statistician interprets "swing state" to mean that π_i is very likely to be between 0.40 and 0.60 and nearly certain to be between 0.35 and 0.65. To recognize this, she uses a $N(0, \sigma^2)$ prior for logit(π_i), with σ such that logit(0.35) and logit(0.65) are 3 standard deviations from the prior mean of 0. Show that the posterior (mean, standard deviation) are (0.183, 0.143) for logit(π_1) and (0.091, 0.060) for logit(π_2). Based on the posterior distributions, show that $P(\pi_1 < 0.50 \mid y_1, n_1) = 0.100$ and $P(\pi_2 < 0.50 \mid y_2, n_2) = 0.066$. Explain why the Bayesian and frequentist approaches give different answers to the question about which state has greater evidence of victory. Summarize what causes this.

10.13 For a binomial distribution with beta prior, show that the marginal distribution of $s = ny$ is the beta-binomial. State its mean and variance.

10.14 For a binomial distribution with beta prior, show how to conduct Bayesian estimation of $\{\pi_i\}$ for c groups in the one-way layout.

10.15 In a diagnostic test for a disease, let D denote the event of having the disease, and let $+$ $(-)$ denote a positive (negative) diagnosis by the test. Let $\pi_1 = P(+ \mid D)$ (the *sensitivity*), $\pi_2 = P(+ \mid D^c)$ (the *false positive rate*), and $\rho = P(D)$ (the *prevalence*). More relevant to a patient who has received a positive diagnosis is $P(D \mid +)$, the *positive predictive value*.

a. Show that $P(D \mid +) = \pi_1\rho/[\pi_1\rho + \pi_2(1-\rho)]$.

b. Suppose $n_i y_i \sim \text{bin}(n_i, \pi_i)$, $i = 1, 2$. When ρ is known, explain how to simulate in a simple manner to obtain a 95% posterior interval for $P(D \mid +)$ based on independent uniform priors for π_1 and π_2. Illustrate using $n_1 = n_2 = 100$, $y_1 = y_2 = 0.95$, and $\rho = 0.005$. Explain the influence of ρ on why $P(D \mid +)$ seems to be so small.

10.16 Consider independent binary observations from two groups with $\pi_i = P(y_i = 1) = 1 - P(y_i = 0)$, and a binary predictor x. For the 2×2 contingency table summarizing the two binomials, let $\text{logit}(\pi_i) = \beta_0 + \beta_1 x_i$, where $x_i = 1$ or $x_i = 0$ according to a subject's group classification. Also, express $\text{logit}(\pi_i) = \beta_0^* + \beta_1^* x_i^*$, where $x_i^* = 1$ when $x_i = 0$ and $x_i^* = 0$ when $x_i = 1$ (i.e., the classification when the group labels are reversed). Let $\theta = \exp(\beta_1)$ and $\theta^* = \exp(\beta_1^*)$ denote the corresponding odds ratios. The ML estimates satisfy $\hat{\beta}_1^* = -\hat{\beta}_1$ and $\hat{\theta}^* = 1/\hat{\theta}$. For a Bayesian solution, denote the means of the posterior distributions of β_1 and β_1^* by $\tilde{\beta}_1$ and $\tilde{\beta}_1^*$ and the means of the posterior distributions of θ and θ^* by $\tilde{\theta}$ and $\tilde{\theta}^*$.

a. Explain why $\tilde{\beta}_1^* = -\tilde{\beta}_1$ but $\tilde{\theta}^* \neq 1/\tilde{\theta}$.

b. Let (L, U) denote the 95% HPD interval from the posterior distribution of θ. Explain why the 95% HPD interval from the posterior distribution of θ^* is not $(1/U, 1/L)$. Explain why such invariance *does* occur for the equal-tail interval and for frequentist inference.

10.17 In the previous exercise, suppose $\hat{\theta} = 0$ and the HPD interval for θ is $(0, U)$. Is $(1/U, \infty)$ the HPD interval or the equal-tail posterior interval for θ^*? Do you think it is a sensible interval? Explain, and give implications for forming posterior intervals for parameters in logistic regression when there is complete or quasi-complete separation.

10.18 Re-do the analysis with prior $\sigma = 1$ for the example in Section 10.3.2, but use (0, 1) coding for the indicator variable x_1 instead of -0.5 and 0.5. Why is the posterior mean for β_1 so different?

10.19 For a Poisson random variable y with mean λ, show that the Jeffreys prior distribution for λ is improper. Using it, find the posterior distribution and indicate whether it is improper.

10.20 For iid Poisson variates $y_1, \ldots, y_n$ with parameter λ, suppose $\lambda \sim$ gamma(μ, k) (Recall Section 4.7.2).

a. Show that the posterior distribution of λ is gamma (i.e., the prior is conjugate), with posterior mean that is a weighted average of the sample mean and the prior mean. Explain how the weights change as n increases. When n is very large, show that the posterior distribution has approximate mean $\bar{y}$ and approximate variance $\bar{y}/n$.

b. Find the posterior predictive distribution.

10.21 Show how the results in (**a**) in the previous exercise generalize to estimating Poisson parameters for c groups in the one-way layout. For fixed gamma hyperparameters, show that the estimate of μ_i for group i is a weighted average of $\bar{y}_i$ and the mean of the gamma prior distribution. When those hyperparameters are estimated using the marginal negative binomial distribution, show that empirical Bayes estimates shrink $\{\bar{y}_i\}$ toward the overall $\bar{y}$.

10.22 Show how results in Section 10.1.3 generalize for n independent multinomial trials and a Dirichlet prior for the multinomial probabilities. (Good (1965) gave one of the first Bayesian analyses with this model, using empirical Bayes and hierarchical approaches. See also Section 11.2.2.)

10.23 Two independent multinomial variates have ordered response categories. Using independent Dirichlet priors, explain how to simulate to find the posterior probability that those two distributions are stochastically ordered. (Altham (1969) found an exact expression for this probability.)

10.24 Refer to the Presidential election data in Section 10.4.3.

a. Use a Bayesian approach to fit model (10.7), with independent $N(0, 100)$ priors for $\{\beta_i\}$. Find corresponding estimates of $\{\pi_i\}$, and evaluate their performance by finding the MSE and comparing with the ordinary sample proportions.

b. Use a Bayesian approach to fit model (10.7), taking the population percentage q_i voting for Obama in the 2008 election (shown in the data file) as prior information by using a $N[\text{logit}(q_i), 1]$ prior for β_i in model (10.7). Find corresponding estimates of $\{\pi_i\}$, and compare their MSE with the values in (**a**).

10.25 In the previous exercise, repeat (**b**), but using an empirical Bayes approach for a $N(\text{logit}(q_i), \sigma^2)$ prior distribution for β_i.

10.26 Conduct Bayesian analogs of the frequentist modeling for the FEV clinical trial data from Exercise 3.31. Compare Bayesian and frequentist results.

10.27 Conduct Bayesian modeling for the smoking prevention data of Section 9.2.3. Compare results to the frequentist results presented there.

10.28 Refer to Exercise 5.34 on horseshoe crabs. Repeat this exercise using Bayesian methods.

10.29 For the endometrial cancer example in Section 10.3.2, fit the logistic model using a hierarchical Bayesian approach with a diffuse inverse chi-squared distribution for the σ^2 hyperparameter. Interpret results.

CHAPTER 11

Extensions of Generalized Linear Models

This final chapter introduces alternatives to maximum likelihood (ML) and Bayes for fitting linear and generalized linear models (GLMs). We also present an extension of the GLM that permits an additive predictor in place of the linear predictor. A complete exposition of these topics is beyond the scope of this book. We aim here merely to present a brief overview and give you references for further study.

Section 11.1 presents alternative ways to estimate model parameters. For the linear model, *M-estimation* methods minimize a function of the residuals, the sum of squared residuals being one special case. Some such estimates are more robust than least squares, because they are less affected by severe outliers or by contamination of the data. *Regularization methods* modify ML to give sensible answers in situations that are unstable because of causes such as collinearity. For the GLM, the *penalized likelihood* regularization method modifies the log-likelihood function by adding a penalty term, resulting in estimates that tend to have smaller variance than ML estimators.

Regularization methods are especially useful when the number p of model parameters is very large. Such datasets are common in genomics, biomedical imaging, functional magnetic resonance imaging, tomography, signal processing, image analysis, market basket data, and portfolio allocation in finance. Sometimes p is even larger than n. Section 11.2 discusses the fitting of GLMs with high-dimensional data, focusing on identifying the usually small subset of the explanatory variables that are truly relevant for modeling $E(y)$.

Another extension of the ordinary GLM replaces the linear predictor by smooth functions of the explanatory variables. Section 11.3 introduces generalizations of the GLM that do this, such as the *generalized additive model*, or that have structure other than modeling the mean response with a linear predictor, such as *quantile regression*

Foundations of Linear and Generalized Linear Models, First Edition. Alan Agresti.

for modeling quantiles of the response distribution and *nonlinear regression* when the response mean is a nonlinear function of parameters.

11.1 ROBUST REGRESSION AND REGULARIZATION METHODS FOR FITTING MODELS

For an ordinary linear model with residuals $\{e_i = y_i - \hat{\mu}_i\}$, the least squares method minimizes $\sum_i e_i^2$. The model fit can be severely affected by observations that have both large leverage and a large residual (recall Section 2.5.5). So that such observations have less influence, we could instead minimize a function that gives less weight to large residuals.

11.1.1 M-Estimation for Robust Regression

An alternative function to minimize is $\sum_i \rho(e_i)$ for an objective function $\rho(e_i)$ that is symmetric with a minimum at 0 but with possibly less than a quadratic increase. This approach is called *M-estimation*. Like least squares, it does not require assuming a distribution for y.

In M-estimation, the estimates $\hat{\boldsymbol{\beta}}$ of the parameters $\boldsymbol{\beta}$ in the linear predictor are the solutions to the equations

$$\frac{\partial}{\partial \hat{\beta}_j}\left[\sum_{i=1}^{n} \rho(e_i)\right] = \sum_{i=1}^{n} \frac{\partial \rho(e_i)}{\partial e_i}\frac{\partial e_i}{\partial \hat{\beta}_j} = 0, \quad j = 1, \dots, p.$$

For the linear model, $\partial e_i/\partial \hat{\beta}_j = -x_{ij}$. The function $\psi(e) = \partial \rho(e)/\partial e$ is called the *influence function*, because it describes the influence of an observation's residual on $\hat{\boldsymbol{\beta}}$. For least squares, the influence increases linearly with the size of the residual. A more robust solution chooses $\rho(e)$ so that $\psi(e)$ is a bounded function.

Let $\boldsymbol{\rho}(\boldsymbol{\beta})$ represent ρ expressed in terms of the population residuals $\{\epsilon_i = y_i - \boldsymbol{x}_i\boldsymbol{\beta}\}$ as $[\rho(y_1 - \boldsymbol{x}_1\boldsymbol{\beta}), \dots, \rho(y_n - \boldsymbol{x}_n\boldsymbol{\beta})]^{\mathrm{T}}$ and satisfying $E[\partial \boldsymbol{\rho}(\boldsymbol{\beta})/\partial \boldsymbol{\beta}] = \mathbf{0}$. Choosing $\rho(\cdot)$ to be strictly convex ensures that a unique estimate exists. A natural choice is the absolute value metric, $\rho(e_i) = |e_i|$. For the null model, this produces the sample median as the estimate of location. But $\rho(e_i)$ is not then strictly convex, the solution may be indeterminate, and the estimator loses considerable efficiency relative to least squares when the normal linear model is adequate. An alternative is $\rho(e) = |e|^p$ for some $1 < p < 2$, although then the influence function is not bounded.

A compromise approach, suggested by Peter Huber in the early literature on M-estimation, takes $\rho(e) = |e|^2$ for small $|e|$ and takes $\rho(e)$ to be a linear function of $|e|$ beyond that point. A popular implementation takes $\rho(e)$ quadratic for $|e| \leq k\hat{\sigma}$, where $k \approx 1.5$ (proposed by Huber) and $\hat{\sigma}$ is a robust estimate of $\sqrt{\mathrm{var}(\epsilon)}$, such as the median absolute residual for the least squares fit divided by 0.67. Smaller values for k protect against a higher proportion of outlying observations, but at a greater loss of efficiency when the normal linear model truly holds. The value $k = 1.345$ provides

95% efficiency under the normal linear model. Other proposals for robust fitting include one by John Tukey for which the influence function is 0 at large absolute values. This completely removes the influence of large outliers.

For a weight function defined by $w(e) = \psi(e)/e$, the estimating equations for the M-estimates $\hat{\boldsymbol{\beta}}$ are

$$\sum_{i=1}^{n} w(e_i)e_i x_{ij} = 0, \quad j = 1, \ldots, p.$$

Finding the solution requires iterative methods, with initial values such as the least squares estimates. At stage t of the iterative process, the estimating equations correspond to those for the iteratively reweighted least squares solution for minimizing $\sum_i w(e_i^{(t)})e_i^2$. That is, for a model matrix $\boldsymbol{X}$ and with $\boldsymbol{W}^{(t)}$ a diagonal matrix having elements $\{w(e_i^{(t)})\}$, their solution is

$$\hat{\boldsymbol{\beta}}^{(t)} = \left[\boldsymbol{X}^{\mathrm{T}}\boldsymbol{W}^{(t)}\boldsymbol{X}\right]^{-1}\boldsymbol{X}^{\mathrm{T}}\boldsymbol{W}^{(t)}\boldsymbol{y}.$$

The asymptotic covariance matrix of the limit $\hat{\boldsymbol{\beta}}$ of this iterative process is

$$\mathrm{var}(\hat{\boldsymbol{\beta}}) = \left(\boldsymbol{X}^{\mathrm{T}}\boldsymbol{X}\right)^{-1}\frac{E[\psi(\epsilon)]^2}{\{E[\psi'(\epsilon)]\}^2}.$$

Substituting the sample analogs $\{\sum_i[\psi(e_i)]^2\}/n$ for $E[\psi(\epsilon)]^2$ and $[\sum_i \psi'(e_i)]/n$ for $E[\psi'(\epsilon)]$ yields an estimated covariance matrix. Fitting is available in software[1].

11.1.2 Penalized-Likelihood Methods

In fitting GLMs, *regularization methods* modify ML to give sensible answers in unstable situations. A popular way to do this adds a term to the log-likelihood function such that the solution of the modified likelihood equations smooths the ordinary estimates. For a model with log-likelihood function $L(\boldsymbol{\beta})$, we maximize

$$L^*(\boldsymbol{\beta}) = L(\boldsymbol{\beta}) - s(\boldsymbol{\beta}),$$

where $s(\cdot)$ is a function such that $s(\boldsymbol{\beta})$ decreases as elements of $\boldsymbol{\beta}$ are smoother in some sense, such as uniformly closer to 0. This smoothing method, referred to as *penalized likelihood*, shrinks the ML estimate toward $\mathbf{0}$. Among its positive features are a reduction in prediction error and existence when the ML estimate is infinite or badly affected by collinearity.

[1]For example, the `rlm` (robust linear modeling) function in the R `MASS` package

A variety of penalized-likelihood methods use the L_q-*norm* smoothing function

$$s(\boldsymbol{\beta}) = \lambda \sum_{j=1}^{p} |\beta_j|^q$$

for some $q \geq 0$ and $\lambda \geq 0$. The explanatory variables should be standardized, as they are treated the same way in the smoothing function, and the degree of smoothing should not depend on the choice of scaling. The response variable should also be standardized, or the intercept should be removed from the smoothing term, because there is no reason to shrink a parameter whose estimate (for the ordinary linear model) is merely the overall sample mean response. The constant λ is called a *smoothing parameter*, because the degree of smoothing depends on it. The choice of λ reflects the bias–variance tradeoff discussed in Section 4.6.2. Increasing λ results in greater shrinkage toward 0 in $\{\hat{\beta}_j\}$ and smaller variance but greater bias.

How well a smoothing method works depends on λ. This is usually chosen by cross-validation. For each λ value in a chosen grid, we fit the model to part of the data and then check the goodness of the predictions for y in the remaining data. With k-fold cross-validation, we do this k times (for k typically about 10), each time leaving out the fraction $1/k$ of the data and predicting those y values using the model fit from the rest of the data. The selected value of λ is the one having the lowest sample mean prediction error for the k runs, for a measure of prediction error such as squared difference between observed and predicted y. We then apply that value with the penalized-likelihood method for all the data.

At each λ, the sample mean prediction error is a random variable. An alternative choice for λ uses a *one standard error rule*, in which the chosen λ has mean prediction error that is one standard error above the minimum, in the direction of greater regularization. Such a choice may be less likely to overfit the model.

Penalized-likelihood estimators have Bayesian connections. With prior pdf proportional to $\exp[-s(\boldsymbol{\beta})]$, the Bayesian posterior pdf is proportional to the penalized-likelihood function. The mode of the posterior distribution then equals the penalized-likelihood estimate.

11.1.3 L_2-Norm Penalty: Ridge Regression

Regularization methods that penalize by a quadratic term, such as $s(\boldsymbol{\beta}) = \lambda \sum_j \beta_j^2$, are called L_2-*norm* penalty methods. For normal linear models, the best known such method is *ridge regression*, which finds the value of $\boldsymbol{\beta}$ that minimizes

$$\sum_{i=1}^{n} \left(y_i - \sum_{j=1}^{p} \beta_j x_{ij} \right)^2 + \lambda \sum_{j=1}^{p} \beta_j^2.$$

Equivalently, this solution minimizes $\|\boldsymbol{y} - \boldsymbol{X}\boldsymbol{\beta}\|^2$ subject to $\sum_j \beta_j^2 \leq \lambda^*$, where a 1–1 inverse correspondence holds between λ and λ^*. The solution for the ridge regression estimate is

$$\tilde{\boldsymbol{\beta}} = \left(\boldsymbol{X}^{\mathrm{T}}\boldsymbol{X} + \lambda \boldsymbol{I}\right)^{-1} \boldsymbol{X}^{\mathrm{T}}\boldsymbol{y},$$

which adds a "ridge" to the main diagonal of $\boldsymbol{X}^{\mathrm{T}}\boldsymbol{X}$ before inverting it. This modification is helpful when the model matrix is ill-conditioned, such as under collinearity. Adding the ridge makes the matrix invertible, even if $\boldsymbol{X}$ does not have full rank. Since $\tilde{\boldsymbol{\beta}}$ is a linear function of $\boldsymbol{y}$, for the ordinary linear model

$$\mathrm{var}(\tilde{\boldsymbol{\beta}}) = \sigma^2 \left(\boldsymbol{X}^{\mathrm{T}}\boldsymbol{X} + \lambda \boldsymbol{I}\right)^{-1} \boldsymbol{X}^{\mathrm{T}}\boldsymbol{X} \left(\boldsymbol{X}^{\mathrm{T}}\boldsymbol{X} + \lambda \boldsymbol{I}\right)^{-1}.$$

The least squares estimate is the limit of $\tilde{\boldsymbol{\beta}}$ as $\lambda \to 0$. As λ increases, the effect is to shrink the least squares estimate toward $\boldsymbol{0}$. For example, when explanatory variables are linearly transformed so that $\boldsymbol{X}$ is orthonormal (i.e., $\boldsymbol{X}^{\mathrm{T}}\boldsymbol{X} = \boldsymbol{I}$), we see that $\tilde{\boldsymbol{\beta}}$ relates to the least squares estimate $\hat{\boldsymbol{\beta}}$ by $\tilde{\boldsymbol{\beta}} = \hat{\boldsymbol{\beta}}/(1+\lambda)$. The ridge regression estimate $\tilde{\boldsymbol{\beta}}$ has the form of the Bayesian posterior mean for the normal linear model presented in Section 10.2.2, when the prior mean for $\boldsymbol{\beta}$ is $\boldsymbol{0}$ and λ here is identified with σ^2/τ^2 in that Bayesian formulation.

11.1.4 L_1-Norm Penalty: The Lasso

Fitting using the L_1-*norm* penalty, for which $s(\boldsymbol{\beta}) = \lambda \sum_j |\beta_j|$, is referred to as the *lasso* ("least absolute shrinkage and selection operator") method (Tibshirani 1996). Equivalently, it maximizes the likelihood function subject to the constraint that $\sum_j |\beta_j| \leq \lambda^*$ for a constant λ^* inversely related to λ. The larger the value of λ, the greater the shrinkage of estimates toward 0. The shrinkage is by a fixed amount, rather than by a fixed proportion as in ridge regression. For λ sufficiently large, this method shrinks some $\hat{\beta}_j$ completely to zero. In constraining $\sum_j |\beta_j| \leq \lambda^*$, the region of acceptable $\{\beta_j\}$ is a region around the origin that is square when $p = 2$. It intersects the contours of the log likelihood, which are elliptical for normal linear models and approximately so for large n with other GLMs, at axes rather than at the interior in which all $\beta_j \neq 0$. Figure 11.1 illustrates. It is informative to plot the penalized estimates as a function

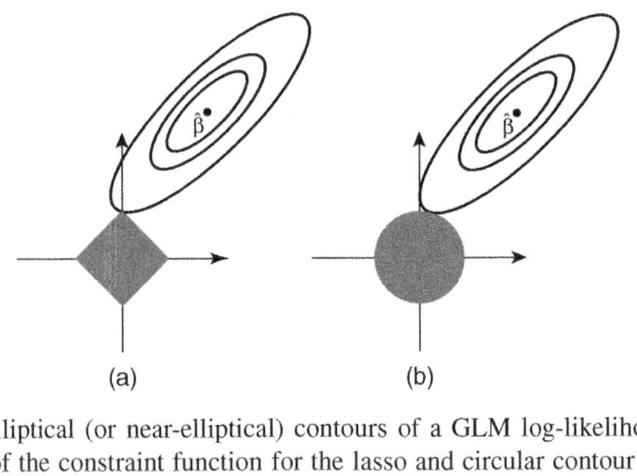

Figure 11.1 Elliptical (or near-elliptical) contours of a GLM log-likelihood function and square contour of the constraint function for the lasso and circular contour of the constraint function for ridge regression. The lasso estimates occur where an ellipse touches the square constraint, often resulting in some $\hat{\beta}_j = 0$. Source: Hastie et al. (2009, p. 71, Figure 3.11), with kind permission from Springer Science+Business Media B.V.

of the permitted value λ^* for $\sum_j |\beta_j|$, to summarize how explanatory variables drop out as λ^* decreases.

Why can shrinking $\{\hat{\beta}_j\}$ toward 0 be effective? In many settings having a large number of explanatory variables, most of them have no effects or very minor effects. An example is genetic association studies, which simultaneously consider each of possibly thousands of genes for the association between the genetic expression levels and the response of interest, such as whether a person has a particular disease. Unless n is extremely large, because of sampling variability the ordinary ML estimates $\{\hat{\beta}_j\}$ tend to be much larger in absolute value than the true values $\{\beta_j\}$. This tendency is exacerbated when we keep only statistically significant variables in a model. Shrinkage toward 0 tends to move $\{\hat{\beta}_j\}$ closer to $\{\beta_j\}$. This is yet another example of the bias–variance tradeoff. Introducing a penalty function results in biased estimates but benefits from reducing the variance.

Penalizing by absolute-value terms makes model fitting more difficult than ridge regression. The estimate of $\boldsymbol{\beta}$ is not linear in $\boldsymbol{y}$, and we need an optimization method to find it. One approach uses the LARS method, to be introduced in Section 11.2.1. A faster method uses coordinate descent, optimizing each parameter separately while holding all the others fixed, and cycling until the estimates stabilize. In a particular cycle t, for explanatory variable x_j, one regresses the residuals $\{y_i - \sum_{k \neq j} \hat{\beta}_k^{(t)} x_{ik}\}$ on $(x_{1j}, \dots, x_{nj})$ to obtain a value which, when reduced in absolute value by an amount dependent on λ, yields the next approximation for $\hat{\beta}_j$. In practice, this is done for a grid of λ values.

Likewise, finding an estimated covariance matrix for lasso estimators is challenging, especially for the parameters having lasso estimates of 0. Tibshirani (1996) noted that the lasso estimate corresponds to a Bayesian posterior mode for the normal linear model when the independent prior distribution for each β_j is a double-exponential (Laplace) distribution, which has pdf

$$g(\boldsymbol{\beta} \mid \sigma^2) = \prod_{j=1}^{p} \frac{\lambda}{2\sqrt{\sigma^2}} e^{-\lambda|\beta_j|/\sqrt{\sigma^2}}.$$

Each component of this prior distribution has a sharp peak at $\beta_j = 0$. Park and Casella (2007) and Hans (2009) used this result as a mechanism for point and interval estimation of $\{\beta_j\}$.

11.1.5 Comparing Penalized Methods, and Generalizations

Ridge regression, the lasso, and other regularization methods are available in software[2]. A disadvantage of ridge regression is that it requires a separate strategy for finding a parsimonious model, because all explanatory variables remain in the model. By contrast, with the lasso, when λ is large, some $\hat{\beta}_j$ shrink to zero, which can help

[2]In R, the `glmnet` and `ridge` packages and the `lm.ridge` function in the `MASS` package provide ridge regression, and the `glmnet` and `lars` packages provide lasso fits.

with model selection. For a factor predictor, the ordinary lasso solution may select individual indicators rather than entire factors, and the solution may depend on the coding scheme, so an alternative *grouped lasso* should be used. A disadvantage of the lasso is that $\{\hat{\beta}_j\}$ are not asymptotically normal and can be highly biased, making inference difficult. Another disadvantage is that the lasso may overly penalize β_j that are truly large. Which of ridge regression and the lasso performs better in terms of bias and variance for estimating the true $\{\beta_j\}$ depends on their values. When p is large but only a few $\{\beta_j\}$ are practically different from 0, the lasso tends to perform better, because many $\{\hat{\beta}_j\}$ may equal 0. When $\{\beta_j\}$ do not vary dramatically in substantive size, ridge regression tends to perform better.

L_0-*norm* penalty regularization takes $s(\boldsymbol{\beta})$ to be proportional to the number of nonzero β_j. This approach has the Akaike information criterion (AIC) and the Bayesian information criterion (BIC) as special cases. This sounds ideal, but optimization for this criterion is impractical with large numbers of variables; for example, the function minimized may not be convex. A compromise method, *SCAD* (smoothly clipped absolute deviation), starts at the origin $\boldsymbol{\beta} = \mathbf{0}$ like an L_1 penalty and then gradually levels off (Fan and Lv 2010). An alternative *elastic net* uses a penalty function that has both L_1 and L_2 terms (Zou and Hastie 2005). It has both ridge regression and the lasso as special cases. Zou (2006) proposed an *adaptive lasso* that can be better for satisfying an *oracle* property, by which asymptotically the method recovers the correct model and has estimators converging to the parameter values at the optimal rate. It uses an adaptive weighted penalty $\sum_j w_j|\beta_j|$ where $w_j = 1/|\hat{\beta}_j|^\gamma$ for a consistent estimator $\hat{\beta}_j$ such as from least squares, and $\gamma > 0$. This has the effect of reducing the penalty when an effect seems to be large.

11.1.6 Example: House Selling Prices Revisited

In Section 4.7.1 we modeled y = the selling price of a house (in thousands of dollars), using as explanatory variables the size of the house, the property tax bill, whether the home is new, the number of bedrooms, and the number of bathrooms. We now illustrate methods of this section by comparing the least squares fit with other methods, for the simple linear model having all the main effects but no interactions. Adjusted for the other variables, the least squares fit shows strong evidence that the mean selling price increases as the house size increases, as the tax bill increases, and for new houses.

```
-------------------------------------------------------------------------------
> attach(Houses) # File Houses.dat at www.stat.ufl.edu/~aa/glm/data
> summary(lm(price ~ size + taxes + new + beds + baths))
Coefficients:
             Estimate  Std. Error  t value  Pr(>|t|)
(Intercept)    4.5258     24.4741    0.185    0.8537
size           0.0683      0.0139    4.904  3.92e-06
taxes          0.0381      0.0068    5.596  2.16e-07
new           41.7114     16.8872    2.470    0.0153
beds         -11.2591      9.1150   -1.235    0.2198
baths         -2.1144     11.4651   -0.184    0.8541
-------------------------------------------------------------------------------
```

For a robust M-estimation fit, we use the Huber influence function mentioned in Section 11.1.1 with $k = 1.345$ and a robust standard deviation estimate. Summaries of effects are similar to least squares, a notable exception being the effect of *new*. The least squares estimated difference of \$41,711 between the mean selling prices of new and older homes, adjusting for the other variables, decreases to \$27,861.

```
-------------------------------------------------------------------------
> library(MASS)
> summary(rlm(price ~ size + taxes + new + beds + baths, psi=psi.huber))
Coefficients: # robust (Huber) fit of linear model
               Value   Std. Error  t value
(Intercept)  11.6233     19.1847    0.6059
size          0.0705      0.0109    6.4533
taxes         0.0341      0.0053    6.3838
new          27.8610     13.2375    2.1047
beds        -16.4034      7.1451   -2.2958
baths         3.9534      8.9873    0.4399
-------------------------------------------------------------------------
```

An alternative parametric check fits the model assuming a gamma distribution for y, which naturally accounts for larger variability in selling prices when the mean is larger. The estimated effect of a new home is also then considerably weaker. The change in the *new* effect for these two fits, relative to least squares, is mainly caused by observation 64 in the data file. This observation, which had a relatively low selling price for a very large house that was not new, was an outlier and influential for least squares but not unusual for the gamma model.

```
-------------------------------------------------------------------------
> summary(glm(price ~ size + taxes + new + beds + baths,
+             family = Gamma(link=identity)))
Coefficients: # fit of gamma GLM with identity link
             Estimate  Std. Error  t value  Pr(>|t|)
(Intercept)   19.1859    13.7759     1.393    0.1670
size           0.0617     0.0125     4.929   3.5e-06
taxes          0.0378     0.0051     7.475   4.0e-11
new           22.6704    19.3552     1.171    0.2444
beds         -19.2618     6.3273    -3.044    0.0030
baths          9.5825     6.4775     1.479    0.1424
-------------------------------------------------------------------------
```

When we implemented the lasso in R with `glmnet`, which operates on the standardized variables, the smoothing parameter value $\lambda = 8.3$ gave the minimum value of cross-validated mean squared error. This fit is not much different from the robust fit but removes *beds* and *baths*, the two predictors that were not significant in the least squares fit. For contrast, we show the coefficients obtained with the much larger value of $\lambda = 23.1$ suggested by the one standard error rule. That fit also removes *new* and has diminished effects of *size* and *taxes*. The first panel of Figure 11.2 shows how the lasso estimates change as λ increases (on the log scale). The *new* estimate decreases from the least squares value of 41.7, becoming 0 when $\log(\lambda) \geq \log(21.0) = 3.0$. For the scaling used, the *size* and *taxes* estimates (which are nonzero for much larger λ values) are too small to appear in the figure. To show this more clearly, the second panel of Figure 11.2 shows the estimates for the standardized variables, for which

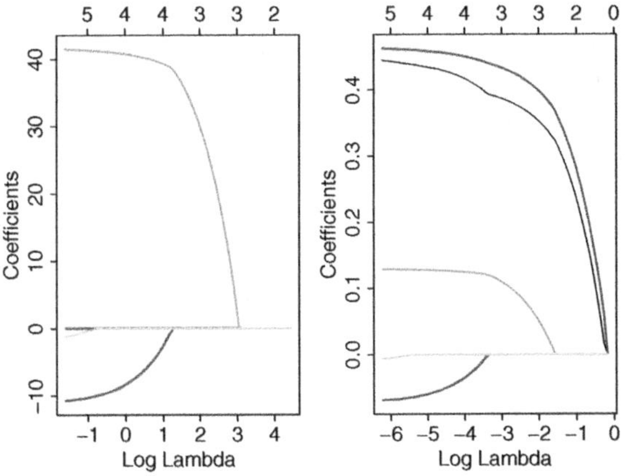

Figure 11.2 Plot of lasso estimates for house selling price data, as function of smoothing parameter $\log(\lambda)$. In the first panel, the least squares estimate of 41.7 for *new* decreases to 0 at $\log(\lambda) = 3.0$. The second panel shows estimates for the standardized variables, from which it is clearer that *size* and *taxes* remain in the model the longest as λ increases. In that panel, the least squares estimates for (size, taxes, new, beds, baths) are (0.45, 0.47, 0.13, −0.07, −0.01), the values of the five curves at the left axis, where λ is essentially 0.

the least-squares estimated effects for (size, taxes, new, beds, baths) are (0.45, 0.47, 0.13, −0.07, −0.01).

```
-------------------------------------------------------------------------
> library(glmnet)
> x <- cbind(size, taxes, new, beds, baths)
> set.seed(1010) # random seed for cross-validation
> cv.glmnet(x,price,alpha=1) # alpha=1 specifies lasso for cross-valid.
$lambda.min # best lambda by 10-fold cross-validation
[1] 8.2883
$lambda.1se # lambda suggested by one standard error rule
[1] 23.06271
> coef(glmnet(x,price, alpha=1, lambda=8.2883))
(Intercept) -5.9947
size         0.0568
taxes        0.0344
new         28.0744
beds          .
baths         .
>  coef(glmnet(x,price, alpha=1, lambda=23.0627))
(Intercept) 22.1646
size         0.0475
taxes        0.0293
new           .
beds          .
baths         .
> fit.lasso <- glmnet(x, price, alpha=1)
> plot(fit.lasso, "lambda")
-------------------------------------------------------------------------
```

For ridge regression, cross-validation suggested using $\lambda = 17.9$. With it, results are not much different from least squares. The fit slightly shrinks the least squares estimates, except for the *new* effect. For $\lambda = 95.3$ from the one standard error rule, the effects of *beds* and *baths* change sign from their least squares values. Keep in mind that for ridge regression and the lasso, results depend greatly on the chosen smoothing parameter λ, and the value chosen for λ in cross-validation will vary considerably according to the seed. One could also report the estimates and standard errors for the standardized variables.

```
-----------------------------------------------------------------------
> cv.glmnet(x,price,alpha=0) # alpha=0 specifies ridge regression
$lambda.min      $lambda.1se
[1] 17.85662     [1] 95.2954
> coef(glmnet(x, price, alpha=0, lambda=95.2954))
(Intercept) -4.4871
size         0.0377
taxes        0.0216
new         41.6077
beds         6.4325
baths       16.9838
-----------------------------------------------------------------------
```

11.1.7 Penalized Likelihood for Logistic Regression

Penalizing a log-likelihood function need not necessarily result in increased bias. One version actually reduces bias of ML estimators. For most models, the ML estimator $\hat{\boldsymbol{\beta}}$ has bias on the order of $1/n$. Firth (1993) penalized the log likelihood in a way that introduces a small bias into the score function but reduces the bias of $\hat{\boldsymbol{\beta}}$ to order $1/n^2$. For the canonical parameter of an exponential family model, Firth's penalized log-likelihood function uses the determinant of the information matrix $\mathcal{J}$,

$$L^*(\boldsymbol{\beta}) = L(\boldsymbol{\beta}) + \frac{1}{2}\log|\mathcal{J}|.$$

The penalized likelihood is proportional to the Bayesian posterior distribution resulting from using the Jeffreys prior distribution. Thus, this penalized ML estimator equals the mode of the posterior distribution induced by the Jeffreys prior.

For logistic regression, Firth noted that the ML estimator is biased away from 0, and the bias correction shrinks the estimator toward 0. When the model matrix is of full rank, $\log|\mathcal{J}|$ is strictly concave. Maximizing the penalized log likelihood yields a maximum penalized-likelihood estimate that always exists and is unique. For the null logistic model and a proportion y of successes in n independent Bernoulli trials, it yields as estimate the *empirical logit*, $\log[(ny + \frac{1}{2})/(n - ny + \frac{1}{2})]$. This corresponds to adding $\frac{1}{2}$ to the success and failure counts. Firth's method is especially appealing for the analysis of data that exhibit complete or quasi-complete separation, because then at least one ordinary ML estimate is infinite or does not exist (Section 5.4.2).

11.1.8 Example: Risk Factors for Endometrial Cancer Revisited

Sections 5.7.1 and 10.3.2 described a study about endometrial cancer that analyzed y = histology of 79 cases (0 = low grade, 1 = high grade), with the explanatory variables x_1 = neovasculation (1 = present, 0 = absent), x_2 = pulsatility index of arteria uterina, and x_3 = endometrium height. For the main-effects model

$$\text{logit}[P(y_i = 1)] = \beta_0 + \beta_1 x_{i1} + \beta_2 x_{i2} + \beta_3 x_{i3},$$

all 13 patients having $x_{i1} = 1$ had $y_i = 1$. So quasi-complete separation occurs, and the ML estimate $\hat{\beta}_1 = \infty$.

```
-----------------------------------------------------------------------
> Endometrial # File Endometrial.dat at www.stat.ufl.edu/~aa/glm/data
   NV PI   EH HG
1   0 13 1.64  0
2   0 16 2.26  0
...
79  0 33 0.85  1
> attach(Endometrial)
> PI2 <- (PI-mean(PI))/sd(PI); EH2 <- (EH-mean(EH))/sd(EH); NV2 <- NV-0.5
> fit.ML <- glm(HG ~ NV2 + PI2 + EH2, family=binomial)
> summary(fit.ML) # ML estimate of NV effect is actually infinite
Coefficients:
             Estimate  Std. Error  z value  Pr(>|z|)
(Intercept)    7.8411    857.8755    0.009    0.9927
NV            18.1856   1715.7509    0.011    0.9915
PI2           -0.4217      0.4432   -0.952    0.3413
EH2           -1.9219      0.5599   -3.433    0.0006
-----------------------------------------------------------------------
```

Table 10.1 showed Bayes estimates, which with standardized x_2 and x_3 shrink $\hat{\beta}_1$ to 9.1 for quite diffuse normal priors ($\sigma = 10$) and to 1.65 for very informative priors ($\sigma = 1$). The maximum penalized-likelihood estimate for β_1 of 2.93 and the 95% profile penalized-likelihood confidence interval of (0.61, 7.85) shrink the ML estimate $\hat{\beta}_1$ and the ordinary profile likelihood interval of (1.28, ∞) considerably toward 0. Results for the other estimates do not change as much.

```
-----------------------------------------------------------------------
> library(logistf)
> fit.penalized <- logistf(HG ~ NV2 + PI2 + EH2, family=binomial)
> summary(fit.penalized)
Confidence intervals and p-values by Profile Likelihood
                coef se(coef) lower 0.95 upper 0.95   Chisq          p
(Intercept)  0.3080   0.8006    -0.9755     2.7888   0.169  6.810e-01
NV2          2.9293   1.5508     0.6097     7.8546   6.798  9.124e-03
PI2         -0.3474   0.3957    -1.2443     0.4045   0.747  3.875e-01
EH2         -1.7243   0.5138    -2.8903    -0.8162  17.759  2.507e-05
-----------------------------------------------------------------------
```

11.2 MODELING WITH LARGE P

High-dimensional data are not well handled by the traditional model-fitting methods presented in this book. In genomics, such applications include classifying tumors by using microarray gene expression or proteomics data or associating protein concentrations with expression of genes or predicting a clinical prognosis by using gene expression data. Generalized linear modeling by ML can be overwhelmed when it needs to detect effects for such applications as differential expression (change between two or more conditions) in many thousands of genes or brain activity in many thousands of locations. We now discuss issues in fitting linear models and GLMs to high-dimensional data in which p is very large, sometimes even with $p > n$. Certain issues are vital yet difficult, such as how to select explanatory variables from an enormous set when nearly all of them are expected to have no effect or a very small effect.

11.2.1 Issues in Variable Selection and Dimension Reduction

In modeling with a very large number of explanatory variables, removing variables that have little if any relevance can ease interpretability and decrease prediction errors. For example, in disease classification, very few of a large number of genes may be associated with the disease. This is reflected by histograms of P-values for testing those effects, which often have appearance similar to the uniform density function that theoretically occurs when the null hypothesis is true. With large p and huge n, ordinary ML fitting may not even be possible and alternative methods may be needed (Toulis and Airoldi 2014). For a binary response, complete or quasi-complete separation often occurs when the number of predictors exceeds a particular point, resulting in some infinite estimates. Even when finite estimates exist, they may be imprecise because of ill-conditioning of the covariance matrix. Moreover, choosing a model that contains a large number of predictors runs the risk of overfitting the data. Future predictions will then tend to be poorer than those obtained with a more parsimonious model.

As in ordinary model selection using ML, variable selection algorithms such as forward selection and backward elimination have pitfalls, especially when p is large. For example, for the set of predictors having no true effect, the maximum sample correlation with the response can be quite large. Also, there can be spurious collinearity among the predictors or spurious correlation between an important predictor and a set of unimportant predictors, because of the high dimensionality[3]. Other criteria exist for identifying an optimal subset of explanatory variables, such as minimizing prediction error or (with AIC) considering models with nearly minimum Kullback–Leibler divergence of the fitted values from true conditional means. With large p, though, it is not feasible to check a high percentage of the possible subsets of predictors, and the danger remains of identifying an effect as important that is actually spurious.

[3]Figure 1 in Fan and Lv (2010) illustrates these issues.

Ordinary variable selection methods such as stepwise procedures are highly discrete: Any particular variable either is or is not selected. Penalized likelihood is more continuous in nature, with some variables perhaps receiving little influence in the resulting prediction equation but not being completely eliminated. Besides providing shrinkage of parameter estimates, some of those methods (L_q-norm with $0 \leq q \leq 1$) also help with variable selection. With the lasso ($q = 1$), many explanatory variables receive zero weight in the prediction equation, the number included depending on the smoothing parameter. A variable can be eliminated, but in a more objective way that does not depend on which variables were previously eliminated.

The variable selection methods for large p fall roughly into two types. One approach adapts dimension-reduction methods, such as stepwise methods and penalized likelihood and regularization using L_q-norm penalties for some q between 0 and 2 and compromise norms. A second approach attempts to identify the relevant effects using standard significance tests but with an adjustment for multiplicity. A fundamental assumption needed for methods to perform well with large p is *sparse structure*, with relatively few elements in $\boldsymbol{\beta}$ being nonzero (Bühlmann et al. 2014).

The first type of variable selection method includes stepwise methods that use regularization procedures. The LARS (least-angle regression) procedure (Efron et al. 2004) for linear models is an adaptation of a forward selection method. Like forward selection, it first adds the predictor having greatest absolute correlation with y, say x_j. This is the variable with the smallest angle between it and the response variable, found for the vectors connecting the origin to the points $\boldsymbol{y}$ and $\boldsymbol{x}_j$ in $\mathbb{R}^n$. The LARS algorithm proceeds from the origin in the x_j direction as long as the angle between the point on that line and the residual between $\boldsymbol{y}$ and that point is smaller than the angle between other predictors and the residual. When some other predictor, say x_k, has as much correlation with the current residual, instead of continuing along the x_j direction, LARS then proceeds in a direction equiangular between x_j and x_k. The algorithm continues in this direction until a third variable x_ℓ earns its way into the "most correlated" set. LARS then proceeds equiangularly between x_j, x_k, and x_ℓ (i.e, along the "least angle direction") until a fourth variable enters, and so on. It smoothly blends in new variables rather than adding them discontinuously.

Advantages of the LARS method are that it is computationally fast and the lasso can be generated in a modified special case. In the published discussion for the Efron et al. article, D. Madigan and G. Ridgeway suggested an extension for logistic regression, and S. Weisberg suggested caution, arguing that any automatic method relying on correlations has potential pitfalls, especially under collinearity. In the rejoinder, the authors discussed possible stopping rules for the algorithm.

An alternative approach that explicitly performs dimension reduction is *principal component analysis*. This method[4] replaces the p predictors by fewer linear combinations of them (the "principal components") that are uncorrelated. The first principal component is the linear combination (using a unit vector) that has the largest possible variance. Each succeeding component has the largest possible variance under the constraint that it is orthogonal to the preceding components. A small number of

[4]Proposed by K. Pearson in 1901 and developed by H. Hotelling in 1933.

principal components often explains a high percentage of the original variability. The components depend on the scaling of the original variables, so when they measure inherently different characteristics they are standardized before beginning the process. Disadvantages, especially with large p, are that it may be difficult to interpret the principal components, and the data may be overfitted, with the derived principal components not explaining variability in another dataset nearly as well. For details, see references in Note 11.3.

The second type of approach searches for effects while adjusting for the number of inferences conducted. This can reduce dramatically the data dimensionality by eliminating the many predictors not having strong evidence of an effect. An approach such as using the *false discovery rate* (FDR) introduced in Section 3.5.3 is especially useful in applications in which a very small proportion of the effects truly are of substantive size. Because of its lessened conservatism and improved power compared with family-wise inference methods such as the Bonferroni, controlling FDR is a sensible strategy to employ in exploratory research involving large-scale testing. A place remains for traditional family-wise inference methods in follow-up validation studies involving the smaller numbers of effects found to be significant in the exploratory studies. Dudoit et al. (2003) surveyed these issues in the context of microarray experiments.

11.2.2 Effect of Large p on Bayesian Methods

Dealing with large p is also challenging for Bayesian inference, perhaps even more so than for frequentist inference. The impact of forming prior distributions for a very large number of parameters may differ from what you intuitively expect. For example, even if you pick a very diffuse prior, the effect may depend strongly on which diffuse prior you choose.

To illustrate, suppose the response distribution is multinomial with p outcome categories and p is very large relative to n, as in a study of the frequency of use of the p words in a language by an author writing in that language. In a particular document, we might observe how many times each word occurs for the n words. Most words would have a count of 0. As in Section 6.1.1, let $\boldsymbol{y}_i = (y_{i1}, \dots, y_{ip})$ represent the multinomial trial for observation i, $i = 1, \dots, n$, where $y_{ij} = 1$ when the response is in category j and $y_{ij} = 0$ otherwise, so $\sum_j y_{ij} = 1$. Let $\pi_{ij} = P(y_{ij} = 1)$, and let $n_j = \sum_i y_{ij}$ denote the total number of observations in category j. Here, for simplicity, we discuss large-p challenges[5] without any reference to explanatory variables, so we will suppress the i subscript and replace π_{ij} by π_j. In practice, similar issues arise when the number of multinomial categories is of any size but the number of explanatory variables is large.

The beta distribution that serves as a conjugate prior distribution for a binomial parameter extends to the *Dirichlet distribution* for multinomial parameters. With hyperparameters $\{\alpha_j\}$, the Dirichlet prior density function is proportional to $\prod_{j=1}^{p} \pi_j^{\alpha_j - 1}$. The posterior density is then also Dirichlet, with parameters $\{n_j + \alpha_j\}$.

[5]Of course, here the actual number of parameters is $p - 1$.

The posterior mean of π_j is $(n_j + \alpha_j)/(n + \sum_k \alpha_k)$. The impact of the prior is essentially to add α_j observations to category j for all j before forming a sample proportion. Most applications use a common value α for $\{\alpha_j\}$, so the impact is to smooth in the direction of the equi-probability model.

The Dirichlet prior with $\alpha = 1$ corresponds to a uniform prior distribution over the probability simplex. This seems diffuse, but it corresponds to adding p observations and then forming sample proportions. This is considerable when p is large relative to n. For example, suppose $n = 100$ but $p = 1000$. The posterior mean of π_j is $(n_j + 1)/(n + p) = (n_j + 1)/1100$. When cell j contains one of the 100 observations, the posterior mean estimate for that cell is 0.0018, shrinking the sample proportion of 0.010 toward the equi-probability value of 0.001. This seems like a reasonable estimate. But what if instead all 100 observations fall in cell j? The posterior mean estimate is then 0.092. This shrinks much more from the sample proportion value of 1.0 than we are likely to believe is sensible. Even though the prior distribution is quite diffuse, it has quite a strong impact on the results. The Jeffreys prior, $\alpha = 1/2$, corresponds to a U-shaped beta density for the binomial case $p = 2$. The shrinkage is then a bit less, but it still gives a posterior mean estimate for π_j of $(n_j + 1/2)/(n + p/2) = (n_j + 1/2)/600$, or 0.1675 when $n_j = n = 100$.

This simplistic example illustrates that the choice of the prior distribution is crucial when p is very large, especially when we depart from the traditional setting in which n is much larger than p. Berger et al. (2013) suggested that the prior distribution should have marginal posterior distributions all close to a common posterior that we'd obtain in the single-parameter case. For instance, we could aim for the posterior distribution of π_j to be approximately a beta distribution with parameters $n_j + \frac{1}{2}$ and $n - n_j + \frac{1}{2}$, which we'd obtain with a Jeffreys prior for the binomial distribution with parameter π_j. We can obtain this by using Dirichlet hyperparameters $\{\alpha_j = 1/p\}$ instead of $\{\alpha_j = 1/2\}$, which is much more diffuse when p is large. This yields a posterior mean for π_j of $(n_j + 1/p)/(n + 1)$. With $n = 100$ observations in $p = 1000$ cells, this is 0.0099 when $n_j = 1$ and is 0.990 when $n_j = 100$.

This approach seems sensible, but even with it, situations exist for which the results may seem inappropriate. When $p = 1000$, suppose we have only $n = 2$ observations, of which $n_j = 1$. The posterior mean for π_j is then 0.334. Would you want to use an estimate that shrinks the sample proportion of 1/2 based on only two observations so little toward the equi-probability value of 0.001? Which prior distribution would you use for such sparse multinomial modeling?

11.3 SMOOTHING, GENERALIZED ADDITIVE MODELS, AND OTHER GLM EXTENSIONS

The models in this text smooth the data rather severely, by producing fitted values satisfying a predictor that is linear in the parameters. In this final section we present frequentist ways of smoothing data that provide more flexibility than linear predictors in GLMs. We also consider alternative models that are nonlinear in the parameters or that describe quantiles instead of mean responses.

11.3.1 Kernel Smoothing

Kernel smoothing, in its basic form, is completely non-model-based. To estimate a mean at a particular point, it smooths the data by using primarily the data at nearby points.

We illustrate with a method that smooths binary response data to portray graphically the form of dependence of y on a quantitative explanatory variable x (Copas 1983). Let $\phi(\cdot)$ denote a symmetric unimodal *kernel function*, such as the standard normal or another bell-shaped pdf. At any x, the kernel-smoothed estimate of $P(y = 1 \mid x)$ is

$$\tilde{\pi}(x) = \frac{\sum_{i=1}^{n} y_i \phi[(x - x_i)/\lambda]}{\sum_{i=1}^{n} \phi[(x - x_i)/\lambda]}, \tag{11.1}$$

where $\lambda > 0$ is a smoothing parameter. At any point x, the estimate $\tilde{\pi}(x)$ is a weighted average of the $\{y_i\}$. For the simple function $\phi(u) = 1$ when $u = 0$ and $\phi(u) = 0$ otherwise, $\tilde{\pi}(x_k)$ simplifies to the sample proportion of successes at $x = x_k$. Then there is no smoothing. When ϕ is proportional to the standard normal pdf, $\phi(u) = \exp(-u^2/2)$, the smoothing approaches this as $\lambda \to 0$. For very small λ, only points near x have much influence. Using mainly very local data produces little bias but high variance. By contrast, as λ increases, data points farther from x also contribute substantially to $\tilde{\pi}(x)$. As λ increases and very distant points receive more weight, the smoothed estimate becomes more like the overall sample proportion. It becomes more highly biased but has smaller variance. As λ grows unboundedly, the smooth function $\tilde{\pi}(x)$ converges to a horizontal line at the level of the overall sample proportion.

For this kernel smoother, the choice of λ is more important in determining $\tilde{\pi}(x)$ than is the choice of ϕ. Copas recommended selecting λ by plotting the resulting function for several values of λ, varying around a value equal to 10 times the average spacing of the x values. The kernel smoothing (11.1) generalizes to incorporate multiple predictors, with a multivariate kernel function such as a multivariate normal pdf.

11.3.2 Nearest-Neighbors Smoothing

In more general contexts than binary regression, smoothers of the kernel type can base estimation at a point on using nearby points. A very simple such method is *nearest-neighbors smoothing*. It is often used for classification, such as by predicting an observation for a subject based on a weighted average of observations for k subjects who have similar values on the explanatory variables.

An advantage of this method is its simplicity, once we select a similarity measure to determine the nearest neighbors. However, the choice of this measure may not be obvious, especially when p is large with possibly some subsets of explanatory variables being highly correlated and some of them being qualitative. More complex smoothers generalize this idea by basing the prediction at a point on a weighted regression using nearby points, such as described next. Such methods have better statistical properties, such as usually lower bias.

11.3.3 The Generalized Additive Model

The GLM generalizes the ordinary linear model by permitting non-normal distributions and modeling functions of the mean. The quasi-likelihood approach (Chapter 8) generalizes GLMs, specifying how the variance depends on the mean without assuming a particular distribution. Another generalization of the GLM replaces the linear predictor by additive smooth functions of the explanatory variables. The GLM structure $g(\mu_i) = \sum_j \beta_j x_{ij}$ generalizes to

$$g(\mu_i) = \sum_{j=1}^{p} s_j(x_{ij}),$$

where $s_j(\cdot)$ is an unspecified smooth function of predictor j. Like GLMs, this model specifies a link function g and a distribution for y. The resulting model is called a *generalized additive model*, symbolized by GAM (Hastie and Tibshirani 1990). The GLM is the special case in which each s_j is a linear function. Also possible is a mixture of explanatory terms of various types: Some s_j may be smooth functions, others may be linear functions as in GLMs, and others may be indicator variables to include qualitative factors.

A useful smooth function is the *cubic spline*. It has separate cubic polynomials over sets of adjacent intervals for an explanatory variable, joined together smoothly at boundaries of those intervals. The boundary points, called *knots*, can be set at evenly spaced points for each predictor or selected according to a criterion involving both smoothness and closeness of the spline to the data. A *smoothing spline* uses knots at the observed predictor values but imposes a smoothing parameter that determines the influence of the integrated squared second derivative of the smoothing function in penalizing the log likelihood. For example, for the normal model with identity link, the fit minimizes a penalized residual sum of squares,

$$\sum_{i=1}^{n} \left[y_i - \sum_{j=1}^{p} s_j(x_{ij}) \right]^2 + \sum_{j=1}^{p} \lambda_j \int [s_j''(x)]^2 dx.$$

Larger smoothing parameter values λ_j result in smoother functions (less "wiggling" and change in the first derivative). In fact, this criterion results in a solution in which each s_j is a cubic spline.

One can select λ_j so that a term s_j in the predictor has an *effective df* value, with higher λ_j corresponding to lower effective *df*. For instance, a smooth function having effective $df = 3$ is similar in overall complexity to a third-degree polynomial, and *df* close to 1 is similar to a straight line. Choosing an effective *df* value or a value for a smoothing parameter determines how smooth the resulting GAM fit looks. It is sensible to try various degrees of smoothing. The goal is not to smooth so much that the fit suppresses interesting patterns yet smooth the data sufficiently so that the data are not overfitted with a highly wiggly function. The smoothing may suggest that a linear model is adequate with a particular link function, or it may suggest ways to improve on linearity.

Using the effective *df* value for each s_j in the additive predictor, we can conduct approximate large-sample inference about those terms. For any model fit, there is a deviance, which reflects the assumed distribution for y. As in comparing GLMs, we can compare deviances for nested GAMs to test whether a particular model gives a significantly better fit than a simpler model.

For fitting a GAM, the *backfitting algorithm* employs a generalization of the Newton–Raphson method that uses local smoothing. The algorithm initializes $\{\hat{s}_j\}$ identically at 0. Then at a particular iteration, it updates the estimate $\hat{s}_j$ by a smoothing of partial residuals $\{y_i - \sum_{k \neq j} \hat{s}_k(x_{ik})\}$ that uses the other estimated smooth functions at that iteration, in turn for $j = 1, \ldots, p$.

An alternative way to smooth the data, without making a distributional assumption for y, employs a type of regression that gives greater weight to nearby observations in predicting the value at a given point; such *locally weighted least squares regression* is often referred to as *lowess* (Cleveland 1979). We prefer GAMs to lowess, because they recognize explicitly the form of the response variable. For instance, with a binary response, lowess can give predicted values below 0 or above 1 at some predictor settings. This cannot happen with a GAM that assumes a binomial random component.

Smoothing methods such as GAMs have the advantage over GLMs of greater flexibility. Using them, we may discover patterns we would miss with ordinary GLMs, and we obtain potentially better predictions of future observations. The smoothness of the function that works well is summarized by its effective df. A disadvantage of GAMs (and other smoothing methods) compared with GLMs is the loss of interpretability for describing the effect of an explanatory variable that has a smooth term in the predictor. Likewise, it is unclear how to apply confidence intervals to effects in a GAM. So it is more difficult to judge when an effect has substantial importance. Thus, when suitable, GLMs are ideal for statistical inference. Also, because any smoothing method has potentially a very large number of parameters, it can require a large n to estimate the functional form accurately.

Even if you plan mainly to use GLMs, a GAM is helpful for exploratory analysis. For instance, for binary responses, scatterplots are not very informative. Plotting the fitted smooth function for a predictor may reveal a general trend without assuming a particular functional relation, as we illustrate in Section 11.3.5.

11.3.4 How Much Smoothing? The Bias–Variance Tradeoff

Smoothing methods have a nonparametric flavor, because they base analyses on a more general structure than a linear predictor. However, in some ways the demands are greater: We need to choose among a potentially infinite number of forms relating the response variable to the explanatory variables, the number of parameters is potentially much larger, and overfitting is a danger.

As discussed in Section 4.6.2, model selection is at the heart of the fundamental statistical tradeoff between bias and variance. Using a particular model has the disadvantage of increasing the potential bias (e.g., a true mean differing from the value corresponding to fitting the model to the population), but it has the advantage that the parsimonious limitation of the parameter space results in decreased variance in estimating characteristics of interest. The methods presented in this chapter provide a compromise. A method typically starts with a model and its likelihood function, but smooths results to adjust for ways an ordinary linear predictor may fail. All smoothing methods require input from the methodologist to control the degree of smoothness imposed on the data in order to deal with the bias–variance tradeoff, whether it be determined by a smoothing parameter in a frequentist approach or a prior distribution in a Bayesian approach.

11.3.5 Example: Smoothing to Portray Probability of Kyphosis

Hastie and Tibshirani (1990, p. 282) described a study to determine risk factors for kyphosis, which is severe forward flexion of the spine following corrective spinal surgery. Figure 11.3 shows this binary outcome y ($1 =$ kyphosis present, $0 =$ absent) plotted against the age in months at the time of the operation. For the youngest and the oldest children, most observations have kyphosis absent.

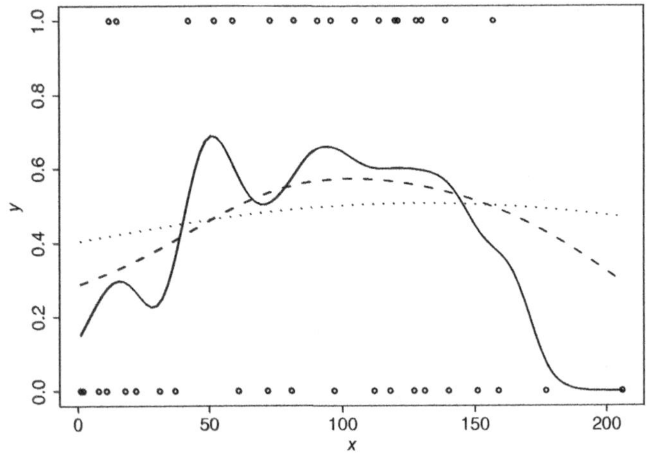

Figure 11.3 Kernel-smoothing estimate of probability of kyphosis as a function of $x =$ age (in months), using standard normal kernel function ϕ and smoothing parameter $\lambda = 25$ (solid curve), 100 (dashed curve), 200 (dotted curve), in equation (11.1).

Figure 11.3 also shows the result of kernel smoothing of the data using the smoother (11.1). The smoothing parameter value $\lambda = 25$ is too low, and the figure is more irregular than the data justify. The higher values of λ give evidence of nonmonotonicity in the relation. In fact, adding a quadratic term to the standard logistic regression model provides an improved fit.

```
-----------------------------------------------------------------------
> Kyphosis # File Kyphosis.dat at www.stat.ufl.edu/~aa/glm/data
      x y
1    12 1
2    15 1
...
40 206 0
> attach(Kyposis)
> plot(x, y)
> k1 <- ksmooth(x, y, "normal", bandwidth=100)
> lines(k1)
> x2 <- x*x
> summary(glm(y ~ x + x2, family=binomial(link=logit)))
              Estimate  Std. Error  z value  Pr(>|z|)
(Intercept)  -2.046255    0.994348   -2.058    0.0396
x             0.060040    0.026781    2.242    0.0250
x2           -0.000328    0.000156   -2.097    0.0360
---
Residual deviance: 48.228  on 37  degrees of freedom
-----------------------------------------------------------------------
```

For fitting a GAM, we treat the data as binomial with a logit link. The default smoothing obtained with the function for GAMs in the VGAM R library falls between a quadratic and cubic in complexity ($df = 2.6$).

```
-----------------------------------------------------------------------
> library(VGAM)
> gam.fit <- vgam(y ~ s(x),family=binomialff(link=logit),data=Kyphosis)
> plot(x, fitted(gam.fit))
> summary(gam.fit)
Residual deviance:  47.948 on 35.358 degrees of freedom
DF for Terms and Approximate Chi-squares for Nonparametric Effects
            Df  Npar Df  Npar Chisq  P(Chi)
(Intercept)  1
s(x)         1      2.6      4.7442  0.1528
-----------------------------------------------------------------------
```

Figure 11.4 shows the fitted values for the 40 observations. This also suggests using a logistic model with a quadratic term[6]. The figure also shows that fit, which is very similar graphically and in the residual deviance.

We can also fit GAMs using the gam and mgcv libraries in R. We next fit models that have successively linear, quadratic, and cubic complexity for the smooth function.

[6]Using a penalized-likelihood approach for GAMs, Eilers and Marx (2002) suggested instead a bell-shaped response curve.

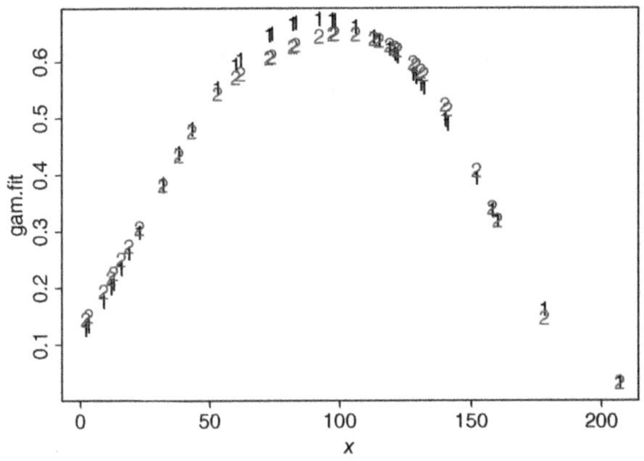

Figure 11.4 Estimates of probability of kyphosis as a function of x = age, using (1) a GAM and (2) logistic regression with quadratic term.

Comparison of deviances shows that quadratic fits better than linear, but cubic is not better than quadratic.

```
-----------------------------------------------------------------------
> library(gam) # R library by Trevor Hastie
> gam.fit1 <- gam(y ~ s(x,1), family=binomial, data=Kyphosis)
> gam.fit2 <- gam(y ~ s(x,2), family=binomial, data=Kyphosis)
> gam.fit3 <- gam(y ~ s(x,3), family=binomial, data=Kyphosis)
> anova(gam.fit1, gam.fit2, gam.fit3)
Analysis of Deviance Table
Model 1: y ~ s(x, 1) # linear complexity
Model 2: y ~ s(x, 2) # quadratic
Model 3: y ~ s(x, 3) # cubic
  Resid. Df  Resid. Dev      Df   Deviance   Pr(>Chi)
1        38      54.504
2        37      49.216  0.9999     5.2880     0.0215
3        36      48.231  1.0002     0.9852     0.3210
-----------------------------------------------------------------------
```

11.3.6 Quantile Regression

Models in this book describe the conditional mean of y as a function of explanatory variables. Alternatively, we could model a quantile. For example, in modeling growth over time for a sample of a biological organism, it might be of interest to estimate the 10th percentile, median, and 90th percentile of the conditional distribution as a function of time. *Quantile regression* models quantiles of a response variable as a function of explanatory variables. M-estimation in regression using $\rho(e_i) = |e_i|$ corresponds to quantile regression for the median.

Like regression fitted by M-estimation, this method can be less severely affected by outliers than is ordinary least squares. When the response conditional distributions are highly skewed with possibly highly nonconstant variance, the method can describe the relationship better than a simple normal model with constant variance. For instance, consider modeling of annual income as a function of the age of a person. We might expect almost no effect at low quantiles, with the effect increasing as the quantile increases, reflecting also increasing variability with age.

Quantile-regression model fitting minimizes a weighted sum of absolute residuals, formulated as a linear programming problem. This is available in software[7]. Why not always use it instead of least squares, since it is less affected by outliers? When the normal linear model truly holds, the least squares estimators are much more efficient.

11.3.7 Nonlinear Regression

In this book, we've focused on predictors that are linear in the parameters. The GAM is one generalization. Another is relevant for applications in which the predictor is naturally nonlinear in parameters. For example, consider the model

$$E(y_i) = \frac{\beta_0}{1 + \exp[-(\beta_1 + \beta_2 x_i)]}.$$

With $\beta_0 = 1$, this is the logistic regression curve for a binary response probability (Chapter 5). For other β_0, it has a symmetric, sigmoidal shape with bounds of 0 and β_0. This model can describe the growth of a tumor or population growth, when the maximum possible size is also a parameter.

A *nonlinear regression model* has the form

$$E(y_i) = f(\boldsymbol{x}_i; \boldsymbol{\beta}),$$

where f is a known function of the explanatory variables and the parameters. With an assumption about the distribution of y_i, inference can use likelihood-based methods. Assuming normality with constant variance σ^2, this again yields the least squares criterion, with $\hat{\boldsymbol{\beta}}$ giving the minimum value of $\sum_i [y_i - f(\boldsymbol{x}_i; \boldsymbol{\beta})]^2$. The likelihood equations are then

$$\sum_{i=1}^{n} [y_i - f(\boldsymbol{x}_i; \boldsymbol{\beta})] \frac{\partial f(\boldsymbol{x}_i; \boldsymbol{\beta})}{\partial \beta_j} = 0, \quad j = 1, \dots, p.$$

Finding the estimates requires an iterative algorithm that starts at initial values $\boldsymbol{\beta}^{(0)}$ for $\hat{\boldsymbol{\beta}}$. Let $\boldsymbol{X}$ denote the model matrix of $\{x_{ij}\}$, and let $\boldsymbol{f}(\boldsymbol{X}; \boldsymbol{\beta})$ be the vector having elements $f(\boldsymbol{x}_i; \boldsymbol{\beta})$. The *Gauss–Newton algorithm*, which for a normal response is equivalent to Fisher scoring, uses the linearization

$$\boldsymbol{f}(\boldsymbol{X}; \boldsymbol{\beta}) = \boldsymbol{f}(\boldsymbol{X}; \boldsymbol{\beta}^{(0)}) + \boldsymbol{G}^{(0)}(\boldsymbol{\beta} - \boldsymbol{\beta}^{(0)}),$$

[7]Such as the `quantreg` package in R.

where the gradient matrix $\boldsymbol{G}^{(0)}$ has elements $\partial f(\boldsymbol{x}_i; \boldsymbol{\beta})/\partial \beta_j$ evaluated at $\boldsymbol{\beta}^{(0)}$. For the initial working residuals $\{e_i^{(0)} = y_i - f(\boldsymbol{x}_i; \boldsymbol{\beta}^{(0)})\}$, the first iteration yields updated estimate

$$\boldsymbol{\beta}^{(1)} = \boldsymbol{\beta}^{(0)} + (\boldsymbol{G}^{(0)\mathrm{T}}\boldsymbol{G}^{(0)})^{-1}\boldsymbol{G}^{(0)\mathrm{T}}\boldsymbol{e}^{(0)}.$$

Each subsequent iteration t regresses the current working residuals $\boldsymbol{e}^{(t)}$ on the current gradient matrix $\boldsymbol{G}^{(t)}$ to find the increment to the working estimate $\boldsymbol{\beta}^{(t)}$. Modifications of the method exist, such as taking smaller increments if needed to decrease the residual sums of squares, or using numerical derivatives rather than computing the gradient matrix. Many nonlinear models have the potential for multiple local maxima of the log likelihood, so it is wise to use a grid of quite different initial values to increase the chance of finding the true least squares estimate.

The linearization is also the basis of standard errors for $\hat{\boldsymbol{\beta}}$. The asymptotic covariance matrix of $\hat{\boldsymbol{\beta}}$ is

$$\mathrm{var}(\hat{\boldsymbol{\beta}}) = \sigma^2 \left(\boldsymbol{G}^{\mathrm{T}}\boldsymbol{G}\right)^{-1},$$

where the gradient matrix $\boldsymbol{G}$ has elements $\partial f(\boldsymbol{x}_i; \boldsymbol{\beta})/\partial \beta_j$ evaluated at $\boldsymbol{\beta}$. This has the same form as $\mathrm{var}(\hat{\boldsymbol{\beta}})$ in formula (2.4) for the ordinary linear model, except that the gradient matrix replaces the model matrix. In practice, we estimate the covariance matrix by substituting $\hat{\boldsymbol{\beta}}$ for $\boldsymbol{\beta}$ in $\boldsymbol{G}$ and by estimating σ^2 by the error mean square $s^2 = \sum_i [y_i - f(\boldsymbol{x}_i; \hat{\boldsymbol{\beta}})]^2/(n-p)$. Nonlinear regression fitting methods and subsequent inference are available in software[8].

CHAPTER NOTES

Section 11.1: Robust Regression and Regularization Methods for Fitting Models

11.1 **Robust regression**: M-estimation evolved out of research by Huber (1964) on robust estimation of a location parameter. Huber and Ronchetti (2009, Chapter 7) presented the regression context of this approach. Rousseeuw (1984) proposed another alternative to least squares, finding the estimate that produces the smallest *median* of the squared residuals, instead of their *mean* (equivalently, sum). Like M-estimation with $\rho(e_i) = |e_i|$, this "least median of squares" method can have low efficiency when the ordinary normal linear model nearly holds. Cantoni and Ronchetti (2001) proposed other robust estimation methods for GLMs. Birkes and Dodge (1993) surveyed alternative fitting methods, including least-absolute-deviations regression, M-estimation, and ridge regression.

11.2 **Ridge regression and lasso**: For more on ridge regression, see Hoerl and Kennard (1970) and Hastie et al. (2009, Section 3.4.1). For the lasso, see Bühlmann and van de Geer (2011, Chapters 2 and 3), Hastie et al. (2009, Section 3.4.2), Izenman (2008), and Tibshirani (1996, and his website `statweb.stanford.edu/~tibs/lasso.html`). James et al. (2013) gave a less technical introduction to such methods,

[8]such as the `nls` function in R.

with extensive R examples. The bootstrap is another possible way to determine standard errors for lasso estimates (Chatterjee and Lahiri 2011). Lockhart et al. (2014) proposed a significance test for the lasso, based on how much of the covariance between y and the model-fitted values can be attributed to a particular predictor when it enters the model. Bühlmann et al. (2014) presented other inference methods, such as tests based on multisample splitting of the data.

Section 11.2: Modeling with Large p

11.3 Penalized likelihood with large p: Fan and Lv (2010) and Tutz (2011) reviewed penalized likelihood methods for variable selection in high dimensions. Fan and Lv noted that the lasso has a tendency to include many false-positive variables when p is large. For details about dimension-reduction methods such as principal component analysis, see Hastie et al. (2009, Chapter 18), Izenman (2008), and James et al. (2013). Bühlmann et al. (2014) presented a brief introductory survey of high-dimensional methods. For multinomial modeling, see Taddy (2013).

11.4 Bayes with large p: For issues in selecting priors when p is large but n may not be, see Berger et al. (2013), Griffin and Brown (2013), Kass and Wasserman (1996, Section 4.2.2), and Polson and Scott (2010). Carvalho et al. (2010) advocated a prior based on multivariate normal scale mixtures. Gelman (2006) argued for using noninformative priors (such as uniform) for variance parameters in hierarchical models. For variable selection issues, see George and McCulloch (1997), George (2000), and Rŏcková and George (2014). Hjort et al. (2010) surveyed nonparametric Bayesian approaches.

Section 11.3: Smoothing, Generalized Additive Models, and Other GLM Extensions

11.5 Smoothing: For smoothing methods, see Fahrmeir et al. (2013, Chapter 8), Fahrmeir and Tutz (2001, Chapter 5), and Faraway (2006, Chapter 11). Green and Silverman (1993), Hastie et al. (2009), Izenman (2008), James et al. (2013), Simonoff (1996), Tutz (2011, Chapters 6 and 10), and Wakefield (2013, Chapters 10–12). For smoothing spatial data , see Fahrmeir et al. (2013, Section 8.2). Albert (2010) presented Bayesian smoothing methods.

11.6 GAMs and penalized-spline regularization: For generalized additive modeling, see Fahrmeir et al. (2013, Chapter 9), Faraway (2006, Chapter 12), Hastie and Tibshirani (1990), Wood (2006), and Yee and Wild (1996). The *generalized additive mixed model* adds random effects to a GAM (Wood 2006, Chapter 6). Eilers and Marx (1996, 2002, 2010) introduced an alternative penalized likelihood approach for splines that provides a way of fitting GAMs as well as a mechanism for regularization. Rather than penalizing by the integrated squared derivative, it penalizes by differences of coefficients of adjacent splines. See also Fahrmeir et al. (2013, Chapter 8).

11.7 Nonlinear and quantile regression: For nonlinear regression methods, see Bates and Watts (1988) and Seber and Wild (1989). For a brief review, see Smyth (2002). For quantile regression and examples, see Davino et al. (2013), Fahrmeir et al. (2013, Chapter 10), and Koenker (2005).

11.8 Functions and images: Methods of this chapter extend to the analysis of more complex types of data, such as functions and images. See, for example, Crainiceanu et al. (2009), Ramsay and Silverman (2005), Di et al. (2009), and `www.smart-stats.org` and the R package `refund`.

EXERCISES

11.1 Show that M-estimation with $\rho(e_i) = |e_i|$ gives the ML solution assuming a Laplace distribution for the response.

11.2 The *breakdown point* of an estimator is the proportion of observations that must be moved toward infinity in order for the estimator to also become infinite. The higher the breakdown point, the more robust the estimator. For estimating the center of a symmetric distribution, explain why the breakdown point is $1/n$ for the sample mean but 0.50 for the sample median. (However, even robust regression methods, such as using $\rho(e_i) = |e_i|$, can have small breakdown points or other unsatisfactory behavior; see Seber and Lee 2003, Sections 3.13.2 and 3.13.3)

11.3 Refer to the equations solved to obtain $\hat{\boldsymbol{\beta}}$ for M-estimation and the expression for var($\hat{\boldsymbol{\beta}}$). Show how they simplify for least squares.

11.4 In M-estimation, let $\rho(x) = 2(\sqrt{1 + x^2/2} - 1)$. Find the influence function, and explain why this gives a compromise between least squares and $\rho(x) = |x|$, having a bounded influence function with a smooth derivative at 0.

11.5 Since the Gauss-Markov theorem says that least squares estimates are "best," are not estimates obtained using M-estimation necessarily poorer? Explain.

11.6 For the saturated model, $E(y_i) = \beta_i$, $i = 1, \ldots, n$, find the ridge-regression estimate of β_i and interpret the impact of λ.

11.7 For the normal linear model, explain how (**a**) the ridge-regression estimates relate to Bayesian posterior means when $\{\beta_j\}$ have independent $N(0, \sigma^2)$ distributions, (**b**) the lasso estimates relate to Bayesian posterior modes when $\{\beta_j\}$ have independent Laplace (double-exponential) distributions with means 0 and common scale parameter.

11.8 Consider the linear model

$$y_i = \beta_1 x_{i1} + \beta_2 x_{i2} + \cdots + \beta_{40} x_{i,40} + \epsilon_i$$

with $\beta_1 = 1$ and $\beta_2 = \cdots = \beta_{40} = 0$, where $x_{ij} = u_i + v_j$ with $\{u_i\}$, $\{v_j\}$, and $\{\epsilon_i\}$ being iid $N(0, 1)$ random variables.

a. Find the correlation between y and x_1, y and x_j for $j \neq 1$, and x_j and x_k for $j \neq k$, and the multiple correlation between y and the set of explanatory variables.

b. Using this model, randomly generate $n = 100$ observations on the 41 variables. Use the lasso to select a model, for a variety of λ smoothing parameter values. Summarize results, and evaluate the effectiveness of this method.

c. Specify alternative values for $\{\beta_j\}$ for which you would not expect the lasso to be effective. Re-generate $\boldsymbol{y}$, and summarize results of using the lasso.

11.9 Refer to the Dirichlet prior distribution introduced for multinomial parameters in Section 11.2.2. Explain why a multivariate normal prior for multinomial logits provides greater flexibility.

11.10 Refer to Copas's kernel smoother (11.1) for binary regression, with $\phi(u) = \exp(-u^2/2)$.

a. To describe how close this estimator falls at a particular x value to a corresponding smoothing in the population, use the delta method to show that an estimated asymptotic variance is

$$\tilde{\pi}(x)[1-\tilde{\pi}(x)]\frac{\sum_i \phi\left[\sqrt{2}(x-x_i)/\lambda\right]}{\left\{\sum_i \phi[(x-x_i)/\lambda]\right\}^2}.$$

Explain why this decreases as λ increases, and explain the implication.

b. As λ increases unboundedly, explain intuitively to what $\tilde{\pi}(x)$ and this estimated asymptotic variance converge.

11.11 When $p > n$, why is backward elimination not a potential method for selecting a subset of explanatory variables?

11.12 Sometimes nonlinear regression models can be converted to ordinary GLMs by employing a link function for the mean response and/or transforming the explanatory variables. Explain how this could be done for the normal-response models (**a**) $E(y_i) = \beta_0 \exp(\beta_1 x_i)$ (an exponential growth model), (**b**) $E(y_i) = 1/(\beta_0 + \beta_1 x_i + \beta_2 x_i^2)$.

11.13 Refer to the form of iterations for the Gauss–Newton algorithm described in Section 11.3.7. Show that an analogous formula

$$\hat{\boldsymbol{\beta}} - \boldsymbol{\beta} = (\boldsymbol{X}^{\mathrm{T}}\boldsymbol{X})^{-1}\boldsymbol{X}^{\mathrm{T}}\boldsymbol{\epsilon}$$

holds for the ordinary linear model, where $\boldsymbol{\epsilon} = \boldsymbol{y} - \boldsymbol{\mu}$ is a "true residual."

11.14 Randomly generate nine observations satisfying a normal linear model by taking $x_i \sim N(50, 20)$ and $y_i = 45.0 + 0.1x_i + \epsilon_i$ with $\epsilon_i \sim N(0, 1)$. Now add to the dataset a contaminated outlying observation 10 having $x_{10} = 100$, $y_{10} = 100$. Fit the normal linear model to the 10 observations using (**a**) least squares, (**b**) Huber's M-estimation. Compare the model parameter estimates and the estimate of σ. Interpret.

11.15 For the data analyzed in Section 11.1.8 on risk factors for endometrial cancer, compare the results shown there with those you obtain using the lasso.

11.16 For the horseshoe crab data introduced in Section 1.5.1 and modeled in Section 7.5, suppose you use graphics to investigate how a female crab's number of male satellites depends on the width (in centimeters) of the carapace shell of the crab. If you plot the response counts of satellites against width, the substantial variability makes it difficult to discern a clear trend. To get a clearer picture, fit a generalized additive model, assuming a Poisson distribution and using the log link. What does this suggest about potentially good predictors and link functions for a GLM?

11.17 For the horseshoe crab dataset, Exercise 5.32 used logistic regression to model the probability that a female crab has at least one male satellite. Plot these binary response data against the crab's carapace width. Also plot a curve based on smoothing the data using a kernel smoother or a generalized additive model, assuming a binomial response and logit link. (This curve shows a roughly increasing trend and is more informative than viewing the binary data alone.)

11.18 For the `Housing.dat` file analyzed in Sections 3.4 and 4.7, use methods of this chapter to describe how the house selling price depends on its size.

11.19 Continue the previous exercise, now using all the explanatory variables.

APPENDIX A

Supplemental Data Analysis Exercises

Note: One purpose of this appendix is to provide exercises for students that are not tied to methodology of a particular text chapter.

1. Using an Internet search or viewing the datasets in the R MASS library (by entering the R commands `library(MASS)` and `data()`), download a dataset relating a quantitative response variable to at least two explanatory variables. Fit a GLM (**a**) using one explanatory variable, and (**b**) using all the explanatory variables in a model-building process. Interpret results.
2. The MASS library of R contains the `Boston` data file, which has several predictors of the per capita crime rate, for 506 neighborhoods in suburbs near Boston. Prepare a four-page report describing a model-building process for these data. Attach edited software output as an appendix.
3. The horseshoe crab dataset `Crabs2.dat` at `www.stat.ufl.edu/~aa/glm/data` comes from a study of factors that affect sperm traits of males. One response variable is total sperm, measured as the log of the number of sperm in an ejaculate. Explanatory variables are the location of the observation, carapace width (centimeters), mass (grams), color (1 = dark, 2 = medium, 3 = light), the operational sex ratio (OSR, the number of males per females on the beach), and a subjective condition number that takes into account mucus, pitting on the prosoma, and eye condition (the higher the better). Prepare a report describing a model-building process for these data. Attach edited software output as an appendix.
4. The `Student survey.dat` file at the text website, a small part of which is shown in Table A.1, shows survey responses of graduate students enrolled in a statistics course in a recent term at the University of Florida. The variables are GE = gender, AG = age in years, HI = high school grade-point average (GPA), CO = college GPA, DH = distance (in miles) of the campus from your home town, DR = distance (in miles) of the classroom from your current residence,

Foundations of Linear and Generalized Linear Models, First Edition. Alan Agresti.

TV = average number of hours per week that you watch TV, SP = average number of hours per week that you participate in sports or have other physical exercise, NE = number of times a week you read a newspaper, AH = number of people you know who have died from AIDS or who are HIV+, VE = whether you are a vegetarian (yes, no), PA = political party affiliation (D = Democrat, R = Republican, I = independent), PI = political ideology (1 = very liberal, 2 = liberal, 3 = slightly liberal, 4 = moderate, 5 = slightly conservative, 6 = conservative, 7 = very conservative), RE = how often you attend religious services (0 = never, 1 = occasionally, 2 = most weeks, 3 = every week), AB = opinion about whether abortion should be legal in the first 3 months of pregnancy (yes, no), AA = support affirmative action (yes, no), LD = belief in life after death (yes, no). Model how political ideology relates to the other variables. Prepare a report, posing a research question and summarizing your graphical analyses, models and interpretations, inferences, checks of assumptions, and overall summary of the relationships.

Table A.1 Part of Data File for Exercise 4

SUBJ	GE	AG	HI	CO	DH	DR	TV	SP	NE	...	LD
1	m	32	2.2	3.5	0	5.0	3	5	0	...	y
2	f	23	2.1	3.5	1200	0.3	15	7	5	...	u
...											
60	f	22	3.4	3.0	650	4	8	6	7	...	y

5. Repeat the previous exercise using college GPA as the response variable.
6. Repeat Exercise 4 using opinion about whether abortion should be legal as the response variable.
7. Repeat Exercise 4 using political party affiliation as the response variable.
8. Refer to Exercise 4. Find an appropriate model for belief in life after death as the response variable, with potential explanatory variables opinion about whether abortion should be legal, how often you attend religious services, political party affiliation, and political ideology.
9. For the `Statewide crime.dat` file at the text website from a *Statistical Abstract of the United States*, a small part of which is in Table A.2, model the statewide violent crime rate with predictors poverty rate, the percent living in metropolitan areas, and percent of high school graduates. Prepare a report in

Table A.2 Part of Data File for Exercise 9

State	Violent	Murder	Metro	High School	Poverty
AK	593	6	65.6	90.2	8.0
AL	430	7	55.4	82.4	13.7
...					
DC	1608	44	100.0	86.4	18.5

which you state a research question you could answer with these data, conduct descriptive and inferential analyses, and provide interpretations and summarize your conclusions.

10. Repeat the previous exercise with the statewide murder rates as the response variable.

11. The `Houses2.dat` file at the text website shows selling prices of homes in Gainesville, Florida in 1996. Write a report describing your model-building analysis, using size, numbers of bathrooms and bedrooms, and whether the house is new as potential explanatory variables.

12. The file `Credit.dat` at the text website, part of which is shown in Table A.3, refers to a sample of subjects randomly selected for an Italian study on the relation between income in millions of lira (the Italian currency at the time of the study) and whether one possesses a travel credit card. Analyze these data.

Table A.3 Part of Data File for Exercise 12

Income	Number sampled	Number with credit card
24	1	0
34	7	1
...		
68	3	3

13. The credit-scoring data file at `www.statistik.lmu.de/service/datenarchiv/kredit/kredit_e.html` includes 20 covariates for 1000 observations. Build a model for credit-worthiness, using as potential predictors: running account, duration of credit, payment of previous credits, intended use, gender, and marital status.

14. According to the *Independent* newspaper (London, March 8, 1994), the Metropolitan Police in London reported 30,475 people as missing in the year ending March 1993. For those of age 13 or less, 33 of 3271 missing males and 38 of 2486 missing females were still missing a year later. For ages 14 to 18, the values were 63 of 7256 males and 108 of 8877 females; for ages 19 and above, the values were 157 of 5065 males and 159 of 3520 females. Analyze by building a model, and interpret. (Thanks to Pat Altham for showing me these data.)

15. The file `Happiness.dat` at the text website shows responses of 18–22 year-olds from a General Social Survey on happiness (categories 1 = very happy, 2 = pretty happy, 3 = not too happy), the total number of traumatic events that happened to the respondent and his/her relatives in the last year, and race (1 = black, 0 = white). Use models to analyze these data.

16. In 2002 the General Social Survey asked "How many people at your work place are close friends?" The 756 responses had a mean of 2.76, standard deviation of 3.65, and a mode of 0. If you plan to build a generalized linear model (GLM) using some explanatory variables for this response, which distribution might be sensible? Why?

17. The file[1] `Fish.dat` at the text website, part of which is shown in Table A.4, reports the results of a study of fish hatching under three environments. Eggs from seven clutches were randomly assigned to three treatments, and the response was whether an egg hatched by day 10. The three treatments were (1) carbon dioxide and oxygen removed, (2) carbon dioxide only removed, and (3) neither removed. Model the probability of hatching for an egg from clutch i in treatment t, allowing for potential overdispersion. Summarize your analyses.

Table A.4 Part of Data File for Exercise 17

clutch	treatment	hatched	n
1	1	0	6
1	2	3	6
...			
7	3	4	20

18. The file[2] `Education.dat` at the text website, shown partly in Table A.5, contains scores on standardized qualitative (reading) and quantitative (math) exams for 890 students taught by 47 teachers working in 14 schools. The goal of the study was to study the effects of the school and of the student's socioeconomic status (1 = high, 0 = low) on the standardized exam scores. Analyze these data.

Table A.5 Part of Data File for Exercise 18

school	teacher	student	ses	test	score
10	10664	271	0	qual	50
10	10664	271	0	quan	31
10	10664	272	0	qual	53
...					
50	50666	1260	1	quan	38

19. Refer to Exercise 9.35 about a salt marsh habitat study. Analyze the data for the 2011 observations alone. Prepare a report describing your analyses. Attach edited software output as an appendix.

20. Refer to Exercise 9.37. The data file[3] `Opacum.dat` at the text website shows data for the marbled salamander (*Ambystoma opacum*). For the embryos that survived to hatching, taking account of the nesting of embryos within jars and jars within families, fit an appropriate model to compare the four rearing environments on the age at hatching. Interpret results.

[1]Data courtesy of Rebecca Hale, University of North Carolina, Asheville.
[2]Data courtesy of Ramon Littell.
[3]Data courtesy of Rebecca Hale, University of North Carolina, Asheville.

21. Refer to the previous exercise. Combining this data file with the data file `Maculatum.dat`, simultaneously model treatment effects on the age of hatching for both species of salamander, and evaluate how the treatment effects differ for the two species.

22. Refer to the previous exercise. For all the embryos of the two species, model the probability of hatching in terms of the treatment and species. Evaluate how the treatment effects differ for the two species.

23. Conduct a Bayesian analysis for one of the exercises assigned for frequentist methods in Chapters 1–9. (Your instructor may assign a particular exercise.)

24. Go to a site with large data files, such as the UCI Machine Learning Repository (`archive.ics.uci.edu/ml`). Find a dataset of interest to you. Use modeling methods to learn something about the data. Summarize your analyses in a report, attaching an appendix showing your use of software.

APPENDIX B

Solution Outlines for Selected Exercises

Note: This appendix contains brief outlines of solutions and hints of solutions for at least a few exercises from each chapter. Many of these are extracts of solutions that were kindly prepared by Jon Hennessy for Statistics 244 at Harvard University in 2013.

Chapter 1

1.1 In the random component, set $\theta_i = \mu_i$, $b(\theta_i) = \theta_i^2/2$, $\phi = \sigma^2$, $a(\phi) = \phi$, and $c(y_i, \phi) = -y_i^2/2\phi - \log(2\pi\phi)$. Use the identity link function.

1.2 **b.** *Hint*: What is the range for a linear predictor, and what is the range of the identity link applied to a binomial probability or to a Poisson mean?

1.5 The predicted number of standard deviation change in y for a standard deviation change in x_i, adjusting for the other explanatory variables.

1.11 Taking $\boldsymbol{\beta} = (\beta_0, \beta_1, \dots, \beta_{c-1})^{\mathrm{T}}$,

$$X = \begin{pmatrix} \mathbf{1}_{n_1} & \mathbf{1}_{n_1} & \mathbf{0}_{n_1} & \cdots & \mathbf{0}_{n_1} \\ \mathbf{1}_{n_2} & \mathbf{0}_{n_2} & \mathbf{1}_{n_2} & \cdots & \mathbf{0}_{n_2} \\ \vdots & \vdots & \vdots & \ddots & \vdots \\ \mathbf{1}_{n_{r-1}} & \mathbf{0}_{n_{r-1}} & \mathbf{0}_{n_{r-1}} & \cdots & \mathbf{1}_{n_{r-1}} \\ \mathbf{1}_{n_r} & -\mathbf{1}_{n_r} & -\mathbf{1}_{n_r} & \cdots & -\mathbf{1}_{n_r} \end{pmatrix}$$

1.18 **b.** Let $\boldsymbol{G} = \boldsymbol{X}_1(\boldsymbol{X}_c^{\mathrm{T}}\boldsymbol{X}_c)^{-1}\boldsymbol{X}_c^{\mathrm{T}}$. Then $\boldsymbol{G}\boldsymbol{X}_c = \boldsymbol{X}_1(\boldsymbol{X}_c^{\mathrm{T}}\boldsymbol{X}_c)^{-1}\boldsymbol{X}_c^{\mathrm{T}}\boldsymbol{X}_c = \boldsymbol{X}_1$.

Foundations of Linear and Generalized Linear Models, First Edition. Alan Agresti.

1.19 **a.** For the model $E(y_{ij}) = \beta_0 + \beta_i + \gamma x_{ij}$, let $\boldsymbol{\beta} = (\beta_0, \beta_1, \ldots, \beta_r, \gamma)^{\mathrm{T}}$, $\boldsymbol{x}_i = (x_{i1}, \ldots, x_{in_i})^{\mathrm{T}}$, and

$$X = \begin{pmatrix} \mathbf{1}_{n_1} & \mathbf{1}_{n_1} & \mathbf{0}_{n_1} & \cdots & \mathbf{0}_{n_1} & \boldsymbol{x}_1 \\ \mathbf{1}_{n_2} & \mathbf{0}_{n_2} & \mathbf{1}_{n_2} & \cdots & \mathbf{0}_{n_2} & \boldsymbol{x}_2 \\ \vdots & \vdots & \vdots & \ddots & \vdots & \vdots \\ \mathbf{1}_{n_r} & \mathbf{0}_{n_r} & \mathbf{0}_{n_r} & \cdots & \mathbf{1}_{n_r} & \boldsymbol{x}_r \end{pmatrix}$$

b. (i) γ, because for each group it is a difference of means at x values one-unit apart. (ii) β_i

c. We can construct $\boldsymbol{X}$ to have identifiable parameters by imposing the constraint $\beta_1 = 0$. For a fixed x, β_i then represents the difference between $E(y)$ in group i and in group 1, and γ represents the change in $E(y)$ per unit increase in x for each group.

Chapter 2

2.3 **a.** The normal equation for β_j is $\sum_i y_i x_{ij} = \sum_i \mu_i x_{ij}$. For β_0, $\sum_i y_i = \sum_i (\beta_0 + \beta_1 x_i)$, so $\hat{\beta}_0 = \bar{y} - \hat{\beta}_1 \bar{x}$. For β_1, $\sum_i y_i x_i = n\bar{x}\beta_0 + \beta_1 \sum_i x_i^2$, so $\hat{\beta}_1 = [\sum_i (x_i - \bar{x})(y_i - \bar{y})]/[\sum_i (x_i - \bar{x})^2]$.

2.8 *Hint*: See McCullagh and Nelder (1989, p. 85) or Wood (2006, p. 13).

2.9 *Hint*: For simplification, express the model with all variables centered, so there is no intercept term, $\hat{\beta}_1$ and $\hat{\beta}_2$ are unchanged, and var$(\hat{\boldsymbol{\beta}})$ is 2×2. Since corr$(\boldsymbol{x}_1, \boldsymbol{x}_2) > 0$, the off-main-diagonal elements of $(\boldsymbol{X}^{\mathrm{T}}\boldsymbol{X})$ are positive.

2.12 $\text{rank}(\boldsymbol{H}) = \text{tr}(\boldsymbol{H}) = \text{tr}[\boldsymbol{X}^{\mathrm{T}}\boldsymbol{X}(\boldsymbol{X}^{\mathrm{T}}\boldsymbol{X})^{-1}] = \text{tr}(\boldsymbol{I}_p) = p = \text{rank}(\boldsymbol{X})$.

2.17 *Hint*: Need $\mathbf{0}$ be a solution?

2.19 **a.** Following an example in Rodgers et al. (1984), let

$$\boldsymbol{X} = \begin{pmatrix} 0 & 1 \\ 0 & 0 \\ 1 & 1 \\ 1 & 0 \end{pmatrix} \quad \text{and} \quad \boldsymbol{W} = \begin{pmatrix} -1 & 5 \\ -5 & 1 \\ 3 & 1 \\ -1 & 3 \end{pmatrix}.$$

The columns of $\boldsymbol{X}$ are uncorrelated and linearly independent. The columns of $\boldsymbol{W}$ are orthogonal and linearly independent. However, the columns of $\boldsymbol{X}$ are not orthogonal and the columns of $\boldsymbol{W}$ are not uncorrelated.

b. If corr$(\boldsymbol{u}, \boldsymbol{v}) = 0$, then $\sum_i (u_i - \bar{u})(v_i - \bar{v}) = 0$. This implies that $\boldsymbol{u}^* = (\boldsymbol{u} - \bar{\boldsymbol{u}})$ and $\boldsymbol{v}^* = (\boldsymbol{v} - \bar{\boldsymbol{v}})$ are orthogonal. If $\bar{u} = 0$ (or equivalently $\bar{v} = 0$), then

$\boldsymbol{u}$ and $\boldsymbol{v}$ are orthogonal since $\sum_i(u_i - \bar{u})(v_i - \bar{v}) = \boldsymbol{u}^{\mathrm{T}}\boldsymbol{v} = 0$. If $\bar{\boldsymbol{u}} \neq \boldsymbol{0}$ and $\bar{\boldsymbol{v}} \neq \boldsymbol{0}$, then there is no guarantee that $\boldsymbol{u}$ and $\boldsymbol{v}$ are orthogonal.

c. If $\boldsymbol{u}^{\mathrm{T}}\boldsymbol{v} = 0$, then the numerator of corr$(\boldsymbol{u}, \boldsymbol{v})$ is $\sum_i(u_i - \bar{u})(v_i - \bar{v}) = (\boldsymbol{u}^{\mathrm{T}}\boldsymbol{v} - n\bar{u}\bar{v}) = -n\bar{u}\bar{v}$. This can only equal 0 if at least one of $\bar{u}$ and $\bar{v}$ equals 0.

2.23 **a.** For the saturated model, $E(y_i) = \beta_i$, $i = 1, \ldots, n$, $\boldsymbol{X} = \boldsymbol{I}_n$ and $C(\boldsymbol{X}) = \mathbb{R}^n$. $C(\boldsymbol{X})^{\perp} = \boldsymbol{0}$. $\boldsymbol{P}_X = \boldsymbol{I}_n$ and $\boldsymbol{I} - \boldsymbol{P}_X = \boldsymbol{0}_{n\times n}$, the matrix of 0's.

b. $\hat{\boldsymbol{\beta}} = \boldsymbol{y}$ and $\hat{\boldsymbol{\mu}} = \hat{\boldsymbol{\beta}} = \boldsymbol{y}$. $s^2 = \sum_i[(y_i - y_i)^2]/(n - n) = 0/0$. The model is not sensible in practice because the prediction $\hat{\mu}_i$ only considers observation y_i and ignores the others. Also, the model is not useful for predicting new values of y.

2.25 $\boldsymbol{a}^{\mathrm{T}} = \boldsymbol{\ell}^{\mathrm{T}}(\boldsymbol{X}^{\mathrm{T}}\boldsymbol{X})^{-1}\boldsymbol{X}^{\mathrm{T}}$

2.29 From (2.6), the leverage for an observation in group i is $1/n_i$.

2.30 **a.** The mean of the leverages is $\mathrm{tr}(\boldsymbol{H})/n = \mathrm{tr}[\boldsymbol{X}(\boldsymbol{X}^{\mathrm{T}}\boldsymbol{X})^{-1}\boldsymbol{X}^{\mathrm{T}}]/n = \mathrm{tr}[\boldsymbol{X}^{\mathrm{T}}\boldsymbol{X}(\boldsymbol{X}^{\mathrm{T}}\boldsymbol{X})^{-1}]/n = \mathrm{tr}(\boldsymbol{I}_p)/n = p/n$.

2.34 Model is not identifiable unless each factor has a constraint such as setting the parameter = 0 at the first or last level.

2.36 *Hint*: $\hat{\mu}_{ij} = \bar{y}_{i.} + \bar{y}_{.j} - \bar{y}_{..}$. See Hoaglin and Welsch (1978).

Chapter 3

3.5 Use $\boldsymbol{I} = \boldsymbol{P}_0 + (\boldsymbol{I} - \boldsymbol{P}_0)$, with $\boldsymbol{P}_0$ the projection matrix for the null model. With $\boldsymbol{\mu}_0 = \mu_0\boldsymbol{1}$ and $\bar{\boldsymbol{y}} = \bar{y}\boldsymbol{1}$, we have $\boldsymbol{P}_0\boldsymbol{\mu}_0 = \boldsymbol{I}\boldsymbol{\mu}_0 = \boldsymbol{\mu}_0$ and $\boldsymbol{P}_0\boldsymbol{y} = \bar{\boldsymbol{y}}$,

$$(\boldsymbol{y} - \boldsymbol{\mu}_0)^{\mathrm{T}}(\boldsymbol{y} - \boldsymbol{\mu}_0) = (\boldsymbol{y} - \boldsymbol{\mu}_0)^{\mathrm{T}}\boldsymbol{P}_0(\boldsymbol{y} - \boldsymbol{\mu}_0) + (\boldsymbol{y} - \boldsymbol{\mu}_0)^{\mathrm{T}}(\boldsymbol{I} - \boldsymbol{P}_0)(\boldsymbol{y} - \boldsymbol{\mu}_0)$$
$$= n(\bar{y} - \mu_0)^2 + (\boldsymbol{y} - \bar{\boldsymbol{y}})^{\mathrm{T}}(\boldsymbol{y} - \bar{\boldsymbol{y}}).$$

By Cochran's Theorem, $(1/\sigma^2)n(\bar{y} - \mu_0)^2 \sim \chi^2_{1,\lambda}$, where $\lambda = (n/\sigma^2)(\mu - \mu_0)^2$, and $(1/\sigma^2)(\boldsymbol{y} - \bar{\boldsymbol{y}})^{\mathrm{T}}(\boldsymbol{y} - \bar{\boldsymbol{y}}) = \frac{1}{\sigma^2}(n - 1)s^2 \sim \chi^2_{n-1}$, and the two quantities are independent. Thus, the test statistic

$$\frac{n(\bar{y} - \mu_0)^2/\sigma^2}{s^2/\sigma^2} \sim \frac{\chi^2_{1,\lambda}}{\chi^2_{n-1}/(n-1)} \sim F_{1,n-1,\lambda}.$$

The equivalent t test uses the signed square root of this test statistic,

$$\frac{\bar{y} - \mu_0}{s/\sqrt{n}} \sim \pm\sqrt{F_{1,n-1,\lambda}} \sim t_{n-1,\lambda}.$$

Under H_0, $\lambda = 0$ and the null distributions are $F_{1,n-1}$ and t_{n-1}.

3.7 **a.** For $\lambda = (1/\sigma^2)\boldsymbol{\mu}^{\mathrm{T}}(\boldsymbol{P}_1 - \boldsymbol{P}_0)\boldsymbol{\mu}$,

$$(\boldsymbol{P}_1 - \boldsymbol{P}_0)\boldsymbol{\mu} = \begin{pmatrix} (\mu_1 - \bar{\mu})\mathbf{1}_{n_1} \\ \vdots \\ (\mu_c - \bar{\mu})\mathbf{1}_{n_c} \end{pmatrix}.$$

Thus, $\lambda = (1/\sigma^2)\boldsymbol{\mu}^{\mathrm{T}}(\boldsymbol{P}_1 - \boldsymbol{P}_0)\boldsymbol{\mu} = (1/\sigma^2)\sum_i n_i(\mu_i - \bar{\mu})^2$.

b. $\lambda = (2n/\sigma^2)0.25\sigma^2 = n/2$. $P(F_{2,3n-3,\lambda} > F_{2,3n-3,0.05})$ equals 0.46 for $n = 10$, 0.94 for $n = 30$, and 0.99 for $n = 50$.

c. $\lambda = 2n\Delta^2$. The powers are 0.05, 0.46, and 0.97 for $\Delta = 0, 0.5, 1.0$.

3.9 **a.** Let $\boldsymbol{y} = \begin{pmatrix} y_1 \\ y_2 \end{pmatrix}$, $\boldsymbol{X} = \begin{pmatrix} \mathbf{1}_{n_1} & \mathbf{1}_{n_1} \\ \mathbf{1}_{n_2} & \mathbf{0}_{n_2} \end{pmatrix}$ and $\boldsymbol{\beta} = \begin{pmatrix} \beta_0 \\ \beta_1 \end{pmatrix}$. Then $E(\boldsymbol{y}) = \boldsymbol{X}\boldsymbol{\beta}$ and $\mu_1 = \beta_0 + \beta_1$ and $\mu_2 = \beta_0$. Thus $\mu_1 - \mu_2 = \beta_1$.

c. $H_0 : \mu_1 = \mu_2$ is $H_0 : \beta_1 = 0$ and $H_1 : \beta_1 \neq 0$. We can use the decomposition

$$\boldsymbol{y}^{\mathrm{T}}\boldsymbol{y} = \boldsymbol{y}^{\mathrm{T}}\boldsymbol{P}_0\boldsymbol{y} + \boldsymbol{y}^{\mathrm{T}}(\boldsymbol{P}_X - \boldsymbol{P}_0)\boldsymbol{y} + \boldsymbol{y}^{\mathrm{T}}(\boldsymbol{I} - \boldsymbol{P}_X)\boldsymbol{y},$$

finding that $(1/\sigma^2)\boldsymbol{y}^{\mathrm{T}}(\boldsymbol{P}_X - \boldsymbol{P}_0)\boldsymbol{y} = (1/\sigma^2)(\bar{y}_1 - \bar{y}_2)^2(\frac{1}{n_1} + \frac{1}{n_2})^{-1} \sim \chi^2_{1,\lambda}$, where $\lambda = 0$ under H_0, and $(1/\sigma^2)\boldsymbol{y}^{\mathrm{T}}(\boldsymbol{I} - \boldsymbol{P}_X)\boldsymbol{y} = (1/\sigma^2)(n_1 + n_2 - 2)s^2 \sim \chi^2_{n_1+n_2-2}$, where s^2 is the pooled variance estimate. The test statistic is thus

$$\frac{(\bar{y}_1 - \bar{y}_2)^2\left(\frac{1}{n_1} + \frac{1}{n_2}\right)^{-1}}{s^2} = \frac{(\bar{y}_1 - \bar{y}_2)^2}{s^2\left(\frac{1}{n_1} + \frac{1}{n_2}\right)} \sim F_{1,n_1+n_2-2}.$$

d. The square root of F gives the t statistic, which has $df = n_1 + n_2 - 2$.

3.10 $(\bar{y}_1 - \bar{y}_2) \pm t_{\alpha/2,n_1+n_2-2}\, s\sqrt{\frac{1}{n_1} + \frac{1}{n_2}}$

3.13 **d.** df values are $r - 1$, $c - 1$, $(r - 1)(c - 1)$, $(N - rc)$. Each mean square = SS/df. For H_0: no interaction, test statistic F = (interaction MS)/(residual MS) has $df_1 = (r - 1)(c - 1)$, $df_2 = N - rc$.

3.15 For models $k = 0$ and 1, substitute $1 - R_k^2 = (1 - R^2_{adj,k})[(n - p_k)/(n - 1)]$ and simplify.

3.17 **a.** Under the general linear hypothesis framework, let $\boldsymbol{\Lambda}$ be a single row of 0s except for 1 in position j and -1 in position k. The F test of H_0: $\boldsymbol{\Lambda}\boldsymbol{\beta} = 0$ gives

$$F = \frac{(\boldsymbol{\Lambda}\hat{\boldsymbol{\beta}})^{\mathrm{T}}[\boldsymbol{\Lambda}(\boldsymbol{X}^{\mathrm{T}}\boldsymbol{X})^{-1}\boldsymbol{\Lambda}^{\mathrm{T}}]^{-1}\boldsymbol{\Lambda}\hat{\boldsymbol{\beta}}}{s^2} = \frac{(\hat{\beta}_j - \hat{\beta}_k)^2}{s^2\boldsymbol{\Lambda}(\boldsymbol{X}^{\mathrm{T}}\boldsymbol{X})^{-1}\boldsymbol{\Lambda}^{\mathrm{T}}} \sim F_{1,n-p}.$$

The denominator simplifies to $[SE_j^2 + SE_k^2 - 2\widehat{\mathrm{cov}}(\hat{\beta}_j, \hat{\beta}_k)]$.

b. Equivalently, the F test compares the model M_1 with p parameters and the simpler model M_0 that replaces columns for $\boldsymbol{x}_{*j}$ and $\boldsymbol{x}_{*k}$ in the model matrix with a single column for $\boldsymbol{x}_{*j} + \boldsymbol{x}_{*k}$.

3.19 **a.** $\Lambda = \begin{pmatrix} 1 & -2 & 1 & 0 & 0 \\ 0 & 1 & -2 & 1 & 0 \\ 0 & 0 & 1 & -2 & 1 \end{pmatrix}$.

b. $df_1 = c - 2$, $df_2 = n - c$.

c. Quantitative model has disadvantage that true relationship may be far from linear in the chosen scores (e.g., nonmonotonic), and qualitative model has disadvantage of lower power for detecting effect if true relationship is close to linear.

3.20 $\boldsymbol{\ell}\hat{\boldsymbol{\beta}} \pm t_{\alpha/2,n-p} s \sqrt{\boldsymbol{\ell}(\boldsymbol{X}^{\mathrm{T}}\boldsymbol{X})^{-1}\boldsymbol{\ell}^{\mathrm{T}}}$.

Special case $(\hat{\beta}_j - \hat{\beta}_k) \pm t_{\alpha/2,n-p}\sqrt{SE_j^2 + SE_k^2 - 2\widehat{\mathrm{cov}}(\hat{\beta}_j, \hat{\beta}_k)}$.

3.21 *Hint*: For the null model, $\bar{y}$ is the estimated linear predictor.

3.22 The actual $P[|y - \bar{y}|/\sigma\sqrt{1 + \frac{1}{n}} \leq 1.96]$

$$= \Phi\left(1.96\sqrt{1+\tfrac{1}{n}} + z_o/\sqrt{n}\right) - \Phi\left(-1.96\sqrt{1+\tfrac{1}{n}} + z_o/\sqrt{n}\right).$$

3.23 Squared partial correlation $= (\mathrm{SSE}_0 - \mathrm{SSE}_1)/\mathrm{SSE}_0$.

3.26 With $\alpha = 0.05$, over most values of c and n, the ratio is on the order 0.96–0.98. For example, you can find the ratio in R using

```
> qtukey(1-alpha, c, c*(n-1))/(sqrt(2)*qt(1 - alpha/
(c*(c-1)), c*(n-1)))
```

3.30 When x_i is uniformly distributed over $(2.0, 4.0)$, $R^2 \approx 0.34$. When x_i is uniformly distributed over $(3.5, 4.0)$, $R^2 \approx 0.05$. The wider the range sampled for the explanatory variable, the larger R^2 tends to be, because in $R^2 = 1 - \mathrm{SSE/TSS}$, SSE tends to be unaffected but TSS tends to increase.

3.28 Expect same $E(\hat{\beta}_1)$ and σ^2 but larger SE for $\hat{\beta}_1$, since from Section 2.1.3, $\mathrm{var}(\hat{\beta}_1) = \sigma^2/[\sum_{i=1}^n (x_i - \bar{x})^2]$.

Chapter 4

4.7 **a.** For observations in group A, since $\partial\mu_A/\partial\eta_i$ is constant, the likelihood equation corresponding to β_1 sets $\sum_A (y_i - \mu_A)/\mu_A = 0$, so $\hat{\mu}_A = \bar{y}_A$. The likelihood equation corresponding to β_0 gives

$$\sum_A \frac{(y_i - \mu_A)}{\mu_A}\left(\frac{\partial\mu_A}{\partial\eta_i}\right) + \sum_B \frac{(y_i - \mu_B)}{\mu_B}\left(\frac{\partial\mu_B}{\partial\eta_i}\right) = 0.$$

The first sum is 0 from the first likelihood equation, and for observations in group B, $\partial\mu_B/\partial\eta_i$ is constant, so the second sum sets $\sum_B(y_i - \mu_B)/\mu_B = 0$, and $\hat{\mu}_B = \bar{y}_B$.

4.9 $W = \sigma^{-2}I$, $\text{var}(\hat{\beta}) = \sigma^2(X^TX)^{-1}$

4.11 *Hint*: If you have difficulty with this exercise, see Section 1.4 of Agresti (2013).

4.12 *Hint*: Construct a large-sample normal interval for $x_0\beta$ and then apply the inverse link function to the endpoints.

4.13

$$D(y;\hat{\mu}_0) - D(y;\hat{\mu}_1) = 2\sum_i y_i(\hat{\mu}_{1i} - \hat{\mu}_{0i}) + \frac{\hat{\mu}_{0i}^2}{2} - \frac{\hat{\mu}_{1i}^2}{2}$$
$$= \sum_i (y_i - \hat{\mu}_{0i})^2 - \sum_i (y_i - \hat{\mu}_{1i})^2.$$

4.14 Note that $\sum_i \hat{\mu}_i = \sum_i y_i$ is the likelihood equation generated by the intercept when the link function is canonical.

4.16 **a.** $d_i = 2[n_i y_i \log(y_i/\hat{\pi}_i) + n_i(1 - y_i)\log[(1 - y_i)/(1 - \hat{\pi}_i)]]$.

4.20 For log likelihood $L(\mu) = -n\mu + (\sum_i y_i)\log(\mu)$, the score is $u = (\sum_i y_i - n\mu)/\mu$, $H = -(\sum_i y_i)/\mu^2$, and the information is n/μ. It follows that the adjustment to $\mu^{(t)}$ in Fisher scoring is $[\mu^{(t)}/n][(\sum_i y_i - n\mu^{(t)})/\mu^{(t)}] = \bar{y} - \mu^{(t)}$, and hence $\mu^{(t+1)} = \bar{y}$. For Newton–Raphson, the adjustment to $\mu^{(t)}$ is $\mu^{(t)} - (\mu^{(t)})^2/\bar{y}$, so that $\mu^{(t+1)} = 2\mu^{(t)} - (\mu^{(t)})^2/\bar{y}$. Note that if $\mu^{(t)} = \bar{y}$, then also $\mu^{(t+1)} = \bar{y}$.

4.22 If the link is not canonical,

$$\frac{\partial\mu_i}{\partial\eta_i} = \frac{\partial\mu_i}{\partial\theta_i}\frac{\partial\theta_i}{\partial\eta_i} = b''(\theta_i)\frac{\partial\theta_i}{\partial\eta_i}$$

$$\frac{\partial L_i}{\partial\beta_j} = \frac{(y_i - \mu_i)}{\text{var}(y_i)}b''(\theta_i)\frac{\partial\theta_i}{\partial\eta_i}x_{ij} = \frac{(y_i - \mu_i)x_{ij}}{a(\phi)}\frac{\partial\theta_i}{\partial\eta_i}.$$

Then, $\partial^2 L_i/\partial\beta_j\partial\beta_k$ depends on y_i, so $\partial^2 L/\partial\beta_j\partial\beta_k \neq E(\partial^2 L/\partial\beta_j\partial\beta_k)$.

4.25 *Hint*: Apply Jensen's inequality to $E[-\log(x)]$, where $P[x = (p_{Mj}/p_j)] = p_j$.

4.27 **a.** If y has standard deviation $\sigma = c\mu$, then using $\log(y) \approx \log(\mu) + (y - \mu)/\mu$, we have $\text{var}[\log(y)] \approx \text{var}(y)/\mu^2 = c^2$.

b. If $\log(y_i) \sim N(\mu_i, \sigma^2)$, then $E(y_i) = e^{\mu_i+\sigma^2/2}$ by the mgf of a normal. Thus, $\log[E(y_i)] = \mu_i + \sigma^2/2 = E[\log(y_i)] + \sigma^2/2$.

c. L_i is also the log-normal fitted median. If $f(\cdot)$ is a monotonic function, $\text{median}[f(x)] = f[\text{median}(x)]$. Then, $e^{\text{median}(\log(y))} = \text{median}(y)$ and e^{L_i} is the ML estimate of the median of the conditional distribution of y_i. The median would often be more relevant because y_i has a skewed distribution.

Chapter 5

5.5 **a.** Let $P(y = 1) = p$ and $P(y = 0) = 1 - p$. By Bayes' Theorem,

$$P(y = 1 \mid x) = \frac{P(x \mid y = 1)p}{P(x \mid y = 1)p + P(x \mid y = 0)(1 - p)}$$

$$= \frac{p(1/\sqrt{2\pi}\sigma)\exp[-(x - \mu_1)^2/2\sigma^2]}{p(1/\sqrt{2\pi}\sigma)\exp[-(x - \mu_1)^2/2\sigma^2] + (1 - p)(1/\sqrt{2\pi}\sigma)\exp[-(x - \mu_0)^2/2\sigma^2]}$$

$$= \frac{\exp\{\log(\frac{p}{1-p}) - \frac{1}{2\sigma^2}[(x - \mu_1)^2 - (x - \mu_0)^2]\}}{1 + \exp\{\log(\frac{p}{1-p}) - \frac{1}{2\sigma^2}[(x - \mu_1)^2 - (x - \mu_0)^2]\}}$$

$$= \frac{\exp[\text{logit}(p) - \frac{1}{2\sigma^2}(\mu_1^2 - \mu_0^2) + \frac{\mu_1-\mu_0}{\sigma^2}x]}{1 + \exp[\text{logit}(p) - \frac{1}{2\sigma^2}(\mu_1^2 - \mu_0^2) + \frac{\mu_1-\mu_0}{\sigma^2}x]}.$$

So, set $\alpha = \text{logit}(p) - (1/2\sigma^2)(\mu_1^2 - \mu_0^2)$ and $\beta = (\mu_1 - \mu_0)/\sigma^2$.

b. When x has a $N(\mu_j, \sigma_j^2)$ distribution, then

$$P(y = 1 \mid x) = \frac{\exp[\text{logit}(p) + \log(\sigma_0/\sigma_1) - (x - \mu_1)^2/2\sigma^2 + (x - \mu_0)^2/2\sigma^2]}{1 + \exp[\text{logit}(p) + \log(\sigma_0/\sigma_1) - (x - \mu_1)^2/2\sigma^2 + (x - \mu_0)^2/2\sigma^2]}$$

Then $\alpha = \text{logit}(p) + \log(\sigma_0/\sigma_1) - \mu_1^2/2\sigma_1^2 - \mu_0^2/2\sigma_0^2$, $\beta = (\mu_1/\sigma_1^2) - (\mu_0/\sigma_0^2)$, and $\gamma = -(1/2)[(1/\sigma_1^2) - (1/\sigma_0^2)]$, where γ is the coefficient for the quadratic term.

5.3

```
> x <- c(1,2,3,4,5,6,7,8)
> y <- c(1,1,0,0,0,0,1,1)
> fit.toy <- glm(y ~ x, family = binomial)
> library(ROCR)
> pred.toy <- prediction(fitted(fit.toy), y)
> perf.toy <- performance(pred.toy,"tpr","fpr")
> plot(perf. toy)
> performance(pred.toy, "auc")
[1] 0.5
```

5.7 In terms of the probability π that $y = 1$ at x_0, $\beta = [\text{logit}(\pi)]/x_0$, $\text{var}(\hat{\beta}) \approx [n\pi(1-\pi)x_0^2]^{-1}$, and the noncentrality goes to 0 as $\beta \to \infty$ (i.e., as $\pi \to 1$). So, the Wald test loses power.

5.8 Condition on the margins of each 2×2 stratum. Let T be total number of successes for treatment 1, summed over strata. P-value is $P(T \geq t_{obs})$ for tables with the given margins, based on hypergeometric probabilities in each stratum. For details, see Agresti (1992 or 2013, Section 7.3.5).

5.9 *Hint*: There are $\binom{6}{3}$ possible data configurations with 3 successes, all equally-likely under H_0. Exact P-value = 0.05.

5.14 **a.** Assuming $\pi_1 = \cdots = \pi_N = \pi$, we can maximize

$$L(\pi) = \sum_{i=1}^{N} y_i \log(\pi) + (n_i - y_i)\log(1-\pi)$$

to show that $\hat{\pi} = (\sum_i y_i)(\sum_i n_i)$. The Pearson statistic for ungrouped data is

$$\begin{aligned} X^2 &= \sum \frac{(\text{observed} - \text{fitted})^2}{\text{fitted}} \\ &= \sum_{i=1}^{N}\sum_{j=1}^{n_i} \frac{(y_{ij} - \hat{\pi})^2}{\hat{\pi}} + \frac{[1 - y_{ij} - (1-\hat{\pi})]^2}{1-\hat{\pi}} \\ &= \sum_{i=1}^{N}\sum_{j=1}^{n_i} \frac{(y_{ij} - \hat{\pi})^2}{\hat{\pi}(1-\hat{\pi})} = \frac{N\hat{\pi}(1-\hat{\pi})}{\hat{\pi}(1-\hat{\pi})} = N, \end{aligned}$$

Because $X^2 = N$, the statistic is completely uninformative.

5.16 **a.** Treating the data as N binomial observations and letting $s_i = \sum_{j=1}^{n_i} y_{ij}$, the kernel of the log likelihood (ignoring the binomial coefficients) is

$$L(\boldsymbol{\pi}) = \sum_{i=1}^{N} s_i \log(\pi_i) + (n_i - s_i)\log(1-\pi_i).$$

Treating the data as $n = \sum_{i=1}^{N} n_i$ Bernoulli observations, the log likelihood is

$$\begin{aligned} L(\boldsymbol{\pi}) &= \sum_{i=1}^{N}\sum_{j=1}^{n_i} y_{ij}\log(\pi_i) + (1-y_{ij})\log(1-\pi_i) \\ &= \sum_{i=1}^{N} s_i \log(\pi_i) + (n_i - s_i)\log(1-\pi_i). \end{aligned}$$

b. For the saturated model case, the two data forms differ. Treating the data as N binomial observations, there are N parameters $\pi_1, \ldots, \pi_N$. Treating the data as n Bernoulli observations, there are n parameters, $\{\pi_{ij}\}$.

c. The difference between deviances of two unsaturated models does not depend on the form of data entry because the log likelihood of the saturated model cancels out when taking the difference between deviances. It depends only on the log likelihoods of the unsaturated models, which from (a) do not depend on the form of data entry.

5.17 **a.** For the ungrouped case, the deviance for M_0 is 16.3 and the deviance for M_1 is 11.0. For the ungrouped case, the deviance for M_0 is 6.3 and the deviance for M_1 is 1.0. The saturated model in the ungrouped case has 12 parameters and the log likelihood of the saturated model is 0 while the saturated model in the grouped case has three parameters.

b. The differences between the deviances is the same ($16.3 - 11.0 = 6.3 - 1.0 = 5.3$). The log likelihoods for the grouped and ungrouped cases only differ by the binomial coefficients. The difference between deviances is double the difference in the log likelihoods. The difference between the log likelihoods for either case is

$$\begin{aligned} D_0 - D_1 &= -2[L(\hat{\boldsymbol{\mu}}_0;\mathbf{y}) - L(\mathbf{y};\mathbf{y})] + 2[L(\hat{\boldsymbol{\mu}}_1;\mathbf{y}) - L(\mathbf{y};\mathbf{y})] \\ &= 2[L(\hat{\boldsymbol{\mu}}_1;\mathbf{y}) - L(\hat{\boldsymbol{\mu}}_0;\mathbf{y})]. \end{aligned}$$

For the grouped case, the binomial coefficients cancel out.

5.22 **a.**

$$\begin{aligned} P(y=1) &= P(U_1 > U_0) = P(\beta_{10} + \beta_{11}x + \epsilon_1 > \beta_{00} + \beta_{01}x + \epsilon_0) \\ &= P\left\{(1/\sqrt{2})(\epsilon_0 - \epsilon_1) < (1/\sqrt{2})[\beta_{10} - \beta_{00} + (\beta_{11} - \beta_{01})x]\right\} \end{aligned} \qquad (11.2)$$

$$= \Phi(\beta_0' + \beta_1' x) \text{ with } \beta_0' = (1/\sqrt{2})(\beta_{10} - \beta_{00}) \text{ and } \beta_1' = (1/\sqrt{2})(\beta_{11} - \beta_{01}).$$

Chapter 6

6.4

$$\frac{\partial \pi_3(x)}{\partial x} = -\frac{[\beta_1 \exp(\alpha_1 + \beta_1 x) + \beta_2 \exp(\alpha_2 + \beta_2 x)]}{[1 + \exp(\alpha_1 + \beta_1 x) + \exp(\alpha_2 + \beta_2 x)]^2}.$$

a. Denominator > 0 and numerator < 0 when $\beta_1 > 0$ and $\beta_2 > 0$.

6.6 **b.**

$$\pi_{ij} = P(u_{ij} > u_{ik}\ \forall j \neq k) = E[P(u_{ij} > u_{ik}\ \forall j \neq k \mid u_{ij})]$$
$$= E\left[\prod_{k\neq j} P(\alpha_k + x_i\beta_k + \epsilon_{ik} < u_{ij} \mid u_{ij})\right] = E\left[\prod_{k\neq j} \Phi(u_{ij} - \alpha_k - x_i\beta_k) \mid u_{ij}\right]$$
$$= \int \phi(u_{ij} - \alpha_j - x_i\beta_j)\prod_{k\neq j} \Phi(u_{ij} - \alpha_k - x_i\beta_k)du_{ij}$$

We can form the likelihood by noting that $\ell = \prod_{i=1}^{N}\prod_{j=1}^{c} \pi_{ij}^{y_{ij}}$.

6.8 For a baseline-category logit model with $\beta_j = j\beta$,

$$\frac{P(y_i = j+1 \mid x_i = u)}{P(y_i = j \mid x_i = u)} = e^{u(j+1)\beta}/e^{uj\beta} = e^{u\beta}.$$

Thus, the odds ratio comparing $x_i = u$ versus $x_i = v$ is

$$\frac{P(y_i = j+1 \mid x_i = u)}{P(y_i = j \mid x_i = u)} \Bigg/ \frac{P(y_i = j+1 \mid x_i = v)}{P(y_i = j \mid x_i = v)} = e^{(u-v)\beta}.$$

Note that the odds ratio does not depend on j (i.e., proportional odds structure for adjacent-category logits).

6.9 See Agresti (2013, Section 16.5) for details.

6.13 **a.** For $j < k$, $\text{logit}[P(y \leq j \mid x = x_i)] - \text{logit}[P(y \leq k \mid x = x_i)] = (\alpha_j - \alpha_k) + (\beta_j - \beta_k)x_i$. This difference of logits cannot be positive since $P(y \leq j) \leq P(y \leq k)$; however, if $\beta_j > \beta_k$ then the difference is positive for large positive x_i, and if $\beta_j < \beta_k$ then the difference is positive for large negative x_i.

b. You need monotone increasing $\{\alpha_j + \beta_j\}$.

6.16 **d.** *Hint*: Show that $\text{var}(T)/\text{var}(S)$ is a squared correlation between two random variables, where with probability π_j the first equals b_j and the second equals $f_j'(\theta)/f_j(\theta)$.

Chapter 7

7.4 $2\sum_i y_i \log(y_i/\bar{y})$, chi-squared with $df = 1$ when μ_1 and μ_2 are large.

7.7 **a.** Use that under H_0, conditional on n the data have a multinomial distribution with equal probabilities.

7.9 For this model, $\partial\mu_i/\partial x_{ij} = \beta_j\mu_i$ and the likelihood equation for the intercept is $\sum_i \mu_i = \sum_i y_i$.

7.11 $\hat{\mu}_{ijk} = n_{i+k}n_{+jk}/n_{++k}$, the same as applying the ordinary independence model to each partial table. The residual $df = \ell(r-1)(c-1)$.

7.16 Baseline-category logit model with additive factor effects for B and C.

7.21 See Greenwood and Yule (1920).

7.23 **a.**

$$f(y;\mu,k) = \frac{\Gamma(y+k)}{\Gamma(k)\Gamma(y+1)}\left(\frac{\mu}{\mu+k}\right)^y\left(\frac{k}{\mu+k}\right)^k$$
$$= \exp\left[y\log\frac{\mu}{\mu+k} + k\log\frac{k}{\mu+k} + \log\Gamma(y+k) - \log\Gamma(k) + \log\Gamma(y+1)\right]$$

Let $\theta = \log[\mu/(\mu+k)]$, $b(\theta) = -\log(1-e^{\theta})$, and $a(\phi) = 1/k$.

b. Letting $x = ya(\phi)$,

$$f(x;\mu,k) = \exp\left\{\frac{x\log[\mu/(\mu+k)] + \log[k/(\mu+k)]}{1/k}\right.$$
$$\left. + \log\Gamma(y+k) - \log\Gamma(k) + \log\Gamma(y+1) + \log k\right\}$$

where the $\log k$ at the end of the equation is the Jacobian.

7.25 The likelihood is proportional to $[k/(\mu+k)]^{nk}[\mu/(\mu+k)]^{\sum_i y_i}$. The log likelihood depends on μ through

$$-nk\log(\mu+k) + \sum_i y_i[\log\mu - \log(\mu+k)].$$

Differentiating with respect to μ, setting equal to 0, and solving for μ yields $\hat{\mu} = \bar{y}$.

7.27 From including the GR term, the likelihood equations imply that the fitted GR marginal totals equal the sample values. For example, the sample had 1040 white females, and necessarily the fitted model will have 1040 white females. The model with AC, AM, CM, AG, AR, GM, GR two-factor terms and no three-factor interaction terms fits well ($G^2 = 19.9$, $df = 19$).

Chapter 8

8.2 For a beta-binomial random variable s, $\text{var}(s) = E[n\pi(1-\pi)] + \text{var}(n\pi) = nE(\pi) - nE(\pi^2) + n[(E\pi)^2 - (E\pi)^2] + n^2\text{var}(\pi) = nE(\pi)[1-E(\pi)] + n(n-1)\text{var}(\pi) = n\mu(1-\mu) + n(n-1)\mu(1-\mu)\theta/(1+\theta) = n\mu(1-\mu)[1+(n-1)\theta/(1+\theta)]$.

8.5 **a.** If $\text{logit}(y_i) = \beta_i + \sigma z$, then $y_i = (e^{\beta_i+\sigma z})/(1+e^{\beta_i+\sigma z})$. Taking $f(\sigma) = (e^{\beta_i+\sigma z})/(1+e^{\beta_i+\sigma z})$ and expanding $f(\sigma)$ around $f(0)$ by Taylor approximation,

$$y_i = \frac{e^{\beta_i}}{1+e^{\beta_i}} + \frac{e^{\beta_i}}{1+e^{\beta_i}}\frac{1}{1+e^{\beta_i}}\sigma z + \frac{e^{\beta_i}(1-e^{\beta_i})}{2(1+e^{\beta_i})^3}\sigma^2 z^2 + \cdots$$

b. Using this approximation and the fact that $E(z) = 0$ and $\text{var}(z) = 1$, we have $E(y_i) \approx \mu_i$ and $\text{var}(y_i) \approx [\mu_i(1-\mu_i)]^2\sigma^2$.

c. The binomial approximation would imply that for a single region $v(\mu_i) = \phi\mu_i(1-\mu_i)$. This approach is inappropriate when $n_i = 1$ since in that case $\phi = 1$. Regardless of n_i, the binomial distribution assumes the small regions are independent, but contiguous regions would likely have dependent results.

8.8 For the null model $\mu_i = \beta$ and $v(\mu_i) = \sigma^2$,

$$u(\beta) = \sum_{i=1}^{n} \frac{\partial \mu_i}{\partial \beta} v(\mu_i)^{-1}(y_i - \mu_i) = \sum_{i=1}^{n} \frac{y_i - \mu_i}{\sigma^2}.$$

Thus $\hat{\beta} = \bar{y}$ and the variance of $\hat{\beta}$ is $V = \sigma^2/n$. A sensible model-based estimate of V is $\hat{V} = (1/n^2)\sum_{i=1}^{n}(y_i - \bar{y})^2$. The actual asymptotic variance of $\hat{\beta}$ is

$$V\left[\sum_{i=1}^{n} \frac{\partial \mu_i}{\partial \beta}\frac{\text{var}(y_i)}{v(\mu_i)^2}\frac{\partial \mu_i}{\partial \beta}\right] V = \frac{\sigma^2}{n}\left(\sum_{i=1}^{n}\frac{\beta}{\sigma^4}\right)\frac{\sigma^2}{n} = \frac{\beta}{n}.$$

To find the robust estimate of the variance that adjusts for model misspecification, we replace $\text{var}(y_i)$ in the expression above with $(y_i - \bar{y})^2$, leading to $[\sum_{i=1}^{n}(y_i - \bar{y})^2]/n^2$.

8.11 The model-based estimator tends to be better when the model holds, and the robust estimator tends to be better when there is severe overdispersion so that the model-based estimator tends to underestimate the actual *SE*.

8.17 **b.** *Hint*: Is it realistic to treat the success probability as identical from shot to shot?

Chapter 9

9.3 *Hint*: Use (2.7) with $\boldsymbol{A} = \boldsymbol{I}_n - \frac{1}{n}\mathbf{1}_n\mathbf{1}_n^{\mathrm{T}}$, for which $\boldsymbol{\mu}^{\mathrm{T}}\boldsymbol{A}\boldsymbol{\mu} = 0$.

9.8 *Hint*: See Exercise 9.8 for correlations for the autoregressive structure.

9.21 *Hints:* The covariance is the same for any pair of cells in the same row, and $\mathrm{var}(\sum_j y_{ij}) = 0$ since y_{i+} is fixed. If $(x_1, \ldots, x_d)$ is multivariate normal with common mean and common variance σ^2 and common correlation ρ for pairs (x_j, x_k), then $[\sum_j (x_j - \bar{x})^2]/\sigma^2(1-\rho)$ is chi-squared with $df = (d-1)$.

9.19 **b.** Given $S_i = 0$, $P(y_{i1} = y_{i2} = 0) = 1$. Given $S_i = 2$, $P(y_{i1} = y_{i2} = 1) = 1$. Given $y_{i1} + y_{i2} = 1$,

$$\begin{aligned} P(y_{i1}, y_{i2} \mid S_i = 1) &= \exp(\beta_1)/[1 + \exp(\beta_1)], \quad y_{i1} = 0, \quad y_{i2} = 1 \\ &= 1/[1 + \exp(\beta_1)], \quad y_{i1} = 1, \quad y_{i2} = 0. \end{aligned}$$

9.20 **a.** *Hint:* Apply the law of large numbers due to A. A. Markov for independent but not identically distributed random variables, or use Chebyshev's inequality.

9.22 $P(y_{ij} = 1 \mid \boldsymbol{u}_i) = \Phi(\boldsymbol{x}_{ij}\boldsymbol{\beta} + \boldsymbol{z}_{ij}\boldsymbol{u}_i)$, so

$$P(y_{ij} = 1) = \int P(z \le \boldsymbol{x}_{ij}\boldsymbol{\beta} + \boldsymbol{z}_{ij}\boldsymbol{u}_i) f(\boldsymbol{u}; \boldsymbol{\Sigma}) d\boldsymbol{u}_i,$$

where z is a standard normal variate that is independent of $\boldsymbol{u}_i$. Since $z - \boldsymbol{z}_{ij}\boldsymbol{u}_i$ has an $N(0, 1 + \boldsymbol{z}_{ij}\boldsymbol{\Sigma}\boldsymbol{z}_{ij}^{\mathrm{T}})$ distribution, the probability in the integrand is $\Phi(\boldsymbol{x}_{ij}\boldsymbol{\beta}[1 + \boldsymbol{z}_{ij}\boldsymbol{\Sigma}\boldsymbol{z}_{ij}^{\mathrm{T}}]^{-1/2})$, which does not depend on u_i, so the integral is the same. The parameters in the marginal model equal those in the GLMM divided by $[1 + \boldsymbol{z}_{ij}\boldsymbol{\Sigma}\boldsymbol{z}_{ij}^{\mathrm{T}}]^{1/2}$, which in the univariate case is $\sqrt{1+\sigma^2}$.

9.24

$$\begin{aligned} \mathrm{cov}(y_{ij}, y_{ik}) &= E[\mathrm{cov}(y_{ij}, y_{ik} \mid u_i)] + \mathrm{cov}[E(y_{ij} \mid u_i), E(y_{ik} \mid u_i)] \\ &= 0 + \mathrm{cov}[\exp(\boldsymbol{x}_{ij}\boldsymbol{\beta} + u_i), \exp(\boldsymbol{x}_{ik}\boldsymbol{\beta} + u_i)]. \end{aligned}$$

The functions in the last covariance term are both monotone increasing functions of u_i, and hence are nonnegatively correlated.

9.31 *Hint*: See Diggle et al. (2002, Sec. 4.6).

Chapter 10

10.1 Given $\sum y_i = n$, $(y_1, \dots, y_c)$ have a multinomial distribution for n trials with probabilities $\{\pi_i = 1/c\}$, and y_i has a binomial distribution with index n and parameter $\pi = 1/c$. The $\{y_i\}$ are exchangeable but not independent.

10.4 Normal with mean

$$\left(\boldsymbol{\Sigma}_0^{-1} + n\boldsymbol{\Sigma}^{-1}\right)^{-1}\left(\boldsymbol{\Sigma}_0^{-1}\boldsymbol{\mu}_0 + n\boldsymbol{\Sigma}^{-1}\bar{\boldsymbol{y}}\right)$$

and covariance matrix $(\boldsymbol{\Sigma}_0^{-1} + n\boldsymbol{\Sigma}^{-1})^{-1}$.

10.5 $E(\sigma^2 \mid \boldsymbol{y}) = ws^2 + (1-w)E(\sigma^2)$, where $w = (n-p)/(n-p+\nu_0-2)$.

10.9

$$E(\tilde{\pi} - \pi)^2 = \left(\frac{n}{n+n^*}\right)^2 \frac{\pi(1-\pi)}{n} + \left(\frac{n^*}{n+n^*}\right)^2 (\mu - \pi)^2.$$

10.11 **a.** ML estimate = 0, confidence interval = (0.0, 0.074).

b. Posterior mean = 1/27 = 0.037, posterior 95% equal-tail interval is (0.001, 0.132), 95% HPD interval is (0, 0.109) where 0.109 is 95th percentile of beta(1, 26) density.

Chapter 11

11.6 $\boldsymbol{X} = \boldsymbol{I}_n$, so

$$\hat{\boldsymbol{\beta}} = (\boldsymbol{X}^{\mathrm{T}}\boldsymbol{X} + \lambda\boldsymbol{I})^{-1}\boldsymbol{X}^{\mathrm{T}}\boldsymbol{y} = \boldsymbol{y}/(1+\lambda).$$

This has greater shrinkage of the ML estimate $\hat{\boldsymbol{\beta}} = \boldsymbol{y}$ as λ increases.

11.9 *Hint*: Can the Dirichlet recognize ordered categories, such as higher correlation between probabilities closer together? Can it recognize hierarchical structure?

11.10 **a.** $\tilde{\pi}(x)$ converges to the overall sample proportion, $\hat{\pi} = (\sum_i y_i)/n$, and the estimated asymptotic variance is approximately $\hat{\pi}(1-\hat{\pi})/n$.

11.11 *Hint*: At the first stage, can you fit the model with ordinary least squares?

References

Agresti, A. 1992. A survey of exact inference for contingency tables. *Stat. Sci.* **7**: 131–153.

Agresti, A. 2010. *Analysis of Ordinal Categorical Data*, 2nd ed. John Wiley & Sons, Inc.

Agresti, A. 2013. *Categorical Data Analysis*, 3rd ed. John Wiley & Sons, Inc.

Agresti, A., J. Booth, J. Hobert, and B. Caffo. 2000. Random-effects modeling of categorical response data. *Sociol. Methodol.* **30**: 27–81.

Agresti, A., B. Caffo, and P. Ohman-Strickland. 2004. Examples in which misspecification of a random effects distribution reduces efficiency, and possible remedies. *Comput. Stat. Data An.* **47**: 639–653.

Aitchison, J., and J. A. Bennett. 1970. Polychotomous quantal response by maximum indicant. *Biometrika* **57**: 253–262.

Aitken, A. C. 1935. On least squares and linear combinations of observations. *Proc. Roy. Soc. Edinburgh*. **55**: 42–48.

Aitkin, M., and D. Clayton. 1980. The fitting of exponential, Weibull, and extreme value distributions to complex censored survival data using GLIM. *Appl. Stat.* **29**: 156–163.

Aitkin, M., and D. B. Rubin. 1985. Estimation and hypothesis testing in finite mixture models, *J. Roy. Stat. Soc.* **B 47**: 67–75.

Aitkin, M., D. Anderson, and J. Hinde. 1981. Statistical modelling of data on teaching styles. *J. Roy. Stat. Soc.* **A 144**: 419–461.

Aitkin, M., B. J. Francis, J. P. Hinde, and R. E. Darnell. 2009. *Statistical Modelling in R*. Oxford: Oxford University Press.

Akaike, H. 1973. Information theory and an extension of the maximum likelihood principle. In *Second International Symposium on Information Theory*, eds B. Petrov and F. Csaki. Budapest: Akademiai Kiado, pp. 267–281.

Albert, J. 2010. Good smoothing. In *Frontiers of Statistical Decision Making and Bayesian Analysis*, eds M.-H. Chen, D. K. Dey, P. Müller, D. Sun, and K. Ye. New York: Springer, pp. 419–436.

Albert, A., and J. A. Anderson. 1984. On the existence of maximum likelihood estimates in logistic models. *Biometrika* **71**: 1–10.

Foundations of Linear and Generalized Linear Models, First Edition. Alan Agresti.

Albert, J., and S. Chib. 1993. Bayesian analysis of binary and polychotomous response data. *J. Am. Stat. Assoc.* **88**: 669–679.

Altham, P. M. E. 1969. Exact Bayesian analysis of a 2×2 contingency table and Fisher's "exact" significance test. *J. Roy. Stat. Soc. B* **31**: 261–269.

Altham, P. M. E. 1978. Two generalizations of the binomial distribution. *Appl. Stat.* **27**: 162–167.

Altham, P. M. E. 1984. Improving the precision of estimation by fitting a model. *J. Roy. Stat. Soc.B* **46**: 118–119.

Amemiya, T. 1984. Tobit models: a survey. *J. Econometrics* **24**: 3–61.

Andersen, E. B. 1980. *Discrete Statistical Models with Social Science Applications*. Amsterdam: North-Holland.

Anderson, J. A. 1975. Quadratic logistic discrimination. *Biometrika* **62**: 149–154.

Anderson, T. W. 2003. *An Introduction to Multivariate Statistical Analysis*, 3rd ed. John Wiley & Sons, Inc.

Anderson, J. A., and P. R. Philips. 1981. Regression, discrimination, and measurement models for ordered categorical variables. *Appl. Stat.* **30**: 22–31.

Anscombe, F. J. 1948. The transformation of Poisson, binomial and negative-binomial data. *Biometrika* **35**: 246–254.

Anscombe, F. J. 1950. Sampling theory of the negative binomial and logarithmic series distributions. *Biometrika* **37**: 358–382.

Aranda-Ordaz, F. J. 1981. On two families of transformations to additivity for binary response data. *Biometrics* **68**: 357–363.

Atkinson, A. C. 1986. Comment: aspects of diagnostic regression analysis. *Stat. Sci.* **1**: 397–402.

Atkinson, A. C., and M. Riani. 2000. *Robust Diagnostic Regression Analysis*. New York: Springer-Verlag.

Azzalini, A. 1994. Logistic regression for autocorrelated data with application to repeated measures. *Biometrika* **81**: 767–775.

Bapat, R. P. 2000. *Linear Algebra and Linear Models*, 2nd ed. Springer.

Bartlett, M. S. 1937. Some examples of statistical methods of research in agriculture and applied biology. *J. Roy. Stat. Soc.* (Suppl 4), 137–183.

Bartlett, M. S. 1947. The use of transformations. *Biometrics* **3**: 39–52.

Bartolucci, F., R. Colombi, and A. Forcina. 2007. An extended class of marginal link functions for modelling contingency tables by equality and inequality constraints. *Stat. Sin.* **17**: 691–711.

Bates, D. M., and D. G. Watts. 1988. *Nonlinear Regression Analysis and Its Application*. John Wiley & Sons, Inc.

Belsley, D. A., E. Kuh, and R. E. Welsch. 1980. *Regression Diagnostics: Identifying Influential Data and Sources of Collinearity*. John Wiley & Sons, Inc.

Benjamini, Y. 2010. Simultaneous and selective inference: current successes and future challenges. *Biomet. J.* **52**: 708–721.

Benjamini, Y., and H. Braun. 2002. John W. Tukey's contributions to multiple comparisons. *Ann. Stat.* **30**: 1576–1594.

Benjamini, Y., and Y. Hochberg. 1995. Controlling the false discovery rate: a practical and powerful approach to multiple testing. *J. Roy. Stat. Soc. B* **57**: 289–300.

Benjamini, Y., and D. Yekutieli. 2001. The control of the false discovery rate in multiple testing under dependency. *Ann. Stat.* **29**: 1165–1188.

Berger, J. 2006. The case for objective Bayesian analysis. *Bayesian Anal.* **1**: 385–402.

Berger, J. O., J. M. Bernardo, and D. Sun. 2013. Overall objective priors. Paper presented at workshop on *Recent Advances in Statistical Inference*, Padova, Italy. See `homes.stat.unipd.it/sites/homes.stat.unipd.it.lauraventura/files/berger.pdf`.

Bergsma, W. P., M. A. Croon, and J. A. Hagenaars. 2009. *Marginal Models: For Dependent, Clustered, and Longitudinal Categorical Data*. New York: Springer.

Berkson, J. 1944. Application of the logistic function to bio-assay. *J. Am. Stat. Assoc.* **39**: 357–365.

Birch, M. W. 1963. Maximum likelihood in three-way contingency tables. *J. Roy. Stat. Soc. B* **25**: 220–233.

Birkes, D., and Y. Dodge. 1993. *Alternative Methods of Regression*. John Wiley & Sons, Inc.

Bishop, Y. M. M., S. E. Fienberg, and P. W. Holland. 1975. *Discrete Multivariate Analysis*. Cambridge, MA: MIT Press.

Bliss, C. I. 1935. The calculation of the dosage–mortality curve. *Ann. Appl. Biol.* **22**: 134–167 (Appendix by R. A. Fisher).

Bliss, C. I. 1953. Fitting the negative binomial distribution to biological data. *Biometrics* **9**: 176–200 (Appendix by R. A. Fisher).

Bock, R. D., and M. Aitkin. 1981. Marginal maximum likelihood estimation of item parameters: application of an EM algorithm. *Psychometrika* **46**: 443–459.

Boole, G. 1854. *An Investigation of the Laws of Thought on Which Are Founded the Mathematical Theories of Logic and Probabilities*. London: Macmillan. See `www.gutenberg.org/files/15114/15114-pdf`.

Booth, J. G., and J. P. Hobert. 1998. Standard errors of prediction in generalized linear mixed models. *J. Am. Stat. Assoc.* **93**: 262–272.

Booth, J. G., G. Casella, H. Friedl, and J. P. Hobert. 2003. Negative binomial loglinear mixed models. *Stat. Modell.* **3**: 179–191.

Box, G. E. P., and D. R. Cox. 1964. An analysis of transformations. *J. Roy. Stat. Soc. B* **26**: 211–252.

Box, G. E. P., and G. C. Tiao. 1973. *Bayesian Inference in Statistical Analysis*. John Wiley & Sons, Inc.

Bradley, E. L. 1973. The equivalence of maximum likelihood and weighted least squares in the exponential family. *J. Am. Stat. Assoc.* **68**: 199–200.

Bradley, R. A., and M. E. Terry. 1952. Rank analysis of incomplete block designs. I. The method of paired comparisons. *Biometrika* **39**: 324–345.

Brazzale, A. R., A. C. Davison, and N. Reid. 2007. *Applied Asymptotics: Case Studies in Small-Sample Statistics*. Cambridge, UK: Cambridge University Press.

Breslow, N. 1984. Extra-Poisson variation in log-linear models. *Appl. Stat.* **33**: 38–44.

Breslow, N., and D. G. Clayton. 1993. Approximate inference in generalized linear mixed models. *J. Am. Stat. Assoc.* **88**: 9–25.

Breslow, N., and N. E. Day. 1980. *Statistical Methods in Cancer Research*, Vol. I. Lyon: International Agency for Research on Cancer (IARC).

Brown, P. J., M. Vannucci, and T. Fearn. 1998. Multivariate Bayesian variable selection and prediction. *J. Roy. Stat. Soc. B* **60**: 627–641.

Bühlmann, P., and S. van de Geer. 2011. *Statistics for High-Dimensional Data*. Springer.

Bühlmann, P., M. Kalisch, and L. Meier. 2014. High-dimensional statistics with a view toward applications in biology. *Annu. Rev. Stat.* **1**: 255–278.

Buonaccorsi, J. P. 2010. *Measurement Error: Models, Methods, and Applications*. CRC Press.

Burnham, K. P., and D. R. Anderson. 2010. *Model Selection and Multimodel Inference: A Practical Information-Theoretic Approach*, 2nd ed. Springer-Verlag.

Cameron, A. C., and P. K. Trivedi. 2013. *Regression Analysis of Count Data*, 2nd ed. Cambridge, UK: Cambridge University Press.

Cantoni, E., and E. Ronchetti. 2001. Robust inference for generalized linear models. *J. Am. Stat. Assoc.* **96**: 1022–1030.

Capanu, M., and B. Presnell. 2008. Misspecification tests for binomial and beta-binomial models. *Stat. Med.* **27**: 2536–2554.

Carey, V., S. L. Zeger, and P. Diggle. 1993. Modelling multivariate binary data with alternating logistic regressions. *Biometrika* **80**: 517–526.

Carlin, B. P., and T. A. Louis. 2009. *Bayesian Methods for Data Analysis*, 3rd ed. CRC Press.

Carvalho, C. M., N. G. Polson, and J. G. Scott. 2010. The horseshoe estimator for sparse signals. *Biometrika* **97**: 465–480.

Casella, G., and R. Berger. 2001. *Statistical Inference*, 2nd ed. Pacific Grove, CA: Wadsworth.

Casella, G., and E. I. George. 1992. Explaining the Gibbs sampler. *Am. Stat.* **46**: 167–174.

Chaloner, K., and I. Verdinelli. 1995. Bayesian experimental design: a review. *Stat. Sci.* **10**: 273–304.

Chatterjee, A., and S. Lahiri. 2011. Bootstrapping lasso estimators. *J. Am. Stat. Assoc.* **106**: 608–625.

Christensen, R. 2011. *Plane Answers to Complex Questions: The Theory of Linear Models*, 4th ed. Springer.

Christensen, R., W. Johnson, A. Branscum, and T. E. Hanson. 2010. *Bayesian Ideas and Data Analysis: An Introduction for Scientists and Statisticians*. Boca Raton, FL: CRC Press.

Cleveland, W. S. 1979. Robust locally weighted regression and smoothing scatterplots. *J. Am. Stat. Assoc.* **74**: 829–836.

Cochran, W. G. 1934. The distribution of quadratic forms in a normal system, with applications to the analysis of covariance. *Math. Proc. Camb. Philos. Soc.* **30**(2): 178–191.

Cochran, W. G. 1940. The analysis of variance when experimental errors follow the Poisson or binomial laws. *Ann. Math. Stat.* **11**: 335–347.

Cochran, W. G. 1950. The comparison of percentages in matched samples. *Biometrika* **37**: 256–266.

Cochran, W. G. 1954. Some methods of strengthening the common χ^2 tests. *Biometrics* **10**: 417–451.

Collett, D. 2005. Binary data. In *Encyclopedia of Biostatistics*, 2nd ed. John Wiley & Sons, Inc., pp. 439–446.

Congdon, P. 2005. *Bayesian Models for Categorical Data*. Hoboken, NJ: John Wiley & Sons, Inc.

Cook, R. D. 1977. Detection of influential observations in linear regression. *Technometrics* **19**: 15–18.

Cook, R. D. 1986. Assessment of local influence. *J. Roy. Stat. Soc. B* **48**: 133–169.

Cook, R. D., and S. Weisberg. 1982. *Residuals and Influence in Regression*. Chapman & Hall.

Cook, R. D., and S. Weisberg. 1997. Graphics for assessing the adequacy of regression models. *J. Amer. Statist. Assoc.* **92**: 490–499.

Copas, J. 1983. Plotting *p* against *x*. *Appl. Stat.* **32**: 25–31.

Copas, J., and S. Eguchi. 2010. Likelihood for statistically equivalent models. *J. Roy. Stat. Soc. B* **72**: 193–217.

Coull, B. A., and A. Agresti. 2000. Random effects modeling of multiple binomial responses using the multivariate binomial logit-normal distribution. *Biometrics* **56**: 73–80.

Cox, D. R. 1958. The regression analysis of binary sequences. *J. Roy. Stat. Soc. B* **20**: 215–242.

Cox, D. R. 1970. *The Analysis of Binary Data* (2nd ed. 1989, by D. R. Cox and E. J. Snell). London: Chapman & Hall.

Cox, D. R., and D. V. Hinkley. 1974. *Theoretical Statistics*. London: Chapman & Hall.

Cox, D. R., and N. Reid. 1987. Parameter orthogonality and approximate conditional inference. *J. Roy. Stat. Soc. B* **49**: 1–39.

Cox, D. R., and E. J. Snell. 1968. A general definition of residuals. *J. Roy. Stat. Soc.* **30**: 248–275.

Crainiceanu, C. M., A.-M. Staicu, and C.-Z. Di. 2009. Generalized multilevel functional regression. *J. Am. Stat. Assoc.* **104**: 1550–1561.

Craiu, R., and J. Rosenthal. 2014. Bayesian computation via Markov chain Monte Carlo. *Annu. Rev. Stat. Appl.* **1**: 179–201.

Crowder, M. J. 1978. Beta-binomial ANOVA for proportions. *Appl. Stat.* **27**: 34–37.

Darroch, J. N., S. L. Lauritzen, and T. P. Speed. 1980. Markov fields and log-linear interaction models for contingency tables. *Ann. Stat.* **8**: 522–539.

Davino, C., M. Furno, and D. Vistocco. 2013. *Quantile Regression: Theory and Applications*. John Wiley & Sons, Inc.

Davison, A. C. 2003. *Statistical Models*. Cambridge University Press.

Davison, A. C., and E. J. Snell. 1991. Residuals and diagnostics. In *Statistical Theory and Modelling: In Honour of Sir David Cox, FRS*, eds D. V. Hinkley, N. Reid, and E. J. Snell. London: Chapman & Hall, pp. 83–106.

Davison, A. C., and C.-L. Tsai. 1992. Regression model diagnostics. *Intern. Stat. Rev.* **60**: 337–353.

Demidenko, E. 2013. *Mixed Models: Theory and Applications with R*, 2nd ed. John Wiley & Sons, Inc.

Dempster, A. P. 1972. Covariance selection. *Biometrics* **28**: 157–175.

Delany, M. F., S. Linda, and C. T. Moore. 1999. Diet and condition of American alligators in 4 Florida lakes. *Proc. Annu. Conf. Southeast. Assoc. Fish and Wildl. Agencies* **53**: 375–389.

Dey, D. K., S. K. Ghosh, and B. K. Mallick (eds). 2000. *Generalized Linear Models: A Bayesian Perspective*. New York: Marcel Dekker.

Di, C.-Z., C. M. Crainiceanu, B. S. Caffo, and N. M. Punjabi. 2009. Multilevel functional principal component analysis. *Ann. Appl. Stat.* **3**: 458–488.

Diggle, P. J., P. Heagerty, K.-Y. Liang, and S. L. Zeger. 2002. *Analysis of Longitudinal Data*, 2nd ed. Oxford University Press.

Dobbie, M., and A. H. Welsh. 2001. Models for zero-inflated count data using the Neyman type A distribution, *Stat. Modell.* **1**: 65–80.

Dobson, A., and A. Barnett. 2008. *An Introduction to Generalized Linear Models*, 3rd ed. CRC Press.

Doss, H. 2010. *Linear Models*. Unpublished notes for STA 6246 at University of Florida.

Draper, D. 1995. Assessment and propagation of model uncertainty. *J. Roy. Stat. Soc. B* **57**: 45–97.

Draper, D., J. S. Hodges, C. L. Mallows, and D. Pregibon. 1993. Exchangeability and data analysis. *J. Roy. Stat. Soc. A* **156**: 9–37.

Draper, N. R., and H. Smith. 1998. *Applied Regression Analysis*, 3rd ed. John Wiley & Sons, Inc.

Dudoit, S., J. P. Shaffer, and J. C. Boldrick. 2003. Multiple hypothesis testing in microarray experiments. *Stat. Sci.* **18**: 71–103.

Dunson, D. B., and A. H. Herring. 2005. Bayesian latent variable models for mixed discrete outcomes. *Biostatistics* **6**: 11–25.

Efron, B., and D. V. Hinkley. 1978. Assessing the accuracy of the maximum likelihood estimator: observed versus expected Fisher information. *Biometrika* **65**: 457–482.

Efron, B., and C. Morris. 1975. Data analysis using Stein's estimator and its generalizations. *J. Am. Stat. Assoc.* **70**: 311–319.

Efron, B., T. Hastie, I. Johnstone, and R. Tibshirani. 2004. Least angle regression. *Ann. Stat.* **32**: 407–499.

Eilers, P. H. C., and B. D. Marx. 1996. Flexible smoothing with *B*-splines and penalties. *Stat. Sci.* **11**: 89–121.

Eilers, P. H. C., and B. D. Marx. 2002. Generalized linear additive smooth structures. *J. Comput. Graph. Stat.* **11**: 758–783.

Eilers, P. H. C., and B. D. Marx. 2010. Splines, knots, and penalties. *Wiley Interdiscip. Rev.: Comput. Stat.* **2**: 637–653.

Fahrmeir, L. 1990. Maximum likelihood estimation in misspecified generalized linear models. *Statistics* **21**: 487–502.

Fahrmeir, L., and H. Kaufmann. 1985. Consistency and asymptotic normality of the maximum likelihood estimator in generalized linear models. *Ann. Stat.* **13**: 342–368.

Fahrmeir, L., and G. Tutz. 2001. *Multivariate Statistical Modelling Based on Generalized Linear Models*, 2nd ed. Springer.

Fahrmeir, L., T. Kneib, S. Lang, and B. Marx. 2013. *Regression: Models, Methods and Applications*. Springer.

Fan, J., and J. Lv. 2010. A selective overview of variable selection in high dimensional feature space. *Stat. Sinica* **20**: 101–148.

Faraway, J. J. 2006. *Extending the Linear Model with R: Generalized Linear, Mixed Effects and Nonparametric Regression Models*. Chapman & Hall/CRC.

Farcomeni, A. 2008. A review of modern multiple hypothesis testing, with particular attention to the false discovery proportion. *Stat. Methods Medic. Res.* **17**: 347–388.

Ferguson, T. S. 1973. A Bayesian analysis of some nonparametric problems. *Ann. Stat.* **1**: 209–230.

Finney, D. J. 1947. The estimation from individual records of the relationship between dose and quantal response. *Biometrika* **34**: 320–334.

Finney, D. J. 1971. *Probit Analysis*, 3rd ed. (earlier editions 1947 and 1952). Cambridge, UK: Cambridge University Press.

Firth, D. 1991. Generalized linear models. In *Statistical Theory and Modelling: In Honour of Sir David Cox, FRS*, eds D. V. Hinkley, N. Reid, and E. J. Snell. London: Chapman & Hall, pp. 55–82.

Firth, D. 1993. Bias reduction of maximum likelihood estimates. *Biometrika* **80**: 27–38.

Fisher, R. A. 1921. On the "probable error" of a coefficient of correlation deduced from a small sample. *Metron* **1**: 3–32.

Fisher, R. A. 1925. *Statistical Methods for Research Workers* (14th ed., 1970). Edinburgh: Oliver & Boyd.

Fisher, R. A. 1935. *The Design of Experiments* (8th ed., 1966). Edinburgh: Oliver & Boyd.

Fitzmaurice, G. M., N. M. Laird, and A. G. Rotnitzky. 1993. Regression models for discrete longitudinal responses. *Stat. Sci.* **8**: 284–299.

Fitzmaurice, G. M., N. M. Laird, and J. H. Ware. 2011. *Applied Longitudinal Analysis*, 2nd ed. John Wiley & Sons, Inc.

Fox, J. 2008. *Applied Regression Analysis and Generalized Linear Models*, 2nd ed. Sage

Frome, E. L. 1983. The analysis of rates using Poisson regression models. *Biometrics* **39**: 665–674.

Gabriel, K. R. 1966. Simultaneous test procedures for multiple comparisons on categorical data. *J. Am. Stat. Assoc.* **61**: 1081–1096.

Galton, F. 1886. Regression towards mediocrity in hereditary stature. *J. Anthropol. Inst. Great Britain and Ireland* **15**: 246–263.

Galton, F. 1888. Co-relations and their measurement. *Proc. Roy. Soc. London* **45**: 135–145.

Gelfand, A. E., and A. F. M. Smith. 1990. Sampling-based approaches to calculating marginal densities. *J. Am. Stat. Assoc.* **85**: 398–409.

Gelman, A. 2006. Prior distributions for variance parameters in hierarchical models. *Bayesian Anal.* **1**: 515–534.

Gelman, A., and J. Hill. 2006. *Data Analysis Using Regression and Multilevel/Hierarchical Models*. Cambridge, UK: Cambridge University Press.

Gelman, A., A. Jakulin, M. G. Pittau, and Y.-S. Su. 2008. A weakly informative default prior distribution for logistic and other regression models. *Ann. Appl. Stat.* **2**: 1360–1383.

Gelman, A., J. B. Carlin, H. S. Stern, D. B. Dunson, A. Vehtari, and D. B. Rubin. 2013. *Bayesian Data Analysis*, 3rd ed. Boca Raton, FL: CRC Press.

Genter, F. C., and V. T. Farewell. 1985. Goodness-of-link testing in ordinal regression models. *Canad. J. Stat.* **13**: 37–44.

George, E. I. 2000. The variable selection problem. *J. Am. Stat. Assoc.* **95**: 1304–1308.

George, E. I., and R. E. McCulloch. 1997. Approaches for Bayesian variable selection. *Stat. Sinica* **7**: 339–373.

Gilchrist, R. 1981. Calculation of residuals for all GLIM models. *GLIM Newsletter* **4**: 26–28.

Godambe, V. P., and C. C. Heyde. 1987. Quasi-likelihood and optimal estimation. *Intern. Stat. Rev.* **55**: 231–244.

Goldstein, H. 2010. *Multilevel Statistical Models*, 4th ed. Hoboken, NJ: John Wiley & Sons, Inc.

Goldstein, H. 2014. Using league table rankings in public policy formulation: statistical issues. *Annu. Rev. Stat. Appl.* **1**: 385–399.

Goldstein, H., and D. J. Spiegelhalter. 1996. League tables and their limitations: statistical issues in comparisons of institutional performance. *J. Roy. Stat. Soc. A* **159**: 385–443.

Good, I. J. 1965. *The Estimation of Probabilities: An Essay on Modern Bayesian Methods*. Cambridge, MA: MIT Press.

Goodman, L. A. 1974. Exploratory latent structure analysis using both identifiable and unidentifiable models. *Biometrika* **61**: 215–231.

Gourieroux, C., A. Monfort, and A. Trognon. 1984. Pseudo maximum likelihood methods: theory. *Econometrica* **52**: 681–700.

Graybill, F. 2000. *Theory and Application of the Linear Model*. Duxbury.

Green, P. J. 1984. Iteratively reweighted least squares for maximum likelihood estimation, and some robust and resistant alternatives. *J. Roy. Stat. Soc. B* **46**: 149–192.

Green, P. J., and B. Silverman. 1993. *Nonparametric Regression and Generalized Linear Models*. Chapman & Hall.

Greene, W. H. 2011. *Econometric Analysis*, 7th ed. Prentice Hall.

Greenwood, M., and G. U. Yule. 1920. An inquiry into the nature of frequency distributions representative of multiple happenings with particular reference to the occurrence of multiple attacks of disease or of repeated accidents. *J. Roy. Stat. Soc. A* **83**: 255–279.

Griffin, J. E., and P. J. Brown. 2013. Some priors for sparse regression modelling. *Bayesian Anal.* **8**: 691–702.

Grizzle, J. E., C. F. Starmer, and G. G. Koch. 1969. Analysis of categorical data by linear models. *Biometrics* **25**: 489–504.

Gueorguieva, R., and J. H. Krystal. 2004. Move over ANOVA. *Arch. Gen. Psychiatry* **61**: 310–317.

Guimarães, P., and R. C. Lindrooth. 2007. Controlling for overdispersion in grouped conditional logit models: a computationally simple application of Dirichlet-multinomial regression. *Econometrics J.* **10**: 439–452.

Haberman, S. J. 1974. *The Analysis of Frequency Data*. Chicago: University of Chicago Press.

Hanley, J. A., and B. J. McNeil. 1982. The meaning and use of the area under a receiving operating characteristic (ROC) curve. *Radiology* **143**: 29–36.

Hans, C. 2009. Bayesian lasso regression. *Biometrika* **96**: 835–845.

Hansen, L. P. 1982. Large sample properties of generalized method of moments estimators. *Econometrica* **50**: 1029–1054.

Harter, H. L. 1974. The method of least squares and some alternatives: Part I. *Intern. Stat. Rev.* **42**: 147–174.

Hartzel, J., A. Agresti, and B. Caffo. 2001. Multinomial logit random effects models. *Stat. Modell.* **1**: 81–102.

Harville, D. A. 1976. Extension of the Gauss–Markov theorem to include the estimation of random effects. *Ann. Stat.* **4**: 384–395.

Harville, D. A. 1977. Maximum likelihood approaches to variance component estimation and to related problems. *J. Am. Stat. Assoc.* **72**: 320–338.

Hastie, T. J., and D. Pregibon. 1991. Generalized linear models. Chapter 6 of *Statistical Models in S*, eds J. M. Chambers and T. J. Hastie. Chapman & Hall/CRC.

Hastie, T., and R. Tibshirani. 1990. *Generalized Additive Models*. London: Chapman & Hall.

Hastie, T., R. Tibshirani, and J. Friedman. 2009. *The Elements of Statistical Learning: Data Mining, Inference, and Prediction*, 2nd ed. New York: Springer.

Hastings, W. K. 1970. Monte Carlo sampling methods using Markov chains and their applications. *Biometrika* **57**: 97–109.

Hauck, W. W., and A. Donner. 1977. Wald's test as applied to hypotheses in logit analysis. *J. Am. Stat. Assoc.* **72**: 851–853.

Hayter, A. J. 1984. A proof of the conjecture that the Tukey–Kramer multiple comparisons procedure is conservative. *Ann. Stat.* **12**: 61–75.

Heagerty, P. J., and S. L. Zeger. 2000. Marginalized multilevel models and likelihood inference. *Stat. Sci.* **15**: 1–19.

Hedeker, D., and R. D. Gibbons. 2006. *Longitudinal Data Analysis*. John Wiley & Sons, Inc.

Heinze, G., and M. Schemper. 2002. A solution to the problem of separation in logistic regression. *Stat. Med.* **21**: 2409–2419.

Henderson, C. R. 1975. Best linear unbiased estimation and prediction under a selection model. *Biometrics* **31**: 423–447.

Hilbe, J. M. 2011. *Negative Binomial Regression*, 2nd ed. Cambridge, UK: Cambridge University Press.

Hinde, J., and C. G. B. Demétrio. 1998. Overdispersion: models and estimation. *Comput. Stat. Data An.* **27**: 151–170.

Hjort, N., C. Holmes, P. Mueller, and S. G. Walker (eds). 2010. *Bayesian Nonparametrics*. Cambridge University Press.

Hoaglin, D. C. 2012. Making sense of coefficients in multiple regression. Unpublished manuscript.

Hoaglin, D. C. 2015. Regressions are commonly misinterpreted. *Stata Journal*, to appear.

Hoaglin, D. C., and R. E. Welsch. 1978. The hat matrix in regression and ANOVA. *Am. Stat.* **32**: 17–22.

Hoaglin, D. C., F. Mosteller, and J. W. Tukey (eds). 1991. *Fundamentals of Exploratory Analysis of Variance*. John Wiley & Sons, Inc.

Hochberg, Y., and A. C. Tamhane. 1987. *Multiple Comparison Procedures*. John Wiley & Sons, Inc.

Hoerl, A. E., and R. Kennard. 1970. Ridge regression: Biased estimation for nonorthogonal problems. *Technometrics* **12**: 55–67.

Hoeting, J. A., D. Madigan, A. E. Raftery, and C. T. Volinsky. 1999. Bayesian model averaging: a tutorial. *Stat. Sci.* **14**: 382–401.

Hoff, P. D. 2009. *A First Course in Bayesian Statistical Methods*. New York: Springer.

Holford, T. R. 1980. The analysis of rates and of survivorship using log-linear models. *Biometrics* **36**: 299–305.

Hosmer, D. W., and S. Lemeshow. 1980. A goodness-of-fit test for multiple logistic regression model. *Commun. Stat. A* **9**: 1043–1069.

Hosmer, D. W., S. Lemeshow, and R. X. Sturdivant. 2013. *Applied Logistic Regression*, 3rd ed. John Wiley & Sons, Inc.

Hotelling, H. 1931. The generalization of Student's ratio. *Ann. Math. Stat.* **2**: 360–378.

Hsu, J. C. 1996. *Multiple Comparisons: Theory and Methods*. Chapman & Hall.

Huber, P. J. 1964. Robust estimation of a location parameter. *Ann. Math. Stat.* **35**: 73–101.

Huber, P. J. 1967. The behavior of maximum likelihood estimates under nonstandard conditions. In *Proceedings of the Fifth Berkeley Symposium on Mathematical Statistics and Probability*, Volume 1: Statistics. Berkeley, CA: University of California Press, pp. 221–233.

Huber, P. J., and E. M. Ronchetti. 2009. *Robust Statistics*, 2nd ed. John Wiley & Sons, Inc.

Ibrahim, J., M.-H. Chen, S. Lipsitz, and A. Herring. 2005. Missing-data methods for generalized linear models: a comparative review. *J. Am. Stat. Assoc.* **100**: 332–346.

Izenman, A. J. 2008. *Modern Multivariate Statistical Techniques: Regression, Classification, and Manifold Learning*. Springer.

James, G., D. Witten, T. Hastie, and R. Tibshirani. 2013. *An Introduction to Statistical Learning, with Applications in R*. Springer.

Johnson, R. A., and D. W. Wichern. 2007. *Applied Multivariate Statistical Analysis*, 6th ed. Prentice Hall.

Johnson, N. L., A. W. Kemp, and S. Kotz. 2005. *Univariate Discrete Distributions*, 3rd ed. New York: John Wiley & Sons, Inc.

Jordan, M. I. 2004. Graphical models. *Stat. Sci.* **19**: 140–155.

Jørgensen, B. 1983. Maximum likelihood estimation and large-sample inference for generalized linear and nonlinear regression models. *Biometrika* **70**: 19–28.

Jørgensen, B. 1987. Exponential dispersion models. *J. Roy. Stat. Soc. B* **49**: 127–162.

Jørgensen, B. 1997. *The Theory of Dispersion Models*. Chapman & Hall.

Karlin, S., and H. M. Taylor. 1975. *A First Course in Stochastic Processes*, 2nd ed. New York: Academic Press.

Kass, R. E., and A. E. Raftery. 1995. Bayes factors. *J. Am. Stat. Assoc.* **90**: 773–795.

Kass, R. E., and L. Wasserman. 1995. A reference Bayesian test for nested hypotheses and its relationship to the Schwarz criterion. *J. Am. Stat. Assoc.* **90**: 928–934.

Kass, R. E., and L. Wasserman. 1996. The selection of prior distributions by formal rules. *J. Am. Stat. Assoc.* **91**: 1343–1370.

Kauermann, G., and R. J. Carroll. 2001. A note on the efficiency of sandwich covariance matrix estimation. *J. Am. Stat. Assoc.* **96**: 1387–1397.

Khuri, A. I. 2009. *Linear Model Methodology*. CRC Press.

Koenker, R. 2005. *Quantile Regression*. Cambridge University Press.

Kosmidis, I. 2014. Improved estimation in cumulative link models. *J. Roy. Stat. Soc. B* **76**: 169–196.

Kosmidis, I., and D. Firth. 2011. Multinomial logit bias reduction via the Poisson log-linear model. *Biometrika* **98**: 755–759.

Kramer, C. Y. 1956. Extension of multiple range tests to group means with unequal numbers of replications. *Biometrics* **12**: 307–310.

Laird, N. M., and J. H. Ware. 1982. Random-effects models for longitudinal data. *Biometrics* **38**: 963–974.

Lambert, D. 1992. Zero-inflated Poisson regression, with an application to defects in manufacturing. *Technometrics* **34**: 1–14.

Lancaster, H. O. 1961. Significance tests in discrete distributions. *J. Am. Stat. Assoc.* **56**: 223–234.

Lang, J. B. 2004. Multinomial-Poisson homogeneous models for contingency tables. *Ann. Stat.* **32**: 340–383.

Lang, J. B. 2005. Homogeneous linear predictor models for contingency tables. *J. Am. Stat. Assoc.* **100**: 121–134.

Lang, J. B., and A. Agresti. 1994. Simultaneously modeling joint and marginal distributions of multivariate categorical responses. *J. Am. Stat. Assoc.* **89**: 625–632.

Lauritzen, S. L. 1996. *Graphical Models*. New York: Oxford University Press.

Lawless, J. F. 1987. Negative binomial and mixed Poisson regression. *Canad. J. Stat.* **15**: 209–225.

Lazarsfeld, P. F., and N. W. Henry. 1968. *Latent Structure Analysis*. Boston: Houghton Mifflin.

Lee, Y., and J. A. Nelder. 1996. Hierarchical generalized linear models. *J. Roy. Stat. Soc. B* **58**: 619–678.

Lee, Y., J. A. Nelder, and Y. Pawitan. 2006. *Generalized Linear Models with Random Effects: Unified Analysis via H-Likelihood*. Boca Raton, FL: Chapman & Hall/CRC.

Legendre, A. M. 1805. Nouvelles méthodes pour la détermination des orbites des comètes. Paris: Courcier (available online at books.google.com/books).

Lehmann, E. L., and J. P. Romano. 2005. *Testing Statistical Hypotheses*, 3rd ed. New York: Springer.

Leonard, T., and J. S. J. Hsu. 1994. The Bayesian analysis of categorical data: a selective review. In *Aspects of Uncertainty: A Tribute to D. V. Lindley*, eds P. R. Freeman and A. F. M. Smith. New York: John Wiley & Sons, Inc., pp. 283–310.

Liang, K. Y., and J. Hanfelt. 1994. On the use of the quasi-likelihood method in teratological experiments. *Biometrics* **50**: 872–880.

Liang, K. Y., and P. McCullagh. 1993. Case studies in binary dispersion. *Biometrics* **49**: 623–630.

Liang, K. Y., and S. L. Zeger. 1986. Longitudinal data analysis using generalized linear models. *Biometrika* **73**: 13–22.

Lindley, D. V., and A. F. M. Smith. 1972. Bayes estimates for the linear model. *J. Roy. Stat. Soc.* **34**: 1–41.

Lindsay, B., C. Clogg, and J. Grego. 1991. Semi-parametric estimation in the Rasch model and related exponential response models, including a simple latent class model for item analysis. *J. Am. Stat. Assoc.* **86**: 96–107.

Lindsey, J. K., and P. M. E. Altham. 1998. Analysis of the human sex ratio by using overdispersion models. *Appl. Stat.* **47**: 149–157.

Lipsitz, S., N. Laird, and D. Harrington. 1991. Generalized estimating equations for correlated binary data: using the odds ratio as a measure of association. *Biometrika* **78**: 153–160.

Littell, R. C., J. Pendergast, and R. Natarajan. 2000. Modelling covariance structure in the analysis of repeated measures data. *Stat. Med.* **19**: 1793–1819.

Little, R. J., and D. B. Rubin. 2002. *Statistical Analysis with Missing Data*, 2nd ed. New York: John Wiley & Sons, Inc.

Liu, Q., and D. A. Pierce. 1994. A note on Gauss–Hermite quadrature. *Biometrika* **81**: 624–629.

Lockhart, R., J. Taylor, R. J. Tibshirani, and R. Tibshirani. 2014. A significance test for the lasso. *Ann. Stat.* **42**: 518–531.

Lovison, G. 2005. On Rao score and Pearson X^2 statistics in generalized linear models. *Stat. Pap.* **46**: 555–574.

Lovison, G. 2014. A note on adjusted responses, fitted values and residuals in generalized linear models. *Stat. Modell.* **14**: 337–359.

Madigan, D., and J. York. 1995. Bayesian graphical models for discrete data. *Intern. Stat. Rev.* **63**: 215–232.

Madsen, H., and P. Thyregod. 2011. *Introduction to General and Generalized Linear Models.* CRC Press.

Mantel, N. 1966. Models for complex contingency tables and polychotomous dosage response curves. *Biometrics* **22**: 83–95.

Mantel, N., and W. Haenszel. 1959. Statistical aspects of the analysis of data from retrospective studies of disease. *J. Natl. Cancer Inst.* **22**: 719–748.

McCullagh, P. 1980. Regression models for ordinal data. *J. Roy. Stat. Soc.* **B 42**: 109–142.

McCullagh, P. 1983. Quasi-likelihood functions. *Ann. Stat.* **11**: 59–67.

McCullagh, P., and J. A. Nelder. 1989. *Generalized Linear Models*, 2nd ed. London: Chapman & Hall.

McCulloch, C. E., and J. M. Neuhaus. 2011. Prediction of random effects in linear and generalized linear models under model misspecification. *Biometrics* **67**: 270–279.

McCulloch, C. E., S. Searle, and J. M. Neuhaus. 2008. *Generalized, Linear, and Mixed Models.* Hoboken, NJ: John Wiley & Sons, Inc.

McFadden, D. 1974. Conditional logit analysis of qualitative choice behavior. In *Frontiers in Econometrics*, ed. P. Zarembka. New York: Academic Press, pp. 105–142.

McKelvey, R. D., and W. Zavoina. 1975. A statistical model for the analysis of ordinal level dependent variables. *J. Math. Sociol.* **4**: 103–120.

McNemar, Q. 1947. Note on the sampling error of the difference between correlated proportions or percentages. *Psychometrika* **12**: 153–157.

Mehta, C. R., and N. R. Patel. 1995. Exact logistic regression: theory and examples. *Stat. Med.* **14**: 2143–2160.

Meng, X.-L. 1997. The EM algorithm and medical studies: a historical link. *Stat. Methods Med. Res.* **6**: 3–23.

Meng, X.-L. 2009. Decoding the H-likelihood. *Stat. Sci.* **24**: 280–293.

Min, Y., and A. Agresti. 2005. Random effects models for repeated measures of zero-inflated count data. *Stat. Modell.* **5**: 1–19.

Molenberghs, G., and G. Verbeke. 2005. *Models for Discrete Longitudinal Data.* New York: Springer.

Monahan, J. F. 2008. *A Primer on Linear Models.* CRC Press.

Moore, D. F., and A. Tsiatis. 1991. Robust estimation of the variance in moment methods for extra-binomial and extra-Poisson variation. *Biometrics* **47**: 383–401.

Morris, C. 1982. Natural exponential families with quadratic variance functions. *Ann. Stat.* **10**: 65–80.

Morris, C. 1983a. Natural exponential families with quadratic variance functions: statistical theory. *Ann. Stat.* **11**: 515–529.

Morris, C. 1983b. Parametric empirical Bayes inference: theory and applications. *J. Am. Stat. Assoc.* **78**: 47–65.

Mosimann, J. E. 1962. On the compound multinomial distribution, the multivariate β-distribution and correlations among proportions. *Biometrika* **49**: 65–82.

Mullahy, J. 1986. Specification and testing of some modified count data models. *J. Econometrics* **3**: 341–365.

Nelder, J., and R. W. M. Wedderburn. 1972. Generalized linear models. *J. Roy. Stat. Soc. A* **135**: 370–384.

Neyman, J. 1949. Contributions to the theory of the χ^2 test. In *Proceedings of the First Berkeley Symposium on Mathematical Statistics and Probability*, ed. J. Neyman. Berkeley, CA: University of California Press, pp. 239–273.

Ntzoufras, I. 2009. *Bayesian Modeling Using WinBUGS*. John Wiley & Sons, Inc.

Ochi, Y., and R. Prentice. 1984. Likelihood inference in a correlated probit regression model. *Biometrika* **71**: 531–543.

O'Hagan, A., and J. Forster. 2004. *Kendall's Advanced Theory of Statistics, Volume 2B: Bayesian Inference*. London: Arnold.

Pace, L., and A. Salvan. 1997. *Principles of Statistical Inference from a Neo-Fisherian Perspective*. Singapore: World Scientific.

Papke, L., and J. Wooldridge. 1996. Econometric methods for fractional response variables with an application to 401(K) plan participation rates. *J. Appl. Economet.* **11**: 619–632.

Park, T., and G. Casella. 2007. The Bayesian lasso. *J. Am. Stat. Assoc.* **103**: 681–686.

Patterson, H. D., and R. Thompson. 1971. Recovery of inter-block information when block sizes are unequal. *Biometrika* **58**: 545–554.

Pearson, K. 1900. On a criterion that a given system of deviations from the probable in the case of a correlated system of variables is such that it can be reasonably supposed to have arisen from random sampling. *Philos. Mag. Ser. 5* **50**: 157–175. (Reprinted in *Karl Pearson's Early Statistical Papers*, ed. E. S. Pearson. Cambridge, UK: Cambridge University Press, 1948.)

Pearson, K. 1920. Notes on the history of correlation. *Biometrika* **13**: 25–45.

Peterson, B., and F. E. Harrell, Jr. 1990. Partial proportional odds models for ordinal response variables. *Appl. Stat.* **39**: 205–217.

Pierce, D. A. 1982. The asymptotic effect of substituting estimators for parameters in certain types of statistics. *Ann. Stat.* **10**: 475–478.

Pierce, D. A., and D. W. Schafer. 1986. Residuals in generalized linear models. *J. Am. Stat. Assoc.* **81**: 977–983.

Plackett, R. L. 1972. The discovery of the method of least squares. *Biometrika* **59**: 239–251.

Polson, N. G., and J. G. Scott. 2010. Shrink globally, act locally: Sparse Bayesian regularisation and prediction. In *Bayesian Statistics 9*, eds J. M. Bernardo, M. J. Bayarri, J. O. Berger, A. P. Dawid, D. Heckerman, A. F. M. Smith, and M. West. Oxford: Clarendon Press, pp. 501–538.

Pregibon, D. 1980. Goodness of link tests for generalized linear models. *Appl. Stat.* **29**: 15–24.

Pregibon, D. 1981. Logistic regression diagnostics. *Ann. Stat.* **9**: 705–724.

Pregibon, D. 1982. Score tests in GLIM with applications. In *Lecture Notes in Statistics*, 14: *GLIM* 82, *Proceedings of the International Conference on Generalised Linear Models*, ed. R. Gilchrist. New York: Springer-Verlag, pp. 87–97.

Prentice, R. 1986. Binary regression using an extended beta-binomial distribution, with discussion of correlation induced by covariate measurement errors. *J. Am. Stat. Assoc.* **81**: 321–327.

Presnell, B., and D. D. Boos. 2004. The IOS test for model misspecification. *J. Am. Stat. Assoc.* **99**: 216–227.

Raftery, A. E. 1995. Bayesian model selection in social research. *Sociol. Methodol.* **25**: 111–163.

Raftery, A. E., D. Madigan, and J. A. Hoeting. 1997. Bayesian model averaging for linear regression models. *J. Am. Stat. Assoc.* **92**: 179–191.

Raiffa, H., and R. Schlaifer. 1961. *Applied Statistical Decision Theory*. Harvard University Press.

Ramsay, J., and B. Silverman. 2005. *Functional Data Analysis*. Springer.

Rao, J. N. K. 2003. *Small Area Estimation*. Hoboken, NJ: John Wiley & Sons, Inc.

Rasch, G. 1961. On general laws and the meaning of measurement in psychology. In *Proceedings of the Fourth Berkeley Symposium on Mathematical Statistics and Probability*, Volume 4, ed. J. Neyman. Berkeley, CA: University of California Press, pp. 321–333.

Rawlings, J. O., S. G. Pantula, and D. A. Dickey. 1998. *Applied Regression Analysis*, 2nd ed. Springer.

Rencher, A. C., and G. B. Schaalje. 2008. *Linear Models in Statistics*, 2nd ed. John Wiley & Sons, Inc.

Ridout, M., J. Hinde, and C. Demetrio. 2001. A score test for testing a zero-inflated Poisson regression model against zero-inflated negative binomial alternatives. *Biometrics* **57**: 219–223.

Robinson, G. K. 1991. That BLUP is a good thing: the estimation of random effects. *Stat. Sci.* **6**: 15–32.

Rŏcková, V., and E. George. 2014. EMVS: the EM approach to Bayesian variable selection. *J. Am. Stat. Assoc.* **109**: 828–846.

Rodgers, J. L., W. A. Nicewander, and L. Toothaker. 1984. Linearly independent, orthogonal, and uncorrelated variables. *Am. Stat.* **38**: 133–134.

Rosenbaum, P. R., and D. B. Rubin. 1983. The central role of the propensity score in observational studies for causal effects. *Biometrika* **70**: 41–55.

Rousseeuw, P. J. 1984. Least median of squares regression. *J. Am. Stat. Assoc.* **79**: 871–880.

Royall, R. M. 1986. Model robust confidence intervals using maximum likelihood estimators. *Intern. Stat. Rev.* **54**: 221–226.

Rubin, D. B. 1974. Estimating causal effects of treatments in randomized and nonrandomized studies. *J. Educ. Psychol.* **66**: 688–701.

Schall, R. 1991. Estimation in generalized linear models. *Biometrika* **78**: 719–727.

Scheffé, H. 1959. *The Analysis of Variance*. John Wiley & Sons, Inc.

Scott, M. A., J. S. Simonoff, and B. D. Marx (eds). 2013. *The SAGE Handbook of Multilevel Modeling*. Los Angeles: Sage.

Schwarz, G. E. 1978. Estimating the dimension of a model. *Ann. Stat.* **6**: 461–464.

Searle, S. 1997. *Linear Models*. John Wiley & Sons, Inc.

Searle, S., G. Casella, and C. E. McCulloch. 1997. *Variance Components*. John Wiley & Sons, Inc.

Seber, G. A. F., and A. J. Lee. 2003. *Linear Regression*, 2nd ed. John Wiley & Sons, Inc.

Seber, G. A. F., and C. J. Wild. 1989. *Nonlinear Regression*. John Wiley & Sons, Inc.

Self, S. G., and K.-Y. Liang. 1987. Asymptotic properties of maximum likelihood estimators and likelihood ratio tests under nonstandard conditions. *J. Am. Stat. Assoc.* **82**: 605–610.

Seneta, E. 1992. On the history of the strong law of large numbers and Boole's inequality. *Historia Mathem.* **19**: 24–39.

Seshadri, V. 1994. *The Inverse Gaussian Distribution: A Case Study in Exponential Families*. Oxford: Oxford University Press.

Simonoff, J. S. 1996. *Smoothing Methods in Statistics*. New York: Springer-Verlag.

Simpson, E. H. 1951. The interpretation of interaction in contingency tables. *J. Roy. Stat. Soc. B* **13**: 238–241.

Skellam, J. G. 1948. A probability distribution derived from the binomial distribution by regarding the probability of success as variable between the sets of trials. *J. Roy. Stat. Soc. B* **10**: 257–261.

Skrondal, A., and S. Rabe-Hesketh. 2004. *Generalized Latent Variable Modeling: Multilevel, Longitudinal, and Structural Equation Models*. Boca Raton, FL: Chapman & Hall/CRC.

Smith, A. F. M. 1973. A general Bayesian linear model. *J. Roy. Stat. Soc.* **35**: 67–75.

Smyth, G. K. 2002. Nonlinear regression. In *Ency. Environmetrics*, Volume 3. John Wiley & Sons, Inc., pp. 1405–1411.

Smyth, G. K. 2003. Pearson's goodness of fit statistic as a score test statistic. In *Science and Statistics: A Festschrift for Terry Speed*, ed. D. R. Goldstein, IMS Lecture Notes–Monograph Series, Vol. 40. Hayward, CA: Institute of Mathematical Statistics, pp. 115–126.

Spiegelhalter, D. J., N. G. Best, B. P. Carlin, and A. van der Linde. 2002. Bayesian measures of model complexity and fit. *J. Roy. Stat. Soc. B* **64**: 583–639.

Spiegelhalter, D. J., N. G. Best, B. P. Carlin, and A. van der Linde. 2014. The deviance information criterion: 12 years on. *J. Roy. Stat. Soc. B* **76**: 485–493.

Stapleton, J. 2009. *Linear Statistical Models*, 2nd ed. John Wiley & Sons, Inc.

Stigler, S. 1981. Gauss and the invention of least squares. *Ann. Stat.* **9**: 465–474.

Stigler, S. 1986. *The History of Statistics: The Measurement of Uncertainty before 1900*. Harvard University Press.

Stone, M. 1974. Cross-validatory choice and assessment of statistical predictions. *J. Roy. Stat. Soc.* **36**: 111–147.

Stroup, W. W. 2012. *Generalized Linear Mixed Models*. CRC Press.

Stukel, T. A. 1988. Generalized logistic models. *J. Am. Stat. Assoc.* **83**: 426–431.

Taddy, M. 2013. Multinomial inverse regression for text analysis. *J. Am. Stat. Assoc.* **108**: 755–770.

Taylor, J. 2013. The geometry of least squares in the 21st century. *Bernoulli* **19**: 1449–1464.

Theil, H. 1969. A multinomial extension of the linear logit model. *Int. Econ. Rev.* **10**: 251–259.

Tibshirani, R. 1996. Regression shrinkage and selection via the lasso. *J. Roy. Stat. Soc. B* **58**: 267–288.

Toulis, P., and E. M. Airoldi. 2014. Stochastic gradient descent methods for principled estimation with massive data sets. Unpublished manuscript.

Touloumis, A., A. Agresti, and M. Kateri. 2013. GEE for multinomial responses using a local odds ratios parameterization. *Biometrics* **69**: 633–640.

Train, K. 2009. *Discrete Choice Methods with Simulation*. Cambridge, UK: Cambridge University Press.

Trivedi, P. K., and D. M. Zimmer. 2007. *Copula Modeling: An Introduction for Practitioners*. Now Publishers, Inc.

Tsiatis, A. A. 1980. A note on the goodness-of-fit test for the logistic regression model. *Biometrika* **67**: 250–251.

Tukey, J. W. 1949. One degree of freedom for non-additivity. *Biometrics* **5**: 232–242.

Tukey, J. W. 1994. *The Collected Works of John W. Tukey*, Vol. 8, ed. H. I. Braun. Chapman & Hall.

Tutz, G. 2011. *Structured Regression for Categorical Data*. Cambridge, UK: Cambridge University Press.

Tutz, G., and W. Hennevogl. 1996. Random effects in ordinal regression models. *Comput. Stat. Data An.* **22**: 537–557.

Tweedie, M. C. K. 1947. Functions of a statistical variate with given means, with special reference to Laplacian distributions. *Proc. Camb. Phil. Soc.* **49**: 41–49.

Varin, C., N. Reid., and D. Firth. 2011. An overview of composite likelihood methods. *Stat. Sinica* **21**: 5–42.

Verbeke, G., and E. Lesaffre. 1998. The effects of misspecifying the random-effects distribution in linear mixed models for longitudinal data. *Comp. Stat. Data Anal.* **23**: 541–566.

Verbeke, G., and G. Molenberghs. 2000. *Linear Mixed Models for Longitudinal Data*. New York: Springer-Verlag.

Wakefield, J. 2013. *Bayesian and Frequentist Regression Methods*. Springer.

Warner, S. L. 1963. Multivariate regression of dummy variates under normality assumptions. *J. Am. Stat. Assoc.* **58**: 1054–1063.

Wedderburn, R. W. M. 1974. Quasi-likelihood functions, generalized linear models, and the Gauss-Newton method. *Biometrika* **61**: 439–447.

Wedderburn, R. W. M. 1976. On the existence and uniqueness of the maximum likelihood estimates for certain generalized linear models. *Biometrika* **63**: 27–32.

Weisberg, S. 2005. *Applied Linear Regression*, 3rd ed. John Wiley & Sons, Inc.

Wherry, R. 1931. A new formula for predicting the shrinkage of the coefficient of multiple correlation. *Ann. Math. Stat.* **2**: 440–457.

White, H. 1980. A heteroskedasticity-consistent covariance matrix estimator and a direct test for heteroskedasticity. *Econometrica* **48**: 817–838.

White, H. 1982. Maximum likelihood estimation of misspecified models. *Econometrica* **50**: 1–26.

Williams, D. A. 1982. Extra-binomial variation in logistic linear models. *Appl. Stat.* **31**: 144–148.

Williams, D. A. 1984. Residuals in generalized linear models. In *Proceedings of the XIIth International Biometric Conference*, pp. 59–68.

Williams, D. A. 1987. Generalized linear model diagnostics using the deviance and single-case deletions. *Appl. Stat.* **36**: 181–191.

Wood, S. 2006. *Generalized Additive Models: An Introduction with R*. Chapman & Hall/CRC.

Yates, F. 1955. The use of transformations and maximum likelihood in the analysis of quantal experiments involving two treatments. *Biometrika* **42**: 382–403.

Yee, T. W., and C. J. Wild. 1996. Vector generalized additive models. *J. R. Stat. Soc. B* **58**: 481–493.

Yule, G. U. 1897. On the theory of correlation. *J. Roy. Stat. Soc.* **60**: 812–854.

Yule, G. U. 1900. On the association of attributes in statistics. *Philos. Trans. Roy. Soc. London Ser. A* **194**: 257–319.

Yule, G. U. 1903. Notes on the theory of association of attributes in statistics. *Biometrika* **2**: 121–134.

Yule, G. U. 1907. On the theory of correlation for any number of variables, treated by a new system of notation. *Proc. Roy. Soc. London, Series A.* **79**: 182–193.

Yule, G. U. 1912. On the methods of measuring association between two attributes. *J. Roy. Stat. Soc.* **75**: 579–642.

Zeger, S. L., and M. R. Karim. 1991. Generalized linear models with random effects: a Gibbs sampling approach. *J. Am. Stat. Assoc.* **86**: 79–86.

Zellner, A., and P. E. Rossi. 1984. Bayesian analysis of dichotomous quantal response models. *J. Economet.* **25**: 365–393.

Zheng, B., and A. Agresti. 2000. Summarizing the predictive power of a generalized linear model. *Stat. Med.* **19**: 1771–1781.

Zou, H. 2006. The adaptive lasso and its oracle properties. *J. Am. Stat. Assoc.* **101**: 1418–1429.

Zou, H., and T. Hastie. 2005. Regularization and variable selection via the elastic net. *J. Roy. Stat. Soc.* **67**: 301–320.

Author Index

Foundations of Linear and Generalized Linear Models, First Edition. Alan Agresti.

Example Index

Foundations of Linear and Generalized Linear Models, First Edition. Alan Agresti.

Subject Index

Foundations of Linear and Generalized Linear Models, First Edition. Alan Agresti.

WILEY SERIES IN PROBABILITY AND STATISTICS

ESTABLISHED BY WALTER A. SHEWHART AND SAMUEL S. WILKS

The ***Wiley Series in Probability and Statistics*** is well established and authoritative. It covers many topics of current research interest in both pure and applied statistics and probability theory. Written by leading statisticians and institutions, the titles span both state-of-the-art developments in the field and classical methods.

Reflecting the wide range of current research in statistics, the series encompasses applied, methodological and theoretical statistics, ranging from applications and new techniques made possible by advances in computerized practice to rigorous treatment of theoretical approaches.

This series provides essential and invaluable reading for all statisticians, whether in academia, industry, government, or research.

† ABRAHAM and LEDOLTER · Statistical Methods for Forecasting
AGRESTI · Analysis of Ordinal Categorical Data, *Second Edition*
AGRESTI · An Introduction to Categorical Data Analysis, *Second Edition*
AGRESTI · Categorical Data Analysis, *Third Edition*
AGRESTI · Foundations of Linear and Generalized Linear Models
ALSTON, MENGERSEN and PETTITT (editors) · Case Studies in Bayesian Statistical Modelling and Analysis
ALTMAN, GILL, and McDONALD · Numerical Issues in Statistical Computing for the Social Scientist
AMARATUNGA and CABRERA · Exploration and Analysis of DNA Microarray and Protein Array Data
AMARATUNGA, CABRERA, and SHKEDY · Exploration and Analysis of DNA Microarray and Other High-Dimensional Data, *Second Edition*
ANDĚL · Mathematics of Chance
ANDERSON · An Introduction to Multivariate Statistical Analysis, *Third Edition*
* ANDERSON · The Statistical Analysis of Time Series
ANDERSON, AUQUIER, HAUCK, OAKES, VANDAELE, and WEISBERG · Statistical Methods for Comparative Studies
ANDERSON and LOYNES · The Teaching of Practical Statistics
ARMITAGE and DAVID (editors) · Advances in Biometry
ARNOLD, BALAKRISHNAN, and NAGARAJA · Records
* ARTHANARI and DODGE · Mathematical Programming in Statistics
AUGUSTIN, COOLEN, DE COOMAN and TROFFAES (editors) · Introduction to Imprecise Probabilities
* BAILEY · The Elements of Stochastic Processes with Applications to the Natural Sciences
BAJORSKI · Statistics for Imaging, Optics, and Photonics
BALAKRISHNAN and KOUTRAS · Runs and Scans with Applications
BALAKRISHNAN and NG · Precedence-Type Tests and Applications

*Now available in a lower priced paperback edition in the Wiley Classics Library.
†Now available in a lower priced paperback edition in the Wiley–Interscience Paperback Series.

BARNETT · Comparative Statistical Inference, *Third Edition*
BARNETT · Environmental Statistics
BARNETT and LEWIS · Outliers in Statistical Data, *Third Edition*
BARTHOLOMEW, KNOTT, and MOUSTAKI · Latent Variable Models and Factor Analysis: A Unified Approach, *Third Edition*
BARTOSZYNSKI and NIEWIADOMSKA-BUGAJ · Probability and Statistical Inference, *Second Edition*
BASILEVSKY · Statistical Factor Analysis and Related Methods: Theory and Applications
BATES and WATTS · Nonlinear Regression Analysis and Its Applications
BECHHOFER, SANTNER, and GOLDSMAN · Design and Analysis of Experiments for Statistical Selection, Screening, and Multiple Comparisons
BEH and LOMBARDO · Correspondence Analysis: Theory, Practice and New Strategies
BEIRLANT, GOEGEBEUR, SEGERS, TEUGELS, and DE WAAL · Statistics of Extremes: Theory and Applications
BELSLEY · Conditioning Diagnostics: Collinearity and Weak Data in Regression
† BELSLEY, KUH, and WELSCH · Regression Diagnostics: Identifying Influential Data and Sources of Collinearity
BENDAT and PIERSOL · Random Data: Analysis and Measurement Procedures, *Fourth Edition*
BERNARDO and SMITH · Bayesian Theory
BHAT and MILLER · Elements of Applied Stochastic Processes, *Third Edition*
BHATTACHARYA and WAYMIRE · Stochastic Processes with Applications
BIEMER, GROVES, LYBERG, MATHIOWETZ, and SUDMAN · Measurement Errors in Surveys
BILLINGSLEY · Convergence of Probability Measures, *Second Edition*
BILLINGSLEY · Probability and Measure, *Anniversary Edition*
BIRKES and DODGE · Alternative Methods of Regression
BISGAARD and KULAHCI · Time Series Analysis and Forecasting by Example
BISWAS, DATTA, FINE, and SEGAL · Statistical Advances in the Biomedical Sciences: Clinical Trials, Epidemiology, Survival Analysis, and Bioinformatics
BLISCHKE and MURTHY (editors) · Case Studies in Reliability and Maintenance
BLISCHKE and MURTHY · Reliability: Modeling, Prediction, and Optimization
BLOOMFIELD · Fourier Analysis of Time Series: An Introduction, *Second Edition*
BOLLEN · Structural Equations with Latent Variables
BOLLEN and CURRAN · Latent Curve Models: A Structural Equation Perspective
BONNINI, CORAIN, MAROZZI and SALMASO · Nonparametric Hypothesis Testing: Rank and Permutation Methods with Applications in R
BOROVKOV · Ergodicity and Stability of Stochastic Processes
BOSQ and BLANKE · Inference and Prediction in Large Dimensions
BOULEAU · Numerical Methods for Stochastic Processes
* BOX and TIAO · Bayesian Inference in Statistical Analysis
BOX · Improving Almost Anything, *Revised Edition*
* BOX and DRAPER · Evolutionary Operation: A Statistical Method for Process Improvement
BOX and DRAPER · Response Surfaces, Mixtures, and Ridge Analyses, *Second Edition*
BOX, HUNTER, and HUNTER · Statistics for Experimenters: Design, Innovation, and Discovery, *Second Edition*
BOX, JENKINS, and REINSEL · Time Series Analysis: Forecasting and Control, *Fourth Edition*

*Now available in a lower priced paperback edition in the Wiley Classics Library.
†Now available in a lower priced paperback edition in the Wiley–Interscience Paperback Series.

BOX, LUCEÑO, and PANIAGUA-QUIÑONES · Statistical Control by Monitoring and Adjustment, *Second Edition*
* BROWN and HOLLANDER · Statistics: A Biomedical Introduction
CAIROLI and DALANG · Sequential Stochastic Optimization
CASTILLO, HADI, BALAKRISHNAN, and SARABIA · Extreme Value and Related Models with Applications in Engineering and Science
CHAN · Time Series: Applications to Finance with R and S-Plus®, *Second Edition*
CHARALAMBIDES · Combinatorial Methods in Discrete Distributions
CHATTERJEE and HADI · Regression Analysis by Example, *Fourth Edition*
CHATTERJEE and HADI · Sensitivity Analysis in Linear Regression
CHEN · The Fitness of Information: Quantitative Assessments of Critical Evidence
CHERNICK · Bootstrap Methods: A Guide for Practitioners and Researchers, *Second Edition*
CHERNICK and FRIIS · Introductory Biostatistics for the Health Sciences
CHILÈS and DELFINER · Geostatistics: Modeling Spatial Uncertainty, *Second Edition*
CHIU, STOYAN, KENDALL and MECKE · Stochastic Geometry and Its Applications, *Third Edition*
CHOW and LIU · Design and Analysis of Clinical Trials: Concepts and Methodologies, *Third Edition*
CLARKE · Linear Models: The Theory and Application of Analysis of Variance
CLARKE and DISNEY · Probability and Random Processes: A First Course with Applications, *Second Edition*
* COCHRAN and COX · Experimental Designs, *Second Edition*
COLLINS and LANZA · Latent Class and Latent Transition Analysis: With Applications in the Social, Behavioral, and Health Sciences
CONGDON · Applied Bayesian Modelling, *Second Edition*
CONGDON · Bayesian Models for Categorical Data
CONGDON · Bayesian Statistical Modelling, *Second Edition*
CONOVER · Practical Nonparametric Statistics, *Third Edition*
COOK · Regression Graphics
COOK and WEISBERG · An Introduction to Regression Graphics
COOK and WEISBERG · Applied Regression Including Computing and Graphics
CORNELL · A Primer on Experiments with Mixtures
CORNELL · Experiments with Mixtures, Designs, Models, and the Analysis of Mixture Data, *Third Edition*
COX · A Handbook of Introductory Statistical Methods
CRESSIE · Statistics for Spatial Data, *Revised Edition*
CRESSIE and WIKLE · Statistics for Spatio-Temporal Data
CSRGO and HORVTH · Limit Theorems in Change Point Analysis
DAGPUNAR · Simulation and Monte Carlo: With Applications in Finance and MCMC
DANIEL · Applications of Statistics to Industrial Experimentation
DANIEL · Biostatistics: A Foundation for Analysis in the Health Sciences, *Eighth Edition*
* DANIEL · Fitting Equations to Data: Computer Analysis of Multifactor Data, *Second Edition*
DASU and JOHNSON · Exploratory Data Mining and Data Cleaning
DAVID and NAGARAJA · Order Statistics, *Third Edition*
DAVINO, FURNO and VISTOCCO · Quantile Regression: Theory and Applications
* DEGROOT, FIENBERG, and KADANE · Statistics and the Law
DEL CASTILLO · Statistical Process Adjustment for Quality Control
DEMARIS · Regression with Social Data: Modeling Continuous and Limited Response Variables

*Now available in a lower priced paperback edition in the Wiley Classics Library.
†Now available in a lower priced paperback edition in the Wiley–Interscience Paperback Series.

DEMIDENKO · Mixed Models: Theory and Applications with R, *Second Edition*
DENISON, HOLMES, MALLICK, and SMITH · Bayesian Methods for Nonlinear Classification and Regression
DETTE and STUDDEN · The Theory of Canonical Moments with Applications in Statistics, Probability, and Analysis
DEY and MUKERJEE · Fractional Factorial Plans
DILLON and GOLDSTEIN · Multivariate Analysis: Methods and Applications
* DODGE and ROMIG · Sampling Inspection Tables, *Second Edition*
* DOOB · Stochastic Processes
DOWDY, WEARDEN, and CHILKO · Statistics for Research, *Third Edition*
DRAPER and SMITH · Applied Regression Analysis, *Third Edition*
DRYDEN and MARDIA · Statistical Shape Analysis
DUDEWICZ and MISHRA · Modern Mathematical Statistics
DUNN and CLARK · Basic Statistics: A Primer for the Biomedical Sciences, *Fourth Edition*
DUPUIS and ELLIS · A Weak Convergence Approach to the Theory of Large Deviations
EDLER and KITSOS · Recent Advances in Quantitative Methods in Cancer and Human Health Risk Assessment
* ELANDT-JOHNSON and JOHNSON · Survival Models and Data Analysis
ENDERS · Applied Econometric Time Series, *Third Edition*
† ETHIER and KURTZ · Markov Processes: Characterization and Convergence
EVANS, HASTINGS, and PEACOCK · Statistical Distributions, *Third Edition*
EVERITT, LANDAU, LEESE, and STAHL · Cluster Analysis, *Fifth Edition*
FEDERER and KING · Variations on Split Plot and Split Block Experiment Designs
FELLER · An Introduction to Probability Theory and Its Applications, Volume I, *Third Edition*, Revised; Volume II, *Second Edition*
FITZMAURICE, LAIRD, and WARE · Applied Longitudinal Analysis, *Second Edition*
* FLEISS · The Design and Analysis of Clinical Experiments
FLEISS · Statistical Methods for Rates and Proportions, *Third Edition*
† FLEMING and HARRINGTON · Counting Processes and Survival Analysis
FUJIKOSHI, ULYANOV, and SHIMIZU · Multivariate Statistics: High-Dimensional and Large-Sample Approximations
FULLER · Introduction to Statistical Time Series, *Second Edition*
† FULLER · Measurement Error Models
GALLANT · Nonlinear Statistical Models
GEISSER · Modes of Parametric Statistical Inference
GELMAN and MENG · Applied Bayesian Modeling and Causal Inference from incomplete-Data Perspectives
GEWEKE · Contemporary Bayesian Econometrics and Statistics
GHOSH, MUKHOPADHYAY, and SEN · Sequential Estimation
GIESBRECHT and GUMPERTZ · Planning, Construction, and Statistical Analysis of Comparative Experiments
GIFI · Nonlinear Multivariate Analysis
GIVENS and HOETING · Computational Statistics
GLASSERMAN and YAO · Monotone Structure in Discrete-Event Systems
GNANADESIKAN · Methods for Statistical Data Analysis of Multivariate Observations, *Second Edition*
GOLDSTEIN · Multilevel Statistical Models, *Fourth Edition*
GOLDSTEIN and LEWIS · Assessment: Problems, Development, and Statistical Issues
GOLDSTEIN and WOOFF · Bayes Linear Statistics
GRAHAM · Markov Chains: Analytic and Monte Carlo Computations

*Now available in a lower priced paperback edition in the Wiley Classics Library.
†Now available in a lower priced paperback edition in the Wiley–Interscience Paperback Series.

GREENWOOD and NIKULIN · A Guide to Chi-Squared Testing
GROSS, SHORTLE, THOMPSON, and HARRIS · Fundamentals of Queueing Theory, *Fourth Edition*
GROSS, SHORTLE, THOMPSON, and HARRIS · Solutions Manual to Accompany Fundamentals of Queueing Theory, *Fourth Edition*
* HAHN and SHAPIRO · Statistical Models in Engineering
HAHN and MEEKER · Statistical Intervals: A Guide for Practitioners
HALD · A History of Probability and Statistics and their Applications Before 1750
† HAMPEL · Robust Statistics: The Approach Based on Influence Functions
HARTUNG, KNAPP, and SINHA · Statistical Meta-Analysis with Applications
HEIBERGER · Computation for the Analysis of Designed Experiments
HEDAYAT and SINHA · Design and Inference in Finite Population Sampling
HEDEKER and GIBBONS · Longitudinal Data Analysis
HELLER · MACSYMA for Statisticians
HERITIER, CANTONI, COPT, and VICTORIA-FESER · Robust Methods in Biostatistics
HINKELMANN and KEMPTHORNE · Design and Analysis of Experiments, Volume 1: Introduction to Experimental Design, *Second Edition*
HINKELMANN and KEMPTHORNE · Design and Analysis of Experiments, Volume 2: Advanced Experimental Design
HINKELMANN (editor) · Design and Analysis of Experiments, Volume 3: Special Designs and Applications
HOAGLIN, MOSTELLER, and TUKEY · Fundamentals of Exploratory Analysis of Variance
* HOAGLIN, MOSTELLER, and TUKEY · Exploring Data Tables, Trends and Shapes
* HOAGLIN, MOSTELLER, and TUKEY · Understanding Robust and Exploratory Data Analysis
HOCHBERG and TAMHANE · Multiple Comparison Procedures
HOCKING · Methods and Applications of Linear Models: Regression and the Analysis of Variance, *Third Edition*
HOEL · Introduction to Mathematical Statistics, *Fifth Edition*
HOGG and KLUGMAN · Loss Distributions
HOLLANDER, WOLFE, and CHICKEN · Nonparametric Statistical Methods, *Third Edition*
HOSMER and LEMESHOW · Applied Logistic Regression, *Second Edition*
HOSMER, LEMESHOW, and MAY · Applied Survival Analysis: Regression Modeling of Time-to-Event Data, *Second Edition*
HUBER · Data Analysis: What Can Be Learned From the Past 50 Years
HUBER · Robust Statistics
† HUBER and RONCHETTI · Robust Statistics, *Second Edition*
HUBERTY · Applied Discriminant Analysis, *Second Edition*
HUBERTY and OLEJNIK · Applied MANOVA and Discriminant Analysis, *Second Edition*
HUITEMA · The Analysis of Covariance and Alternatives: Statistical Methods for Experiments, Quasi-Experiments, and Single-Case Studies, *Second Edition*
HUNT and KENNEDY · Financial Derivatives in Theory and Practice, *Revised Edition*
HURD and MIAMEE · Periodically Correlated Random Sequences: Spectral Theory and Practice
HUSKOVA, BERAN, and DUPAC · Collected Works of Jaroslav Hajek—with Commentary
HUZURBAZAR · Flowgraph Models for Multistate Time-to-Event Data

*Now available in a lower priced paperback edition in the Wiley Classics Library.
†Now available in a lower priced paperback edition in the Wiley–Interscience Paperback Series.

JACKMAN · Bayesian Analysis for the Social Sciences
† JACKSON · A User's Guide to Principle Components
JOHN · Statistical Methods in Engineering and Quality Assurance
JOHNSON · Multivariate Statistical Simulation
JOHNSON and BALAKRISHNAN · Advances in the Theory and Practice of Statistics: A Volume in Honor of Samuel Kotz
JOHNSON, KEMP, and KOTZ · Univariate Discrete Distributions, *Third Edition*
JOHNSON and KOTZ (editors) · Leading Personalities in Statistical Sciences: From the Seventeenth Century to the Present
JOHNSON, KOTZ, and BALAKRISHNAN · Continuous Univariate Distributions, Volume 1, *Second Edition*
JOHNSON, KOTZ, and BALAKRISHNAN · Continuous Univariate Distributions, Volume 2, *Second Edition*
JOHNSON, KOTZ, and BALAKRISHNAN · Discrete Multivariate Distributions
JUDGE, GRIFFITHS, HILL, LÜTKEPOHL, and LEE · The Theory and Practice of Econometrics, *Second Edition*
JUREK and MASON · Operator-Limit Distributions in Probability Theory
KADANE · Bayesian Methods and Ethics in a Clinical Trial Design
KADANE AND SCHUM · A Probabilistic Analysis of the Sacco and Vanzetti Evidence
KALBFLEISCH and PRENTICE · The Statistical Analysis of Failure Time Data, *Second Edition*
KARIYA and KURATA · Generalized Least Squares
KASS and VOS · Geometrical Foundations of Asymptotic Inference
† KAUFMAN and ROUSSEEUW · Finding Groups in Data: An Introduction to Cluster Analysis
KEDEM and FOKIANOS · Regression Models for Time Series Analysis
KENDALL, BARDEN, CARNE, and LE · Shape and Shape Theory
KHURI · Advanced Calculus with Applications in Statistics, *Second Edition*
KHURI, MATHEW, and SINHA · Statistical Tests for Mixed Linear Models
* KISH · Statistical Design for Research
KLEIBER and KOTZ · Statistical Size Distributions in Economics and Actuarial Sciences
KLEMELÄ · Smoothing of Multivariate Data: Density Estimation and Visualization
KLUGMAN, PANJER, and WILLMOT · Loss Models: From Data to Decisions, *Third Edition*
KLUGMAN, PANJER, and WILLMOT · Loss Models: Further Topics
KLUGMAN, PANJER, and WILLMOT · Solutions Manual to Accompany Loss Models: From Data to Decisions, *Third Edition*
KOSKI and NOBLE · Bayesian Networks: An Introduction
KOTZ, BALAKRISHNAN, and JOHNSON · Continuous Multivariate Distributions, Volume 1, *Second Edition*
KOTZ and JOHNSON (editors) · Encyclopedia of Statistical Sciences: Volumes 1 to 9 with Index
KOTZ and JOHNSON (editors) · Encyclopedia of Statistical Sciences: Supplement Volume
KOTZ, READ, and BANKS (editors) · Encyclopedia of Statistical Sciences: Update Volume 1
KOTZ, READ, and BANKS (editors) · Encyclopedia of Statistical Sciences: Update Volume 2
KOWALSKI and TU · Modern Applied U-Statistics
KRISHNAMOORTHY and MATHEW · Statistical Tolerance Regions: Theory, Applications, and Computation

*Now available in a lower priced paperback edition in the Wiley Classics Library.
†Now available in a lower priced paperback edition in the Wiley–Interscience Paperback Series.

KROESE, TAIMRE, and BOTEV · Handbook of Monte Carlo Methods
KROONENBERG · Applied Multiway Data Analysis
KULINSKAYA, MORGENTHALER, and STAUDTE · Meta Analysis: A Guide to Calibrating and Combining Statistical Evidence
KULKARNI and HARMAN · An Elementary Introduction to Statistical Learning Theory
KUROWICKA and COOKE · Uncertainty Analysis with High Dimensional Dependence Modelling
KVAM and VIDAKOVIC · Nonparametric Statistics with Applications to Science and Engineering
LACHIN · Biostatistical Methods: The Assessment of Relative Risks, *Second Edition*
LAD · Operational Subjective Statistical Methods: A Mathematical, Philosophical, and Historical Introduction
LAMPERTI · Probability: A Survey of the Mathematical Theory, *Second Edition*
LAWLESS · Statistical Models and Methods for Lifetime Data, *Second Edition*
LAWSON · Statistical Methods in Spatial Epidemiology, *Second Edition*
LE · Applied Categorical Data Analysis, *Second Edition*
LE · Applied Survival Analysis
LEE · Structural Equation Modeling: A Bayesian Approach
LEE and WANG · Statistical Methods for Survival Data Analysis, *Fourth Edition*
LEPAGE and BILLARD · Exploring the Limits of Bootstrap
LESSLER and KALSBEEK · Nonsampling Errors in Surveys
LEYLAND and GOLDSTEIN (editors) · Multilevel Modelling of Health Statistics
LIAO · Statistical Group Comparison
LIN · Introductory Stochastic Analysis for Finance and Insurance
LINDLEY · Understanding Uncertainty, *Revised Edition*
LITTLE and RUBIN · Statistical Analysis with Missing Data, *Second Edition*
LLOYD · The Statistical Analysis of Categorical Data
LOWEN and TEICH · Fractal-Based Point Processes
MAGNUS and NEUDECKER · Matrix Differential Calculus with Applications in Statistics and Econometrics, *Revised Edition*
MALLER and ZHOU · Survival Analysis with Long Term Survivors
MARCHETTE · Random Graphs for Statistical Pattern Recognition
MARDIA and JUPP · Directional Statistics
MARKOVICH · Nonparametric Analysis of Univariate Heavy-Tailed Data: Research and Practice
MARONNA, MARTIN and YOHAI · Robust Statistics: Theory and Methods
MASON, GUNST, and HESS · Statistical Design and Analysis of Experiments with Applications to Engineering and Science, *Second Edition*
McCULLOCH, SEARLE, and NEUHAUS · Generalized, Linear, and Mixed Models, *Second Edition*
McFADDEN · Management of Data in Clinical Trials, *Second Edition*
* McLACHLAN · Discriminant Analysis and Statistical Pattern Recognition
McLACHLAN, DO, and AMBROISE · Analyzing Microarray Gene Expression Data
McLACHLAN and KRISHNAN · The EM Algorithm and Extensions, *Second Edition*
McLACHLAN and PEEL · Finite Mixture Models
McNEIL · Epidemiological Research Methods
MEEKER and ESCOBAR · Statistical Methods for Reliability Data
MEERSCHAERT and SCHEFFLER · Limit Distributions for Sums of Independent Random Vectors: Heavy Tails in Theory and Practice
MENGERSEN, ROBERT, and TITTERINGTON · Mixtures: Estimation and Applications

*Now available in a lower priced paperback edition in the Wiley Classics Library.
†Now available in a lower priced paperback edition in the Wiley–Interscience Paperback Series.

MICKEY, DUNN, and CLARK · Applied Statistics: Analysis of Variance and Regression, *Third Edition*

* MILLER · Survival Analysis, *Second Edition*

MONTGOMERY, JENNINGS, and KULAHCI · Introduction to Time Series Analysis and Forecasting

MONTGOMERY, PECK, and VINING · Introduction to Linear Regression Analysis, *Fifth Edition*

MORGENTHALER and TUKEY · Configural Polysampling: A Route to Practical Robustness

MUIRHEAD · Aspects of Multivariate Statistical Theory

MULLER and STOYAN · Comparison Methods for Stochastic Models and Risks

MURTHY, XIE, and JIANG · Weibull Models

MYERS, MONTGOMERY, and ANDERSON-COOK · Response Surface Methodology: Process and Product Optimization Using Designed Experiments, *Third Edition*

MYERS, MONTGOMERY, VINING, and ROBINSON · Generalized Linear Models. With Applications in Engineering and the Sciences, *Second Edition*

NATVIG · Multistate Systems Reliability Theory With Applications

† NELSON · Accelerated Testing, Statistical Models, Test Plans, and Data Analyses

† NELSON · Applied Life Data Analysis

NEWMAN · Biostatistical Methods in Epidemiology

NG, TAIN, and TANG · Dirichlet Theory: Theory, Methods and Applications

OKABE, BOOTS, SUGIHARA, and CHIU · Spatial Tesselations: Concepts and Applications of Voronoi Diagrams, *Second Edition*

OLIVER and SMITH · Influence Diagrams, Belief Nets and Decision Analysis

PALTA · Quantitative Methods in Population Health: Extensions of Ordinary Regressions

PANJER · Operational Risk: Modeling and Analytics

PANKRATZ · Forecasting with Dynamic Regression Models

PANKRATZ · Forecasting with Univariate Box-Jenkins Models: Concepts and Cases

PARDOUX · Markov Processes and Applications: Algorithms, Networks, Genome and Finance

PARMIGIANI and INOUE · Decision Theory: Principles and Approaches

* PARZEN · Modern Probability Theory and Its Applications

PEÑA, TIAO, and TSAY · A Course in Time Series Analysis

PESARIN and SALMASO · Permutation Tests for Complex Data: Applications and Software

PIANTADOSI · Clinical Trials: A Methodologic Perspective, *Second Edition*

POURAHMADI · Foundations of Time Series Analysis and Prediction Theory

POURAHMADI · High-Dimensional Covariance Estimation

POWELL · Approximate Dynamic Programming: Solving the Curses of Dimensionality, *Second Edition*

POWELL and RYZHOV · Optimal Learning

PRESS · Subjective and Objective Bayesian Statistics, *Second Edition*

PRESS and TANUR · The Subjectivity of Scientists and the Bayesian Approach

PURI, VILAPLANA, and WERTZ · New Perspectives in Theoretical and Applied Statistics

† PUTERMAN · Markov Decision Processes: Discrete Stochastic Dynamic Programming

QIU · Image Processing and Jump Regression Analysis

* RAO · Linear Statistical Inference and Its Applications, *Second Edition*

RAO · Statistical Inference for Fractional Diffusion Processes

RAUSAND and HØYLAND · System Reliability Theory: Models, Statistical Methods, and Applications, *Second Edition*

*Now available in a lower priced paperback edition in the Wiley Classics Library.

†Now available in a lower priced paperback edition in the Wiley–Interscience Paperback Series.

RAYNER, THAS, and BEST · Smooth Tests of Goodnes of Fit: Using R, *Second Edition*
RENCHER and SCHAALJE · Linear Models in Statistics, *Second Edition*
RENCHER and CHRISTENSEN · Methods of Multivariate Analysis, *Third Edition*
RENCHER · Multivariate Statistical Inference with Applications
RIGDON and BASU · Statistical Methods for the Reliability of Repairable Systems
* RIPLEY · Spatial Statistics
* RIPLEY · Stochastic Simulation
ROHATGI and SALEH · An Introduction to Probability and Statistics, *Second Edition*
ROLSKI, SCHMIDLI, SCHMIDT, and TEUGELS · Stochastic Processes for Insurance and Finance
ROSENBERGER and LACHIN · Randomization in Clinical Trials: Theory and Practice
ROSSI, ALLENBY, and McCULLOCH · Bayesian Statistics and Marketing
† ROUSSEEUW and LEROY · Robust Regression and Outlier Detection
ROYSTON and SAUERBREI · Multivariate Model Building: A Pragmatic Approach to Regression Analysis Based on Fractional Polynomials for Modeling Continuous Variables
* RUBIN · Multiple Imputation for Nonresponse in Surveys
RUBINSTEIN and KROESE · Simulation and the Monte Carlo Method, *Second Edition*
RUBINSTEIN and MELAMED · Modern Simulation and Modeling
RUBINSTEIN, RIDDER, and VAISMAN · Fast Sequential Monte Carlo Methods for Counting and Optimization
RYAN · Modern Engineering Statistics
RYAN · Modern Experimental Design
RYAN · Modern Regression Methods, *Second Edition*
RYAN · Sample Size Determination and Power
RYAN · Statistical Methods for Quality Improvement, *Third Edition*
SALEH · Theory of Preliminary Test and Stein-Type Estimation with Applications
SALTELLI, CHAN, and SCOTT (editors) · Sensitivity Analysis
SCHERER · Batch Effects and Noise in Microarray Experiments: Sources and Solutions
* SCHEFFE · The Analysis of Variance
SCHIMEK · Smoothing and Regression: Approaches, Computation, and Application
SCHOTT · Matrix Analysis for Statistics, *Second Edition*
SCHOUTENS · Levy Processes in Finance: Pricing Financial Derivatives
SCOTT · Multivariate Density Estimation: Theory, Practice, and Visualization
* SEARLE · Linear Models
† SEARLE · Linear Models for Unbalanced Data
† SEARLE · Matrix Algebra Useful for Statistics
† SEARLE, CASELLA, and McCULLOCH · Variance Components
SEARLE and WILLETT · Matrix Algebra for Applied Economics
SEBER · A Matrix Handbook For Statisticians
† SEBER · Multivariate Observations
SEBER and LEE · Linear Regression Analysis, *Second Edition*
† SEBER and WILD · Nonlinear Regression
SENNOTT · Stochastic Dynamic Programming and the Control of Queueing Systems
* SERFLING · Approximation Theorems of Mathematical Statistics
SHAFER and VOVK · Probability and Finance: It's Only a Game!
SHERMAN · Spatial Statistics and Spatio-Temporal Data: Covariance Functions and Directional Properties
SILVAPULLE and SEN · Constrained Statistical Inference: Inequality, Order, and Shape Restrictions
SINGPURWALLA · Reliability and Risk: A Bayesian Perspective

*Now available in a lower priced paperback edition in the Wiley Classics Library.
†Now available in a lower priced paperback edition in the Wiley–Interscience Paperback Series.

SMALL and MCLEISH · Hilbert Space Methods in Probability and Statistical Inference
SRIVASTAVA · Methods of Multivariate Statistics
STAPLETON · Linear Statistical Models, *Second Edition*
STAPLETON · Models for Probability and Statistical Inference: Theory and Applications
STAUDTE and SHEATHER · Robust Estimation and Testing
STOYAN · Counterexamples in Probability, *Second Edition*
STOYAN and STOYAN · Fractals, Random Shapes and Point Fields: Methods of Geometrical Statistics
STREET and BURGESS · The Construction of Optimal Stated Choice Experiments: Theory and Methods
STYAN · The Collected Papers of T. W. Anderson: 1943–1985
SUTTON, ABRAMS, JONES, SHELDON, and SONG · Methods for Meta-Analysis in Medical Research
TAKEZAWA · Introduction to Nonparametric Regression
TAMHANE · Statistical Analysis of Designed Experiments: Theory and Applications
TANAKA · Time Series Analysis: Nonstationary and Noninvertible Distribution Theory
THOMPSON · Empirical Model Building: Data, Models, and Reality, *Second Edition*
THOMPSON · Sampling, *Third Edition*
THOMPSON · Simulation: A Modeler's Approach
THOMPSON and SEBER · Adaptive Sampling
THOMPSON, WILLIAMS, and FINDLAY · Models for Investors in Real World Markets
TIERNEY · LISP-STAT: An Object-Oriented Environment for Statistical Computing and Dynamic Graphics
TROFFAES and DE COOMAN · Lower Previsions
TSAY · Analysis of Financial Time Series, *Third Edition*
TSAY · An Introduction to Analysis of Financial Data with R
TSAY · Multivariate Time Series Analysis: With R and Financial Applications
UPTON and FINGLETON · Spatial Data Analysis by Example, Volume II: Categorical and Directional Data
† VAN BELLE · Statistical Rules of Thumb, *Second Edition*
VAN BELLE, FISHER, HEAGERTY, and LUMLEY · Biostatistics: A Methodology for the Health Sciences, *Second Edition*
VESTRUP · The Theory of Measures and Integration
VIDAKOVIC · Statistical Modeling by Wavelets
VIERTL · Statistical Methods for Fuzzy Data
VINOD and REAGLE · Preparing for the Worst: Incorporating Downside Risk in Stock Market Investments
WALLER and GOTWAY · Applied Spatial Statistics for Public Health Data
WEISBERG · Applied Linear Regression, *Fourth Edition*
WEISBERG · Bias and Causation: Models and Judgment for Valid Comparisons
WELSH · Aspects of Statistical Inference
WESTFALL and YOUNG · Resampling-Based Multiple Testing: Examples and Methods for *p*-Value Adjustment
* WHITTAKER · Graphical Models in Applied Multivariate Statistics
WINKER · Optimization Heuristics in Economics: Applications of Threshold Accepting
WOODWORTH · Biostatistics: A Bayesian Introduction
WOOLSON and CLARKE · Statistical Methods for the Analysis of Biomedical Data, *Second Edition*
WU and HAMADA · Experiments: Planning, Analysis, and Parameter Design Optimization, *Second Edition*

*Now available in a lower priced paperback edition in the Wiley Classics Library.
†Now available in a lower priced paperback edition in the Wiley–Interscience Paperback Series.

WU and ZHANG · Nonparametric Regression Methods for Longitudinal Data Analysis
YAKIR · Extremes in Random Fields
YIN · Clinical Trial Design: Bayesian and Frequentist Adaptive Methods
YOUNG, VALERO-MORA, and FRIENDLY · Visual Statistics: Seeing Data with Dynamic Interactive Graphics
ZACKS · Examples and Problems in Mathematical Statistics
ZACKS · Stage-Wise Adaptive Designs
* ZELLNER · An Introduction to Bayesian Inference in Econometrics
ZELTERMAN · Discrete Distributions—Applications in the Health Sciences
ZHOU, OBUCHOWSKI, and McCLISH · Statistical Methods in Diagnostic Medicine, *Second Edition*

*Now available in a lower priced paperback edition in the Wiley Classics Library.
†Now available in a lower priced paperback edition in the Wiley–Interscience Paperback Series.